Lecture Notes in Mathematics 1807

Editors:
J.-M. Morel, Cachan
F. Takens, Groningen
B. Teissier, Paris

T0226273

Springer
Berlin
Heidelberg
New York
Hong Kong
London
Milan
Paris
Tokyo

V. D. Milman G. Schechtman (Eds.)

Geometric Aspects of Functional Analysis

Israel Seminar 2001–2002

2001–2002

 Springer

Editors

Vitali D. Milman

Department of Mathematics
Tel Aviv University
Ramat Aviv
69978 Tel Aviv
Israel

e-mail: milman@post.tau.ac.il

Gideon Schechtman

Department of Mathematics
The Weizmann Institute of Science
P.O. Box 26
76100 Rehovot
Israel

e-mail: gideon@wisdom.weizmann.ac.il
http://www.wisdom.weizmann.ac.il/˜gideon

Cataloging-in-Publication Data applied for

Bibliographic information published by Die Deutsche Bibliothek

Die Deutsche Bibliothek lists this publication in the Deutsche Nationalbibliografie;
detailed bibliographic data is available in the Internet at http://dnb.ddb.de

Mathematics Subject Classification (2000): 46-06, 46B07, 52-06 60-06

ISSN 0075-8434
ISBN 978-3-540-00485-1 Springer-Verlag Berlin Heidelberg New York

Springer-Verlag Berlin Heidelberg New York a member of BertelsmannSpringer
Science + Business Media GmbH

http://www.springer.de

© Springer-Verlag Berlin Heidelberg 2003

Typesetting: Camera-ready TEX output by the author

SPIN: 10904442   41/3142/ du - 543210 - Printed on acid-free paper

Preface

During the last two decades the following volumes containing the proceedings of the Israel Seminar in Geometric Aspects of Functional Analysis appeared

1983-1984   Published privately by Tel Aviv University
1985-1986   Springer Lecture Notes, Vol. 1267
1986-1987   Springer Lecture Notes, Vol. 1317
1987-1988   Springer Lecture Notes, Vol. 1376
1989-1990   Springer Lecture Notes, Vol. 1469
1992-1994   Operator Theory: Advances and Applications, Vol. 77, Birkhauser
1994-1996   MSRI Publications, Vol. 34, Cambridge University Press
1996-2000   Springer Lecture Notes, Vol. 1745.

Of these, the first six were edited by Lindenstrauss and Milman, the seventh by Ball and Milman and the last by the two of us.

As in the previous volumes, the current one reflects general trends of the Theory. The connection between Probability and Convexity continues to broaden and deepen and a number of papers of this collection reflect this fact. There is a renewed interest (and hope for solution) in the old and fascinating slicing problem (also known as the hyperplane conjecture). Several papers in this volume revolve around this conjecture as well as around some related topics as the distribution of functionals, regarded as random variables on a convex set equipped with its normalized Lebesgue measure. Some other papers deal with more traditional aspects of the Theory like the concentration phenomenon. Finally, the volume contains a long paper on approximating convex sets by randomly chosen polytopes which also contains a deep study of floating bodies, an important subject in Classical Convexity Theory.

All the papers here are original research papers and were subject to the usual standards of refereeing.

As in previous proceedings of the GAFA Seminar, we also list here all the talks given in the seminar as well as talks in related workshops and conferences. We believe this gives a sense of the main directions of research in our area.

We are grateful to Ms. Diana Yellin for taking excellent care of the typesetting aspects of this volume.

Vitali Milman
Gideon Schechtman

Table of Contents

A Note on Simultaneous Polar and Cartesian Decomposition

F. Barthe[1]*, M. Csörnyei[2]** and A. Naor[3]***

[1] CNRS-Université de Marne-la-Vallée, Equipe d'analyse et de Mathématiques appliquées, Cité Descartes, Champs-sur-Marne, 77454, Marne-la-Vallée, Cedex 2, France *barthe@math.univ-mlv.fr*

[2] Department of Mathematics, University College London, Gower Street, London, WC1E 6BT, United Kingdom *mari@math.ucl.ac.uk*

[3] Institute of Mathematics, Hebrew University, Jerusalem 91904, Israel *naor@math.huji.ac.il*

Summary. We study measures on $\mathbb{R}^n$ which are product measures for the usual Cartesian product structure of $\mathbb{R}^n$ as well as for the polar decomposition of $\mathbb{R}^n$ induced by a convex body. For finite atomic measures and for absolutely continuous measures with density $d\mu/dx = e^{-V(x)}$, where V is locally integrable, a complete characterization is presented.

1 Introduction

A subset $K \subset \mathbb{R}^n$ is called a star-shaped body if it is star-shaped with respect to the origin, compact, has non-empty interior, and for every $x \neq 0$ there is a unique $r > 0$ such that $x/r \in \partial K$. We denote this r by $\|x\|_K$ ($\|\cdot\|_K$ is the Minkowski functional of K). Note that $\|x\|_K$ is automatically continuous (if x_n tends to $x \neq 0$, then for every subsequence x_{n_k} such that $\|x_{n_k}\|_K$ converges to r, the compactness ensures that $x/r \in \partial K$, so that $r = \|x\|_K$ by the uniqueness assumption). Any star-shaped body $K \subset \mathbb{R}^n$ induces a polar product structure on $\mathbb{R}^n \setminus \{0\}$ through the identification

$$x \mapsto \left(\|x\|_K, \frac{x}{\|x\|_K} \right).$$

In this note we study the measures on $\mathbb{R}^n$, $n \geq 2$ which are product measures with respect to the Cartesian coordinates, and the above polar decomposition.

In measure theoretic formulation, we will be interested in the measures μ on $\mathbb{R}^n$ which are product measures with respect to the product structures

* Partially supported by EPSRC grant 64 GR/R37210.

** Supported by the Hungarian National Foundation for Scientific Research, grant # F029768.

*** Supported in part by the Binational Science Foundation Israel-USA, the Clore Foundation and the EU grant HPMT-CT-2000-00037. This work is part of a Ph.D. thesis being prepared under the supervision of Professor Joram Lindenstrauss.

$\mathbb{R}^n = \mathbb{R} \times \cdots \times \mathbb{R} = \mathbb{R}^+ \cdot \partial K$. Here $\times$ is the usual Cartesian product and for $R \subset \mathbb{R}^+$, $\Omega \in \partial K$, the polar product is by definition $R \cdot \Omega = \{r\omega; \ r \in R \text{ and } \omega \in \Omega\}$. We adopt similar notation for product measures: $\otimes$ will be used for Cartesian-product measures and $\odot$ for polar-product measures. With this notation, we say that μ has a simultaneous product decomposition with respect to K if there are measures $\mu_1, \ldots \mu_n$ on $\mathbb{R}$ such that $\mu = \mu_1 \otimes \cdots \otimes \mu_n$, and there is a measure τ on $\mathbb{R}^+$ and a measure ν on ∂K such that $\mu = \tau \odot \nu$ (in what follows, all measures are Borel). Notation like A^k or $\prod_i A_i$ always refers to the Cartesian product.

For probability measures one can formulate the notion of simultaneous product decomposition as follows. A measure μ on $\mathbb{R}^n$ has a simultaneous product decomposition with respect to K if and only if there are independent real valued random variables $X_1, \ldots, X_n$ such that if we denote $X = (X_1, \ldots, X_n)$ then $\mu(A) = P(X \in A)$ and $X/\|X\|_K$ is independent of $\|X\|_K$.

The standard Gaussian measure on $\mathbb{R}^n$ is obviously a Cartesian product. A consequence of its rotation invariance is that it is also a polar-product measure for the usual polar structure induced by the Euclidean ball. Many characterizations of the Gaussian distribution have been obtained so far. The motivations for such characterizations arise from several directions. Maxwell proved that the Gaussian measure is the only rotation invariant product probability measure on $\mathbb{R}^3$, and deduced that this is the distribution of the velocities of gas particles. The classical Cramer and Bernstein characterizations of the Gaussian measure, as well as the numerous related results that appeared in the literature arose from various probabilistic and statistical motivations. We refer to the book [Br] and the references therein for a detailed account. The more modern characterization due to Carlen [C] arose from the need to characterize the equality case in a certain functional inequality.

To explain the motivation for the present paper, we begin by noting that the Gaussian density is in fact one member of a wider family of measures with simultaneous product decomposition, involving bodies other than the Euclidean ball. They will be easily introduced after setting notation. The cone measure on the boundary of K, denoted by μ_K is defined as:

$$\mu_K(A) = \text{vol}([0,1] \cdot A).$$

This measure is natural when studying the polar decomposition of the Lebesgue measure with respect to K, i.e. for every integrable $f : \mathbb{R}^n \to \mathbb{R}$, one has

$$\int_{\mathbb{R}^n} f(x)\, dx = \int_0^{+\infty} nr^{n-1} \int_{\partial K} f(r\omega)\, d\mu_K(\omega) dr.$$

For the particular case $K = B_p^n = \{x \in \mathbb{R}^n; \ \|x\|_p \leq 1\}$, where $\|x\|_p = (\sum_{i=1}^n |x_i|^p)^{1/p}$, a fundamental result of Schechtman and Zinn [SZ1] (see also Rachev and Rüschendorff [RR]), gives a concrete representation of μ_K:

Theorem 1. *Let g be a random variable with density $e^{-|t|^p}/(2\Gamma(1+1/p))$, $t \in \mathbb{R}$. If $g_1, ..., g_n$ are i.i.d. copies of g, set:*

$$S = \left(\sum_{i=1}^{n} |g_i|^p \right)^{1/p},$$

and consider the random vector:

$$Z = \left(\frac{g_1}{S}, ..., \frac{g_n}{S} \right) \in \mathbb{R}^n.$$

Then the random vector Z is independent of S. Moreover, for every measurable $A \subset \partial B_p^n$ we have:

$$\frac{\mu_{B_p^n}(A)}{\mathrm{vol}(B_p^n)} = P(Z \in A).$$

The independence of Z and S, that is the simultaneous product decomposition, turns out to be very useful for probabilistic as well as geometric purposes ([SZ2],[NR],[N],[BN]). One might hope that such a statement holds true for other norms and other densities. The aim of this note is to show that the ℓ_p^n norm is in fact characterized by this property, although we will show that such an independence result holds for other measures. Section 2 is devoted to absolutely continuous measures. Section 3 presents a classification for finite atomic measures, when K is convex. As the reader will see there are more examples. Some of them, however, are not interesting and we will discard them by suitable assumptions. For example: a constant random variable is independent of any other. This observation allows us to produce several measures with simultaneous product decomposition. Any random variable X with values in the half-line $\{x; \; x_1 > 0\}$ works. Its law is clearly a Cartesian product measure, and $X/\|X\|_K$ is constant regardless of K, so it is independent of $\|X\|_K$. Similarly, if X has independent components and takes values in only one sphere $r\partial K$ it has a simultaneous product decomposition.

Let λ and λ_0 denote the Lebesgue measure and the counting measure on $\mathbb{R}^n$, respectively. For any (not necessarily finite) measure μ on $\mathbb{R}^n$ with a simultaneous product decomposition with respect to a star-shaped body K, $\mu = \mu_1 \otimes \cdots \otimes \mu_n = \tau \odot \nu$ and a function $f : \mathbb{R}^n \to \mathbb{R}$, we say that f has a simultaneous product decomposition with respect to μ and K, if there are some functions $f_j \in L_1(\mu_j)$, $g \in L_1(\nu)$ and $h \in L_1(\tau)$ such that

$$f(x) = \prod_{i=1}^{n} f_i(x_i) = g\left(\frac{x}{\|x\|_K} \right) \cdot h(\|x\|_K)$$

μ almost everywhere. It is immediate to see that for any countable set $S \subset \mathbb{R}^n$ an atomic measure $f 1_S d\lambda_0$ has a simultaneous product decomposition with respect to K if and only if f has a simultaneous product decomposition with respect to λ_0 and K. Analogously (see Lemma 2 below), an absolutely continuous measure $f d\lambda$ has a simultaneous product decomposition with respect to K if and only if its density function f has a simultaneous product decomposition with respect to λ and K. From this it is immediate to see that if $f_1 d\lambda$ and $f_2 d\lambda$ (resp. $f_1 d\lambda_0$ and $f_2 d\lambda_0$) have simultaneous product decompositions with respect to K and $f_1 f_2 \in L_1(\lambda)$ then $f_1 f_2 d\lambda$ (resp. $f_1 f_2 d\lambda_0$) has a simultaneous product decomposition with respect to K. Similarly, if $f d\lambda$ (resp. $f d\lambda_0$) has a simultaneous product decomposition with respect to K and $f^\alpha \in L_1(\lambda)$ (resp. $f^\alpha \in L_1(\lambda_0)$) then $f^\alpha d\lambda$ (resp. $f^\alpha d\lambda_0$) has a simultaneous product decomposition with respect to K.

If K is not assumed to be convex, many different examples may be produced: take two sets of positive numbers $\{x_1, \ldots, x_N\}$ and $\{y_1, \ldots, y_N\}$ and consider $\mu_1 = \sum \delta_{x_i}$ and $\mu_2 = \sum \delta_{y_i}$. If one assumes that the numbers y_i/x_j are all different then the measure $\mu_1 \otimes \mu_2$ is supported on points (x_j, y_i) all having different directions. So there are several origin-star-shaped bodies K such that $\mu_1 \otimes \mu_2$ is supported on the boundary of K. For such K's, $\mu_1 \otimes \mu_2$ admits a polar decomposition.

Finally, if μ has simultaneous product decomposition, and $\epsilon_1, \ldots, \epsilon_n \in \{-1, 1\}$ then the restriction of μ to $\{x; \ x_i \epsilon_i > 0\}$ still has this property. This remark allows us to restrict the study to the positive orthant $(0, +\infty)^n$ (and one has to glue pieces together at the end). For some positive numbers $a_1, a_2, \ldots, a_n$ let $T = T_{a_1, \ldots, a_n} : (0, +\infty)^n \to (0, +\infty)^n$ denote the linear bijection $(x_1, \ldots, x_n) \mapsto (x_1/a_1, \ldots, x_n/a_n)$. If μ is supported in the positive orthant and it has a simultaneous product decomposition with respect to K, then $\mu \circ T$ has a simultaneous product decomposition with respect to $T(K)$. We will show that if μ is an absolutely continuous measure with density of the form $e^{-V(x)}$ where V is locally integrable, and if μ has a simultaneous product decomposition with respect to a star-shaped body K, then there are some positive numbers $a_1, \ldots, a_n$ and there is a $p > 0$ such that $K \cap (0, \infty)^n = T(B_p^n \cap (0, \infty)^n)$. We will also show that if an atomic measure has a simultaneous product decomposition with respect to a convex body K, and it is not supported on a sphere rK for some $r > 0$, then for some $a_1, a_2, \ldots, a_n$ we have $K \cap (0, \infty)^n = T(B_\infty^n \cap (0, \infty)^n)$.

2 Absolutely Continuous Measures

As in the classical characterizations of the Gaussian measure, the assumption that the measure is absolutely continuous reduces the characterization problem to a solution of a functional equation which holds almost everywhere (with respect to the Lebesgue measure). Unless we add some smoothness assumptions on the densities, the next step is to apply a smoothing procedure. Of course, after "guessing" the family of solutions of the equations, we must come up with a smoothing procedure which sends each member of this family to another member of the family. The classical Cramer and Bernstein characterizations use Fourier transform techniques (see [Br]), while Carlen [C] applies the heat semi-group. The particular form of the equation we will derive will force us to use yet another smoothing procedure.

When μ is absolutely continuous the following (easy) characterization holds:

Lemma 2. *Assume that μ is an absolutely continuous measure on $\mathbb{R}^n$. Then it has a simultaneous product decomposition with respect to K if and only if there are locally integrable non-negative functions $f_1, \ldots, f_n$ defined on $\mathbb{R}$, g defined on ∂K (locally integrable with respect to μ_K) and h on $(0, \infty)$ such that*

$$\frac{d\mu}{dx}(x) = \prod_{i=1}^{n} f_i(x_i) = g\left(\frac{x}{\|x\|_K}\right) \cdot h(\|x\|_K),$$

is Lebesgue almost everywhere.

Proof. Assume that μ has a simultaneous product decomposition with respect to K. In the above notation, write $\mu = \mu_1 \otimes \cdots \otimes \mu_n = \tau \odot \nu$. For every measurable $B \subset \partial K$:

$$\nu(B) = \mu(\mathbb{R}^+ \cdot B) = \int_{R^+ \cdot B} \frac{d\mu}{dx}(x) dx$$

$$= \int_B \left(\int_0^\infty n \cdot r^{n-1} \frac{d\mu}{dx}(r\omega) dr\right) d\mu_K(\omega).$$

Similarly, for every measurable $A \subset \mathbb{R}^+$:

$$\tau(A) = \mu(A \cdot \partial K) = \int_A n \cdot \left(\int_{\partial K} \frac{d\mu}{dx}(r\omega) d\mu_K(\omega)\right) dr.$$

This shows that both τ and ν are absolutely continuous. Similarly $\mu_1, \ldots, \mu_n$ are absolutely continuous.

Now, for every measurable $A \subset \mathbb{R}^+$, $B \subset \partial K$, $C_1, \ldots C_n \subset \mathbb{R}$:

$$\mu(A \cdot B) = \tau(A)\nu(B) = \int_0^\infty \int_{\partial K} \frac{d\tau}{dr}(r) \cdot \frac{d\nu}{d\mu_K}(\omega) dr d\mu_K(\omega)$$

$$= \int_{A \cdot B} \frac{1}{n \cdot \|x\|_K^{n-1}} \cdot \frac{d\tau}{dr}(\|x\|_K) \cdot \frac{d\nu}{d\mu_K}\left(\frac{x}{\|x\|_K}\right) dx,$$

and

$$\mu(C_1 \times \ldots \times C_n) = \int_{C_1 \times \ldots \times C_n} \prod_{i=1}^{n} \frac{d\mu_i}{dx_i}(x_i) dx.$$

Since the product Borel σ-algebras on $\mathbb{R}^+ \cdot \partial K$ and $\mathbb{R} \times \ldots \times \mathbb{R}$ (n times) coincide, this shows that:

$$\frac{d\mu}{dx}(x) = \prod_{i=1}^{n} \frac{d\mu_i}{dx_i}(x_i) = \frac{1}{n \cdot \|x\|_K^{n-1}} \cdot \frac{d\tau}{dr}(\|x\|_K) \cdot \frac{d\nu}{d\mu_K}\left(\frac{x}{\|x\|_K}\right),$$

is Lebesgue almost everywhere. The reverse implication is even simpler. $\square$

Fix some $p > 0$, $b_1, \ldots b_n > -1$ and $a_1, \ldots a_n > 0$. Let $X_1, \ldots, X_n$ be independent random variables, such that the density of X_i is:

$$\frac{p a_i^{(b_i+1)/p}}{2\Gamma\left(\frac{b_i+1}{p}\right)} \cdot |t|^{b_i} e^{-a_i|t|^p}.$$

Note that:

$$\prod_{i=1}^n |x_i|^{b_i} e^{-a_i|x_i|^p} =$$

$$= \left(\sum_{i=1}^n a_i|x_i|^p\right)^{\frac{1}{p}\sum_{i=1}^n b_i} \cdot e^{-\sum_{i=1}^n a_i|x_i|^p} \prod_{i=1}^n \left[\frac{|x_i|}{\left(\sum_{j=1}^n a_j|x_j|^p\right)^{1/p}}\right]^{b_i}.$$

Hence, if we denote $X = (X_1, \ldots, X_n)$ then by Lemma 2, $X/\left(\sum_{i=1}^n a_i|X_i|^p\right)^{1/p}$ and $\left(\sum_{i=1}^n a_i|X_i|^p\right)^{1/p}$ are independent. Moreover, if $b_1 = \ldots = b_n = 0$ and $a_1 = \ldots = a_n = 1$ then it follows from the proof of Lemma 2 that $X/\|X\|_p$ generates the cone measure on the sphere of ℓ_p^n. We have therefore obtained a generalization of Theorem 1.

The main goal of this section is to prove that the above densities are the only way to obtain a measure with a simultaneous product decomposition with respect to a star-shaped body $K \subset \mathbb{R}^n$ (and that K must then be a weighted ℓ_p^n ball). In solving the functional equation of Lemma 2 we will require a smoothing procedure. Clearly, we require a way to smooth a function such that a function of the form $c|t|^b e^{-a|t|^p}$ is transformed to a function of the same form. Let $\psi : \mathbb{R} \to \mathbb{R}$ be locally integrable. For any infinitely differentiable $\rho : (0, \infty) \to [0, \infty)$ which is compactly supported in $(0, \infty)$ define:

$$(\rho \star \psi)(x) = \int_0^\infty \rho(t)\psi\left(\frac{x}{t}\right)dt = \int_0^\infty x\rho(sx)\psi\left(\frac{1}{s}\right)ds.$$

It is easy to verify that $\rho \star \psi$ is infinitely differentiable on $(-\infty, 0) \cup (0, \infty)$. Fix some $\epsilon < 1/2$ and let $\rho_\epsilon : (0, \infty) \to [0, 1/(2\epsilon)]$ be any infinitely differentiable function such that $\rho_\epsilon(t) = 1/(2\epsilon)$ when $|t - 1| \leq \epsilon$ and $\rho_\epsilon(t) = 0$ when $|t - 1| > \epsilon + \epsilon^2$. Now, for $x > 0$ (and similarly when $x < 0$):

$$\left|(\rho_\epsilon \star \psi)(x) - \frac{1}{2x\epsilon}\int_{x/(1+\epsilon+\epsilon^2)}^{x/(1-\epsilon-\epsilon^2)} \psi(u)du\right| =$$

$$= \frac{1}{x}\int_{x/(1+\epsilon+\epsilon^2)}^{x/(1-\epsilon-\epsilon^2)} \left|\frac{1}{2\epsilon} - \left(\frac{x}{u}\right)^2 \rho_\epsilon\left(\frac{x}{u}\right)\right| |\psi(u)|du \leq$$

$$\leq \frac{1}{x}\int_{x/(1+\epsilon)}^{x/(1-\epsilon)} |\psi(u)|du + \frac{1}{2\epsilon x}\left[\int_{x/(1+\epsilon+\epsilon^2)}^{x/(1+\epsilon)} |\psi(u)|du + \int_{x/(1-\epsilon)}^{x/(1-\epsilon-\epsilon^2)} |\psi(u)|du\right],$$

which, by the Lebesgue density theorem, implies that $\lim_{\epsilon \to 0} \rho \star \psi = \psi$ almost everywhere. Since, $\lim_{\epsilon \to 0} \int_0^\infty \rho_\epsilon(t)dt = 1$, the same holds for $\beta_\epsilon = \rho_\epsilon / \int_0^\infty \rho_\epsilon$.

Since for every function of the form $f(t) = c|t|^b e^{-a|t|^p}$, the function $\exp(\rho_\epsilon \star (\log f))$ has the same form, and the above smoothing procedure allows us to prove our main result. In what follows $\varepsilon(x) \in \{-1, 1\}$ denotes the sign of x (any convention for the sign of zero will do).

Theorem 3. *Let $K \subset \mathbb{R}^n$ be a star-shaped body. Assume that μ is an absolutely continuous probability measure on $\mathbb{R}^n$ which has a simultaneous product decomposition with respect to K. Assume in addition that $\log(\frac{d\mu}{dx})$ is locally integrable. Then there is some $p > 0$ and there are $b_1, \ldots, b_n > -1$ and $r, a_1(1), a_1(-1), c_1(1), c_1(-1) \ldots, a_n(1), a_n(-1), c_n(1), c_n(-1) > 0$ such that:*

$$K = \left\{ x \in \mathbb{R}^n; \ \sum_{i=1}^n a_i(\varepsilon(x_i))|x_i|^p \leq r \right\},$$

and

$$d\mu(x) = \prod_{i=1}^n c_i(\varepsilon(x_i))|x_i|^{b_i} e^{-a_i(\varepsilon(x_i))|x_i|^p} \, dx_i.$$

Conversely, for K and μ as above, μ has a simultaneous product decomposition with respect to K.

We will require the following elementary lemma:

Lemma 4. *Fix $\alpha, \alpha' > 0$. Let $f : (0, \infty) \to \mathbb{R}$ be a continuous function such that for every $x > 0$,*

$$f(\alpha x) = 2f(x) \qquad \text{and} \qquad f(\alpha' x) = \frac{3}{2}f(x).$$

Then for every $x > 0$, $f(x) = f(1)x^{\frac{\log 2}{\log \alpha}}$ (if $\alpha = 1$, $f(x) = 0$ for all x).

Proof. We may assume that f is not identically zero. Then $\alpha \neq \alpha'$ and $\alpha, \alpha' \neq 1$. Consider the set

$$P = \left\{ \beta > 0; \ \text{there is} \ \ c_\beta > 0 \ \ \text{s.t.} \ \ f(\beta x) = c_\beta f(x), \quad \text{for all} \quad x > 0 \right\}.$$

It is a multiplicative subgroup of $(0, \infty)$. By classical results, P is either dense in $(0, \infty)$ or discrete. Assume first that it is dense. Fix some x_0 such that $f(x_0) \neq 0$. For any $\beta \in P$, $c_\beta = f(\beta x_0)/f(x_0)$. By continuity of f and by the density of P it follows that for every $x, \beta > 0$ one has

$$f(\beta x) = \frac{f(\beta x_0)}{f(x_0)} f(x).$$

So, if for some β, $f(\beta x_0) = 0$ then f is identically zero. Therefore, f does not vanish. We can choose $x_0 = 1$ and setting $g = f/f(1)$, we have that for every

$x, y > 0$, $g(xy) = g(x)g(y)$. It is well known that the continuity of g ensures that it is a power function.

To finish, let us note that P cannot be discrete. Indeed, if P is discrete, since it contains α and $\alpha' \neq \alpha$, it is of the form $\{T^h; h \in \mathbb{Z}\}$ for some positive $T \neq 1$. So there are $k, k' \in \mathbb{Z} \setminus \{0\}$ such that $\alpha = T^k$ and $\alpha' = T^{k'}$. Our hypothesis is that for all $x > 0$

$$2f(x) = f(\alpha x) = f(T^k x) = c_T^k f(x),$$
$$\frac{3}{2}f(x) = f(\alpha' x) = f(T^{k'} x) = c_T^{k'} f(x).$$

For an x such that $f(x) \neq 0$ we get $2 = c_T^k$ and $\frac{3}{2} = c_T^{k'}$. It follows that $3^k = 2^{k+k'}$. This is impossible because $k \neq 0$. $\square$

Proof of Theorem 3. Using the notation and the result of Lemma 2,

$$\frac{d\mu}{dx}(x) = \prod_{i=1}^{n} f_i(x_i) = g\left(\frac{x}{\|x\|_K}\right) \cdot h(\|x\|_K).$$

For $i = 1, \ldots, n$ denote $F_i = \log f_i$. Denote also $G = \log g$ and $H = \log h$. Let $\varphi : \mathbb{R}^n \to \mathbb{R}$ be a compactly supported continuous function. For every $t > 0$:

$$\sum_{i=1}^{n} \int_{\mathbb{R}^n} \varphi(tx)F_i(x_i)dx = \int_{\mathbb{R}^n} \varphi(tx)G\left(\frac{x}{\|x\|_K}\right)dx + \int_{\mathbb{R}^n} \varphi(tx)H\left(\|x\|_K\right)dx.$$

Changing variables this translates to

$$\sum_{i=1}^{n} \int_{\mathbb{R}^n} \varphi(y)F_i\left(\frac{y_i}{t}\right)dy = \int_{\mathbb{R}^n} \varphi(y)G\left(\frac{y}{\|y\|_K}\right)dy + \int_{\mathbb{R}^n} \varphi(y)H\left(\frac{\|y\|_K}{t}\right)dy.$$

Fix some $\epsilon > 0$. Multiplying by β_ϵ and integrating, we get,

$$\sum_{i=1}^{n} \int_{\mathbb{R}^n} \varphi(y)(\beta_\epsilon \star F_i)(y_i)dy$$
$$= \int_{\mathbb{R}^n} \varphi(y)G\left(\frac{y}{\|y\|_K}\right)dy + \int_{\mathbb{R}^n} \varphi(y)(\beta_\epsilon \star H)(\|y\|_K)dy.$$

Denote $\phi_i = \beta_\epsilon \star F_i$ and $\eta = \beta_\epsilon \star H$. By the above identity for almost every $y \in \mathbb{R}^n$:

$$\sum_{i=1}^{n} \phi_i(y_i) = G\left(\frac{y}{\|y\|_K}\right) + \eta(\|y\|_K).$$

Since ϕ_i and η are continuous on $\mathbb{R} \setminus \{0\}$, we can change G on a set of measure zero such that the latter identity holds for every $y \in \mathbb{R}^n$ with non-zero coordinates.

Fix some $y \in \mathbb{R}^n$ with non-zero coordinates. For every $\lambda > 0$, one has

$$\sum_{i=1}^{n} \phi_i(\lambda y_i) = G\left(\frac{y}{\|y\|_K}\right) + \eta(\lambda \|y\|_K).$$

Since both sides of the equation are differentiable in λ, taking derivatives at $\lambda = 1$ yields:

$$\sum_{i=1}^{n} \chi_i(y_i) = \zeta(\|y\|_K), \tag{1}$$

where for simplicity, we write $\chi_i(t) = t\phi_i'(t)$ and $\zeta(t) = t\eta'(t)$. From this, we shall deduce that ζ, χ_i are power functions. This can be proved by differentiation along the boundary of K, under smoothness assumptions. Since we want to deal with general star shaped bodies, we present now another reasoning.

Note that $\lim_{t \to 0} \chi_i(t)$ exists. Indeed since η is smooth on $(0, \infty)$ and $\|\cdot\|_K$ is continuous, the above equation for $y^t = \sum_{j \neq i} e_j + te_i$ gives

$$\lim_{t \to 0} \chi_i(t) = \zeta\left(\left\|\sum_{j \neq i} e_j\right\|_K\right) - \sum_{j \neq i} \chi_i(1).$$

Similarly, ζ may be extended by continuity at 0. Hence, (1) holds on $\mathbb{R}^n$. Applying (1) to λe_i, gives

$$\chi_i(\lambda) = \zeta(|\lambda| \cdot \|\varepsilon(\lambda)e_i\|_K) + \gamma_i,$$

for some constant γ_i. Plugging this into (1) for $y_i \geq 0$ we obtain an equation in ζ only:

$$\sum_{i=1}^{n} \zeta(y_i \|e_i\|_K) + \gamma_i = \zeta(\|y\|_K).$$

Choosing $y = \lambda y_1 e_1 / \|e_1\|_K + \lambda y_2 e_2 / \|e_2\|_K$, with $\lambda, y_1, y_2 > 0$, we get

$$\zeta(\lambda y_1) + \zeta(\lambda y_2) + \sum_{i=1}^{n} \gamma_i + (n-2)\zeta(0) = \zeta\left(\lambda \left\|y_1 \frac{e_1}{\|e_1\|_K} + y_2 \frac{e_2}{\|e_2\|_K}\right\|_K\right).$$

Differentiating in λ at $\lambda = 1$ and setting $f(t) = t\zeta'(t)$, $t > 0$,

$$f(y_1) + f(y_2) = f\left(\left\|y_1 \frac{e_1}{\|e_1\|_K} + y_2 \frac{e_2}{\|e_2\|_K}\right\|_K\right). \tag{2}$$

For $y_1 = y_2 = t > 0$, we obtain $2f(t) = f(\alpha t)$, with $\alpha = \left\|\frac{e_1}{\|e_1\|_K} + \frac{e_2}{\|e_2\|_K}\right\|_K$. Combining this relation with (2) gives

$$\frac{1}{2}f(y_1 \alpha) + f(y_2) = f\left(\left\|y_1 \frac{e_1}{\|e_1\|_K} + y_2 \frac{e_2}{\|e_2\|_K}\right\|_K\right).$$

For $y_1 = t/\alpha$ and $y_2 = t > 0$ we get $\frac{3}{2} f(t) = f(t\alpha')$, with $\alpha' = \|\frac{1}{\alpha} \frac{e_1}{\|e_1\|_K} + \frac{e_2}{\|e_2\|_K}\|_K$. Lemma 4 ensures that $f(t) = f(1)t^p$ with $p = \log 2 / \log \alpha \neq 0$, and with the convention that $p = 0$ if $\alpha = 1$ (in this case $f(1) = 0$). It follows that

$$\zeta(t) = \zeta'(1)t^p/p + \zeta(0)$$

if $p \neq 0$ and $\zeta(t) = \zeta(0)$ otherwise. Integrating again, we get an expression for $\eta = \beta_\epsilon \star \log h$. Letting ϵ tend to zero shows that there are constants $a, b, c \in \mathbb{R}$ such that for a.e. $t > 0$, $h(t) = ct^b e^{-at^p}$ (this is valid even if $p = 0$).

Next we find an expression of the functions f_i. We start with the relation

$$t\phi'(t) = \chi_i(t) = \zeta(|t| \cdot \|\varepsilon(t)e_i\|_K) + \gamma_i = a_i(\varepsilon(t))|t|^p + b_i,$$

for some constants $a_i(1), a_i(-1), b_i$. Thus for $t \neq 0$, $\phi'(t) = a_i(\varepsilon(t))|t|^{p-1}\varepsilon(t) + b_i/t$. Integrating (with different constants on $(-\infty, 0)$ and on $(0, \infty)$) and taking the limit $\epsilon \to 0$ as before we arrive at $f_i(t) = c_i(\varepsilon(t))|t|^{b_i} e^{-a_i|t|^p}$, for almost every t and for some constants $c_i(1), c_i(-1)$. For f_i to have a finite integral, p has to be non-zero. Our initial equation reads as: for a.e. x,

$$\prod_{i=1}^{n} c_i(\varepsilon(x_i))|x_i|^{b_i} e^{-a_i|x_i|^p} = c\|x\|_K^b e^{-a\|x\|_K^p} g\left(\frac{x}{\|x\|_K}\right).$$

By continuity this holds on $\mathbb{R}^n \setminus \{x; \prod_{i=1}^{n} x_i = 0\}$. For such an x and $\lambda > 0$, the equation becomes

$$\prod_{i=1}^{n} c_i(\varepsilon(x_i)) \lambda^{\sum_{i=1}^{n} b_i} e^{-\lambda^p \sum_{i=1}^{n} a_i(\varepsilon(x_i))|x_i|^p} \prod_{i=1}^{n} c_i|x_i|^{b_i}$$

$$= c\lambda^b \|x\|_K^b e^{-a\lambda^p \|x\|_K^p} g\left(\frac{x}{\|x\|_K}\right).$$

This clearly implies that $a\|x\|_K^p = \sum_{i=1}^{n} a_i(\varepsilon(x_i))|x_i|^p$. Since μ is a probability measure, necessarily $a, a_i(1), a_i(-1) > 0$ and $b_i > -1$. Thus K is determined. The boundedness of K forces $p > 0$. The proof is complete. $\square$

We now pass to the case of μ being an infinite measure. In this case, every star-shaped body gives rise to a measure with a simultaneous product decomposition. Indeed, for every $b_1, \ldots, b_n > -1$ in $\mathbb{R}$, and every star-shaped body K, the measure $d\mu(x) = \prod_{i=1}^{n} |x_i|^{b_i} dx$ admits such a decomposition, due to the identity:

$$\prod_{i=1}^{n} |x_i|^{b_i} = \prod_{i=1}^{n} \left(\frac{|x_i|}{\|x\|_K}\right)^{b_i} \cdot \|x\|_K^{\sum_{i=1}^{n} b_i}.$$

We can however prove that the above example is the only additional case. For simplicity we work with measures on $(0, \infty)^n$.

Theorem 5. *Let $K \subset \mathbb{R}^n$ be a star-shaped body. Assume that μ is an absolutely continuous measure on $(0, \infty)^n$ which has a simultaneous product decomposition with respect to K. Assume in addition that $\log(\frac{d\mu}{dx})$ is locally integrable. Then one of the following assertions holds:*
1) *There are $p, r > 0$, $b_1, \ldots, b_n \in \mathbb{R}$, $c \geq 0$ and $a_1, \ldots, a_n \neq 0$ all having the same sign such that:*

$$K \cap (0, \infty)^n = \left\{ x \in (0, \infty)^n; \ \sum_{i=1}^{n} |a_i| \cdot |x_i|^p \leq r \right\},$$

and

$$d\mu = c \prod_{i=1}^{n} \left(x_i^{b_i} e^{-a_i x_i^p} 1_{\{x_i > 0\}} dx_i \right).$$

2) *K is arbitrary and there are $b_1, \ldots, b_n \in \mathbb{R}$ and $c > 0$ such that*

$$d\mu = c \prod_{i=1}^{n} \left(x_i^{b_i} 1_{\{x_i > 0\}} dx_i \right).$$

Conversely if K and μ satisfy 1) or 2) then μ has a simultaneous product decomposition with respect to K.

Proof. This result follows from the proof of Theorem 3. The writing is simpler since we work on $(0, \infty)^n$. We present the modifications. If in the argument $p = 0$, then $f_i(t) = c_i t^{b_i}$ and we are done. If $p \neq 0$ then the argument provides $a, a_1, \ldots, a_n$ such that whenever $x_i > 0$

$$a\|x\|_K^p = \sum_{i=1}^{n} a_i x_i^p.$$

If $a = 0$ then $a_i = 0$ for all i's, $f_i(t) = c_i t^{b_i}$ and there is no constraint on K. If $a \neq 0$ then the previous relation gives $a_i = a\|e_i\|_K^p$, so the a_i's are not zero and have same sign. Since $\|x\|_K = \left(\sum_{i=1}^{n} \frac{a_i}{a} |x_i|^p \right)^{1/p}$, the set $K \cap (0, \infty)^n$ is a weighted ℓ_p^n-ball. By boundedness $p > 0$. As before $f_i(t) = c_i t^{b_i} e^{-a_i t^p}$. This ends the proof. □

3 Atomic Measures

In this section we focus on finite atomic measures $\sum_{P \in S} \alpha_P \delta_P$, where δ_P is the Dirac measure at P and $S \subset \mathbb{R}^n$ is countable. For convenience, we write $\mu(P)$ for $\mu(\{P\})$. We also restrict ourselves to convex sets K. The following result deals with measures which are not supported on a sphere. Measures which concentrate on a sphere, when K is convex are much easier to classify and we leave this to the reader (one of the μ_i's has to be a Dirac mass).

Theorem 6. *Assume that $K \subset \mathbb{R}^n$ is convex, symmetric and contains the origin in its interior, and that μ is a finite (and non-zero) atomic measure on $(0, \infty)^n$, which admits a simultaneous polar and Cartesian decomposition with respect to K: $\mu = \mu_1 \otimes \cdots \otimes \mu_n = \tau \odot \nu$. Assume in addition that τ is not a Dirac measure. Then the following assertions hold:*
a) There are $\lambda_1, \ldots, \lambda_n > 0$ such that $K \cap [0, \infty)^n = \prod_{i=1}^{n} [0, \lambda_i]$.
b) There are $c, r, \alpha_1, \ldots \alpha_n > 0$, $0 < q < 1$ and $D = \prod_{i=1}^{n} \{ r\lambda_i q^k; \ k \in \mathbb{N} \}$ such that :

$$\mu(x) = \begin{cases} c \prod_{i=1}^{n} x_i^{\alpha_i} & \text{if } x \in D \\ 0 & \text{if } x \notin D \end{cases}$$

Conversely, if K and μ satisfy a) and b) then μ has a simultaneous product decomposition with respect to K.

As a matter of illustration, we check that conditions *a)* and *b)* ensure simultaneous Cartesian and polar product decompositions. Let $x \in (0, \infty)^n$. Since the set D is a Cartesian product, it is clear that μ defined in *b)* is a Cartesian-product measure. Next, write $x = \rho\omega$ with $\rho > 0$ and $\omega \in \partial K$; then

$$\mu(x) = \mu(\rho\omega) = c\,\rho^{\sum_{i=1}^{n} \alpha_i} \prod_{i=1}^{n} \omega_i^{\alpha_i} \delta_D(\rho\omega),$$

so we just need to check that D is a polar product set in order to show that the latter is the product of a quantity depending only on ρ times a quantity depending on ω. But this is easy: if $x \in D$ then for each i, one has $x_i = r\lambda_i q^{k_i}$. By a), $\|x\|_K = r\, q^{\min_i k_i} \in T = \{ rq^k; \ k \in \mathbb{N} \}$ and

$$\frac{x}{\|x\|_K} = \left(\lambda_i q^{k_i - \min_j k_j} \right)_{i=1}^{n} \in \Omega = \left\{ (\lambda_i q^{h_i})_{i=1}^{n}; \ h_i \geq 0 \ \text{ and} \prod_i h_i = 0 \right\}.$$

This shows that $D \subset T \cdot \Omega$. The converse inclusion is easily checked. Hence μ has a simultaneous product decomposition.

The rest of this section is devoted to the proof of the necessary condition in Theorem 6. From now on we assume that μ and K satisfy the assumption of the theorem. We begin with some notation. If λ is a measure on a measurable space (Ω, Σ), let $\mathcal{M}_\lambda = \{ x \in \Omega; \lambda(x) = \sup_{\omega \in \Omega} \lambda(\omega) \}$. Clearly, if λ is a finite measure then $|\mathcal{M}_\lambda| < \infty$. We also put

$$\mathcal{M}_\lambda^2 = \left\{ x \in \Omega; \lambda(x) = \sup_{\omega \in \Omega \setminus \mathcal{M}_\lambda} \lambda(\omega) \right\}.$$

When λ is countably supported we define $\mathrm{supp}(\lambda) = \{ \omega \in \Omega; \lambda(\omega) > 0 \}$.

Returning to the setting of Theorem 6, we clearly have that $\mathrm{supp}(\mu) = \mathrm{supp}(\mu_1) \times \cdots \times \mathrm{supp}(\mu_n) = \mathrm{supp}(\tau) \cdot \mathrm{supp}(\nu)$ and $\mathcal{M}_\mu = \mathcal{M}_{\mu_1} \times \cdots \times \mathcal{M}_{\mu_n} = \mathcal{M}_\tau \cdot \mathcal{M}_\nu$.

The next lemma will be used for a measure other than μ of the main theorem. This is why the fact that τ in the product decomposition is not Dirac is specifically stated as an hypothesis there.

Lemma 7. *Define the slope function $s_1(x) = \frac{x_1}{\|(x_2,\ldots,x_n)\|_2}$. If τ is not a Dirac measure, then s_1 does not attain its minimum on $\mathrm{supp}(\mu)$.*

Proof. Assume that $\mu(x) > 0$. Then $\mu_1(x_1),\ldots\mu_n(x_n) > 0$, $\nu(x/\|x\|_K) > 0$ and $\tau(\|x\|_K) > 0$. Since τ is not a Dirac measure, there is $r > 0$, $r \neq \|x\|_K$ such that $\tau(r) > 0$. Let $y = \frac{r}{\|x\|_K}x$. Then $\mu(y) = \tau(r)\nu(x/\|x\|_K) > 0$, so that for every $i = 1,\ldots,n$, $\mu_i(\frac{r}{\|x\|_K}x_i) > 0$. Therefore for

$$u = \left(\frac{r}{\|x\|_K}x_1, x_2, \ldots, x_n\right),$$

$$v = \left(x_1, \frac{r}{\|x\|_K}x_2, \frac{r}{\|x\|_K}x_3, \ldots, \frac{r}{\|x\|_K}x_n\right),$$

$\mu(u) > 0$ and $\mu(v) > 0$. But $s_1(u) = \frac{r}{\|x\|_K}s_1(x)$, $s_1(v) = \frac{\|x\|_K}{r}s_1(x)$ and either $\frac{r}{\|x\|_K} < 1$ or $\frac{\|x\|_K}{r} < 1$. □

Corollary 8. *Under the assumptions of Theorem 6, $|\mathrm{supp}(\mu)| = \infty$ and $|\mathcal{M}_\tau| = 1$.*

Proof. The first assertion is obvious, and the second assertion follows since $\mu|_{\mathcal{M}_\mu}$ has a simultaneous product decomposition. Indeed $\mu|_{\mathcal{M}_\mu}$ also has a finite support, and therefore s_1 attains its minimum on it. It follows that the radial measure $\tau|_{\mathcal{M}_\mu}$ has to be a Dirac measure. □

Put $\mathcal{M}_\tau = \{r\}$.

Lemma 9. $\mathrm{supp}(\mu) \subset \{x; \|x\|_K \leq r\}$.

Proof. If $\mathrm{supp}(\mu) \not\subset \{x; \|x\|_K \leq r\}$ then since τ is a finite measure there is $R > r$ such that $\tau(R) > 0$ is maximal on (r, ∞). For every $i = 1,\ldots,n$, let $M_i = \max \mathcal{M}_{\mu_i} > 0$. Now, $x = (M_1, M_2, \ldots, M_n) \in \mathcal{M}_\mu$, so that $\|x\|_K = r$. Put $y = \frac{R}{r}x$. Clearly $\mu(y) = \tau(R)\nu(x/\|x\|_K)$ is maximal on the set $\{x; \|x\|_K > r\}$. For every $i = 1,\ldots,n$ define

$$x^i = \left(M_1, \ldots, M_{i-1}, \frac{R}{r}M_i, M_{i+1}, \ldots, M_n\right).$$

Note that for every $j = 1,\ldots,n$, $M_j = \max \mathcal{M}_{\mu_j} < \frac{R}{r}M_j = y_j$ so that $y_j \notin \mathcal{M}_{\mu_j}$. It follows that for every $i = 1,\ldots,n$, $\mu(x^i) > \mu(y)$, so that $\|x^i\|_K \leq r$. Now, using the convexity of K we have

$$r = \|x\|_K = \left\|\frac{1}{n-1+\frac{R}{r}} \sum_{i=1}^n x^i\right\|_K$$

$$\leq \frac{1}{n-1+\frac{R}{r}} \sum_{i=1}^n \|x^i\|_K \leq \frac{nr}{n-1+\frac{R}{r}} < r,$$

which is a contradiction. □

Since τ isn't Dirac, there is some $r' \in \mathcal{M}_\tau^2$. By Lemma 9, $r' < r$.

Lemma 10. *For every $i = 1, \ldots, n$, $\inf \operatorname{supp}(\mu_i) = 0$.*

Proof. As in Lemma 7, we will study the function:

$$s_i(x) = \frac{x_i}{\left\| \sum_{j \neq i} x_j e_j \right\|_2}.$$

Since Lemma 9 implies in particular that $\operatorname{supp}(\mu)$ is bounded, our claim will follow once we show that $\inf_{x \in \operatorname{supp}(\mu)} s_i(x) = 0$. Let $\sigma_i = \inf_{x \in \operatorname{supp}(\mu)} s_i(x)$ and assume that $\sigma_i > 0$. For every $\epsilon > 0$ there is $x \in \operatorname{supp}(\mu)$ such that $s_i(x) \leq (1 + \epsilon)\sigma_i$. From the proof of Lemma 7 it follows that for every $\rho \in \operatorname{supp}(\tau)$ there are $u, v \in \operatorname{supp}(\tau)$ such that $s_i(u) = \frac{\rho}{\|x\|_K} s_i(x)$ and $s_i(v) = \frac{\|x\|_K}{\rho} s_i(x)$. Hence, $\min\{\frac{\rho}{\|x\|_K}, \frac{\|x\|_K}{\rho}\} \geq \frac{1}{1+\epsilon}$. If $\|x\|_K = r$ take $\rho = r'$. Otherwise, $\|x\|_K \leq r'$, in which case take $\rho = r$. In both cases we get that $r \leq (1 + \epsilon)r'$, which is a contradiction when ϵ is small enough. $\square$

In what follows we will continue to use the notation $M_i = \max \mathcal{M}_{\mu_i}$, and we will also put $m_i = \min \mathcal{M}_{\mu_i}$. Let $x = (M_1, \ldots, M_n)$, $x' = (m_1, \ldots, m_n)$.

Corollary 11. *For every $J \subset \{1, \ldots, n\}$:*

$$\left\| \sum_{i \in J} M_i e_i \right\|_K \leq r.$$

Proof. By Lemma 10 for every $\epsilon > 0$ and $i = 1, \ldots, n$ there is $z_i \in \operatorname{supp}(\mu_i)$ with $z_i < \epsilon$. Now:

$$\sum_{i \in J} M_i e_i + \sum_{i \notin J} z_i e_i \in \operatorname{supp}(\mu),$$

so that by Lemma 9 we get:

$$\left\| \sum_{i \in J} M_i e_i + \sum_{i \notin J} z_i e_i \right\|_K \leq r.$$

The result follows by taking $\epsilon \to 0$. $\square$

Corollary 12. $\prod_{i=1}^n [0, M_i] \subset rK$.

Lemma 13. *Let J be a non-empty subset of $\{1, \ldots, n\}$. Then:*

$$\left\| \sum_{i \in J} M_i e_i + \sum_{i \notin J} \frac{r'}{r} m_i e_i \right\|_K = r.$$

Proof. Denote $y = \frac{r'}{r}x'$. Since $x' \in M_\mu$, $\|x'\|_K = r$. Now, $\mu(y) = \tau(r')\nu(x'/\|x'\|_K)$, and because $r' \in M_r^2$ and $x'/\|x'\|_K \in M_\nu$ we deduce that $\mu(y)$ is maximal on the set $\{x; \|x\|_K \neq r\}$. But since $r' < r$, for any $j = 1, \ldots, n$ one has $y_j = \frac{r'}{r}m_j < m_j = \min M_{\mu_j}$, so that $y_j \notin M_{\mu_j}$. It follows that since $J \neq \emptyset$,

$$\mu\left(\sum_{i \in J} M_i e_i + \sum_{i \notin J} \frac{r'}{r}m_i e_i\right) = \left(\prod_{i \in J} \mu_i(M_i)\right)\left(\prod_{i \notin J} \mu_i(y_i)\right) > \prod_{i=1}^{n} \mu_i(y_i) = \mu(y),$$

so that $\left\|\sum_{i \in J} M_i e_i + \sum_{i \notin J} \frac{r'}{r}m_i e_i\right\|_K = r$. $\square$

We can now prove the first part of Theorem 6:

Proposition 14. $K \cap [0, \infty)^n = \prod_{i=1}^{n} \left[0, \frac{M_i}{r}\right]$.

Proof. We set $Q = \prod_{i=1}^{n} [0, M_i]$. Let $1 \leq i \leq n$. Since $0 < m_i r'/r < M_i$ the point

$$P_i = M_i e_i + \sum_{j \neq i} \frac{r'}{r}m_j e_j$$

lies in the interior of the facet $Q \cap \{x; x_i = M_i\}$ of Q. It is also a boundary point of rK by Lemma 13. As guaranteed by Corollary 12, $Q \subset rK$, so that any supporting hyperplane of rK at P_i is a supporting hyperplane of Q at this point. Therefore at P_i the convex set rK admits $\{x; x_i = M_i\}$ as a (unique) supporting hyperplane. It follows that $rK \subset \{x; x_i \leq M_i\}$. This is true for every $1 \leq i \leq n$ and the proof is complete. $\square$

We now pass to the proof of the final assertion of Theorem 6. We have proved that there are real numbers $t_i = r/M_i > 0$, $i = 1 \ldots n$ such that for every $x \in [0, \infty)^n$, $\|x\|_K = \max_{1 \leq i \leq n} t_i x_i$. Moreover for every $x \in \operatorname{supp}(\mu)$, $\|x\|_K \leq r$ and $\mu(r/t_1, \ldots, r/t_n) > 0$. By replacing K with rK we may assume that $r = 1$. Moreover, as we remarked in Section 1, for any $a_1, \ldots, a_n > 0$, μ admits a simultaneous product decomposition if and only if $\mu \circ T_{a_1,\ldots,a_n}$ admits a simultaneous product decomposition. Therefore, by composing μ with a suitable $T_{a_1,\ldots,a_n}$, we can assume that $M_1 = \cdots = M_n = 1$ and $t_1 = \cdots = t_n = 1$. In addition, by replacing the measures μ, μ_i, ν, τ by $\mu/\mu(1, \ldots, 1)$, $\mu_i/\mu_i(1), \nu/\nu(1, \ldots, 1), \tau/\tau(1)$ we can assume that $\mu(1, \ldots, 1) = \mu_i(1) = \nu(1, \ldots, 1) = \tau(1) = 1$.

Lemma 15. *If* $p, q \in \operatorname{supp}(\tau)$ *and* $q > p$ *then* $\frac{p}{q}, pq \in \operatorname{supp}(\tau)$.

Proof. By the product property

$$0 < \tau(p) = \mu(p, \ldots, p) = \prod_{i=1}^{n} \mu_i(p),$$

so that for every $i = 1, \ldots, n$, $p \in \text{supp}(\mu_i)$. Similarly, $q \in \text{supp}(\mu_i)$. Hence, since $p < q$:

$$0 < \mu(p, q, \ldots, q) = \tau(q)\nu\left(\frac{p}{q}, 1, \ldots, 1\right),$$

so that $\nu(\frac{p}{q}, 1, \ldots, 1) > 0$. Now, using again the fact that $p < q$ we have:

$$\mu_1\left(\frac{p}{q}\right) = \mu\left(\frac{p}{q}, 1, \ldots, 1\right) = \nu\left(\frac{p}{q}, 1, \ldots, 1\right) > 0.$$

This shows that $\frac{p}{q} \in \text{supp}(\mu_1)$. Similarly, for every i, $\frac{p}{q} \in \text{supp}(\mu_i)$. Hence,

$$0 < \prod_{i=1}^{n} \mu_i\left(\frac{p}{q}\right) = \mu\left(\frac{p}{q}, \ldots, \frac{p}{q}\right) = \tau\left(\frac{p}{q}\right).$$

This shows that $\frac{p}{q} \in \text{supp}(\tau)$. Now, since the remark preceding Lemma 15 implies that $p \leq 1$,

$$\mu_1(pq) \prod_{i=2}^{n} \mu_i(q) = \mu(pq, q, \ldots, q) = \tau(q)\nu(p, 1, \ldots, 1)$$

$$= \tau(q)\mu(p, 1, \ldots, 1) = \tau(q)\mu_1(p) > 0.$$

Hence, $pq \in \text{supp}(\mu_1)$. Similarly, for every i, $pq \in \text{supp}(\mu_i)$, so that:

$$0 < \mu(pq, \ldots, pq) = \tau(pq),$$

which shows that $pq \in \text{supp}(\tau)$. □

Lemma 16. *For every $i = 1, \ldots, n$, $\text{supp}(\mu_i) = \text{supp}(\tau)$.*

Proof. In the proof of Lemma 15 we have seen that $\text{supp}(\tau) \subset \text{supp}(\mu_i)$. We show the other inclusion. First, note that $\inf \text{supp}(\tau) = 0$. Indeed for every $\epsilon > 0$, Lemma 10 ensures the existence of $x_i \in \text{supp}(\mu_i)$ such that $x_i < \epsilon$. Now,

$$0 < \prod_{i=1}^{n} \mu_i(x_i) = \tau\left(\max_{1 \leq i \leq n} x_i\right)\nu\left(\frac{x}{\|x\|_K}\right).$$

So $\inf \text{supp}(\tau) \leq \epsilon$.

Take any $p \in \text{supp}(\mu_i)$. There is some $q \in \text{supp}(\tau)$ such that $q < p$. By the proof of Lemma 15, for every j, $q \in \text{supp}(\mu_j)$; therefore:

$$0 < \mu_i(p) \prod_{j \neq i} \mu_j(q) = \mu\left(pe_i + \sum_{j \neq i} qe_j\right) = \tau(p)\nu\left(e_i + \sum_{j \neq i} \frac{q}{p}e_j\right),$$

so that $p \in \text{supp}(\tau)$. □

Lemma 17. *For every $i = 1, \ldots, n$ and for every $p, q \in \text{supp}(\mu_i)$,*

$$\mu_i(pq) = \mu_i(p)\mu_i(q).$$

Proof. Note that since $p \in \text{supp}(\tau)$, $p \leq 1$. Hence, using the fact that $q \in \text{supp}(\mu_j)$ we have:

$$0 < \mu_i(pq)\prod_{j\neq i}\mu_j(q) = \mu\left(pqe_i + \sum_{j\neq i}qe_j\right) = \tau(q)\nu\left(pe_i + \sum_{j\neq i}e_j\right)$$

$$= \tau(q)\mu_i(p) = \mu_i(p)\mu(q,\ldots,q) = \mu_i(p)\mu_i(q)\prod_{j\neq i}\mu_j(q).$$

□

Lemma 18. *Assume that $A \subset (0,1]$, $A \neq \{1\}$, $A \neq \emptyset$, has the property that xy and x/y are in A whenever $x, y \in A$ and $x \leq y$. Let $f : A \to (0,\infty)$ be a function such that if $x, y \in A$ then $f(xy) = f(x)f(y)$ and:*

$$\sum_{a\in A} f(a) < \infty.$$

Then there are $\alpha > 0$ and $0 < q < 1$ such that $f(a) = a^\alpha$ and $A = \{q^n\}_{n=0}^\infty$.

Proof. For any $a \in A \setminus \{1\}$ and $n \in \mathbb{N}$, $a^n \in A$ and $f(a^n) = f(a)^n$. Since $\sum_{n=1}^\infty f(a^n) < \infty$, $f(a) < 1$. Now, if $a, b \in A \setminus \{1\}$ and $\frac{a^n}{b^m} < 1$, $\frac{a^n}{b^m} \in A$, so that $1 > f\left(\frac{a^n}{b^m}\right) = \frac{f(a)^n}{f(b)^m}$. We have shown that for every $n, m \in \mathbb{N}$ and $a, b \in A \setminus \{1\}$:

$$\frac{n}{m} > \frac{\log b}{\log a} \implies \frac{n}{m} > \frac{\log f(b)}{\log f(a)}.$$

Hence, $\frac{\log b}{\log a} \geq \frac{\log f(b)}{\log f(a)}$ for every $a, b \in A \setminus \{1\}$. By symmetry, there is $\alpha \in \mathbb{R}$ such that for every $a \in A$, $\frac{\log f(a)}{\log a} = \alpha$. This proves the first assertion ($\alpha > 0$ since $f(a) < 1$).

Put $B = \{-\log a; a \in A\}$. Clearly:

$$a, b \in B \implies a + b, |a - b| \in B.$$

Since $f(a) = a^\alpha$ and $\sum_{a\in A} f(a) < \infty$, for every $x > 0$ there are only finitely many $a \in A$ with $a \geq x$. In other words, for every $x > 0$ there are only finitely many $b \in B$ with $b \leq x$. In particular, if we let $p = \inf B \setminus \{0\}$ then $p > 0$ and $p \in B$. Now, for every $n = 0, 1, 2, \ldots$, $np \in B$. We claim that $B = \{0, p, 2p, 3p, \ldots\}$. Indeed, if $x \in B \setminus \{0, p, 2p, 3p, \ldots\}$ then there is an integer n such that $0 < |x - np| < p$. But, $|x - np| \in B$, and this contradicts the definition of p. Finally, for $q = e^{-p}$, $A = \{1, q, q^2, \ldots\}$.   □

Remark. All the assumptions in Lemma 18 are necessary. Apart from the trivial examples such as $A = (0,1]$ and $A = (0,1] \cap \mathbb{Q}$ we would like to point out the more interesting example $A = \{2^n 3^m; m, n \in \mathbb{Z} \text{ and } 2^n 3^m \leq 1\}$, $f(2^n 3^m) = 2^{\alpha m} 3^{\beta m}$ where α and β are distinct real numbers (of course in this case the condition $\sum_{a \in A} f(a) < \infty$ is not satisfied).

Proof of Theorem 6. Assertion $a)$ is given by Proposition 14. To prove $b)$ fix some $1 \leq i \leq n$ and define: $A = \mathrm{supp}(\mu_i)$. By Lemma 16 and Lemma 9, $A = \mathrm{supp}(\tau) \subset (0,1]$. Additionally, Lemma 15 implies that if $x, y \in A$, $x \leq y$ then $xy, \frac{x}{y} = \in A$. Clearly for every $x \in A$, $\mu_i(x) > 0$ and since μ_i is a finite measure, $\sum_{a \in A} \mu_i(a) < \infty$. An application of Lemma 17 gives that for every $x, y \in A$, $\mu_i(xy) = \mu_i(x)\mu_i(y)$. Now, Lemma 18 implies that there are $\alpha_i > 0$ and $0 < q < 1$ such that $\mu_i(a) = a^{\alpha_i}$ and $A = \{q^n\}_{n=0}^\infty$. So, $\mathrm{supp}(\tau) = \{q^k; \ k \in \mathbb{N}\}$ and by Lemma 16, one gets $\mathrm{supp}(\mu_i) = \{q^k; \ k \in \mathbb{N}\}$. Moreover, $\mu_i(q^k) = q^{k\alpha_i}$. This concludes the proof of the theorem. $\square$

4 Concluding Remarks

In this section we list some remarks and open problems that arise from the results of the previous two sections.

1) There are examples when K is allowed to be unbounded (of course in this case it is no longer a body). Indeed the "unit ball" of ℓ_p^n for non-positive p gives such a decomposition with $f_i(t) = |t|^{b_i} \exp(-|t|^p)$.

2) Theorem 3 does not cover the case of the uniform measure on $B_\infty^n = [-1,1]^n$, which clearly has simultaneously the Cartesian and the polar decomposition with respect to $K = B_\infty^n$. It is the natural limit case of the examples with the densities $\exp(-|t|^p)$. Under strong conditions on the density and its support, results can be obtained which encompass measures supported on the cube. It would be very nice to get rid of the conditions. It seems that one of the necessary steps would be to understand the structure of sets in $\mathbb{R}^n$ which are products with respect to the Cartesian structure and for the polar structure generated by a convex set K. This is a problem of independent interest.

3) The classification of simultaneous product measures, without additional hypothesis, is a very challenging problem. Note that our results may be used. Indeed if μ has simultaneous product decomposition, then its absolutely continuous part has it too. Similarly, if a singular measure has the property, then its atomic part has it too, so Theorem 6 applies. The main obstacle seems to be dealing with singular continuous measures.

Acknowledgement. Part of this work was carried out while the first and last-named authors were visiting University College London. They are grateful to their UCL hosts, Professors K. Ball and D. Preiss, for their invitation.

References

[BN] Barthe, F., Naor, A.: Hyperplane projections of the unit ball of ℓ_p^n. Discrete Comput. Geom., to appear

[Br] Bryc, W. (1995): The normal distribution. Characterizations with applications. Lecture Notes in Statistics, **100**, Springer Verlag, New York

[C]   Carlen, E.A. (1991): Superadditivity of Fisher's information and logarithmic Sobolev inequalities. J. Funct. Anal., **101(1)**, 194–211

[N]   Naor, A. (2001): The surface measure and cone measure on the sphere of ℓ_p^n. Preprint

[NR] Naor, A., Romik, D.: Projecting the surface measure of the sphere of ℓ_p^n. Ann. Inst. H. Poincaré Probab. Statist., to appear

[RR] Rachev, S.T., Rüschendorf, L. (1991): Approximate independence of distributions on spheres and their stability properties. Ann. Probab., **19(3)**, 1311–1337

[SZ1] Schechtman, G., Zinn, J. (1990): On the volume of intersection of two L_p^n balls. Proc. Amer. Math. Soc., **110(1)**, 217–224

[SZ2] Schechtman, G., Zinn, J. (2000): Concentration on the ℓ_p^n ball. Geometric Aspects of Functional Analysis, Lecture Notes in Math., **1745**, Springer Verlag, Berlin, 245–256

Approximating a Norm by a Polynomial*

Alexander Barvinok

Department of Mathematics, University of Michigan, Ann Arbor, MI 48109-1109, USA *barvinok@umich.edu*

Summary. We prove that for any norm $\| \cdot \|$ in the d-dimensional real vector space V and for any odd $n > 0$ there is a non-negative polynomial $p(x)$, $x \in V$ of degree $2n$ such that

$$p^{\frac{1}{2n}}(x) \leq \|x\| \leq \binom{n+d-1}{n}^{\frac{1}{2n}} p^{\frac{1}{2n}}(x).$$

Corollaries and polynomial approximations of the Minkowski functional of a convex body are discussed.

1 Introduction and the Main Result

Our main motivation is the following general question. Let us fix a norm $\| \cdot \|$ in a finite-dimensional real vector space V (or, more generally, the Minkowski functional of a convex body in V). Given a point $x \in V$, how fast can one compute or approximate $\|x\|$? For example, various optimization problems can be posed this way. As is well known (see, for example, Lecture 3 of [B]), any norm in V can be approximated by an ℓ_2 norm in V within a factor of $\sqrt{\dim V}$. From the computational complexity point of view, an ℓ_2 norm of x is just the square root of a positive definite quadratic form p in x and hence can be computed "quickly", that is, in time polynomial in $\dim V$ for any $x \in V$ given by its coordinates in some basis of V. Note that we do not count the time required for "preprocessing" the norm to obtain the quadratic form p, as we consider the norm fixed and not a part of the input. It turns out that by employing higher degree forms p, we can improve the approximation: for any $c > 0$, given an $x \in V$, one can approximate $\|x\|$ within a factor of $c\sqrt{\dim V}$ in time polynomial in $\dim V$. This, and some other approximation results follow easily from our main theorem.

Theorem 1.1. *Let V be a d-dimensional real vector space and let $\| \cdot \| : V \longrightarrow \mathbb{R}$ be a norm in V. For any odd integer $n > 0$ there exists a homogeneous polynomial $p : V \longrightarrow \mathbb{R}$ of degree $2n$ such that $p(x) \geq 0$ (in fact, p is the sum of squares of homogeneous polynomials of degree n) and*

$$p^{\frac{1}{2n}}(x) \leq \|x\| \leq \binom{n+d-1}{n}^{\frac{1}{2n}} p^{\frac{1}{2n}}(x)$$

for all $x \in V$.

* This research was partially supported by NSF Grant DMS 9734138.

We observe that the approximation factor $\binom{n+d-1}{n}^{1/2n}$ approaches 1 as long as $n/d \longrightarrow \infty$, see also Section 2.2.

We prove Theorem 1.1 in Section 3. In Section 2, we discuss some corollaries of Theorem 1.1. In Section 4, we show how to extend Theorem 1.1 to Minkowski functionals of arbitrary convex bodies.

1.1 Best Approximating Polynomials

Let us fix a norm $\| \cdot \|$ in a finite-dimensional vector space V and a positive integer n. One can ask what is the smallest possible constant $C = C(\| \cdot \|, n)$ for which there exists a polynomial p of degree $2n$ such that

$$p^{\frac{1}{2n}}(x) \leq \|x\| \leq Cp^{\frac{1}{2n}}(x) \quad \text{for all} \quad x \in V. \tag{1.1.1}$$

Moreover, what is the value of

$$C(d,n) = \sup_{\substack{\|\cdot\| \text{ is a norm in } V \\ \text{and } \dim V = d}} C(\| \cdot \|, n).$$

Theorem 1.1 asserts that $C(d,n) \leq \binom{n+d-1}{n}^{1/2n}$ for odd n. It is not known whether the equality holds except in the case of $n = 1$ when indeed $C(d,1) = \sqrt{d}$ (see, for example, Lecture 3 of [B]). The following simple observation can be useful to determine what a best approximating polynomial may look like.

Suppose that there is a set of finitely many non-negative polynomials p_i with $\deg p_i = 2n$ which satisfy (1.1.1). Thus we have

$$p_i(x) \leq \|x\|^{2n} \leq C^{2n}p_i(x) \quad \text{for all} \quad x \in V \quad \text{and for every} \quad p_i. \tag{1.1.2}$$

Then any convex combination p of polynomials p_i satisfies (1.1.2) and hence (1.1.1). Suppose now that the normed space V possesses a compact group G of linear isometries. If a polynomial p satisfies (1.1.1) then, for any $g \in G$, the polynomial $p_g(x) = p(gx)$ satisfies (1.1.1) with the same constant C and hence, by averaging over G, we can choose a G-invariant polynomial p which satisfies (1.1.1). Hence if a norm $\|\cdot\|$ is invariant under the action of a compact group, we can always choose an invariant best approximating polynomial.

2 Corollaries and Remarks

2.1 Approximation by Polynomials of a Fixed Degree

Let us fix an n in Theorem 1.1. Then, as d grows, the value of $p^{1/2n}(x)$ approximates $\| \cdot \|$ within a factor of $c_n\sqrt{d}$, where $c_n \approx (n!)^{-1/2n} \approx \sqrt{e/n}$. Since for any fixed n, computation of $p(x)$ takes a $d^{O(n)}$ time, for any $c > 0$ we obtain a polynomial time algorithm to approximate $\|x\|$ within a factor of $c\sqrt{d}$ (again, we do not count the time required for preprocessing, that is, to find the polynomial p).

One may wonder whether a significantly better approximation factor, for example $C(d, n) = O(d^{1/2n})$, can be achieved in Theorem 1.1 (that would also agree with the $\sqrt{d}$ bound in the classical case of $n = 1$). This, however, does not seem to be the case as the following example shows. Let $\| \cdot \|$ be the ℓ_1 norm in $\mathbb{R}^d$, that is

$$\|x\| = \sum_{i=1}^{d} |\xi_i| \quad \text{for} \quad x = (\xi_1, \ldots, \xi_d).$$

The norm $\| \cdot \|$ is invariant under signed permutations of the coordinates

$$(\xi_1, \ldots, \xi_d) \longmapsto (\pm\xi_{i_1}, \ldots \pm \xi_{i_d})$$

and hence, as discussed in Section 1.1, we can choose an invariant best approximating polynomial p. Hence p is a symmetric polynomial in $\xi_1^2, \ldots, \xi_d^2$. In particular, for $n = 3$, we have

$$p(x) = \alpha_d \sum_{i=1}^{d} \xi_i^6 + \beta_d \sum_{1 \leq i \neq j \leq d} \xi_i^2 \xi_j^4 + \gamma_d \sum_{1 \leq i < j < k \leq d} \xi_i^2 \xi_j^2 \xi_k^2$$

for some real α_d, β_d and γ_d. Since we must have $0 \leq p(x) \leq \|x\|^6$, by substituting $x = (1, 0, \ldots, 0)$, $x = (1, 1, 0, \ldots, 0)$ and $x = (1, 1, 1, 0, \ldots, 0)$ we get

$$0 \leq \alpha_d \leq 1, \quad 0 \leq 2\alpha_d + 2\beta_d \leq 64 \quad \text{and} \quad 0 \leq 3\alpha_d + 6\beta_d + \gamma_d \leq 729,$$
$$\text{which implies that} \quad \alpha_d \leq 1, \quad \beta_d \leq 32 \quad \text{and} \quad \gamma_d \leq 735.$$

Substituting $x = (1, \ldots, 1)$, we observe that $\|x\| = d$ and that $p(x) = O(d^3)$. Therefore, we must have $C(d, 3) \geq C(\ell_1, 3) \geq c\sqrt{d}$ for some absolute constant $c > 0$.

2.2 Linear Growth of the Degree

If we allow n to grow linearly with d, we can get a constant factor approximation. Indeed, if we choose $n = \gamma d$ for some $\gamma > 0$ in Theorem 1.1, for large d we have

$$C_0(\gamma) = \binom{n + d - 1}{n}^{\frac{1}{2n}} \approx \exp\left\{ \frac{1}{2} \ln \frac{\gamma + 1}{\gamma} + \frac{1}{2\gamma} \ln(\gamma + 1) \right\}. \quad (2.2.1)$$

Thus $p^{1/2n}(x)$ approximates $\| \cdot \|$ within a factor of $C_0(\gamma)$ depending on γ alone. In particular, if $\gamma \longrightarrow \infty$, the approximation factor approaches 1. Since for any fixed $\gamma > 0$, computation of $p(x)$ takes $2^{O(d)}$ time, for any constant

$c > 1$ we can get an algorithm of $2^{O(d)}$ complexity approximating the value of $\|x\|$ within a factor of c.

The anonymous referee noticed that a different constant factor approximation can be achieved via the following construction. Let V^* be the dual space and let

$$B^* = \left\{ f \in V^* : \quad f(x) \leq 1 \quad \text{for all} \quad x \in V \quad \text{such that} \quad \|x\| \leq 1 \right\}$$

be the unit ball of the dual norm in V^*. As is known, (see, for example, Lemma 4.10 of [P]), for any $0 < \delta < 1$ one can choose a set N of $|N| \leq (1 + 2/\delta)^d$ points in B^* which form a δ-net (in the dual norm). Given an integer n, let us define the polynomial p by

$$p(x) = \frac{1}{|N|} \sum_{f \in N} f^{2n}(x).$$

Let us fix a $\gamma > 0$. It is not hard to show that if $\delta = \delta(\gamma)$ is chosen in the optimal way, as long as $n = \gamma d$, the value of $p^{1/2n}(x)$ approximates $\|x\|$ within a constant factor $C_1(\gamma)$. Interestingly, for $\gamma \longrightarrow \infty$, the asymptotics of (2.2.1) and $C_1(\gamma)$ coincide:

$$C_0(\gamma), C_1(\gamma) = 1 + \frac{\ln \gamma}{2\gamma}(1 + o(1)) \quad \text{as} \quad \gamma \longrightarrow \infty.$$

However, for small γ, the bound $C_0(\gamma)$ of (2.2.1) is substantially better than the one obtained for $C_1(\gamma)$ using this construction. We have

$$C_0(\gamma) = \sqrt{e/\gamma}(1 + o(1)) \quad \text{for} \quad \gamma \longrightarrow 0 \quad \text{as opposed to}$$
$$C_1(\gamma) = 3^{\frac{1}{2\gamma}\left(1 + o(1)\right)} \quad \text{for} \quad \gamma \longrightarrow 0.$$

2.3 Approximating by Other Computable Functions

It is possible that one can achieve a better approximation by employing a wider class of computable functions. For example, the ℓ_1 norm which appears to be resistant to polynomial approximations (cf. Section 2.1) is very easy to compute. A natural candidate would be the class of functions which are sums of $p_i^{1/2n}$ for different polynomials p_i. In particular, the ℓ_1 norm itself is a function of this type.

3 Proof of Theorem 1.1

Let B be the unit ball of $\|\cdot\|$, so

$$B = \left\{ x \in V : \|x\| \leq 1 \right\}.$$

Hence B is a convex compact set containing the origin in its interior and symmetric about the origin.

Let V^* be the dual space of all linear functions $f : V \longrightarrow \mathbb{R}$ and let $B^* \subset V^*$ be the polar of B:

$$B^* = \left\{ f \in V^* : f(x) \leq 1 \quad \text{for all} \quad x \in B \right\}.$$

Hence B^* is a convex compact set symmetric about the origin. Using the standard duality argument, we can write

$$\|x\| = \max_{f \in B^*} f(x). \tag{3.1}$$

Let

$$W = V^{\otimes n} = \underbrace{V \otimes \ldots \otimes V}_{n \text{ times}} \quad \text{and} \quad W^* = \left(V^{\otimes n}\right)^* = \underbrace{V^* \otimes \ldots \otimes V^*}_{n \text{ times}}$$

be the n-th tensor powers of V and V^* respectively.

For vectors $x \in V$ and $f \in V^*$ let

$$x^{\otimes n} = \underbrace{x \otimes \ldots \otimes x}_{n \text{ times}} \quad \text{and} \quad f^{\otimes n} = \underbrace{f \otimes \ldots \otimes f}_{n \text{ times}}$$

denote the n-th tensor power $x^{\otimes n} \in W$ and $f^{\otimes n} \in W^*$ respectively.

By (3.1), we can write

$$\|x\|^n = \max_{f \in B^*} \left(f(x)\right)^n = \max_{f \in B^*} f^{\otimes n}(x^{\otimes n}). \tag{3.2}$$

Let D be the convex hull of $f^{\otimes n}$ for $f \in B^*$:

$$D = \text{conv}\left\{ f^{\otimes n} : f \in B^* \right\}.$$

Then D is a convex compact subset of W^*, symmetric about the origin (we use that n is odd). From (3.2) we can write

$$\|x\|^n = \max_{f \in B^*} f^{\otimes n}(x^{\otimes n}) = \max_{g \in D} g(x^{\otimes n}). \tag{3.3}$$

Let us estimate the dimension of D. There is a natural action of the symmetric group S_n in W^* which permutes the factors V^*, so that

$$\sigma(f_1 \otimes \ldots \otimes f_n) = f_{\sigma^{-1}(1)} \otimes \ldots \otimes f_{\sigma^{-1}(n)}.$$

Let $\text{Sym}(W^*) \subset W^*$ be the *symmetric part* of W^*, that is, the invariant subspace of that action. As is known, the dimension of $\text{Sym}(W^*)$ is that of the space of homogeneous polynomials of degree n in d real variables (cf., for example, Lecture 6 of [FH]). Next, we observe that $f^{\otimes n} \in \text{Sym}(W^*)$ for all $f \in V^*$ and, therefore,

$$\dim D \le \dim \mathrm{Sym}(W^*) = \binom{n+d-1}{n} \tag{3.4}$$

Let E be the John ellipsoid of D in the affine hull of D, that is the (unique) ellipsoid of the maximum volume inscribed in D. As is known, (see, for example, Lecture 3 of [B])

$$E \subset D \subset (\sqrt{\dim D})E.$$

Combining this with (3.3), we write

$$\max_{g \in E} g(x^{\otimes n}) \le \|x\|^n \le (\sqrt{\dim D}) \max_{g \in E} g(x^{\otimes n})$$

and, by (3.4),

$$\max_{g \in E} g(x^{\otimes n}) \le \|x\|^n \le \binom{n+d-1}{n}^{\frac{1}{2}} \max_{g \in E} g(x^{\otimes n}). \tag{3.5}$$

Let

$$q(x) = \max_{g \in E} g(x^{\otimes n}).$$

We claim that $p(x) = q^2(x)$ is a homogeneous polynomial in x of degree $2n$. Indeed, let us choose a basis $e_1, \ldots, e_d$ in V. Then W acquires the basis

$$e_{i_1 \ldots i_n} = e_{i_1} \otimes \ldots \otimes e_{i_n} \quad \text{for} \quad 1 \le i_1, \ldots, i_n \le d.$$

Geometrically, V and V^* are identified with $\mathbb{R}^d$ and W and W^* are identified with $\mathbb{R}^{d^n}$. Let $K \subset W^*$ be the Euclidean unit ball defined by the inequality

$$K = \left\{ h \in W^* : \sum_{1 \le i_1, \ldots, i_n \le d} h^2(e_{i_1 \ldots i_n}) \le 1 \right\}.$$

Since E is an ellipsoid, there is a linear transformation $T : W^* \longrightarrow W^*$ such that $T(K) = E$. Let $T^* : W \longrightarrow W$ be the conjugate linear transformation and let $y = T^*(x^{\otimes n})$. Hence the coordinates $y_{i_1 \ldots i_n}$ of y with respect to the basis $\{e_{i_1 \ldots i_n}\}$ are polynomials in x of degree n. Then

$$q(x) = \max_{g \in E} g(x^{\otimes n}) = \max_{h \in K} T(h)(x^{\otimes n}) = \max_{h \in K} h(T^*(x^{\otimes n}))$$

$$= \max_{h \in K} h(y) = \sqrt{\sum_{1 \le i_1, \ldots, i_n \le d} y_{i_1 \ldots i_n}^2}.$$

Hence we conclude that $p(x) = q^2(x)$ is a homogeneous polynomial in x of degree $2n$, which is non-negative for all $x \in V$ (moreover, $p(x)$ is a sum of squares). From (3.5), we conclude that

$$p^{\frac{1}{2n}}(x) \le \|x\| \le \binom{n+d-1}{n}^{\frac{1}{2n}} p^{\frac{1}{2n}}(x),$$

as claimed.

4 An Extension to Minkowski Functionals

There is a version of Theorem 1.1 for Minkowski functionals of convex bodies which are not necessarily symmetric about the origin.

Theorem 4.1. *Let V be a d-dimensional real vector space, let $B \subset V$ be a convex compact set containing the origin in its interior and let $\|x\| = \inf\{\lambda > 0 : x \in \lambda B\}$ be its Minkowski functional. For any odd integer $n > 0$ there exist a homogeneous polynomial $p : V \longrightarrow \mathbb{R}$ of degree $2n$ and a homogeneous polynomial $r : V \longrightarrow \mathbb{R}$ of degree n such that $p(x) \geq 0$ and*

$$\left(r(x) + \sqrt{p(x)}\right)^{\frac{1}{n}} \leq \|x\| \leq \left(r(x) + \binom{n+d-1}{n}\sqrt{p(x)}\right)^{\frac{1}{n}}$$

for all $x \in V$.

Proof. The proof follows the proof of Theorem 1.1 with some modifications. Up to (3.4) no essential changes are needed (note, however, that now we have to use that n is odd in (3.2)). Then, since the set D is not necessarily symmetric about the origin, we can only find an ellipsoid E (centered at the origin) of W^* and a point $w \in D$, such that

$$E \subset D - w \subset (\dim D)E,$$

see, for example, Lecture 3 of [B]. Then (3.5) transforms into

$$\max_{g \in E} g(x^{\otimes n}) \leq \|x\|^n - w(x^{\otimes n}) \leq \binom{n+d-1}{n}\max_{g \in E} g(x^{\otimes n}).$$

Denoting

$$p(x) = \left(\max_{g \in E} g(x^{\otimes n})\right)^2 \quad \text{and} \quad r(x) = w(x^{\otimes n})$$

we proceed as in the proof of Theorem 1.1. □

Acknowledgement. The author is grateful to the anonymous referee for useful comments and interesting questions.

References

[B]   Ball, K. (1997): An elementary introduction to modern convex geometry. Flavors of Geometry, Math. Sci. Res. Inst. Publ., **31**, Cambridge University Press, Cambridge, 1–58

[FH]   Fulton, W., Harris, J. (1991): Representation Theory. A First Course. Graduate Texts in Mathematics, **129**, Springer-Verlag, New York

[P]   Pisier, G. (1989): The Volume of Convex Bodies and Banach Space Geometry. Cambridge Tracts in Mathematics, **94**, Cambridge University Press, Cambridge

Concentration of Distributions of the Weighted Sums with Bernoullian Coefficients*

S.G. Bobkov

School of Mathematics, University of Minnesota, 127 Vincent Hall, 206 Church St. S.E., Minneapolis, MN 55455, USA *bobkov@math.umn.edu*

Summary. For non-correlated random variables, we study a concentration property of the distributions of the weighted sums with Bernoullian coefficients. The obtained result is used to derive an "almost surely version" of the central limit theorem.

Let $X = (X_1, \ldots, X_n)$ be a vector of n random variables with finite second moments such that, for all k, j,

$$\mathbf{E}\, X_k X_j = \delta_{kj} \tag{1}$$

where δ_{kj} is Kronecker's symbol. It is known that, for growing n, the distribution functions

$$F_\theta(x) = \mathbf{P}\left\{ \sum_{k=1}^{n} \theta_k X_k \le x \right\}, \quad x \in \mathbf{R},$$

of the weighted sums of (X_k), with coefficients $\theta = (\theta_1, \ldots, \theta_n)$ satisfying $\theta_1^2 + \ldots + \theta_n^2 = 1$, form a family possessing a certain concentration property with respect to the uniform measure σ_{n-1} on the unit sphere S^{n-1}. Namely, most of F_θ's are close to the average distribution

$$F(x) = \int_{S^{n-1}} F_\theta(x)\, d\sigma_{n-1}(\theta)$$

in the sense that, for each $\delta > 0$, there is an integer n_δ such that if $n \ge n_\delta$ one can select a set of coefficients $\Theta \subset S^{n-1}$ of measure $\sigma_{n-1}(\Theta) \ge 1 - \delta$ such that $d(F_\theta, F) \le \delta$, for all $\theta \in \Theta$. This property was first observed by V.N. Sudakov [S] who stated it for the Kantorovich–Rubinshtein distance $d(F_\theta, F) = \int_{-\infty}^{+\infty} |F_\theta(x) - F(x)|\, dx$, with a proof essentially relying on the isoperimetric theorem on the sphere. A different approach to this result was suggested by H. von Weizsäcker [W] (cf. also [D-F]). V.N. Sudakov also considered "Gaussian coefficients" in which case, as shown in [W], there is a rather general infinite dimensional formulation. An important special situation where the random vector X is uniformly distributed over a centrally symmetric convex body in $\mathbf{R}^n$ was recently studied, for the uniform distance

---

* Supported by an NSF grant and EPSRC Visiting Fellowship.

$\sup_x |F_\theta(x) - F(x)|$, by M. Antilla, K. Ball, and I. Perissinaki [A-B-P], see also [B] for refinements and extensions to log-concave distributions. One can find there quantitative versions of Sudakov's theorem, while in the general case, the following statement proven in [B] holds true: under (1), for all $\delta > 0$,

$$\sigma_{n-1}\{L(F_\theta, F) \geq \delta\} \leq 4n^{3/8}\, e^{-n\delta^4/8}. \tag{2}$$

Here $L(F_\theta, F)$ stands for the Lévy distance defined as the minimum over all $\delta \geq 0$ such that $F(x - \delta) - \delta \leq F_\theta(x) \leq F(x + \delta) + \delta$, for all $x \in \mathbf{R}$. As well as the Kantorovich–Rubinshtein distance d, the metric L is responsible for the weak convergence, and there is a simple relation $d(F_\theta, F) \leq 6L(F_\theta, F)^{1/2}$ (so one can give an appropriate estimate for d on the basis of (2)).

The aim of this note is to show that a property similar to (2) still holds with respect to very small pieces of the sphere. As a basic example, we consider coefficients of the special form $\theta = \frac{1}{\sqrt{n}}\varepsilon$ where $\varepsilon = (\varepsilon_1, \dots, \varepsilon_n)$ is an arbitrary sequence of signs ± 1. Thus, consider the weighted sums

$$S_\varepsilon = \frac{1}{\sqrt{n}}\sum_{k=1}^{n}\varepsilon_k X_k$$

together with their distribution functions $F_\varepsilon(x) = \mathbf{P}\{S_\varepsilon \leq x\}$ and the corresponding average distribution

$$F(x) = \int_{\{-1,1\}^n} F_\varepsilon(x)\, d\mu_n(\varepsilon) = \frac{1}{2^n}\sum_{\varepsilon_k = \pm 1}\mathbf{P}\left\{\frac{\varepsilon_1 X_1 + \dots + \varepsilon_n X_n}{\sqrt{n}} \leq x\right\}. \tag{3}$$

Here and throughout, μ_n stands for the normalized counting measure on the discrete cube $\{-1, 1\}^n$. We prove:

Theorem 1. *Under* (1), *for all* $\delta > 0$,

$$\mu_n\{\varepsilon : L(F_\varepsilon, F) \geq \delta\} \leq Cn^{1/4}\, e^{-cn\delta^8}, \tag{4}$$

where C and c are certain positive numerical constants.

Note that the condition (1) is invariant under rotations, i.e., it is fulfilled for random vectors $U(X)$ with an arbitrary linear unitary operator U in $\mathbf{R}^n$. Being applied to such vectors, the inequality (4) will involve the average $F = F^U$ which of course depends on U. However, under mild integrability assumptions on the distribution of X, all these F^U (not just most of them) turn out to be close to the one appearing in Sudakov's theorem as the typical distribution for the uniformly distributed (on the sphere) or suitably squeezed Gaussian coefficients. In particular, one can give an analogue of (4) with a certain distribution F not depending on the choice of the basis in $\mathbf{R}^n$. On the other hand, some additional natural assumptions lead to the following version of the central limit theorem. We will denote by μ_∞ the canonical infinite product measure $\mu_1 \otimes \mu_1 \otimes \dots$ on the product space $\{-1, 1\}^\infty$.

Theorem 2. *Let $\{X_{n,k}\}_{k=1}^n$ be an array of random variables satisfying (1) for all n and such that in probability, as $n \to \infty$,*

a) $\dfrac{\max\{|X_{n,1}|,\ldots,|X_{n,n}|\}}{\sqrt{n}} \to 0,$

b) $\dfrac{X_{1,1}^2+\ldots+X_{n,n}^2}{n} \to 1.$

Then, for μ_∞-almost all sequences $\{\varepsilon_k\}_{k\geq 1}$ of signs,

$$\frac{1}{\sqrt{n}} \sum_{k=1}^n \varepsilon_k X_{n,k} \to N(0,1), \quad as \quad n \to \infty.$$

If we consider the sum $\frac{1}{\sqrt{n}} \sum_{k=1}^n \varepsilon_k X_{n,k}$ with ε_k regarded as independent Bernoullian random variables which are independent of all $X_{n,k}$, then the above statement will become much weaker and will express just the property that the average distribution F defined by (3) for the random vector $(X_{1,1},\ldots,X_{n,n})$ is close to $N(0,1)$ (here is actually a step referring to the assumptions a) and b)). In addition to this property, we need to have a sufficiently good closeness (in spaces of finite dimension) of most of F_ε's to F and thus to the normal law.

Both the assumption a) and b) are important for the conclusion of Theorem 2. Under a), the property b) is necessary. To see that a) cannot be omitted, assume that the underlying probability space $(\Omega, \mathbf{P})$ is non-atomic and take a partition $A_{n,1},\ldots,A_{n,n}$ of Ω consisting of the sets of $\mathbf{P}$-measure $1/n$. Then, the array $X_{n,k} = \sqrt{n}\,1_{A_{n,k}}$, $1 \leq k \leq n$, satisfies (1), and

$$\frac{\max\{|X_{n,1}|,\ldots,|X_{n,n}|\}}{\sqrt{n}} = 1, \qquad \frac{X_{1,1}^2 + \ldots + X_{n,n}^2}{n} = 1,$$

so, the property b) is fulfilled, while a) is not. On the other hand, for any sign sequence $(\varepsilon_1,\ldots,\varepsilon_n)$, the random variable $\frac{1}{\sqrt{n}} \sum_{k=1}^n \varepsilon_k X_{n,k}$ takes only the two values ± 1, so it cannot be approximated by the standard normal distribution. Note, however, that Theorem 1 still holds in this degenerate case, with the μ_n-typical distribution F having two equal atoms at ± 1.

It might be worthwhile also noting that in general it is not possible to state Theorem 2 for any prescribed coefficients, say, for $\varepsilon_k = 1$ – similarly to the case of independent variables, even if, for each n, $\{X_{n,k}\}$ are bounded, symmetrically distributed and pairwise independent. For example, start from a sequence of independent Bernoullian random variables $\xi_1,\ldots,\xi_d$ (with $\mathbf{P}\{\xi_k = \pm 1\} = \frac{1}{2}$) and construct a double index sequence $X_{n,(k,j)} = \xi_k\xi_j$, $1 \leq k < j \leq d$. The collection $\{X_{n,(k,j)}\}$, of cardinality $n = d(d-1)/2$, satisfies the basic correlation condition (1), and since $|X_{n,(k,j)}| = 1$, both the assumption a) and b) are fulfilled. Nevertheless, in probability, as $d \to \infty$,

$$\frac{1}{\sqrt{n}} \sum_{1 \leq k < j \leq d} X_{n,(k,j)} = \frac{1}{2\sqrt{n}} \left(\sum_{k=1}^d \xi_k \right)^2 - \frac{d}{2\sqrt{n}} \to \frac{\zeta^2 - 1}{\sqrt{2}}$$

where $\zeta \in N(0,1)$.

We turn to the proof of Theorem 1. To this task, we first study the concentration property of the family $\{F_\varepsilon\}$ on the level of their characteristic functions

$$f_\varepsilon(t) = \mathbf{E}\, e^{itS_\varepsilon}, \quad t \in \mathbf{R}.$$

Concentration of $\{f_\varepsilon\}$ around its μ_n-mean

$$f(t) = \int f_\varepsilon(t)\, d\mu_n(\varepsilon) = \int_{-\infty}^{+\infty} e^{itx}\, dF(x)$$

can be then converted, with the help of standard facts from Fourier analysis, into the concentration of distributions in the form (4). This route somewhat different than that of [A-B-P] or [B] has apparently to be chosen in view of a specific form of concentration on the discrete cube.

With every complex-valued function f on $\{-1,1\}^n$, we connect the length of the discrete gradient $|\nabla f|$ defined by

$$|\nabla f(\varepsilon)|^2 = \sum_{k=1}^{n} \left| \frac{f(\varepsilon) - f(s_k(\varepsilon))}{2} \right|^2, \quad \varepsilon \in \{-1,1\}^n,$$

where $s_k(\varepsilon)$ is the neighbour of ε along kth coordinate, i.e., $(s_k(\varepsilon))_j = \varepsilon_j$ for $j \neq k$, and $(s_k(\varepsilon))_k = -\varepsilon_k$. Set $\|\nabla f\|_\infty = \max_\varepsilon |\nabla f(\varepsilon)|$.

Lemma 1. *For every f such that $\|\nabla f\|_\infty \leq \sigma$,*

$$\mu_n \left\{ \left| f - \int f\, d\mu_n \right| \geq h \right\} \leq 4e^{-h^2/(4\sigma^2)}, \quad h > 0.$$

This Gaussian bound is standard. It can be obtained using the so-called modified logarithmic Sobolev inequalities, see e.g. [B-G], [L]. In fact, for real-valued f, a sharper estimate holds true,

$$\mu_n \left\{ \left| f - \int f\, d\mu_n \right| \geq h \right\} \leq 2e^{-h^2/(2\sigma^2)},$$

while in general the latter can be applied separately to the real and the imaginary part of f to yield the inequality of Lemma 1.

Lemma 2. *Under (1), for every $t \in \mathbf{R}$,*

$$\|\nabla f_\varepsilon(t)\|_\infty \leq \frac{|t| + t^2}{\sqrt{n}}.$$

Proof. Using the equality $f_\varepsilon(t) - f_{s_k(\varepsilon)}(t) = \mathbf{E}\, e^{itS_\varepsilon}(1 - e^{-2it\,\varepsilon_k X_k/\sqrt{n}})$, we may write

$$|\nabla f_\varepsilon(t)| = \sup \left| \mathbf{E}\, e^{itS_\varepsilon} \sum_{k=1}^{n} a_k \frac{1 - e^{-2it\varepsilon_k X_k/\sqrt{n}}}{2} \right|$$

$$\leq \sup \mathbf{E} \left| \sum_{k=1}^{n} a_k \frac{1 - e^{-2it\varepsilon_k X_k/\sqrt{n}}}{2} \right|,$$

where the supremum runs over all complex numbers $a_1, \ldots, a_n$ such that $|a_1|^2 + \ldots + |a_n|^2 = 1$. Using the estimate $|e^{i\alpha} - 1 - i\alpha| \leq \frac{1}{2}\alpha^2$ ($\alpha \in \mathbf{R}$) and the assumption $\mathbf{E}X_k^2 = 1$, we can continue to get

$$|\nabla f_\varepsilon(t)| \leq \frac{|t|}{\sqrt{n}} \sup \mathbf{E} \left| \sum_{k=1}^{n} a_k \varepsilon_k X_k \right| + \frac{t^2}{n} \sup \mathbf{E} \sum_{k=1}^{n} |a_k| X_k^2$$

$$= \frac{|t|}{\sqrt{n}} \sup \mathbf{E} \left| \sum_{k=1}^{n} a_k \varepsilon_k X_k \right| + \frac{t^2}{\sqrt{n}}.$$

It remains to note that, by Schwarz' inequality and (1), $(\mathbf{E}\,|\sum_{k=1}^{n} a_k \varepsilon_k X_k|)^2 \leq \mathbf{E}\,|\sum_{k=1}^{n} a_k \varepsilon_k X_k|^2 = 1$.

We also need the following observation due to H. Bohman [Bo].

Lemma 3. *Given characteristic functions φ_1 and φ_2 of the distribution functions F_1 and F_2, respectively, if $|\varphi_1(t) - \varphi_2(t)| \leq \lambda|t|$, for all $t \in \mathbf{R}$, then, for all $x \in \mathbf{R}$ and $a > 0$,*

$$F_1(x - a) - \frac{2\lambda}{a} \leq F_2(x) \leq F_1(x + a) + \frac{2\lambda}{a}.$$

The particular case $a = \sqrt{2\lambda}$ gives an important relation

$$\frac{1}{2} L(F_1, F_2)^2 \leq \sup_{t>0} \left| \frac{\varphi_1(t) - \varphi_2(t)}{t} \right|. \tag{5}$$

Proof of Theorem 1. Fix a number $h > 0$. For $0 < t \leq \frac{2}{h}$, by Lemma 2, $\|\nabla f_\varepsilon(t)\|_\infty \leq \frac{t+t^2}{\sqrt{n}} \leq \frac{t}{\sqrt{n}}(1 + \frac{2}{h})$, so that, by Lemma 1 applied to the function $\varepsilon \to f_\varepsilon(t)$, we get

$$\mu_n \left\{ \varepsilon : \left| \frac{f_\varepsilon(t) - f(t)}{t} \right| \geq h \right\} \leq 4 e^{-nh^4/4(h+2)^2}. \tag{6}$$

In the case $t > \frac{2}{h}$, this inequality is immediate, since $|f_\varepsilon(t) - f(t)| \leq 2 < th$, for all ε. Thus, we have the estimate (6) for all t separately, but in order to apply Lemma 3, we need a similar bound holding true for the supremum over all $t > 0$. To this end, apply (6) to the points $t_r = rh^2$, $r = 1, 2, \ldots, N = [\frac{2}{h}] + 1$, to get

$$\mu_n \left\{ \varepsilon : \max_{1 \le r \le N} \left| \frac{f_\varepsilon(t_r) - f(t_r)}{t_r} \right| \ge h \right\} \le 4N e^{-nh^4/4(h+2)^2}. \tag{7}$$

Since $\mathbf{E}S_\varepsilon = 0$, $\mathbf{E}S_\varepsilon^2 = 1$, we have $|f_\varepsilon'(t)| \le 1$, $f_\varepsilon'(0) = 0$, $|f_\varepsilon''(t)| \le 1$, and similarly for f. Therefore, $|f_\varepsilon(t) - f(t)| \le t^2 \le th$, for all ε, as soon as $0 \le t \le h$. In case $h \le t \le \frac{2}{h}$, since $t_N \ge \frac{2}{h}$, one can pick an index $r = 1, \ldots, N-1$ such that $t_r < t \le t_{r+1}$. Assuming that $\left| \frac{f_\varepsilon(t_r) - f(t_r)}{t_r} \right| < h$, and recalling that $t_{r+1} - t_r = h^2$, we may write

$$|f_\varepsilon(t) - f(t)| \le |f_\varepsilon(t) - f_\varepsilon(t_r)| + |f_\varepsilon(t_r) - f(t_r)| + |f(t_r) - f(t)|$$
$$< 2|t - t_r| + t_r h \le 2h^2 + t_r h < 3th.$$

Consequently, (7) implies

$$\mu_n \left\{ \sup_{t>0} \left| \frac{f_\varepsilon(t) - f(t)}{t} \right| \ge 3h \right\} \le 4N e^{-nh^4/4(h+2)^2}$$
$$\le 4 \left(\frac{2}{h} + 1 \right) e^{-nh^4/4(h+2)^2}.$$

Therefore, by (5),

$$\mu_n \left\{ \frac{1}{2} L(F_\varepsilon, F)^2 \ge 3h \right\} \le 4 \left(\frac{2}{h} + 1 \right) e^{-nh^4/4(h+2)^2}.$$

Replacing $6h$ with δ^2 and noticing that only $0 < \delta \le 1$ should be taken into consideration, one easily arrives at the estimate $\mu_n\{L(F_\varepsilon, F) \ge \delta\} \le \frac{C}{\delta^2} e^{-cn\delta^8}$ with some positive numerical constants C and c. On the other hand, in the latter inequality, we may restrict ourselves to values $\delta > c_1 n^{-1/8}$ which make the bound $\frac{C}{\delta^2} e^{-cn\delta^8}$ less than 1, and then we arrive at the desired inequality (4).

Theorem 1 has been proved, and we may state its immediate consequence:

Corollary 1. *Under* (1), *for at least* 2^{n-1} *sequences* $\varepsilon = (\varepsilon_1, \ldots, \varepsilon_n)$ *of signs,* $L(F_\varepsilon, F) \le C \left(\frac{\log n}{n} \right)^{1/8}$, *where* C *is a universal constant.*

Let us now turn to the second task: approximation of the μ_n-typical F by more canonical distributions. Namely, denote by G the distribution function of the random variable $\zeta \frac{|X|}{\sqrt{n}}$ where ζ is a standard normal random variable independent of the Euclidean norm $|X| = (X_1^2 + \ldots + X_n^2)^{1/2}$. Clearly, G represents a mixture of a family of centered Gaussian measures on the line and has characteristic function

$$g(t) = \mathbf{E}e^{-t^2|X|^2/(2n)}, \quad t \in \mathbf{R}, \tag{8}$$

while F has characteristic function

$$f(t) = \mathbf{E} \prod_{k=1}^{n} \cos\left(\frac{tX_k}{\sqrt{n}}\right). \tag{9}$$

In order to bound the Lévy distance $L(F, G)$, the following general elementary observation, not using the condition (1), can be applied.

Lemma 4. *Assume* $\mathbf{E}|X|^2 \le n$. *For all* $\alpha > 0$ *and* $|t| \le \frac{1}{2\alpha}$,

$$|f(t) - g(t)| \le \frac{1}{9}\alpha^2 t^4 + 2\mathbf{P}\left\{\frac{\max\{|X_1|, \ldots, |X_n|\}}{\sqrt{n}} > \alpha\right\}.$$

Proof. By Taylor's expansion, in the interval $|s| \le \frac{1}{2}$, we have $\cos(s) = e^{-\frac{s^2}{2} - u(s)}$ with u satisfying $0 \le u(s) \le \frac{s^4}{9}$. Therefore, provided that $|\frac{X_k}{\sqrt{n}}| \le \alpha$, for all $k \le n$, and $\alpha|t| \le \frac{1}{2}$,

$$\prod_{k=1}^{n} \cos\left(\frac{tX_k}{\sqrt{n}}\right) = \exp\left\{-\frac{t^2|X|^2}{2n} - \sum_{k=1}^{n} u\left(\frac{tX_k}{\sqrt{n}}\right)\right\}$$

with $0 \le \sum_{k=1}^{n} u(\frac{tX_k}{\sqrt{n}}) \le \frac{1}{9}\max_k |\frac{tX_k}{\sqrt{n}}|^2 \sum_{k=1}^{n} |\frac{tX_k}{\sqrt{n}}|^2 \le \frac{\alpha^2 t^4}{9}\frac{|X|^2}{n}$. So,

$$e^{-\frac{t^2|X|^2}{2n}} \ge \prod_{k=1}^{n} \cos\left(\frac{tX_k}{\sqrt{n}}\right) \ge e^{-\frac{t^2|X|^2}{2n} - \frac{\alpha^2 t^4}{9}\frac{|X|^2}{n}}.$$

Taking the expectations and using $|\prod_{k=1}^{n} \cos(\frac{tX_k}{\sqrt{n}}) - e^{-t^2|X|^2/(2n)}| \le 2$ for the complementary event $\frac{\max\{|X_1|, \ldots, |X_n|\}}{\sqrt{n}} > \alpha$, we thus get

$$|f(t) - g(t)| \le 2\mathbf{P}\left\{\frac{\max\{|X_1|, \ldots, |X_n|\}}{\sqrt{n}} > \alpha\right\} + \mathbf{E}\, e^{-\frac{t^2|X|^2}{2n}}\left(1 - e^{-\frac{\alpha^2 t^4}{9}\frac{|X|^2}{n}}\right).$$

The last term is bounded by $\mathbf{E}(1 - e^{-\frac{\alpha^2 t^4}{9}\frac{|X|^2}{n}}) \le 1 - e^{-\frac{\alpha^2 t^4}{9}\frac{\mathbf{E}|X|^2}{n}} \le \frac{\alpha^2 t^4}{9}$ where we applied Jensen's inequality together with the assumption $\mathbf{E}|X|^2 \le n$. Lemma 4 follows.

Via the inequality of Lemma 4, with mild integrability assumptions on the distribution of X, one can study a rate of closeness of F and thus of F_ε to the distribution function G. One can start, for instance, with the moment assumption

$$\mathbf{E}|X_k|^4 \le \beta, \quad 1 \le k \le n, \tag{10}$$

implying $\mathbf{P}\{\frac{\max\{|X_1|, \ldots, |X_n|\}}{\sqrt{n}} > \alpha\} \le \frac{\beta}{\alpha^4 n}$, so that, by Lemma 4,

$$|f(t) - g(t)| \le \frac{1}{9}\alpha^2 t^4 + \frac{2\beta}{\alpha^4 n}, \quad \text{as soon as } |t| \le \frac{1}{2\alpha}.$$

Minimizing the right-hand side over all $\alpha > 0$, we obtain that

$$|f(t) - g(t)| \leq \frac{\beta^{1/3}|t|^{16/3}}{3n^{1/3}}, \quad \text{provided that} \quad |t| \leq \frac{n^{1/4}}{24\beta^{1/2}}.$$

Now apply Zolotarev's estimate, [Z], [P], to get

$$L(F, G) \leq \frac{1}{\pi} \int_0^T \left| \frac{f(t) - g(t)}{t} \right| dt + 2e \frac{\log T}{T} \quad (T > 1.3)$$

$$\leq \frac{\beta^{1/3} T^{16/3}}{16\pi n^{1/3}} + 2e \frac{\log T}{T}, \quad \text{if} \ \ 1.3 < T \leq \frac{n^{1/4}}{24\beta^{1/2}}.$$

Taking $T = \frac{n^{1/19}}{\beta^{1/19}}$ and using $\beta \geq 1$, we will arrive at the estimate of the form

$$L(F, G) \leq C \frac{\beta^{1/19} + \log n}{n^{1/19}}, \quad n \geq C\beta^{37/15},$$

up to some numerical constant C. Higher moments or exponential integrability assumption improve this rate of convergence, but it seems, with the above argument, the rate of Corollary 1 cannot be reached.

On the other hand, the closeness of G to the normal distribution function Φ requires some additional information concerning the rate of convergence of $\frac{X_1^2 + ... + X_n^2}{n}$ to 1. For example, the property $\text{Var}(|X|^2) \leq O(n)$ guarantees a rate of the form $L(G, \Phi) = O(n^{-c})$ with a certain power $c > 0$. Thus, together with the moment assumption (10), one arrives at the bound $L(F, \Phi) = O(n^{-c})$.

Finally, let us note that G is determined via the distribution of the Euclidean norm $|X|$, so it is stable under the choice of the basis in $\mathbf{R}^n$. The condition (10) is stated for the canonical basis in $\mathbf{R}^n$, and the appropriate basis free assumption may read as

$$\sup_{\theta \in S^{n-1}} \mathbf{E} |\langle \theta, X \rangle|^p \leq \beta_p, \quad p > 2. \tag{11}$$

Then, at the expense of the rate of closeness, one may formulate an analogue of Theorem 1 for the distribution G in the place of F and with respect to an arbitrary basis in $\mathbf{R}^n$. The inequality (11) includes many interesting classes of distributions such as log-concave probability measures satisfying (1), for example.

Proof of Theorem 2. Denote by f_n and g_n the characteristic functions defined for the random vectors $(X_{n,1}, \ldots, X_{n,n})$ according to formulas (9) and (8), respectively. Also, according to (3), denote by $F^{(n)}$ the corresponding average distribution functions.

In view of the assumption a), one can select a sequence $\alpha_n \downarrow 0$ such that $\mathbf{P}\{\frac{\max\{|X_{1,1}|, \ldots, |X_{n,n}|\}}{\sqrt{n}} > \alpha_n\} \to 0$, as $n \to \infty$. Then, by Lemma 4, for all $t \in \mathbf{R}$, $|f_n(t) - g_n(t)| \to 0$, as $n \to \infty$. On the other hand, the condition b) readily implies $g_n(t) \to e^{-t^2/2}$, so $f_n(t) \to e^{-t^2/2}$. Thus, $L(F^{(n)}, \Phi) \to 0$.

Now, given an infinite sequence $\varepsilon \in \{-1, 1\}^\infty$, denote by $T_n(\varepsilon)$ its projection $(\varepsilon_1, \ldots, \varepsilon_n)$. It remains to show that $L(F_{T_n(\varepsilon)}, F^{(n)}) \to 0$, for μ_∞-almost all ε. Fix any small number $p > 0$, and take a sequence $\delta_n \to 0^+$ such that

$$\sum_{n=1}^\infty C n^{1/4} e^{-cn\delta_n^8} \leq p,$$

where C and c are numerical constants from Theorem 1 (δ_n may depend on p). Then the application of (4) yields

$$\mu_\infty\{\varepsilon : L(F_{T_n(\varepsilon)}, F^{(n)}) > \delta_n, \text{ for some } n \geq 1\}$$
$$\leq \sum_{n=1}^\infty \mu_\infty\{L(F_{T_n(\varepsilon)}, F^{(n)}) > \delta_n\}$$
$$= \sum_{n=1}^\infty \mu_n\{\varepsilon = (\varepsilon_1, \ldots, \varepsilon_n) : L(F_\varepsilon, F^{(n)}) > \delta_n\} \leq p.$$

Therefore, $L(F_{T_n(\varepsilon)}, F^{(n)}) \leq \delta_n$, for all $n \geq 1$ and for all ε except for a set of μ_∞-measure at most p. That is,

$$\mu_\infty\left\{\varepsilon : \sup_{n \geq 1}\left(L(F_{T_n(\varepsilon)}, F^{(n)}) - \delta_n\right) \leq 0\right\} \geq 1 - p. \tag{12}$$

But since $\delta_n \to 0$,

$$\sup_{n \geq 1}\left(L(F_{T_n(\varepsilon)}, F^{(n)}) - \delta_n\right) \geq \limsup_{n \to \infty}\left(L(F_{T_n(\varepsilon)}, F^{(n)}) - \delta_n\right)$$
$$= \limsup_{n \to \infty} L(F_{T_n(\varepsilon)}, F^{(n)}).$$

Consequently, (12) implies $\mu_\infty\{\limsup_{n \to \infty} L(F_{T_n(\varepsilon)}, F^{(n)}) = 0\} \geq 1 - p$. The probability on the left does not depend on p, and letting $p \to 0$ finishes the proof.

Acknowledgement. We would like to thank B. Zegarlinski and A. Grigoryan for the invitation to visit to the Imperial College, London, where this work was partly done.

References

[A-B-P] Antilla, M., Ball, K., Perissinaki, I. (1998): The central limit problem for convex bodies. Preprint

[B] Bobkov, S.G. On concentration of distributions of random weighted sums. Ann. Probab., to appear

[B-G]   Bobkov, S.G., Götze, F. (1999): Exponential integrability and transportation cost related to logarithmic Sobolev inequalities. J. Funct. Anal., **163**, No. 1, 1–28

[Bo]   Bohman, H. (1961): Approximate Fourier analysis of distribution functions. Arkiv Mat., **4**, No. 2-3, 99–157

[D-F]   Diaconis, P., Freedman, D. (1984): Asymptotics of graphical projection pursuit. Ann. Stat., **12**, No. 3, 793–815

[Le]   Ledoux, M. (1999): Concentration of measure and logarithmic Sobolev inequalities. Séminaire de Probabilités, XXXIII, Lecture Notes in Math., **1709**, 120–216

[P]   Petrov, V.V. (1987): Limit theorems for sums of independent random variables. Moscow, Nauka (in Russian)

[S]   Sudakov, V.N. (1978): Typical distributions of linear functionals in finite-dimensional spaces of higher dimension. Soviet Math. Dokl., **19**, No. 6, 1578–1582. Translated from: Dokl. Akad. Nauk SSSR, **243**, No. 6

[W]   von Weizsäcker, H. (1997): Sudakov's typical marginals, random linear functionals and a conditional central limit theorem. Probab. Theory Rel. Fields, **107**, 313–324

[Z]   Zolotarev, B.M. (1971): An estimate for the difference between distributions in Lévy metric. Proceedings of Steklov's Math. Institute, **112**, 224–231 (in Russian)

Spectral Gap and Concentration for Some Spherically Symmetric Probability Measures*

S.G. Bobkov

School of Mathematics, University of Minnesota, 127 Vincent Hall, 206 Church St. S.E., Minneapolis, MN 55455, USA *bobkov@math.umn.edu*

Summary. We study the spectral gap and a related concentration property for a family of spherically symmetric probability measures.

This note appeared in an attempt to answer the following question raised by V. Bogachev: How do we effectively estimate the spectral gap for the exponential measures μ on the Euclidean space $\mathbf{R}^n$ with densities of the form $\frac{d\mu(x)}{dx} = ae^{-b|x|}$?

By the spectral gap, we mean here the best constant $\lambda_1 = \lambda_1(\mu)$ in the Poincaré-type inequality

$$\lambda_1 \int_{\mathbf{R}^n} |u(x)|^2 \, d\mu(x) \le \int_{\mathbf{R}^n} |\nabla u(x)|^2 \, d\mu(x) \tag{1}$$

with u being an arbitrary smooth (or, more generally, locally Lipschitz) function on $\mathbf{R}^n$ such that $\int u(x) \, d\mu(x) = 0$. Although it is often known that $\lambda_1 > 0$, in many problems of analysis and probability, one needs to know how the dimension n reflects on this constant. One important case, the canonical Gaussian measure $\mu = \gamma_n$, with density $(2\pi)^{-n/2} e^{-|x|^2/2}$, provides an example with a dimension-free spectral gap $\lambda_1 = 1$. This fact can already be used to recover a dimension-free concentration phenomenon in Gauss space.

To unite both the Gaussian and the exponential cases, we consider a spherically symmetric probability measure μ on $\mathbf{R}^n$ with density

$$\frac{d\mu(x)}{dx} = \rho(|x|), \quad x \in \mathbf{R}^n,$$

assuming that $\rho = \rho(t)$ is an arbitrary log-concave function on $(0, +\infty)$, that is, the function $\log \rho(t)$ is concave on its support interval. In order that μ be log-concave itself (cf. [Bor2] for a general theory of log-concave measures), ρ has also to be non-increasing in $t > 0$. However, this will not be required.

It is a matter of normalization, if we assume that μ satisfies

$$\int_{\mathbf{R}^n} \langle x, \theta \rangle^2 \, d\mu(x) = |\theta|^2, \quad \text{for all } \theta \in \mathbf{R}^n. \tag{2}$$

---
* Supported in part by an NSF grant.

As usual, $\langle \cdot, \cdot \rangle$ and $|\cdot|$ denote the scalar product and the Euclidean norm, respectively. Since μ is symmetrically invariant, this normalization condition may also be written as $\int x_1^2 \, d\mu(x) = 1$, or $\int |x|^2 \, d\mu(x) = n$. We prove:

Theorem 1. *Under* (2), *the optimal value of* λ_1 *in* (1) *satisfies* $\frac{1}{13} \leq \lambda_1 \leq 1$.

Returning to the exponential measure $d\mu(x) = a \, e^{-b|x|} dx$, $b > 0$, we thus obtain that λ_1 is of order b^2/n.

Using Theorem 1 and applying Gromov–Milmans's theorem on concentration under Poincaré-type inequalities, one may conclude that all the considered measures share a dimension-free concentration phenomenon:

Theorem 2. *Under* (2), *given a measurable set* A *in* $\mathbf{R}^n$ *of measure* $\mu(A) \geq \frac{1}{2}$, *for all* $h > 0$,

$$1 - \mu(A^h) \leq 2e^{-ch}, \tag{3}$$

where c *is a certain positive universal constant.*

Here, we use $A^h = \{x \in \mathbf{R}^n : \text{dist}(A, x) < h\}$ to denote an h-neighborhood of A with respect to the Euclidean distance.

Note that, in polar coordinates, every spherically symmetric measure μ with density $\rho(|x|)$ represents a product measure, i.e., it may be viewed as the distribution of $\xi\theta$, where θ is a random vector uniformly distributed over the unit sphere S^{n-1}, and where $\xi > 0$ is an independent of θ random variable with distribution function

$$\mu\{|x| \leq t\} = n\omega_n \int_0^t s^{n-1}\rho(s) \, ds, \quad t > 0 \tag{4}$$

(ω_n is the volume of the unit ball in $\mathbf{R}^n$). For example, one can take $(\mathbf{R}^n, \mu)$ for the underlying probability space and put $\xi(x) = |x|$, $\theta(x) = \frac{x}{|x|}$. It is a classical fact that $\lambda_1(S^{n-1}) = n - 1$. To reach Theorems 1-2, our task will be therefore to estimate $\lambda_1(\xi)$ from below and to see in particular that the values of ξ are strongly concentrated around its mean $\mathbf{E}\xi$ which is of order $\sqrt{n}$. When ρ is log-concave, the density $q(t) = n\omega_n t^{n-1}\rho(t)$ of ξ is log-concave, as well. Of course, this observation is not yet enough to reach the desired statements, since it "forgets" about an important factor t^{n-1}. As a first step, we will need the following one-dimensional:

Lemma 1. *Given a positive integer* n, *if a random variable* $\xi > 0$ *has density* $q(t)$ *such that the function* $q(t)/t^{n-1}$ *is log-concave on* $(0, +\infty)$, *then*

$$\text{Var}(\xi) \leq \frac{1}{n}(\mathbf{E}\xi)^2. \tag{5}$$

As usual, $\text{Var}(\xi) = \mathbf{E}\xi^2 - (\mathbf{E}\xi)^2$ and $\mathbf{E}\xi$ denote the variance and the expectation of a random variable ξ.

For $\xi(x) = |x|$ as above, with distribution given by (4), in view of the normalization condition (2), we have $\mathbf{E}\xi^2 = n$, so the bound (5) yields a dimension-free inequality

$$\mathrm{Var}(\xi) \leq 1. \tag{6}$$

Lemma 1 represents a particular case of a theorem due to R.E. Barlow, A.W. Marshall, and F. Proshan (cf. [B-M-P], p. 384, and [Bor1]) which states the following: If a random variable $\eta > 0$ has a distribution with increasing hazard rate (in particular, if η has a log-concave density), then its normalized moments $\lambda_a = \frac{1}{\Gamma(a+1)} \mathbf{E}\eta^a$ satisfy a reverse Lyapunov's inequality

$$\lambda_a^{b-c} \lambda_c^{a-b} \leq \lambda_b^{a-c}, \qquad a \geq b \geq c \geq 1, \quad c \text{ integer.} \tag{7}$$

Indeed, putting $a = n+1$, $b = n$, $c = n - 1$ $(n \geq 2)$, we get

$$\mathbf{E}\eta^{n+1}\, \mathbf{E}\eta^{n-1} \leq \left(1 + \frac{1}{n}\right)(\mathbf{E}\eta^n)^2. \tag{8}$$

If the random variable ξ has density $q(t) = t^{n-1}p(t)$ with p log-concave on $(0, +\infty)$, and η has density $p(t)/\int_0^{+\infty} p(t)\, dt$, the above inequality becomes $\mathbf{E}\xi^2 \leq (1 + \frac{1}{n})(\mathbf{E}\xi)^2$ which is exactly (5).

When $n = 1$, the latter is equivalent to the well-known Khinchine-type inequality $\mathbf{E}\eta^2 \leq 2\,(\mathbf{E}\eta)^2$. More generally, one has

$$\mathbf{E}\eta^a \leq \Gamma(a+1)\,(\mathbf{E}\eta)^a, \qquad a \geq 1,$$

which is known to hold true in the class of all random variables $\eta > 0$ with log-concave densities. This fact cannot formally be deduced from (7) because of the assumption $c \geq 1$. It was obtained in 1961 by S. Karlin, F. Proshan, and R.E. Barlow [K-P-B] as an application of their study of the so-called totally positive functions (similar to [B-M-P] – with techniques and ideas going back to the work of I.J. Schoenberg [S]).

To make the proof of Theorem 1 more self-contained, we would like to include a different argument leading to the inequality (7) for a related function:

Lemma 2. *Given a log-concave random variable $\eta > 0$, the function $\lambda_a = \frac{1}{a^a}\mathbf{E}\eta^a$ is log-concave in $a > 0$. Equivalently, it satisfies (7), for all $a \geq b \geq c > 0$.*

Again putting $a = n + 1$, $b = n$, $c = n - 1$, we obtain $\mathbf{E}\eta^{n+1}\,\mathbf{E}\eta^{n-1} \leq C_n(\mathbf{E}\eta^n)^2$ with constant $C_n = \frac{(n+1)^{n+1}(n-1)^{n-1}}{n^{2n}}$ which is a little worse than that of (8). On the other hand, one can easily see that $C_n \leq 1 + \frac{1}{n} + \frac{1}{n^3}$, so, we get, for example, the constant $\frac{2}{n}$ in Lemma 1 (and this leads to the lower bound $\frac{1}{25}$ in Theorem 1).

Finally, it might also be worthwhile to mention here the following interesting immediate consequence of Lemmas 1-2. Given an integer $d \geq 1$ and an arbitrary sequence of probability measures $(\mu_n)_{n \geq d}$ on $\mathbf{R}^n$ (from the class we are considering), their projections to the coordinate subspace $\mathbf{R}^d$ must converge, as $n \to \infty$, to the standard Gaussian measure on $\mathbf{R}^d$.

A second step to prove Theorem 1 is based on the following statement ([B1], Corollary 4.3):

Lemma 3. *If a random variable ξ has distribution ν with log-concave density on the real line, then*

$$\frac{1}{12\,\mathrm{Var}(\xi)} \leq \lambda_1(\nu) \leq \frac{1}{\mathrm{Var}(\xi)}.$$

Together with (6) for $\xi(x) = |x|$, we thus get

$$\lambda_1(\nu) \geq \frac{1}{12}. \tag{9}$$

Proof of Theorem 1. We may assume that $n \geq 2$. As before, denote by ν the distribution of the Euclidean norm $\xi(x) = |x|$ under μ, and by σ_{n-1} the normalized Lebesgue measure on the unit sphere S^{n-1}. To prove the Poincaré-type inequality (1), take a smooth bounded function u on $\mathbf{R}^n$ and consider another smooth bounded function $v(r, \theta) = u(r\theta)$ on the product space $(0, +\infty) \times \mathbf{R}^n$. Under the product measure $\nu \times \sigma_{n-1}$, v has the same distribution as u has under μ.

By (9), the measure ν satisfies the Poincaré-type inequality on the line,

$$\mathrm{Var}_\nu(g) \leq 12 \int_0^{+\infty} |g'(r)|^2 d\nu(r),$$

where $g = g(r)$ is an arbitrary absolutely continuous function on $(0, +\infty)$. In particular, for $g(r) = v(r, \theta)$ with fixed $\theta \in S^{n-1}$, we get

$$\int_0^{+\infty} v(r, \theta)^2\, d\nu(r) \leq \left(\int_0^{+\infty} v(r, \theta)\, d\nu(r) \right)^2 + 12 \int_0^{+\infty} \left| \frac{\partial v}{\partial r} \right|^2 d\nu(r).$$

Now, $\frac{\partial v}{\partial r} = \langle \nabla u(r\theta), \theta \rangle$, so $\left| \frac{\partial v}{\partial r} \right| \leq |\nabla u(r\theta)|$. Integrating the above inequality over σ_{n-1}, we get

$$\int_{\mathbf{R}^n} u(x)^2\, d\mu(x) \leq \int_{S^{n-1}} w(\theta)^2\, d\sigma_{n-1}(\theta) + 12 \int_{\mathbf{R}^n} |\nabla u(x)|^2\, d\mu(x), \tag{10}$$

where $w(\theta) = \int_0^{+\infty} v(r, \theta)\, d\nu(r)$. For this function, which is well-defined and smooth on the whole space $\mathbf{R}^n$, the average over σ_{n-1} is exactly the average of u over μ. Hence, by the Poincaré inequality on the unit sphere,

$$\int_{S^{n-1}} w(\theta)^2\, d\sigma_{n-1}(\theta) \leq \left(\int_{\mathbf{R}^n} u(x)\, d\mu(x) \right)^2 + \frac{1}{n} \int_{S^{n-1}} |\nabla w(\theta)|^2 d\sigma_{n-1}(\theta). \tag{11}$$

(The classical Riemannian version of the Poincaré inequality is formulated for the "inner" gradient $\nabla_{S^{n-1}} w(\theta)$ on the unit sphere which is the projection of the usual gradient $\nabla w(\theta)$ onto the subspace orthogonal to θ. In this case the constant $\frac{1}{n}$ in (11) should be replaced with $\frac{1}{n-1}$.)

Since $\nabla w(\theta) = \int_0^{+\infty} r \, \nabla u(r\theta) \, d\nu(r)$, we have $|\nabla w(\theta)| \leq \int_0^{+\infty} r \, |\nabla u(r\theta)| \, d\nu(r)$. Hence, by the Cauchy–Bunyakovski inequality,

$$|\nabla w(\theta)|^2 \leq \int_0^{+\infty} r^2 \, d\nu(r) \int_0^{+\infty} |\nabla u(r\theta)|^2 \, d\nu(r) = n \int_0^{+\infty} |\nabla u(r\theta)|^2 \, d\nu(r),$$

where we used the normalization condition $\mathbf{E}\xi^2 = n$. Together with (10) and (11), this estimate yields

$$\int_{\mathbf{R}^n} u(x)^2 \, d\mu(x) \leq \left(\int_{\mathbf{R}^n} u(x) \, d\mu(x) \right)^2 + 13 \int_{\mathbf{R}^n} |\nabla u(x)|^2 \, d\mu(x),$$

that is, the Poincaré-type inequality (1) with the lower bound $\lambda_1 \geq 1/13$.

The upper bound is trivial and follows by testing (1) on linear functions. This finishes the proof.

As already mentioned, the fact that (1) implies a concentration inequality, namely,

$$1 - \mu(A^h) \leq C e^{-c\sqrt{\lambda_1}\, h}, \quad h > 0, \quad \mu(A) \geq \frac{1}{2}, \tag{12}$$

where C and c are certain positive universal constants, was proved by M. Gromov and V.D. Milman, see [G-M]. They formulated it in the setting of a compact Riemannian manifold, but the assertion remains to hold in many other settings, e.g., for an arbitrary metric space (see e.g. [A-S], [B-L], [L]). The best possible constant in the exponent in (12) is $c = 2$ ([B2]), but this is not important for the present formulation of Theorem 1.

Remark. We do not know how to adapt the argument in order to prove, for all smooth u with μ-mean zero, a stronger inequality in comparison with (1),

$$c \int_{\mathbf{R}^n} |u(x)| \, d\mu(x) \leq \int_{\mathbf{R}^n} |\nabla u(x)| \, d\mu(x), \tag{13}$$

called sometimes a Cheeger-type inequality. On the shifted indicator functions $u = 1_A - \mu(A)$, (13) turns into an equivalent isoperimetric inequality for the μ-perimeter, $\mu^+(A) \geq 2c\,\mu(A)(1 - \mu(A))$. One deep conjecture ([K-L-S]) asserts that, for some universal $c > 0$, this isoperimetric inequality holds true under the isotropic condition (2) in the class of all log-concave measures μ. However, the hypothesis remains open even in the weaker forms such as Poincaré and concentration inequalities. And as we saw, already the particular case of a symmetrically log-concave measure leads to a rather sophisticated one-dimensional property such as Lemma 1.

Proof of Lemma 2. Let p be the probability density of η on $(0, +\infty)$. We apply the one-dimensional Prékopa–Leindler theorem (see [Pr1-2], [Le], or [Pi] for a short proof): given $t, s > 0$ with $t + s = 1$ and non-negative measurable functions u, v, w on $(0, +\infty)$ satisfying $w(tx + sy) \geq u^t(x)v^s(y)$, for all $x, y > 0$, we have

$$\int_0^{+\infty} w(z)\, dz \geq \left(\int_0^{+\infty} u(x)\, dx \right)^t \left(\int_0^{+\infty} v(y)\, dy \right)^s. \tag{14}$$

Let $a > b > c > 0$ and $b = ta + sc$. Since

$$\sup_{tx+sy=z} x^a y^c = a^{ta} c^{sc} \left(\frac{z}{ta + sc} \right)^{ta+sc},$$

the inequality (14) applies to $u(x) = (\frac{x}{a})^a p(x)$, $v(y) = (\frac{y}{c})^c p(y)$, and $w(z) = (\frac{z}{b})^b p(z)$. This is exactly what we need.

Remark. The multidimensional Prékopa–Leindler theorem yields a similar statement: For any random vector $(\eta_1, \ldots, \eta_n)$ in $\mathbf{R}_+^n$ with log-concave distribution, the function $\varphi(a_1, \ldots, a_n) = \mathbf{E}\left(\frac{\eta_1}{a_1}\right)^{a_1} \ldots \left(\frac{\eta_n}{a_n}\right)^{a_n}$ is log-concave on $\mathbf{R}_+^n$.

References

[A-S]   Aida, S., Stroock, D. (1994): Moment estimates derived from Poincaré and logarithmic Sobolev inequalities. Math. Res. Lett., **1**, No. 1, 75–86

[B-M-P] Barlow, R.E., Marshall, A.W., Proshan, F. (1963): Properties of probability distributions with monotone hazard rate. Ann. Math. Stat., **34**, 375–389

[B1]   Bobkov, S.G. (1999): Isoperimetric and analytic inequalities for log-concave probability distributions. Ann. Probab., **27**, No. 4, 1903–1921

[B2]   Bobkov, S.G. (1999): Remarks on Gromov-Milman's inequality. Vestnik of Syktyvkar University, **3**, Ser.1, 15–22 (in Russian)

[B-L]   Bobkov, S.G., Ledoux, M. (1997): Poincaré's inequalities and Talagrand's concentration phenomenon for the exponential distribution. Probab. Theory Rel. Fields, **107**, 383–400

[Bor1]   Borell, C. (1973): Complements of Lyapunov's inequality. Math. Ann., **205**, 323–331

[Bor2]   Borell, C. (1974): Convex measures on locally convex spaces. Ark. Math., **12**, 239–252

[G-M]   Gromov, M., Milman, V.D. (1983): A topological application of the isoperimetric inequality. Amer. J. Math., **105**, 843–854

[K-L-S] Kannan, R., Lovász, L., Simonovits, M. (1995): Isoperimetric problems for convex bodies and a localization lemma. Discrete and Comput. Geom., **13**, 541–559

[K-P-B] Karlin, S., Proshan, F., Barlow, R.E. (1961): Moment inequalities of Polya frequency functions. Pacific J. Math., **11**, 1023–1033

[L]   Ledoux, M. (1999). Concentration of measure and logarithmic Sobolev inequalities. Séminaire de Probabilités, XXXIII, Lecture Notes in Math., **1709**, Springer, 120–216

[Le]   Leindler, L. (1972): On a certain converse of Hölder's inequality II. Acta Sci. Math. Szeged, **33**, 217–223

[Pi]   Pisier, G. (1989): The Volume of Convex Bodies and Banach Space Geometry. Cambridge Tracts in Math., **94**, Cambridge University Press, Cambridge

[Pr1]   Prékopa, A. (1971): Logarithmic concave measures with applications to stochastic programming. Acta Sci. Math. Szeged, **32**, 301–316

[Pr2]   Prékopa, A. (1973): On logarithmic concave measures and functions. Acta Sci. Math. Szeged, **34**, 335–343

[S]   Schoenberg, I.J. (1951): On Pólya frequency functions I. The totally positive functions and their Laplace transforms. J. d'Analyse Math., **1**, 331–374

On the Central Limit Property of Convex Bodies

S.G. Bobkov[1]* and A. Koldobsky[2]**

[1] School of Mathematics, University of Minnesota, 127 Vincent Hall, 206 Church St. S.E., Minneapolis, MN 55455, USA *bobkov@math.umn.edu*
[2] Department of Mathematics, Mathematical Sciences Building, University of Missouri, Columbia, MO 65211, USA *koldobsk@math.missouri.edu*

Summary. For isotropic convex bodies K in $\mathbf{R}^n$ with isotropic constant L_K, we study the rate of convergence, as n goes to infinity, of the average volume of sections of K to the Gaussian density on the line with variance L_K^2.

Let K be an isotropic convex body in $\mathbf{R}^n$, $n \geq 2$, with volume one. By the isotropy assumption we mean that the baricenter of K is at the origin, and there exists a positive constant L_K so that, for every unit vector θ,

$$\int_K \langle x, \theta \rangle^2 \, dx = L_K^2.$$

Introduce the function

$$f_K(t) = \int_{S^{n-1}} \mathrm{vol}_{n-1}\big(K \cap H_\theta(t)\big) \, d\sigma(\theta), \quad t \in \mathbf{R},$$

expressing the average $(n-1)$-dimensional volume of sections of K by hyperplanes $H_\theta(t) = \{x \in \mathbf{R}^n : \langle x, \theta \rangle = t\}$ perpendicular to $\theta \in S^{n-1}$ at distance $|t|$ from the origin (and where σ is the normalized uniform measure on the unit sphere).

When the dimension n is large, the function f_K is known to be very close to the Gaussian density on the line with mean zero and variance L_K^2. Being general and informal, this hypothesis needs to be formalized and verified, and precise statements may depend on certain additional properties of convex bodies. For some special bodies K, several types of closeness of f_K to Gaussian densities were recently studied in [B-V], cf. also [K-L]. To treat the general case, the following characteristic σ_K^2 associated with K turns out to be crucial:

$$\sigma_K^2 = \frac{\mathrm{Var}(|X|^2)}{n L_K^4}.$$

Here X is a random vector uniformly distributed over K, and $\mathrm{Var}(|X|^2)$ denotes the variance of $|X|^2$. In particular, we have the following statement which is proved in this note.

---

* Supported in part by the NSF grant DMS-0103929.
** Supported in part by the NSF grant DMS-9996431.

Theorem 1. *For all* $0 < |t| \le c\sqrt{n}$,

$$\left| f_K(t) - \frac{1}{\sqrt{2\pi}L_K} e^{-t^2/(2L_K^2)} \right| \le C \left[\frac{\sigma_K L_K}{t^2\sqrt{n}} + \frac{1}{n} \right], \qquad (1)$$

where c and C are positive numerical constants.

Using Bourgain's estimate $L_K \le c\log(n)\, n^{1/4}$ ([Bou], cf. also [D], [P]) the right-hand side of (1) can be bounded, up to a numerical constant, by

$$\frac{\sigma_K \log n}{t^2 n^{1/4}} + \frac{1}{n},$$

which is small for large n up to the factor σ_K. Let us look at the behavior of this quantity in some canonical cases.

For the n-cube $K = [-\frac{1}{2}, \frac{1}{2}]^n$, by the independence of coordinates, $\sigma_K^2 = \frac{4}{5}$.

For K's the normalized ℓ_1^n balls,

$$\sigma_K^2 = 1 - \frac{2(n+1)}{(n+3)(n+4)} \to 1, \quad \text{as} \quad n \to \infty.$$

Normalization condition refers to $\mathrm{vol}_n(K) = 1$, but a slightly more general definition $\sigma_K^2 = \frac{n\mathrm{Var}(|X|^2)}{(\mathbf{E}|X|^2)^2}$ makes this quantity invariant under homotheties and simplifies computations.

For K's the normalized Euclidean balls,

$$\sigma_K^2 = \frac{4}{n+4} \to 0, \quad \text{as} \quad n \to \infty.$$

Thus, σ_K^2 can be small and moreover, in the space of any fixed dimension, the Euclidean balls provide the minimum (cf. Theorem 2 below).

The property that σ_K^2 is bounded by an absolute constant for all ℓ_p^n balls simultaneously was recently observed by K. Ball and I. Perissinaki [B-P] who showed for these bodies that the covariances $\mathrm{cov}(X_i^2, X_j^2) = \mathbf{E}X_i^2 X_j^2 - \mathbf{E}X_i^2 \mathbf{E}X_j^2$ are non-positive. Since in general $\mathrm{Var}(|X|^2) = \sum_{i=1}^n \mathrm{Var}(X_i^2) + \sum_{i \ne j} \mathrm{cov}(X_i^2, X_j^2)$, the above property together with the Khinchine-type inequality implies

$$\mathrm{Var}(|X|^2) \le \sum_{i=1}^n \mathrm{Var}(X_i^2) \le \sum_{i=1}^n \mathbf{E}X_i^4 \le CnL_K^4.$$

The result was used in [A-B-P] to study the closeness of random distribution functions $F_\theta(t) = \mathbf{P}\{\langle X, \theta \rangle \le t\}$, for most of θ on the sphere, to the normal distribution function with variance L_K^2. This randomized version of the central limit theorem originates in the paper by V. N. Sudakov [S], cf. also [D-F], [W]. The reader may find recent related results in [K-L], [Bob],

[N-R], [B-H-V-V]. It has become clear since the work [S] that, in order to get closeness to normality, the convexity assumption does not play a crucial role, and one rather needs a dimension-free concentration of $|X|$ around its mean. Clearly, the strength of concentration can be measured in terms of the variance of $|X|^2$, for example.

Nevertheless, the question on whether or not the quantity σ_K^2 can be bounded by a universal constant in the general convex isotropic case is still open, although it represents a rather weak form of Kannan-Lovász-Simonovits' conjecture about Cheeger-type isoperimetric constants for convex bodies [K-L-S]. For isotropic K, the latter may equivalently be expressed as the property that, for any smooth function g on $\mathbf{R}^n$, for some absolute constant C,

$$\int_K \left| g(x) - \int_K g(x)\,dx \right| dx \le CL_K \int_K |\nabla g(x)|\,dx. \qquad (2)$$

By Cheeger's theorem, the above implies a Poincaré-type inequality

$$\int_K \left| g(x) - \int_K g(x)\,dx \right|^2 dx \le 4(CL_K)^2 \int_K |\nabla g(x)|^2\,dx$$

which for $g(x) = |x|^2$ becomes $\mathrm{Var}(|X|^2) \le 16nC^2 L_K^4$, that is, $\sigma_K^2 \le 16C^2$.

To bound an optimal C in (2), R. Kannan, L. Lovász, and M. Simonovits considered in particular the geometric characteristic

$$\chi(K) = \int_K \chi_K(x)\,dx$$

where $\chi_K(x)$ denotes the length of the longest interval lying in K with center at x. By applying the localization lemma of [L-S], they proved that (2) holds true with $CL_K = 2\chi(K)$. Therefore, $\sigma_K L_K \le 8\chi(K)$, and thus the right-hand side of (1) can also be bounded, up to a constant, by

$$\frac{\chi(K)}{t^2\sqrt{n}} + \frac{1}{n}.$$

To prove Theorem 1, we need the following formula which also appears in [B-V, Lemma 1.2].

Lemma 1. *For all* t,

$$f_K(t) = \frac{\Gamma\left(\frac{n}{2}\right)}{\sqrt{\pi}\,\Gamma\left(\frac{n-1}{2}\right)} \int_{K \cap \{|x| \ge |t|\}} \frac{1}{|x|} \left(1 - \frac{t^2}{|x|^2}\right)^{\frac{n-3}{2}} dx.$$

For completeness, we prove it below (with a somewhat different argument).

Proof. We may assume $t \geq 0$. Denote by $\lambda_{\theta,t}$ the Lebesgue measure on $H_\theta(t)$. Then

$$\lambda_t = \int_{S^{n-1}} \lambda_{\theta,t} \, d\sigma(\theta)$$

is a positive measure on $\mathbf{R}^n$ such that $f_K(t) = \lambda_t(K)$. This measure has density that is invariant with respect to rotations, i.e.,

$$\frac{d\lambda_t}{dx} = p_t(|x|),$$

where p_t is a function on $[t, \infty)$. To find the function p_t, note first that, for every $r > t$,

$$\lambda_t\big(B(0,r)\big) = \int_{B(0,r)} p_t(|x|) \, dx = |S^{n-1}| \int_t^r p_t(s) s^{n-1} \, ds,$$

where $B(0,r)$ is the Euclidean ball with center at the origin and radius r, and $|S^{n-1}| = \frac{2\pi^{n/2}}{\Gamma(n/2)}$ is the surface area of the sphere S^{n-1}. On the other hand, since the section of $B(0,r)$ by the hyperplane $H_\theta(t)$ is the Euclidean ball in $\mathbf{R}^{n-1}$ of radius $(r^2 - t^2)^{1/2}$, we have

$$\lambda_t\big(B(0,r)\big) = \int_{S^{n-1}} \lambda_{\theta,t}\big(B(0,r)\big) \, d\sigma(\theta) = \frac{\pi^{(n-1)/2}}{\Gamma(1 + (n-1)/2)}(r^2 - t^2)^{(n-1)/2}.$$

Taking the derivatives by r, we see that for every $r \geq t$,

$$\frac{n-1}{2}(r^2 - t^2)^{(n-1)/2} \, 2r = \frac{2\pi^{1/2} \, \Gamma\left(\frac{n-1}{2}\right)}{\Gamma\left(\frac{n}{2}\right)} \, p_t(r) r^{n-1},$$

which implies

$$p_t(r) = \frac{\Gamma\left(\frac{n}{2}\right)}{\sqrt{\pi} \, \Gamma\left(\frac{n-1}{2}\right)} \frac{(r^2 - t^2)^{(n-3)/2}}{r^{n-2}}.$$

Since $f_K(t) = \lambda_t(K)$, the result follows.

Proof of Theorem 1. Let $t > 0$. By the Cauchy-Schwarz inequality,

$$\int_K \big||x|^2 - nL_K^2\big| \, dx \leq \left(\int_K \big||x|^2 - nL_K^2\big|^2 \, dx\right)^{1/2} = \sqrt{n} \, \sigma_K L_K^2,$$

so

$$\int_K \big||x| - \sqrt{n}L_K\big| \, dx = \int_K \frac{\big||x|^2 - nL_K^2\big|}{|x| + \sqrt{n}L_K} \, dx \leq \sigma_K L_K. \tag{3}$$

By Stirling's formula,

$$\lim_{n \to \infty} \frac{\sqrt{2\pi}}{\sqrt{n}} \frac{\Gamma(n/2)}{\sqrt{\pi}\Gamma((n-1)/2)} = 1.$$

so that the constants $c_n = \frac{\Gamma(n/2)}{\sqrt{\pi}\Gamma((n-1)/2))}$ appearing in Lemma 1 are $O(\sqrt{n})$.

Now, on the interval $[t, \infty)$ consider the function

$$g_n(z) = \frac{1}{z}\left(1 - \frac{t^2}{z^2}\right)^{(n-3)/2}.$$

Its derivative

$$g_n'(z) = \frac{t^2(n-3)}{z^4}\left(1 - \frac{t^2}{z^2}\right)^{(n-5)/2} - \frac{1}{z^2}\left(1 - \frac{t^2}{z^2}\right)^{(n-3)/2}$$

represents the difference of two non-negative terms. Both of them are equal to zero at t, tend to zero at infinity and each has one critical point, the first at $z = t\sqrt{n-1}/2$, and the second at $z = t\sqrt{n-1}/\sqrt{2}$. Therefore,

$$\max_{z \in [t, \infty)} |g_n'(z)| \le \frac{16}{t^2(n-1)}.$$

This implies that, for every $x \in K$, $|x| \ge t$, if $\sqrt{n}L_K \ge t$, then

$$|g_n(|x|) - g_n(\sqrt{n}L_K)| \le \frac{16}{t^2(n-1)}\big||x| - \sqrt{n}L_K\big|,$$

and by (3),

$$\int_{K_t} |g_n(|x|) - g_n(\sqrt{n}L_K)|\, dx \le \frac{16\sigma_K L_K}{t^2(n-1)}, \qquad (4)$$

where $K_t = K \cap \{|x| \ge t\}$.

Now, writing

$$f_K(t) = c_n \int_{K_t} g_n(|x|)\, dx$$

$$= c_n g_n(\sqrt{n}L_K)\mathrm{vol}_n(K_t) + c_n \int_{K_t} (g_n(|x|) - g_n(\sqrt{n}L_K))\, dx$$

and applying (4), we see that, for all $t \le \sqrt{n}L_K$,

$$|f_K(t) - c_n g_n(\sqrt{n}L_K)\mathrm{vol}_n(K_t)| \le \frac{C\sigma_K L_K}{t^2\sqrt{n}},$$

where C is a numerical constant. This gives

$$\big|f_K(t) - c_n g_n(\sqrt{n}L_K)\big| \le c_n g_n(\sqrt{n}L_K)\big(1 - \mathrm{vol}_n(K_t)\big) + \frac{C\sigma_K L_K}{t^2\sqrt{n}}. \qquad (5)$$

Recall that $L_K \ge c$, for some universal $c > 0$ (the worst situation is attained at Euclidean balls, cf. eg. [Ba]). Therefore (5) is fulfilled under $t \le c\sqrt{n}$.

To further bound the first term on the right-hand side of (5), note that $g_n(z) \le 1/z$, so $c_n g_n(\sqrt{n}L_K) \le C_0$, for some numerical C_0. Also, if $t \le c\sqrt{n}$,

$$1 - \mathrm{vol}_n(K_t) \le \mathrm{vol}_n\big(B(0,t)\big) = \omega_n t^n \le \left(\frac{c_0}{\sqrt{n}}\right)^n (c\sqrt{n})^n < 2^{-n},$$

where ω_n denotes the volume of the unit ball in $\mathbf{R}^n$, and where $c_0 c$ can be made less than $1/2$ by choosing a proper c. This also shows that the first term in (5) will be dominated by the second one. Indeed, the inequality $C_0 2^{-n} \le \frac{C \sigma_K L_K}{t^2 \sqrt{n}}$ immediately follows from $t \le c\sqrt{n}$ and the lower bound on σ_K given in Theorem 2.

Thus,

$$\left| f_K(t) - c_n g_n(\sqrt{n}L_K) \right| \le \frac{C \sigma_K L_K}{t^2 \sqrt{n}},$$

and we are left with the task of comparing $c_n g_n(\sqrt{n}L_K)$ with the Gaussian density on the line. This is done in the following elementary

Lemma 2. *If* $0 \le t \le \sqrt{n}L_K$, *for some absolute* C,

$$\left| \frac{\Gamma\left(\frac{n}{2}\right)}{\sqrt{\pi}\,\Gamma\left(\frac{n-1}{2}\right)} \left(1 - \frac{t^2}{nL_K^2}\right)^{(n-3)/2} \frac{1}{\sqrt{n}L_K} - \frac{1}{\sqrt{2\pi}L_K} e^{-t^2/2L_K^2} \right| \le \frac{C}{n}.$$

Proof. Using the fact that L_K is bounded from below, multiplying the above inequality by $\sqrt{2\pi}L_K$ and replacing $u = t^2/(2L_K^2)$, we are reduced to estimating

$$\left| \frac{\sqrt{2}\,\Gamma\left(\frac{n}{2}\right)}{\sqrt{n}\,\Gamma\left(\frac{n-1}{2}\right)} \left(1 - \frac{2u}{n}\right)^{\frac{n-3}{2}} - e^{-u} \right| \le \left| e^{-u} - \frac{\sqrt{2}\,\Gamma\left(\frac{n}{2}\right)}{\sqrt{n}\,\Gamma\left(\frac{n-1}{2}\right)} e^{-u} \right|$$

$$+ \frac{\sqrt{2}\,\Gamma\left(\frac{n}{2}\right)}{\sqrt{n}\,\Gamma\left(\frac{n-1}{2}\right)} \left| e^{-u} - \left(1 - \frac{2u}{n}\right)^{\frac{n-3}{2}} \right|.$$

In order to estimate the first summand, use the asymptotic formula for the Γ-function, $\Gamma(x) = x^{x-1} e^{-x} \sqrt{2\pi x}\,(1 + \frac{1}{12x} + O(\frac{1}{x^2}))$, as $x \to +\infty$, to get

$$\sqrt{\frac{2}{n}}\,\frac{\Gamma\left(\frac{n}{2}\right)}{\Gamma\left(\frac{n-1}{2}\right)} = \frac{\left(\frac{n}{2}\right)^{(n-3)/2} e^{-n/2} \sqrt{\pi n}\,\left(1 + \frac{1}{6n} + O(\frac{1}{n^2})\right)}{\left(\frac{n-1}{2}\right)^{(n-3)/2} e^{-(n-1)/2} \sqrt{\pi(n-1)}\,\left(1 + \frac{1}{6(n-1)} + O(\frac{1}{n^2})\right)}$$

$$= e^{-1/2} \left(\frac{n}{n-1}\right)^{\frac{n}{2}-1} \left(1 + O\left(\frac{1}{n^2}\right)\right).$$

Since, by Taylor, $\left(\frac{n}{n-1}\right)^{\frac{n}{2}-1} = e^{(-\frac{n}{2}+1)\log(1-\frac{1}{n})} = e^{1/2}\left(1 + O\left(\frac{1}{n}\right)\right)$, the first summand is $O(\frac{1}{n})$ uniformly over $u \ge 0$.

To estimate the second summand, recall that $0 \le u \le n/2$. The function $\psi_n(u) = e^{-u} - \left(1 - \frac{2u}{n}\right)^{\frac{n-3}{2}}$ satisfies $\psi_n(0) = 0$, $\psi_n(n/2) = e^{-n/2}$, and the point $u_0 \in [0, n/2]$ where $\psi'_n(u_0) = 0$ (if it exists) satisfies $\left(1 - \frac{2u_0}{n}\right)^{\frac{n-5}{2}} = \frac{n}{n-3} e^{-u_0}$ (when $n \geq 4$). Hence, $\psi_n(u_0) = \frac{2u_0 - 3}{n-3} e^{-u_0} = O(\frac{1}{n})$, and thus $\sup_u \psi_n(u) = O(\frac{1}{n})$. This proves Lemma 2.

Remark. Returning to the inequality (1) of Theorem 1, it might be worthwhile to note that, in the range $|t| \geq c\sqrt{n}$, the function f_K satisfies, for some absolute $C > 0$, the estimate

$$f_K(t) \leq \frac{C}{|t|} e^{-t^2/(CnL_K^2)} \leq \frac{C}{c\sqrt{n}},$$

and in this sense it does not need to be compared with the Gaussian distribution in this range. Indeed, it follows immediately from the equality in Lemma 1 that

$$f_K(t) \leq C\sqrt{n} \max_{z \geq |t|} g_n(z) \, \mathbf{P}\{|X| \geq |t|\},$$

where X denotes a random vector uniformly distributed over K. When $n \geq 3$, in the interval $z \geq |t|$, the function $g_n(z) = \frac{1}{z}(1 - \frac{t^2}{z^2})^{(n-3)/2}$ attains its maximum at the point $z_0 = |t|\sqrt{n-2}$ where it takes the value $g_n(z_0) \leq \frac{1}{|t|\sqrt{n-2}}$. Hence,

$$C\sqrt{n} \max_{z \geq |t|} g_n(z) \leq \frac{C'}{|t|} \leq \frac{C'}{c\sqrt{n}}.$$

On the other hand, the probability $\mathbf{P}\{|X| \geq |t|\}$ can be estimated with the help of Alesker's ψ_2-estimate, [A],

$$\mathbf{E} e^{|X|^2/(C''nL_K^2)} \leq 2.$$

We finish this note with a simple remark on the extremal property of the Euclidean balls in the minimization problem for σ_K^2.

Theorem 2. $\sigma_K^2 \geq \frac{4}{n+4}$.

Proof. The distribution function $F(r) = \mathrm{vol}_n(\{x \in K : |x| \leq r\})$ of the random vector X uniformly distributed in K has density

$$F'(r) = r^{n-1} \left| S^{n-1} \cap \frac{1}{r}K \right| = |S^{n-1}| r^{n-1} \sigma\left(\frac{1}{r}K\right), \quad r > 0.$$

We only use the property that $q(r) = |S^{n-1}| \sigma(\frac{1}{r}K)$ is non-increasing in $r > 0$. Clearly, this function can also be assumed to be absolutely continuous so that we can write

$$q(r) = n \int_r^{+\infty} \frac{p(s)}{s^n} \, ds, \quad r > 0,$$

for some non-negative measurable function p on $(0, +\infty)$.

We have

$$1 = \int_0^\infty dF(r) = \int_0^\infty r^{n-1} q(r)\, dr = n \iint_{0<r<s} r^{n-1} \frac{p(s)}{s^n}\, dr ds = \int_0^\infty p(s)\, ds.$$

Hence, p represents a probability density of a positive random variable, say, ξ. Similarly, for every $\alpha > -n$,

$$\mathbf{E}|X|^\alpha = \int_0^\infty r^{\alpha+n-1} q(r)\, dr = \frac{n}{n+\alpha} \int_0^\infty s^\alpha p(s)\, ds = \frac{n}{n+\alpha} \mathbf{E}\xi^\alpha.$$

Therefore,

$$\begin{aligned}
\mathrm{Var}(|X|^2) &= \frac{n}{n+4} \mathbf{E}\xi^4 - \left(\frac{n}{n+2} \mathbf{E}\xi^2 \right)^2 \\
&= \frac{4n}{(n+4)(n+2)^2} (\mathbf{E}\xi^2)^2 + \frac{n}{n+4} \mathrm{Var}(\xi^2) \\
&\geq \frac{4n}{(n+4)(n+2)^2} (\mathbf{E}\xi^2)^2.
\end{aligned}$$

One can conclude that

$$\sigma_K^2 = n \frac{\mathrm{Var}(|X|^2)}{(\mathbf{E}|X|^2)^2} \geq n \frac{\frac{4n}{(n+4)(n+2)^2} (\mathbf{E}\xi^2)^2}{\left(\frac{n}{n+2} \mathbf{E}\xi^2 \right)^2} = \frac{4}{n+4}.$$

Theorem 2 follows.

Acknowledgement. We would like to thank V. D. Milman for stimulating discussions.

References

[A]   Alesker, S. (1995): ψ_2-estimate for the Euclidean norm on a convex body in isotropic position. Geom. Aspects Funct. Anal. (Israel 1992-1994), Oper. Theory Adv. Appl., **77**, 1–4

[A-B-P] Antilla, M., Ball, K., Perissinaki, I. (1998): The central limit problem for convex bodies. Preprint

[Ba]   Ball, K. (1988): Logarithmically concave functions and sections of convex sets. Studia Math., **88**, 69–84

[B-P]   Ball, K., Perissinaki, I. (1998): Subindependence of coordinate slabs in ℓ_p^n balls. Israel J. of Math., **107**, 289–299

[Bob]   Bobkov, S.G. On concentration of distributions of random weighted sums. Ann. Probab., to appear

[Bou]   Bourgain, J. (1991): On the distribution of polynomials on high dimensional convex sets. Lecture Notes in Math., **1469**, 127–137

[B-H-V-V]   Brehm, U., Hinow, P., Vogt, H., Voigt, J. Moment inequalities and central limit properties of isotropic convex bodies. Preprint

[B-V]   Brehm, U., Voigt, J. (2000): Asymptotics of cross sections for convex bodies. Beiträge Algebra Geom., **41**, 437–454

[D]   Dar, S. (1995): Remarks on Bourgain's problem on slicing of convex bodies. Geom. Aspects of Funct. Anal., Operator Theory: Advances and Applications, **77**, 61–66

[D-F]   Diaconis, P., Freedman, D. (1984): Asymptotics of graphical projection pursuit. Ann. Stat., **12(3)**, 793–815

[K-L-S]   Kannan, R., Lovász, L., Simonovits, M. (1995): Isoperimetric problems for convex bodies and a localization lemma. Discrete and Comput. Geom., **13**, 541–559

[K-L]   Koldobsky, A., Lifshitz, M. (2000): Average volume of sections of star bodies. Geom. Aspects of Funct. Anal. (Israel Seminar 1996-2000), Lecture Notes in Math., **1745**, 119–146

[L-S]   Lovász, L., Simonovits, M. (1993): Random walks in a convex body and an improved volume algorithm. Random Structures and Algorithms, **4(3)**, 359–412

[N-R]   Naor, A., Romik, D. Projecting the surface measure of the sphere of ℓ_p^n. Annales de L'Institut Henri Poincaré, to appear

[P]   Paoris, G. (2000): On the isotropic constant of non-symmetric convex bodies. Geom. Aspects of Funct. Anal. (Israel Seminar 1996-2000), Lecture Notes in Math., **1745**, 239–244

[S]   Sudakov, V.N. (1978): Typical distributions of linear functionals in finite-dimensional spaces of higher dimensions. Soviet Math. Dokl., **19(6)**, 1578–1582. Translated from Dokl. Akad. Nauk SSSR, **243(6)**

[W]   von Weizsäcker, H. (1997): Sudakov's typical marginals, random linear functionals and a conditional central limit theorem. Probab. Theory Rel. Fields, **107**, 313–324

On Convex Bodies and Log-Concave Probability Measures with Unconditional Basis[*]

S.G. Bobkov[1] and F.L. Nazarov[2]

[1] School of Mathematics, University of Minnesota, Minneapolis, MN 55455, USA
  bobkov@math.umn.edu
[2] Department of Mathematics, Michigan State University, East Lansing,
  MI 48824-1027, USA *fedja@math.msu.edu*

1 Introduction

We consider here two asymptotic properties of finite dimensional convex bodies which generate a norm with an unconditional basis. For definiteness, such a basis is taken to be the canonical basis in $\mathbf{R}^n$. Thus, assume we are given a convex set $K \subset \mathbf{R}^n$ of volume $\mathrm{vol}_n(K) = 1$ which, together with every point $x = (x_1, \ldots, x_n)$, contains the parallelepiped with the sides $[-|x_j|, |x_j|]$, $1 \le j \le n$. In addition, K is supposed to be in isotropic position, which is equivalent to the property that the integrals

$$\int_K x_j^2 \, dx = L_K^2, \quad 1 \le j \le n, \tag{1.1}$$

do not depend on j.

The isotropic constant L_K is known to satisfy $c_1 \le L_K \le c_2$, for some universal $c_1, c_2 > 0$. Hence, for the Euclidean norm $|x| = (x_1^2 + \ldots + x_n^2)^{1/2}$ we have

$$c_1 n \le \int_K |x|^2 \, dx \le c_2 n$$

and similarly, the average value of $|x|$ over K is about $\sqrt{n}$.

Consider the linear functional

$$f(x) = \frac{x_1 + \ldots + x_n}{\sqrt{n}}.$$

By (1.1), its L_2-norm over K is exactly $\|f\|_2 = L_K$. As in the case of any other linear functional, L_p-norms satisfy $\|f\|_p \le Cp\|f\|_2$ for every $p \ge 1$ and some absolute C. Up to a universal constant, this property can equivalently be expressed as one inequality $\|f\|_{\psi_1} \le C\|f\|_2$ for the Orlicz norm corresponding to the Young function $\psi_1(t) = e^{|t|} - 1$, $t \in \mathbf{R}$. For the concrete functional f introduced above, this can be sharpened in terms of the Young function $\psi_2(t) = e^{|t|^2} - 1$.

---

[*] Supported in part by NSF grants.

Theorem 1.1. $\|f\|_{\psi_2} \leq C$, *for some universal* C.

The proof might require some information on the distribution of the Euclidean norm of a point x over K. Indeed, if we observe $x = (x_1, \ldots, x_n)$ as a random vector uniformly distributed in K, and if $(\varepsilon_1, \ldots, \varepsilon_n)$ is an arbitrary collection of signs, then $(\varepsilon_1 x_1, \ldots, \varepsilon_n x_n)$ has the same uniform distribution (by the assumption that the canonical basis is unconditional). In particular,

$$f(x, \varepsilon) = \frac{\varepsilon_1 x_1 + \ldots + \varepsilon_n x_n}{\sqrt{n}}$$

has the same distribution as $f(x)$. But with respect to the symmetric Bernoulli measure $\mathbf{P}_\varepsilon$ on the discrete cube $\{-1, 1\}^n$, there is a subgaussian inequality

$$\mathbf{P}_\varepsilon \{|f(x, \varepsilon)| \geq t\} \leq 2 e^{-nt^2/(2|x|^2)}, \quad t \geq 0.$$

Taking the expectation over K, we arrive at

$$\text{vol}_n\{x \in K : |f(x)| \geq t\} \leq 2 \int_K e^{-nt^2/(2|x|^2)}\, dx. \tag{1.2}$$

This is how the distribution of the norm $|x|$ can be involved in the study of the distribution of $f(x)$. The statement of Theorem 1.1 is equivalent to the assertion that the tails of f admit a subgaussian bound

$$\text{vol}_n\{x \in K : |f(x)| \geq t\} \leq C e^{-ct^2}.$$

Hence, it suffices to prove such a bound for the integral in (1.2) taken over a sufficiently big part of K. The function $e^{-nt^2/(2|x|^2)}$ under the integral sign has the desired subgaussian behaviour on the part of K where $|x|/\sqrt{n} \leq$ const. To control large deviations of $|x|/\sqrt{n}$, we prove:

Theorem 1.2. *There exist universal* $t_0 > 0$ *and* $c > 0$ *such that, for all* $t \geq t_0$,

$$\text{vol}_n \left\{ x \in K : \frac{|x|}{\sqrt{n}} \geq t \right\} \leq e^{-ct\sqrt{n}}. \tag{1.3}$$

For the "normalized" ℓ_1^n-ball, this inequality was proved by G. Schechtman and J. Zinn in [S-Z1], see also [S-Z2] for related results on deviations of the Euclidean norm and other Lipschitz functions on the ℓ_p^n-balls.

Note that too large t may be ignored in (1.3), since we always have $|x| \leq Cn$, for all $x \in K$ (V.D. Milman, A. Pajor, [M-P]). Therefore, for $t > C\sqrt{n}$, the left hand side is zero. For $t \leq C\sqrt{n}$, the inequality implies

$$\text{vol}_n \left\{ x \in K : \frac{|x|}{\sqrt{n}} \geq t \right\} \leq e^{-ct^2/C},$$

which means that the $L_{\psi_2}(K)$-norm of the Euclidean norm is bounded by its L_2-norm, up to a universal constant. Thus, Theorem 1.2 can also be viewed as a sharpening, for isotropic convex sets with an unconditional basis, of a result of S. Alesker [A]. We do not know whether the unconditionality assumption is important for the conclusion such as (1.3). On the other hand, Theorem 1.2 as well as Theorem 1.1 (under an extra condition on the support) can be extended to all isotropic log-concave probability measures which are invariant under transformations $(x_1, \ldots, x_n) \to (\pm x_1, \ldots, \pm x_n)$, cf. Propositions 5.1 and 6.1 below.

Using Theorem 1.2, one may estimate the integral in (1.2) as follows:

$$\int_K e^{-nt^2/(2|x|^2)} \, dx = \int_{|x| \leq t_0 \sqrt{n}} + \int_{|x| \geq t_0 \sqrt{n}} \leq e^{-t^2/(2t_0^2)} + e^{-ct_0\sqrt{n}}$$
$$\leq 2\,e^{-t^2/(2t_0^2)}$$

provided that $t \leq \text{const}\, n^{1/4}$. Hence, we obtain the desired subgaussian bound for relatively "small" t. To treat the values $t \geq \text{const}\, n^{1/4}$, one needs to involve some other arguments which are discussed in section 6.

2 Preliminaries (the case of bodies)

Here we collect some useful, although basically known, facts about the sets K with the canonical unconditional basis as in section 1. It is reasonable to associate with K its normalized part in the positive octant $\mathbf{R}_+^n = [0, +\infty)^n$,

$$K^+ = 2K \cap \mathbf{R}_+^n.$$

Thus, if $x = (x_1, \ldots, x_n)$ is viewed as a random vector uniformly distributed in K, then the vector $(2|x_1|, \ldots, 2|x_n|)$ is uniformly distributed in K^+.

The set K^+ has the properties:

a) $\text{vol}_n(K^+) = 1$;

b) for all $x \in K^+$ and $y \in \mathbf{R}_+^n$ with $y_j \leq x_j$, $1 \leq j \leq n$, we have $y \in K^+$;

c) $\int_{K^+} x_j^2 \, dx = 4L_K^2$, for all $1 \leq j \leq n$.

Proposition 2.1. $L_K^2 \leq \frac{1}{2}$.

Proof. With every point $x = (x_1, \ldots, x_n)$, the set K^+ contains the parallelepiped $\prod_{j=1}^n [0, x_j]$. So $\prod_{j=1}^n x_j \leq 1$, for every $x \in K$. Since both the sets K^+ and $V = \{x \in \mathbf{R}_+^n : \prod_{j=1}^n x_j \geq 1\}$ are convex and do not intersect each other (excluding the points on the boundaries), there exists a separating hyperplane. But any hyperplane touching the boundary of V has equation $\lambda_1 x_1 + \ldots + \lambda_n x_n = n$ with some $\lambda_j > 0$ such that $\prod_{j=1}^n \lambda_j = 1$. Therefore, $K^+ \subset \{x \in \mathbf{R}_+^n : \frac{\lambda_1 x_1 + \ldots + \lambda_n x_n}{n} \leq 1\}$, and so, by the geometric-arithmetic inequality,

$$1 \geq \int_{K+} \frac{\lambda_1 x_1 + \ldots + \lambda_n x_n}{n} \, dx \geq \left(\prod_{j=1}^n \int_{K+} x_j \, dx \right)^{1/n}.$$

By a Khinchine-type inequality,

$$\int_{K+} x_j \, dx \geq \frac{1}{\sqrt{2}} \left(\int_{K+} x_j^2 \, dx \right)^{1/2} = \sqrt{2} \, L_K, \tag{2.1}$$

according to the property c). Thus, $1 \geq \sqrt{2} \, L_K$.

Remark 2.1. It is a well-known fact that, in the class of all measurable sets K in $\mathbf{R}^n$ of volume one, the integral $\int_K |x|^2 \, dx$ is minimized for the normalized Euclidean ball B_n with center at the origin. Therefore, for isotropic K, we always have $L_K \geq L_{B_n}$ which leads to the optimal dimension-free lower bound

$$L_K \geq \frac{1}{\sqrt{2\pi e}}. \tag{2.2}$$

More generally, in the class of all probability densities q on $\mathbf{R}^n$ attaining maximum at the origin, the quantity $q^2(0) \int |x|^2 q(x) \, dx$ is minimized for the indicator function of B_n. This property was observed by D. Hensley [H] who assumed additionally that q is log-concave and symmetric, and later K. Ball [Ba] gave a shorter argument not using log-concavity and symmetry. In the one-dimensional case, the property reads as

$$q(0) \left(\int_{\mathbf{R}} t^2 q(t) \, dt \right)^{1/2} \geq \frac{1}{2\sqrt{3}}. \tag{2.3}$$

Remark 2.2. The inequality (2.1) is a particular case of the following theorem due to S. Karlin, F. Proschan, and R.E. Barlow [K-P-B]: Given a positive random variable ξ with a log-concave density on $(0, +\infty)$, for all real $s > 1$

$$\mathbf{E} \, \xi^s \leq \Gamma(s+1) \, (\mathbf{E} \, \xi)^s.$$

Equality is achieved if and only if ξ has an exponential distribution, that is, when $\mathrm{Prob}\{\xi > t\} = e^{-\lambda t}$, $t > 0$, for some parameter $\lambda > 0$.

Proposition 2.2. *For every hyperspace H in $\mathbf{R}^n$,*

$$\mathrm{vol}_{n-1}(K \cap H) \geq \frac{1}{\sqrt{6}}.$$

Moreover, if K is invariant under permutations of coordinates, then every section $K_j = K \cap \{x_j = 0\}$, $1 \leq j \leq n$, satisfies $\mathrm{vol}_{n-1}(K_j) \geq 1$.

Proof. If $H = \{x \in \mathbf{R}^n : \langle \theta, x \rangle = 0\}$, $|\theta| = 1$, apply (2.3) to the density $q(t)$ of the linear function $x \to \langle \theta, x \rangle$ over K: then we get

$$\text{vol}_{n-1}(K \cap H) L_K \geq \frac{1}{2\sqrt{3}}.$$

This inequality holds true for any symmetric isotropic convex set K of volume one. In our specific case, it remains to apply Proposition 2.1.

For the second statement, given a non-empty set $\pi \subset \{1, \ldots, n\}$, denote by K_π^+ the section of K by the $(n - |\pi|)$-dimensional subspace $\{x : x_j = 0,$ for all $j \in \pi\}$. Write the Steiner decomposition

$$\text{vol}_n \left(K^+ + r[0, 1]^n\right) = \sum_{k=0}^{n} a_k(K^+) r^k, \quad r > 0,$$

where $a_k = \sum_{|\pi|=k} \text{vol}_{n-k}(K_\pi^+)$ with the convention that $a_0 = \text{vol}_n(K^+) = 1$. By the Brunn-Minkowski inequality, $\text{vol}_n \left(K^+ + r[0, 1]^n\right) \geq (1 + r)^n$, so the coefficient $a_1(K^+)$ in front of r should satisfy $a_1 \geq n$. That is,

$$\sum_{j=1}^{n} \text{vol}_{n-1}(K_j^+) \geq n,$$

where $K_j^+ = K^+ \cap \{x_j = 0\}$. Since all these $(n-1)$-dimensional volumes are equal to each other, and $\text{vol}_{n-1}(K_j) = \text{vol}_{n-1}(K_j^+)$, the conclusion follows.

Proposition 2.3. *For all $\alpha_1, \ldots, \alpha_n \geq 0$,*

$$\text{vol}_n\{x \in K^+ : x_1 \geq \alpha_1, \ldots, x_n \geq \alpha_n\} \leq e^{-c(\alpha_1 + \ldots + \alpha_n)}$$

with $c = 1/\sqrt{6}$. If K is invariant under permutations of coordinates, one may take $c = 1$.

Proof. The function $u(\alpha_1, \ldots, \alpha_n) = \text{vol}_n\{x \in K^+ : x_1 \geq \alpha_1, \ldots, x_n \geq \alpha_n\}$ is log-concave on $\mathbf{R}_+^n$, $u(0) = 1$, and

$$\left. \frac{\partial u(\alpha)}{\partial \alpha_j} \right|_{\alpha=0} = -\text{vol}_{n-1}(K_j) \leq -c,$$

according to Proposition 2.2. These properties easily imply the desired inequality.

Actually, Proposition 2.3 can be sharpened by applying the Brunn-Minkowski inequality in its full volume. The latter implies that the function $u^{1/n}$ is concave on K^+ which is a slightly stronger property than just log-concavity. Hence, with the same argument, we have the inequality

$$\text{vol}_n^{1/n}\{x \in K^+ : x_1 \geq \alpha_1, \ldots, x_n \geq \alpha_n\} \leq 1 - \frac{c(\alpha_1 + \ldots + \alpha_n)}{n}$$

holding true for all $(\alpha_1, \ldots, \alpha_n) \in K^+$ with $c = 1/\sqrt{6}$. Since the right hand side of this inequality must be non-negative, an immediate consequence of such a refinement is:

Proposition 2.4. *For all* $(x_1, \ldots, x_n) \in K^+$,

$$x_1 + \ldots + x_n \leq \sqrt{6}\, n.$$

Equivalently, for all $(x_1, \ldots, x_n) \in K$, $|x_1| + \ldots + |x_n| \leq \frac{\sqrt{6}}{2}\, n$.

Thus, the normalized ℓ^1-ball in $\mathbf{R}^n$ is the largest set within the class of all K's which we consider (up to a universal enlarging factor). One may wonder therefore whether or not it is true that the cube would be the smallest one. The question turns out simple as one can see from the proof of the following:

Proposition 2.5. *The set* K *contains the cube* $[-\frac{1}{\sqrt{2}} L_K, \frac{1}{\sqrt{2}} L_K]^n$ *which in turn contains* $[-\frac{1}{2\sqrt{\pi e}}, \frac{1}{2\sqrt{\pi e}}]^n$.

Proof. The baricenter $v = \mathrm{bar}(K^+)$ must belong to K^+, so K^+ contains parallelepiped $\prod_{j=1}^n [0, v_j]$ with $v_j = \int_{K^+} x_j \, dx$. Hence the first statement immediately follows from the Khinchine-type inequality (2.1). The second one is based on the lower bound (2.2).

3 Log-Concave Measures

Here we extend Propositions 2.1–2.3 to log-concave measures. Let μ be a probability measure on $\mathbf{R}^n$ with a log-concave density $p(x)$, $x \in \mathbf{R}^n$, such that

a) $p(0) = 1$;

b) $p(\pm x_1, \ldots, \pm x_n)$ does not depend on the choice of signs;

c) $\int x_j^2 \, d\mu(x) = \int x_j^2 p(x) \, dx = L_\mu^2$ does not depend on $j = 1, \ldots, n$.

The case of the indicator density $p(x) = 1_K(x)$ reduces to the previous section. As in the body case, we associate with μ its squeezed restriction μ^+ to the positive octant $\mathbf{R}_+^n$: this measure has density

$$p^+(x) = p\left(\frac{1}{2} x\right), \quad x \in \mathbf{R}_+^n.$$

If $x = (x_1, \ldots, x_n)$ is distributed according to μ, then the vector $(2|x_1|, \ldots, 2|x_n|)$ is distributed according to μ^+. The function p^+ is log-concave, is non-increasing in each coordinate, and satisfies

$$\int_{\mathbf{R}_+^n} x_j^2 \, d\mu^+(x) = 4L_\mu^2, \quad 1 \leq j \leq n.$$

Proposition 3.1. $L_\mu \leq C$, *for some absolute* C.

Proof. Since p^+ is non-increasing, for every $x \in \mathbf{R}_+^n$,

$$1 \geq \int_0^{x_1} \cdots \int_0^{x_n} p^+(y)\, dy \geq p^+(x) \int_0^{x_1} \cdots \int_0^{x_n} dy = p^+(x) \prod_{j=1}^n x_j.$$

Hence,

$$u(x) \equiv -\log p^+(x) \geq \log \prod_{j=1}^n x_j \equiv v(x).$$

Note that u is convex, while v is a concave function. Therefore, there must exist an affine function ℓ such that $u(x) \geq \ell(x) \geq v(x)$, for all $x \in \mathbf{R}_+^n$. This function can be chosen to be tangent to v at some point $a = (a_1, \ldots, a_n)$ with positive coordinates. That is, we may take

$$\ell(x) = v(a) + \langle \nabla v(a), x - a \rangle = \log \prod_{j=1}^n a_j + \sum_{j=1}^n \frac{x_j - a_j}{a_j}.$$

Setting $\lambda_j = \frac{1}{a_j}$, the inequality $u(x) \geq \ell(x)$ becomes

$$p^+(x) \leq e^n \prod_{j=1}^n \lambda_j\, e^{-\lambda_j x_j}, \quad x \in \mathbf{R}_+^n.$$

In particular, since $p^+(0) = 1$, we have $\prod_{j=1}^n \lambda_j \geq e^{-n}$. Hence,

$$\int_{\mathbf{R}_+^n} \prod_{j=1}^n x_j\, p^+(x)\, dx \leq \int_{\mathbf{R}_+^n} \prod_{j=1}^n x_j \left(e^n \prod_{j=1}^n \lambda_j\, e^{-\lambda_j x_j} \right) dx = e^n \prod_{j=1}^n \frac{1}{\lambda_j} \leq e^{2n}.$$

On the other hand, with respect to μ^+,

$$\left\| \prod_{j=1}^n x_j \right\|_1 \geq \left\| \prod_{j=1}^n x_j \right\|_0 = \prod_{j=1}^n \|x_j\|_0 \geq c^n \prod_{j=1}^n \|x_j\|_2 = (2c)^n L_\mu^n,$$

where we have used a Khinchine-type inequality $\|g\|_0 = \lim_{p \to 0+} \|g\|_p \geq c \|g\|_2$ for linear functions g with respect to log-concave measures (which is actually valid for any norm, cf. [L]). Proposition 3.1 follows with $C = e^2/(2c)$.

Proposition 3.2. *For every hyperspace* H *in* $\mathbf{R}^n$,

$$\int_H p(x)\, dx \geq \frac{1}{e\sqrt{6}}.$$

If p *is invariant under permutations of coordinates, then* $\int_{\{x_j=0\}} p(x)\, dx \geq \frac{1}{e}$, *for every* $1 \leq j \leq n$.

There is a way to prove this statement without appealing to Proposition 3.1. In turn, starting from Proposition 3.2, one can easily obtain Proposition 3.1 with $C = e\sqrt{3}$. Indeed, the reverse one-dimensional Hensley inequality (for the class of all symmetric log-concave probability densities q on the line, cf. [H], Lemma 4) asserts that

$$q(0) \left(\int_{\mathbf{R}} t^2 \, dx \right)^{1/2} \leq \frac{1}{\sqrt{2}} \tag{3.1}$$

(equality is achieved at $q(t) = e^{-2|t|}$). If we take any hyperspace $H = \{x \in \mathbf{R}^n : \langle \theta, x \rangle = 0\}$, $|\theta| = 1$, and apply this inequality to the density $q(t)$ of the distribution of the linear function $\langle \theta, x \rangle$ under the measure μ, then we arrive exactly at

$$\int_H p(x) \, dx \, L_\mu \leq \frac{1}{\sqrt{2}}.$$

Hence, the lower bound $\int_H p(x) \, dx \geq 1/(e\sqrt{6})$ would lead to $L_\mu \leq e\sqrt{3}$, while in the case where μ is invariant under permutations of coordinates we would similarly obtain the estimate $L_\mu \leq e/\sqrt{2}$.

Proposition 3.2 will be derived from a more general:

Lemma 3.1. *For any log-concave probability density p on $\mathbf{R}^n$ such that $p(0) = 1$ and $p(\pm x_1, \ldots, \pm x_n)$ does not depend on the choice of signs,*

$$\prod_{j=1}^n \int_{\{x_j=0\}} p(x) \, dx \geq e^{-n}. \tag{3.2}$$

It is interesting that the constant $1/e$ appearing on the right is asymptotically optimal. Indeed, for the density

$$p(x) = \exp\left\{ -2n!^{1/n} \max_{j \leq n} |x_j| \right\},$$

for every $j \leq n$, we have $\int_{\{x_j=0\}} p(x) \, dx = \frac{n!^{1/n}}{n} \to \frac{1}{e}$, as $n \to \infty$.

As in this example, when a density p is invariant under permutations of coordinates, all $(n-1)$-dimensional integrals $\int_{\{x_j=0\}} p(x) \, dx$ coincide, so, by (3.2), these integrals must be greater or equal to $1/e$. In the general case, we may only conclude that $\max_j \int_{\{x_j=0\}} p(x) \, dx \geq 1/e$. On the other hand, the combination of the two Hensley's inequalities (2.3) and (3.1) immediately implies that, for any symmetric log-concave isotropic density p on $\mathbf{R}^n$ and for any two hyperspaces H_1, H_2, we have $\int_{H_1} p(x) \, dx \leq \sqrt{6} \int_{H_2} p(x) \, dx$. Hence,

$$\min_H \int_H p(x) \, dx \geq \frac{1}{\sqrt{6}} \max_j \int_{\{x_j=0\}} p(x) \, dx \geq \frac{1}{e\sqrt{6}}.$$

Thus, Lemma 3.1 implies Proposition 3.2.

Proof of Lemma 3.1. Given a measurable set A in $\mathbf{R}^n$, an inequality due to L. H. Loomis and H. Whitney asserts ([L-W], [B-Z]) that

$$\prod_{j=1}^{n} \text{vol}_{n-1}(A_j) \geq \text{vol}_n(A)^{n-1},$$

where A_j is the projection of A to the hyperspace $x_j = 0$. As a matter of fact, being applied to $A = K$, the above yields yet another proof of the second part of Proposition 2.2.

Loomis-Whitney's inequality admits a certain functional formulation. Namely, given a measurable function $g \geq 0$ on $\mathbf{R}^n$, not identically zero, consider the family $A(t) = \{x : g(x) > t\}$, $t > 0$. Define on $\mathbf{R}^{n-1}$ the functions

$$g_j(x_1, \ldots, x_{j-1}, x_{j+1}, \ldots, x_n) = \sup_{x_j} g_j(x_1, \ldots, x_{j-1}, x_j, x_{j+1}, \ldots, x_n)$$

together with $A_j(t) = \{x : g_j(x) > t\}$, $t > 0$. Then $A_j(t)$ are projections of $A(t)$, so

$$\text{vol}_n\big(A(t)\big)^{n-1} \leq \prod_{j=1}^{n} \text{vol}_{n-1}\big(A_j(t)\big).$$

Put $\varphi_j(t) = \text{vol}_{n-1}(A_j(t))$, $\varphi(t) = \text{vol}_n(A(t))$. Raising the above to the power $1/n$, integrating over $t > 0$ and applying Hölder's inequality, we get

$$\int_0^{+\infty} \varphi(t)^{(n-1)/n}\, dt \leq \int_0^{+\infty} \prod_{j=1}^{n} \varphi_j(t)^{1/n}\, dt \leq \prod_{j=1}^{n} \left(\int_0^{+\infty} \varphi_j(t)\, dt \right)^{1/n}$$

$$= \left(\prod_{j=1}^{n} \int_{\mathbf{R}^{n-1}} g_j(x)\, dx \right)^{1/n}.$$

In order to bound from below the first integral, we use the property that $\varphi(t)$ is non-increasing in $t > 0$. For such functions, for all $\alpha \in (0,1]$, there is a simple inequality (cf. [B-Z])

$$\left(\int_0^{+\infty} \varphi(t)^{\alpha}\, dt \right)^{1/\alpha} \geq \int_0^{+\infty} \varphi(t^{\alpha})\, dt.$$

But the right hand side is exactly $\int_{\mathbf{R}^n} g(x)^{1/\alpha}\, dx$, and for $\alpha = \frac{n-1}{n}$, we thus get

$$\prod_{j=1}^{n} \int_{\mathbf{R}^{n-1}} g_j(x)\, dx \geq \left(\int_{\mathbf{R}^n} g(x)^{n/(n-1)}\, dx \right)^{n-1}.$$

This is the desired functional form yielding the original inequality on indicator functions $g = 1_A$. For $g = p$, the supremum in the definition of g_j is attained at $x_j = 0$, and the functional inequality becomes

$$\prod_{j=1}^{n} \int_{\{x_j=0\}} p(x)\, dx \geq \left(\int_{\mathbf{R}^n} p(x)^{n/(n-1)}\, dx \right)^{n-1}.$$

The right hand side can further be estimated using the log-concavity of p. Namely, since $p(0) = 1$, for every $t \in (0,1)$ and $x \in \mathbf{R}^n$, we have $p(tx)^{1/t} \geq p(x)$. Integrating over x, we get $\int_{\mathbf{R}^n} p(x)^{1/t}\, dx \geq t^n$ which for $t = \frac{n-1}{n}$ gives

$$\int_{\mathbf{R}^n} p(x)^{n/(n-1)}\, dx \geq \left(\frac{n-1}{n} \right)^n, \quad n \geq 2.$$

It remains to note that $\left(\frac{n-1}{n} \right)^{n(n-1)} \geq e^{-n}$.

Lemma 3.1 follows. As a consequence, we get an analogue of Proposition 2.3:

Proposition 3.3. *For all* $\alpha_1, \ldots, \alpha_n \geq 0$,

$$\mu^+ \{ x \in \mathbf{R}_+^n : x_1 \geq \alpha_1, \ldots, x_n \geq \alpha_n \} \leq e^{-c(\alpha_1 + \ldots + \alpha_n)}$$

with $c = \frac{1}{e\sqrt{6}}$. *If* μ *is invariant under permutations of coordinates, one may take* $c = 1/e$.

4 Decreasing Rearrangement

For any vector $x = (x_1, \ldots, x_n)$ in $\mathbf{R}^n$, its coordinates can be written in the decreasing order,

$$X_1 \geq X_2 \geq \ldots \geq X_n.$$

In particular, $X_1 = \max_j x_j$, $X_n = \min_j x_j$. When x is observed as a random vector with uniform distribution in K^+ or more generally with distribution μ^+, the distribution of the random vector $(X_1, \ldots, X_n)$ can be studied on the basis of Propositions 2.3 and 3.3, respectively. In particular, we have:

Proposition 4.1. *For any* $\alpha \geq 0$, $1 \leq k \leq n$,

$$\mu^+ \{ x \in \mathbf{R}_+^n : X_k \geq \alpha \} \leq C_n^k\, e^{-c k \alpha},$$

where $c > 0$ *is a numerical constant.*

One may always take $c = 1/(e\sqrt{6})$ but the constant can be improved for special situations. For example, $c = 1/e$, when μ^+ is invariant under permutations of coordinates, and moreover $c = 1$ when μ^+ is uniform distribution on K^+ which is invariant under permutations of coordinates.

We denote by C_n^k the usual combinatorial coefficients $\frac{n!}{k!(n-k)!}$.

Proof. Since

$$\{x \in \mathbf{R}_+^n : X_k \geq \alpha\} = \cup_{n \geq j_1 > \ldots > j_k \geq 1}\{x \in \mathbf{R}_+^n : x_{j_1} \geq \alpha, \ldots, x_{j_k} \geq \alpha\},$$

we get

$$\mu^+\{X_k \geq \alpha\} \leq \sum_{n \geq j_1 > \ldots > j_k \geq 1} \mu^+\{x_{j_1} \geq \alpha, \ldots, x_{j_k} \geq \alpha\} \leq C_n^k e^{-c\,k\alpha},$$

where we applied Proposition 3.3 (or, respectively, Proposition 2.3) on the last step.

The combinatorial argument easily extends to yield a more general:

Proposition 4.2. *For any collection of indices* $1 \leq k_1 < \ldots < k_r \leq n$, *and for all* $\alpha_1, \ldots, \alpha_r \geq 0$,

$$\mu^+\{X_{k_1} \geq \alpha_1, \ldots, X_{k_r} \geq \alpha_r\} \leq \frac{n!\, e^{-c\,(k_1\alpha_1 + (k_2 - k_1)\alpha_2 \ldots + (k_r - k_{r-1})\alpha_r)}}{k_1!(k_2 - k_1)! \ldots (k_r - k_{r-1})!(n - k_r)!},$$

where $c > 0$ *is a numerical constant.*

Let us now illustrate one of the possible applications to large deviations, say, for ℓ^1-norm $\|x\|_1 = \sum_{k=1}^n |x_k|$ under the measure μ. For all numbers $\alpha_1, \ldots, \alpha_n \geq 0$,

$$\mu\left\{\|x\|_1 \geq \sum_{k=1}^n \alpha_k\right\} = \mu^+\left\{\sum_{k=1}^n x_k \geq 2\sum_{k=1}^n \alpha_k\right\} = \mu^+\left\{\sum_{k=1}^n X_k \geq 2\sum_{k=1}^n \alpha_k\right\}$$

$$\leq \sum_{k=1}^n \mu^+\{X_k \geq 2\alpha_k\} \leq \sum_{k=1}^n C_n^k e^{-2c\,k\,\alpha_k}$$

where we applied Proposition 4.1 on the last step. Using $C_n^k \leq \left(\frac{ne}{k}\right)^k$, we thus get

$$\mu\left\{c\,\|x\|_1 \geq \sum_{k=1}^n \alpha_k\right\} \leq \sum_{k=1}^n e^{-k\left(2\alpha_k - \log\frac{ne}{k}\right)}.$$

Now, take $\alpha_k = \frac{1}{2}\log\frac{ne}{k} + t\frac{n}{k(\log n + 1)}$ which is almost an optimal choice. Then, $\sum_{k=1}^n \alpha_k \leq n(1 + t)$, and we arrive at:

Proposition 4.3. *For any* $t \geq 0$,

$$\mu\left\{\frac{c\,\|x\|_1}{n} \geq 1 + t\right\} \leq n\exp\left\{-2t\frac{n}{\log n + 1}\right\}.$$

The right hand side converges to zero for any fixed $t > 0$. In particular, for large n, we have $\|x\|_1 \leq 2n/c$ with μ-probability almost one. In probabilistic language, this means that the random variables $\|x\|_1/n$ are stochastically bounded as $n \to \infty$. Since $L^1(\mu)$-norm of $\|x\|_1/n$ is about 1, this property cannot be deduced from the usual exponential bound for norms under log-concave measures (cf. [Bo]).

5 Euclidean Norm. Proof of Theorem 1.2

As in the proof of Proposition 4.3, for all $\alpha_1, \dots, \alpha_n \geq 0$, we similarly obtain that

$$\mu\left\{|x|^2 \geq \sum_{k=1}^n \alpha_k^2\right\} = \mu^+\left\{\sum_{k=1}^n X_k^2 \geq 4\sum_{k=1}^n \alpha_k^2\right\}$$

$$\leq \sum_{k=1}^n \mu^+\{X_k \geq 2\alpha_k\} \leq \sum_{k=1}^n C_n^k \, e^{-2ck\alpha_k}$$

where again we applied Proposition 4.1 on the last step. Using $C_n^k \leq \left(\frac{ne}{k}\right)^k$, we thus get

$$\mu\left\{c^2\,|x|^2 \geq \sum_{k=1}^n \alpha_k^2\right\} \leq \sum_{k=1}^n e^{-k\left(2\alpha_k - \log\frac{ne}{k}\right)}.$$

Now, take $\alpha_k = \frac{1}{2}\log\frac{ne}{k} + t\,\frac{\sqrt{n}}{k}$. Then, $\sum_{k=1}^n \alpha_k^2 \leq 4nt^2$, for all $t \geq 2$, so

$$\mu\left\{\frac{c\,|x|}{\sqrt{n}} \geq 2t\right\} \leq n\,e^{-2t\sqrt{n}}.$$

In a more compact form:

Proposition 5.1. *For any $t \geq 4$,*

$$\mu\left\{x \in \mathbf{R}^n : \frac{c\,|x|}{\sqrt{n}} \geq t\right\} \leq e^{-\frac{1}{2}t\sqrt{n}}.$$

As in Proposition 4.1, we may take $c = 1/(e\sqrt{6})$ in general, and $c = 1/\sqrt{6}$ in the body case. As explained in section 1, the above inequality implies:

Proposition 5.2. *For every number $C \geq 56$, in the interval $0 \leq t \leq Cn^{1/4}$,*

$$\mu\left\{x \in \mathbf{R}^n : \left|\frac{x_1 + \dots + x_n}{\sqrt{n}}\right| \geq t\right\} \leq 2\exp\left\{-\frac{t^2}{8\,C^{4/3}}\right\}.$$

Indeed, applying Proposition 5.1 with $c = \frac{1}{7} < \frac{1}{e\sqrt{6}}$, we get

$$\mu\left\{\frac{1}{\sqrt{n}}\left|\sum_{j=1}^n x_j\right| \geq t\right\} = \mu \otimes \mathbf{P}_\varepsilon\left\{\frac{1}{\sqrt{n}}\left|\sum_{j=1}^n \varepsilon_j x_j\right| \geq t\right\}$$

$$\leq \int e^{-nt^2/(2|x|^2)}\,d\mu(x) = \int_{|x|\leq t_0\sqrt{n}} + \int_{|x|\geq t_0\sqrt{n}}$$

$$\leq e^{-t^2/(2t_0^2)} + e^{-\frac{1}{14}t_0\sqrt{n}},$$

for every t_0 provided that $ct_0 \geq 4$, that is, $t_0 \geq 28$. By the assumption on t, the last term is bounded by $e^{-t_0\,t^2/(14\,C^2)}$. It remains to take (the optimal) $t_0 = (7C^2)^{1/3}$.

6 Theorem 1.1 for Log-Concave Measures

In order to involve the region $t \geq Cn^{1/4}$ in Proposition 5.2, an extra condition on the measure μ is required. One important property distinguishing the case where μ is the uniform distribution on K from the general measure case is indicated in Proposition 2.4: for all $x \in K$,

$$|x_1| + \ldots + |x_n| \leq An \tag{6.1}$$

with $A = \sqrt{6}/2$. It is therefore natural to assume that the measure μ is supported on a convex set satisfying (6.1) for some $A = A(\mu)$. In this case Theorem 1.1 admits a corresponding extension:

Proposition 6.1. $\|f\|_{L_{\psi_2}(\mu)} \leq C\sqrt{A(\mu)}$, where C is a numerical constant.

Note that in terms of the linear functional

$$f(x) = \frac{x_1 + \ldots + x_n}{\sqrt{n}}$$

the quantity $A(\mu)$ is described as $1/\sqrt{n}\,\|f\|_{L_\infty(\mu)}$. Thus, Proposition 6.1 relates L_{ψ_2}-norm to L_∞-norm of f via $\|f\|_{L_{\psi_2}(\mu)} \leq C/\sqrt{n}\,\sqrt{\|f\|_{L_\infty(\mu)}}$. This inequality is not linear in f which is due to the basic assumption $p(0) = 1$ on the density p of μ. Without this condition, Proposition 6.1 can be formulated as follows:

Corollary 6.1. *Let μ be a probability measure on $\mathbf{R}^n$ with a log-concave density p such that, for all $x \in \mathbf{R}^n$, $p(\pm x_1, \ldots, \pm x_n)$ does not depend on the choice of signs, and $\int_{\mathbf{R}^n} x_j^2\, p(x)\, dx$ does not depend on $j = 1, \ldots, n$. Then, for some universal C,*

$$\|f\|^2_{L_{\psi_2}(\mu)} \leq \frac{C}{\sqrt{n}}\,\|f\|_{L_2(\mu)}\,\|f\|_{L_\infty(\mu)}.$$

Let us return to the original assumption $p(0) = 1$. Then $A(\mu)$ is always separated from zero. Indeed, since the density $p(x)$ is bounded by 1, we have

$$1 = \int_{|x_1|+\ldots+|x_n| \leq An} p(x)\, dx \leq \mathrm{vol}_n\{x \in \mathbf{R}^n : |x_1| + \ldots + |x_n| \leq An\}$$
$$= \frac{(2An)^n}{n!}.$$

Hence, $A \geq \frac{n!^{1/n}}{2n} \geq \frac{1}{2e}$.

While the first applications are based upon Proposition 4.1, the proof of Proposition 6.1 uses a more general Proposition 4.2. The estimate given in it can be simplified as follows: using a general bound $m! \geq \left(\frac{m}{e}\right)^m$ and the fact that the function $x \rightarrow \left(\frac{ne}{x}\right)^x$ increases in $0 < x \leq n$, we get

$$\frac{n(n-1)\ldots(n-k_r+1)}{k_1!(k_2-k_1)!\ldots(k_r-k_{r-1})!} \leq \prod_{j=1}^{r}\left(\frac{ne}{k_j-k_{j-1}}\right)^{k_j-k_{j-1}} \leq \prod_{j=1}^{r}\left(\frac{ne}{k_j}\right)^{k_j}$$

with the convention that $k_0 = 0$ on the middle step. Hence, for all $\alpha_1,\ldots,\alpha_r \geq 0$,

$$\mu^+\{X_{k_1} \geq \alpha_1,\ldots,X_{k_r} \geq \alpha_r\} \leq$$
$$\prod_{j=1}^{r}\left(\frac{ne}{k_j}\right)^{k_j} e^{-c\,(k_1\alpha_1+(k_2-k_1)\alpha_2\ldots+(k_r-k_{r-1})\alpha_r)}.$$

From now on, the indices k_j will be assumed to be the powers of 2. Thus let $\ell = [\log_2 n]$ (the integer part), and let S be any non-empty subset of $\{0,1,\ldots,\ell\}$. From the previous inequality, for any collection $\alpha_k \geq 0$ indexed by $k \in S$,

$$\mu^+\{X_{2^k} \geq \alpha_k, \text{ for all } k \in S\} \leq \prod_{k \in S}\left(\frac{ne}{2^k}\right)^{2^k} \exp\left\{-c\sum_{k \in S}2^{k-1}\alpha_k\right\}.$$

The choice $\alpha_k = \beta_k + \frac{2}{c}\log\frac{ne}{2^k}$ leads to:

Lemma 6.1. *For any non-empty subset S of $\{0,1,\ldots,\ell\}$ and any collection $\beta = (\beta_k)_{k \in S}$ of non-negative numbers,*

$$\mu^+\left\{X_{2^k} \geq \beta_k + \frac{2}{c}\log\frac{ne}{2^k}, \text{ for all } k \in S\right\} \leq \exp\left\{-c\sum_{k \in S}2^{k-1}\beta_k\right\}.$$

As before, one may take $c = 1/(e\sqrt{6})$. In view of the assumption (6.1), the measure μ^+ is supported by

$$x_1 + \ldots + x_n \leq 2An$$

so, only $\beta_k < 2An$ can be of interest in Lemma 6.1. Assume moreover that each β_k also represents a power of 2. The couples (S,β) with these properties will be called blocks, and we say that a vector $x \in \mathbf{R}_+^n$ is controlled by a block (S,β) if

$$X_{2^k} \geq \beta_k + \frac{2}{c}\log\frac{ne}{2^k}, \quad \text{for all } k \in S.$$

Lemma 6.2. *The total number of blocks does not exceed, $e^{2\log 2n\log(2\log 4An)}$.*

Indeed, given a non-empty $S \subset \{0,1,\ldots,\ell\}$, the number of admissible functions β on S is equal to $[\log_2 2An]^{|S|}$. Hence, the number of all blocks is equal to

$$\sum_{S}[\log_2 2An]^{|S|} = \sum_{r=1}^{\ell+1}C_{\ell+1}^r[\log_2 2An]^r = (1 + [\log_2 2An])^{[\log_2 n]+1} - 1$$

from which the desired bound easily follows.

Combining Lemma 6.1 with Lemma 6.2 and using $c = 1/(e\sqrt{6}) > 1/7$, we thus obtain that

$$\mu^+\left\{x \in \mathbf{R}_+^n : x \text{ is controlled by a block } (S, \beta) \text{ with } \sum_{k \in S} 2^{k-1}\beta_k \geq \frac{1}{8} t\sqrt{n}\right\}$$

$$\leq e^{2\log 2n \log(2\log 4An)} e^{-\frac{1}{8.7} t\sqrt{n}}. \tag{6.2}$$

Lemma 6.3. *Given $t > 0$, assume that a vector $x \in \mathbf{R}_+^n$ is not controlled by any block (S, β) with $\sum_{k \in S} 2^{k-1}\beta_k \geq \frac{1}{8} t\sqrt{n}$. Then, with some absolute constant $B > 0$,*

$$\mathbf{P}_\varepsilon\left\{\left|\frac{\varepsilon_1 x_1 + \ldots + \varepsilon_n x_n}{\sqrt{n}}\right| \geq t\right\} \leq 2 e^{-t^2/B}.$$

Proof. It is also possible that x is not controlled by any block (S, β) at all: by the very definition, this holds if and only if

$$X_{2^k} < 1 + \frac{2}{c} \log \frac{ne}{2^k}, \quad \text{for all } 0 \leq k \leq \ell.$$

But then

$$|x|^2 = \sum_{j=1}^n X_j^2 \leq \sum_{k=0}^\ell X_{2^k}^2 2^k < \sum_{k=0}^\ell \left(1 + \frac{2}{c}\log\frac{ne}{2^k}\right)^2 2^k \leq Bn,$$

for some absolute constant B. Therefore, for all $t > 0$,

$$\mathbf{P}_\varepsilon\left\{\left|\frac{\varepsilon_1 x_1 + \ldots + \varepsilon_n x_n}{\sqrt{n}}\right| \geq t\right\} \leq 2 e^{-nt^2/2|x|^2} \leq 2 e^{-t^2/2B},$$

and the statement follows.

In the other case, there is a maximal block controlling the given vector x. Namely, introduce (the canonical) set

$$S = \left\{k = 0, 1, \ldots, \ell : X_{2^k} \geq 1 + \frac{2}{c} \log \frac{ne}{2^k}\right\},$$

and for each $k \in S$, denote by β_k the maximal power of 2 not exceeding $X_{2^k} - \frac{2}{c}\log\frac{ne}{2^k}$. In particular,

$$\beta_k \leq X_{2^k} - \frac{2}{c}\log\frac{ne}{2^k} < 2\beta_k, \tag{6.3}$$

and, by the assumption of the lemma,

$$\sum_{k \in S} 2^{k-1}\beta_k < \frac{1}{8} t\sqrt{n}. \tag{6.4}$$

Define a new vector $(Y_j)_{1 \le j \le n}$ approximating $(X_j)_{1 \le j \le n}$ in a certain sense. First put

$$\alpha_k = \left(X_{2^k} - \left(1 + \frac{2}{c} \log \frac{ne}{2^k} \right) \right)^+, \quad 0 \le k \le \ell,$$

so that $\alpha_k = 0$ outside S and $0 \le \alpha_k \le 2\beta_k - 1 < 2\beta_k$, for all $k \in S$, according to (6.3). Let $Y_j = (X_j - \alpha_k)^+$, for $2^k \le j < 2^{k+1}$ $(0 \le k \le \ell)$. Then, clearly $0 \le Y_j \le X_j \le Y_j + \alpha_k$, and by (6.4),

$$\sum_{j=1}^n X_j - Y_j \le \sum_{k=0}^{\ell} 2^k \alpha_k = \sum_{k \in S} 2^k \alpha_k \le \sum_{k \in S} 2^{k+1} \beta_k < \frac{1}{2} t \sqrt{n}.$$

Hence,

$$\mathbf{P}_\varepsilon \left\{ \frac{1}{\sqrt{n}} \left| \sum_{j=1}^n \varepsilon_j x_j \right| \ge t \right\} = \mathbf{P}_\varepsilon \left\{ \frac{1}{\sqrt{n}} \left| \sum_{j=1}^n \varepsilon_j X_j \right| \ge t \right\}$$

$$\le \mathbf{P}_\varepsilon \left\{ \frac{1}{\sqrt{n}} \left| \sum_{j=1}^n \varepsilon_j Y_j \right| \ge \frac{t}{2} \right\}.$$

It remains to observe that, for $2^k \le j < 2^{k+1}$, we have $Y_j \le Y_{2^k} \le 1 + \frac{2}{c} \log \frac{ne}{2^k}$, so $\sum_{j=1}^n Y_j^2 \le \sum_{k=0}^{\ell} \left(1 + \frac{2}{c} \log \frac{ne}{2^k} \right)^2 2^k \le Bn$. Lemma 6.3 follows.

Proof of Proposition 6.1. We need to get a subgaussian bound of the form $\mu\{|f| \ge t\} \le c_1 e^{-c_2 t^2 / A}$, for some absolute $c_1, c_2 > 0$. By the assumption (6.1) on the support of μ, we may assume $t \le A\sqrt{n}$.

Put $C = (\sigma A)^{3/4}$ with a positive universal constant σ to be determined later on. Since necessarily $A \ge 1/(2e)$, we assume $\left(\frac{\sigma}{2e} \right)^{3/4} \ge 56$ so that to apply Proposition 5.2 in the interval $0 \le t \le Cn^{1/4}$: it then gives

$$\mu\{x \in \mathbf{R}^n : |f(x)| \ge t\} \le 2e^{-t^2/(8\sigma A)}.$$

The right hand side is of the desired order both in t and A in that interval.

Now, let $t \ge Cn^{1/4}$. Define $\Omega_0(t)$ to be the collection of all vectors $x \in \mathbf{R}^n$ which are controlled by a block (S, β) with $\sum_{k \in S} 2^{k-1} \beta_k \ge \frac{1}{8} t \sqrt{n}$. Let $\Omega_1(t) = \mathbf{R}_+^n \setminus \Omega_0(t)$. In terms of $f(x, \varepsilon) = \frac{\varepsilon_1 x_1 + \ldots + \varepsilon_n x_n}{\sqrt{n}}$, we may write

$$\mu\{|f| > t\} = \mu^+ \otimes \mathbf{P}_\varepsilon \{(x, \varepsilon) : |f(x, \varepsilon)| > 2t\}$$

$$= \int_{\Omega_0} \mathbf{P}_\varepsilon \{|f(x, \varepsilon)| > 2t\} \, d\mu^+(x) + \int_{\Omega_1} \mathbf{P}_\varepsilon \{|f(x, \varepsilon)| > 2t\} \, d\mu^+(x).$$

The second integral does not exceed $2e^{-t^2/B}$ with some numerical B (Lemma 6.3). The first integral can be bounded, according to (6.2), by

$$\mu^+ \left(\Omega_0(t) \right) \le e^{-\frac{1}{56} t \sqrt{n} + \Delta_n(A)},$$

where $\Delta_n(A) = 2\log(2n)\log(2\log(4An))$. Thus, for the values $Cn^{1/4} \le t \le A\sqrt{n}$, it suffices to show that

$$e^{-\frac{1}{56}\,t\sqrt{n}+\Delta_n(A)} \le e^{-t^2/(112\,A)}$$

(note that if $A\sqrt{n} < Cn^{1/4}$, we are done). Equivalently,

$$\frac{1}{112\,A}\,t^2 - \frac{1}{56}\,t\sqrt{n} + \Delta_n(A) \le 0.$$

Since $t \le A\sqrt{n}$, the above is implied by $\Delta_n(A) \le \frac{1}{112}\,t\sqrt{n}$. In view of $t \ge Cn^{1/4} = (\sigma A)^{3/4}n^{1/4}$, the latter is equivalent to

$$\Delta_n(A) \le \frac{1}{112}\,(\sigma A)^{3/4}n^{1/4}.$$

Clearly, if σ is sufficiently large, the above inequality holds true for all $A \ge \frac{1}{2e}$ and $n \ge 1$. Summarizing, we may write the following estimate for all $t > 0$:

$$\mu\{|f| > t\} \le \max\left\{2e^{-t^2/(8\sigma A)},\, 2e^{-t^2/B} + e^{-t^2/(112\,A)}\right\}.$$

This gives the desired result.

References

[A]   Alesker, S. (1995): ψ_2-estimate for the Euclidean norm on a convex body in isotropic position. Geom. Aspects Funct. Anal. (Israel 1992-1994), Oper. Theory Adv. Appl., **77**, 1–4

[Ba]   Ball, K.M. (1988): Logarithmically concave functions and sections of convex bodies. Studia Math., **88**, 69–84

[Bo]   Borell, C. (1974): Convex measures on locally convex spaces. Ark. Math., **12**, 239–252

[B-Z]   Burago, Yu.D., Zalgaller, V.A. (1988): Geometric Inequalities. Springer-Verlag, Berlin

[H]   Hensley, D. (1980): Slicing convex bodies – bounds for slice area in terms of the body's covariance. Proc. Amer. Math. Soc., **79(4)**, 619–625

[K-P-B] Karlin, S., Proschan, F., Barlow, R.E. (1961): Moment inequalities of Pólya frequency functions. Pacific J. Math., **11**, 1023–1033

[L]   Latala, R. (1996): On the equivalence between geometric and arithmetic means for log-concave measures. In: Convex Geometric Analysis (Berkeley, CA, 1996), pp. 123–127; Math. Sci. Res. Inst. Publ., **34**, Cambridge Univ. Press, Cambridge, 1999

[L-W]   Loomis, L.H., Whitney, H. (1949): An inequality related to the isoperimetric inequality. Bull. Amer. Math. Soc., **55**, 961–962

[M-P]   Milman, V.D., Pajor, A. (1989): Isotropic position and inertia ellipsoids and zonoids of the unit ball of a normed n-dimensional space. Geom. Aspects of Funct. Anal. (1987–88), Lecture Notes in Math., **1376**, 64–104

[S-Z1]   Schechtman, G., Zinn, J. (1990): On the volume of the intersection of two L_p^n balls. Proc. of the Amer. Math. Soc., **110**, 217–224

[S-Z2]   Schechtman, G., Zinn, J. (2000): Concentration on the ℓ_p^n ball. GAFA, Israel Seminar 1996-2000, Lecture Notes in Math., Springer, **1745**, 247–256

Random Lattice Schrödinger Operators with Decaying Potential: Some Higher Dimensional Phenomena

J. Bourgain

Institute for Advanced Study, Princeton, NJ 08540, USA *bourgain@math.ias.edu*

Summary. We consider lattice Schrödinger operators on $\mathbb{Z}^d$ of the form $H_\omega = \Delta + V_\omega$ where Δ denotes the usual lattice Laplacian on $\mathbb{Z}^d$ and V_ω is a random potential $V_\omega(n) = \omega_n v_n$. Here $\{\omega_n | n \in \mathbb{Z}^d\}$ are independent Bernoulli or normalized Gaussian variables and $(v_n)_{n \in \mathbb{Z}^d}$ is a sequence of weights satisfying a certain decay condition. In what follows, we will focus on some results related to absolutely continuous (ac)-spectra and proper extended states that, roughly speaking, distinguish $d > 1$ from $d = 1$ (but are unfortunately also far from satisfactory in this respect). There will be two parts. The first part is a continuation of [Bo], thus $d = 2$. We show that the results on ac spectrum and wave operators from [Bo], where we assumed $|v_n| < C|n|^{-\alpha}, \alpha > \frac{1}{2}$, remain valid if $(v_n|n|^\varepsilon)$ belongs to $\ell^3(\mathbb{Z}^2)$, for some $\varepsilon > 0$. This fact is well-known to be false if $d = 1$.

The second part of the paper is closely related to [S]. We prove for $d \geq 5$ and letting $V_\omega(n) = \kappa \omega_n |n|^{-\alpha} (\alpha > \frac{1}{3})$ existence of (proper) extended states for $H_\omega = \Delta + \bar{V}_\omega$, where $\bar{V}_\omega$ is a suitable renormalization of V_ω (involving only deterministic diagonal operators with decay at least $|n|^{-2\alpha}$). Since in 1D for $\alpha < \frac{1}{2}$, ω a.s. all extended states are in $\ell^2(\mathbb{Z})$, this is again a higher dimensional phenomenon. It is likely that the method may be made to work for all $\alpha > 0$. But even so, this is again far from the complete picture since it is conjectured that $H_\omega = \Delta + \omega_n \delta_{nn'}$ has a component of ac spectrum if $d \geq 3$.

I On Random Schrödinger Operators on $\mathbb{Z}^2$

1 Introduction

The present paper is a continuation of [Bo].

In [Bo] we considered spectral issues for lattice Schrödinger operators on $\mathbb{Z}^2$ of the form

$$H_\omega = \Delta + V_\omega \tag{1.1}$$

where Δ is the lattice Laplacian on $\mathbb{Z}^2$, i.e.

$$\Delta(n, n') = 1 \quad \text{if } |n_1 - n_1'| + |n_2 - n_2'| = 1$$
$$= 0 \text{ otherwise}$$

and V_ω is a random potential

$$V_\omega(n) = \omega_n v_n \tag{1.2}$$

with $\{v_n | n \in \mathbb{Z}^2\} \subset \mathbb{R}_+$ satisfying a certain decay condition and where $\{\omega_n | n \in \mathbb{Z}^2\}$ are independent Bernoulli or normalized Gaussians (this restrictive distribution hypothesis for the random variables - more specifically the L^{ψ_2}-tail distribution - is of importance here).

The results obtained in [Bo] are the following

Theorem 1. *Fix $\tau > 0$ and denote $I = \{E \in [-4, 4] | \tau < |E| < 4 - \tau\}$.*
Fix $\rho > \frac{1}{2}$ and assume

$$\sup_n |v_n|\, |n|^\rho < \kappa = \kappa(\tau, \rho). \tag{1.3}$$

Let H_ω be defined by (1.1), (1.2).
For κ sufficiently small and ω outside a set of small measure

(1) *H_ω has only a.c. spectrum on I*
(2) *Denoting $E_0(I)$ and $E(I)$ the spectral projections for Δ and $H = H_\omega$*
   resp., the wave operators $W_\pm(H, \Delta)E_0(I), W_\pm(\Delta, H)E(I)$ exist and establish unitary eigenvalue of $\Delta E_0(I)$ and $HE(I)$.

Theorem 2. *Assume again $\rho > \frac{1}{2}$ and instead of (1.3)*

$$\sup_n |v_n|\, |n|^\rho < \infty.$$

Then, for almost all ω

(1) *a.c. spectrum $H_\omega \supset [-4, 4]$*
(2) *The wave operators $W_\pm(H, \Delta)E_0([-4, 4])$ and generalized wave operators*
   $W_\pm(\Delta, H)E([-4, 4])$ exist.

As shown in [Bo], Theorem 2 follows from Theorem 1 and the existence of generalized wave operators $W_\pm(H, H + P)$, whenever P is a finite rank perturbation of the self-adjoint operator H (cf. [Ka]).

As mentioned, our aim in this paper is to focus on results that distinguish the one-dimensional and higher dimensional setting. For $d = 1$, the spectral theory of random Schrödinger operators has been extensively studied over the past decades (relying heavily on the transfer matrix formalism - a method not available in higher dimension). In particular, if we let for instance

$$V_\omega = \omega_n |n|^{-\alpha}$$

with $\{\omega_n | n \in \mathbb{Z}\}$ i.i.d. variables, uniformly distributed in $[-1, 1]$, it is known that, almost surely, H_ω has pure point spectrum for $0 < \alpha < \frac{1}{2}$ and a.c. spectrum for $\frac{1}{2} < \alpha$ (cf [Si], [K-L-S]). Also, Theorems 1 and 2 stated above hold for $d = 1$ (replacing the interval $[-4, 4]$ by $[-2, 2]$ of course) with analogous (slightly simpler) proofs and the crucial decay exponent $\frac{1}{2}$ for the potential remains the same. Our purpose here is to impose conditions on the potential that does depend on dimension. We consider again only the case $d = 2$.

Compared with [Bo], the geometric properties (in particular curvature) of the level sets of $\hat{\Delta}$ will play a more important role (if $d > 2$, there are additional difficulties related to vanishing curvature in this respect which we don't intend to explore here).

Theorem 3. *Assume*

$$V_\omega(n) = \omega_n |n|^{-\varepsilon} v_n \tag{1.4}$$

where $\{\omega_n | n \in \mathbb{Z}^2\}$ are as in Theorems 1 and 2, $\varepsilon > 0$ is fixed and

$$\|v\|_{\ell^3(\mathbb{Z}^2)} < \kappa \tag{1.5}$$

with $\kappa = \kappa(\tau, \varepsilon) > 0$ small enough.
Then the statement of Theorem 1 holds.

Theorem 4. *Replacing in Theorem 3 condition (1.5) by*

$$\|v\|_{\ell^3(\mathbb{Z}^2)} < \infty \tag{1.6}$$

the conclusion of Theorem 2 holds.

2 Preliminaries

Recall that for $f \in \ell^2(\mathbb{Z}^2)$

$$\Delta f(n) = \int_{\mathbb{T}^2} 2(\cos 2\pi \xi_1 + \cos 2\pi \xi_2) \hat{f}(\xi) e^{-2\pi i n . \xi} d\xi$$

where

$$\hat{f}(\xi) = (\mathcal{F}f)(\xi) = \sum_{n \in \mathbb{Z}^2} f(n) e^{2\pi i n . \xi}.$$

Hence the free resolvent $R_0(z) = (\Delta - z)^{-1}$ is obtained by applying the Fourier multiplier

$$\frac{1}{m(\xi) - z}$$

with

$$m(\xi) = 2(\cos 2\pi \xi_1 + \cos 2\pi \xi_2). \tag{2.1}$$

Thus

$$R_0(z) = \mathcal{F}^{-1} \frac{1}{m(\xi) - z} \mathcal{F}.$$

Fixing $\tau > 0$ and $\tau < |\lambda| < 4 - \tau$, the equation

$$m(\xi) = \lambda$$

represents a smooth curve Γ_λ with non-vanishing curvature. As in [Bo], denote σ_λ the arclength-measure of Γ_λ. Thus

$$|\hat{\sigma}_\lambda(n)| < C(1 + |n|)^{-\frac{1}{2}}. \tag{2.2}$$

From the theory of Fourier transforms of measures supported by smooth hyper-surfaces with non-vanishing curvature, we obtain therefore

Lemma 2.3 *Let μ be a measure supported by Γ_λ such that $\mu \ll \sigma_\lambda$ and $\frac{d\mu}{d\sigma_\lambda} \in L^2(\Gamma_\lambda, d\sigma_\lambda)$. Then*

$$\|\hat{\mu}\|_{\ell^6(\mathbb{Z}^2)} \le C \left\| \frac{d\mu}{d\sigma_\lambda} \right\|_2. \tag{2.4}$$

Remark. Lemma 2.3 is a standard fact from harmonic analysis (see [St]). Estimates (2.2), (2.4) are obviously dimension dependent and were not involved in [Bo].

For the proof of Theorems 3 and 4, we proceed exactly as in [Bo]. Thus the proof of Theorem 3 is perturbative and the main issue is to control the Born-series expansion

$$R(z) = (H - z)^{-1} = \sum_{s \ge 0} (-1)^s \left[R_0(z)V \right]^s R_0(z). \tag{2.5}$$

To achieve this, we rely on the basic estimates stated as Lemmas 3.18 and 3.48 of [Bo]. Let us recall them.

Denote

$$V_0 = V_0 \mathcal{X}_{\{0\}} \text{ and } V_k = V \mathcal{X}_{[2^{k-1} \le |n| < 2^k]}$$

the dyadic restrictions of V.

By $C^{(\delta)}$ we denote a function in ξ-space satisfying a bound

$$|C^{(\delta)}(\xi)| < \left[|m(\xi) - \lambda| + \delta \right]^{-1/2} \ (0 < \delta < 1) \tag{2.6}$$

(here and in the sequel, λ always assumed to satisfy $\tau < |\lambda| < 4 - \tau$).

Lemma 3.18 from [Bo] is the following statement

Lemma 2.7 *One has the operator norm estimate on $\ell^2(\mathbb{Z}^2)$*

$$\left\| C_2^{(\delta_2)} \mathcal{F} V_\ell \mathcal{F}^{-1} C_1^{(\delta_1)} \right\| < (\kappa 2^{-\ell})^c \left(\log \frac{1}{\delta_1} + \log \frac{1}{\delta_2} \right)^A \tag{2.8}$$

except for ω in a set of measure at most

$$\exp \left\{ - (\kappa^{-1} 2^\ell)^c \left(\log \frac{1}{\delta_1} + \log \frac{1}{\delta_2} \right) \right\} \tag{2.9}$$

($c, A > 0$ are constants independent of δ_1, δ_2, ℓ).

Denoting ρ_λ the restriction operator to Γ_λ, Lemma 3.48 in [Bo] states

Lemma 2.10

$$\left\| \rho_\lambda \mathcal{F} V_\ell \mathcal{F}^{-1} C^{(\delta)} \right\|_{L^2_\xi \to L^2(\Gamma_\lambda, d\sigma_\lambda)} < (\kappa 2^{-\ell})^c \left(\log \frac{1}{\delta} \right)^A \tag{2.11}$$

except for ω in a set of measure at most

$$\exp \left\{ - (\kappa^{-1} 2^\ell)^c \log \frac{1}{\delta} \right\}. \tag{2.12}$$

Summarizing in [Bo], Theorems 1 and 2 are derived from the estimates (2.8)–(2.12) in Lemmas 2.7, 2.10. To prove Theorems 3 and 4, it will suffice to establish these lemmas replacing the assumption (1.3) on V by (1.4), (1.5).

From assumption (2.6) on $C^{(\delta)}$ and representation as average on the level sets of $m(\xi)$, the left side of (2.8), (2.11) is clearly captured by an estimate on

$$\left\| \rho_{\lambda_2} \mathcal{F} V_\ell \mathcal{F}^{-1} \right\|_{L^2(\Gamma_{\lambda_1}, d\sigma_{\lambda_1}) \to L^2(\Gamma_{\lambda_2}, d\sigma_{\lambda_2})} \tag{2.13}$$

(with $\frac{\tau}{2} < |\lambda_1|, |\lambda_2| < 4 - \frac{\tau}{2}$),
and it suffices to prove that for fixed λ_1, λ_2

$$\mathbb{E}_\omega \left[(2.13) \right] < (\kappa 2^{-\ell})^c. \tag{2.14}$$

Finally, also recall the entropy bound (1.13), (1.14) in [Bo], known as the 'dual Sudakov inequality' (due to [P-T]). As in [Bo], the following particular setting is the one we need. Consider a linear operator $S : \mathbb{R}^d \to \ell_m^\infty$ and denote for fixed $t > 0$ by $\mathcal{N}(t)$ the minimal number of balls in ℓ_m^∞ of radius t needed to cover the set $\{ Sx \,|\, x \in \mathbb{R}^d, \|x\|_2 \leq 1 \}$. Then the following inequality holds

$$\log \mathcal{N}(t) < C(\log m) t^{-2} \|S\|_{\ell_d^2 \to \ell_m^\infty}^2 \tag{2.15}$$

where C is a universal constant, (see inequality (4.2) in [Bo]).

3 Proof of Theorems 3 and 4

Again, from the stability of the a.c. spectrum under finite rank perturbations, it suffices to prove Theorem 3. From the discussion in the previous section, the result will follow from (2.14). This inequality will be derived by combining the argument from [Bo] (section 4) with Lemma 2.3.

We start by applying (2.15) considering the operator

$$S : L^2(\Gamma_\lambda, d\sigma_\lambda) \to \ell^\infty(\mathbb{Z}^2) : \mu \mapsto \hat\mu|_{|n| \sim 2^\ell}. \tag{3.1}$$

Thus in (2.15), $m \sim 4^\ell$. The domain may clearly be replaced by a finite dimensional Hilbert space (notice that the estimate (2.15) does not depend on d). Obviously

$$\|S\|_{L^2(d\sigma_\lambda) \to \ell^\infty(\mathbb{Z}^2)} \leq C \tag{3.2}$$

and in fact, from (2.4)

$$\|S\|_{L^2(d\sigma_\lambda) \to \ell^6(\mathbb{Z}^2)} \leq C. \tag{3.3}$$

From (2.15), (3.2)

$$\log \mathcal{N}(t) < C\ell t^{-2}. \tag{3.4}$$

Recalling the definition of $\mathcal{N}(t)$ and taking also (3.3) into account, this means that for each $t > 0$, there is a set $\mathcal{E}_t \subset \ell^\infty_{|n|\sim 2^\ell}$ of vectors ξ with the following properties

$$\log |\mathcal{E}_t| < C\ell t^{-2} \tag{3.5}$$

$$\max_{\substack{\mu \in L^2(\Gamma_\lambda) \\ \|\frac{d\mu}{d\sigma_\lambda}\|_2 \leq 1}} \min_{\xi \in \mathcal{E}_t} \max_{|n|\sim 2^\ell} |\hat\mu(n) - \xi_n| < t \tag{3.6}$$

$$\max_{\xi \in \mathcal{E}_t} \|\xi\|_6 < C. \tag{3.7}$$

Next, taking t of the form $2^{-r}, r \in \mathbb{Z}_+$, we may then obtain sets $\mathcal{F}_r \subset \mathcal{E}_{2^{-r-1}} - \mathcal{E}_{2^{-r}}$ s.t

$$\|\xi\|_\infty < 2^{-r+1} \text{ and } \|\xi\|_6 < C \text{ for } \xi \in \mathcal{F}_r \tag{3.8}$$

and for each $\mu \in L^2(\Gamma_\lambda), \|\frac{d\mu}{d\sigma_\lambda}\| \leq 1$, there is a representation

$$S\mu = \sum_r \xi^{(r)} \text{ for some } \xi^{(r)} \in \mathcal{F}_r. \tag{3.9}$$

We use here (3.6), (3.7).

Proceeding further as in [Bo], (2.13) equals

$$\sup \left| \sum_{|n|\sim 2^\ell} V_\omega(n)\hat\mu_1(n)\hat\mu_2(n) \right| = \sup 2^{-\varepsilon\ell} \left| \sum_{|n|\sim 2^\ell} \omega_n v_n \hat\mu_1(n)\hat\mu_2(n) \right| \tag{3.10}$$

where the sup is taken over all pairs $(\mu_1, \mu_2) \in L^2(\Gamma_{\lambda_1}) \times L^2(\Gamma_{\lambda_2}), \|\frac{d\mu_i}{d\sigma_{\lambda_i}}\|_2 \leq 1$.

Introducing the families $\mathcal{F}_r^{(i)}(r \in \mathbb{Z}_+)$ for $\Gamma_{\lambda_i}(i = 1,2)$, decomposition (3.9) and convexity reduces (3.10) to the following expression

$$\left| \sum_{|n|\sim 2^\ell} \omega_n v_n \hat\mu_1(n)\hat\mu_2(n) \right| \leq \sum_{r_1, r_2 \in \mathbb{Z}_+} \max_{\substack{\xi' \in \mathcal{F}_{r_1}^{(1)} \\ \xi'' \in \mathcal{F}_{r_2}^{(2)}}} \left| \sum_{|n|\sim 2^\ell} \omega_n v_n \xi'_n \xi''_n \right|. \tag{3.11}$$

Take ω-expectation of each of the terms. Since $\{\omega_n | n \in \mathbb{Z}^2\}$ are independent Bernoulli or Gaussians, Dudley's L^{ψ_2}-estimate applies. Thus

$$\mathbb{E}_\omega \left[\max_{\mathcal{F}_{r_1}^{(1)} \times \mathcal{F}_{r_2}^{(2)}} \left| \sum_{|n| \sim 2^\ell} \omega_n v_n \xi'_n \xi''_n \right| \right]$$

$$\leq C \left(\log |\mathcal{F}_{r_1}^{(1)}| + \log |\mathcal{F}_{r_2}^{(2)}| \right)^{1/2} \left\{ \max_{\mathcal{F}_{r_1}^{(1)} \times \mathcal{F}_{r_2}^{(2)}} \left[\sum_{|n| \sim 2^\ell} |v_n|^2 |\xi'_n|^2 |\xi''_n|^2 \right]^{1/2} \right\}.$$

$$(3.12)$$

By (3.5) and construction, $|\mathcal{F}_r| \leq |\mathcal{E}_{2^{-r-1}}| \, |\mathcal{E}_{2^{-r}}|$, hence

$$\log |\mathcal{F}_r| < C\ell 4^{-r}. \tag{3.13}$$

By (3.8) and Hölder's inequality, we get for $\xi' \in \mathcal{F}_{r_1}^{(1)}, \xi'' \in \mathcal{F}_{r_2}^{(2)}$

$$\left(\sum_{|n| \sim 2^\ell} |v_n|^2 |\xi'_n|^2 |\xi''_n|^2 \right)^{1/2} \leq \|v\|_3 \, \|\xi' . \xi''\|_6 < C \|v\|_3 \min(2^{-r_1}, 2^{-r_2}). \tag{3.14}$$

Substitution of (3.13), (3.14) in (3.12) gives

$$C\sqrt{\ell}(2^{-r_1} + 2^{-r_2}) \min(2^{-r_1}, 2^{-r_2}) \|v\|_3 < C\sqrt{\ell} \|v\|_3. \tag{3.15}$$

Invoking also the obvious bound

$$\max_{\mathcal{F}_{r_1}^{(1)} \times \mathcal{F}_{r_2}^{(2)}} \left| \sum_{|n| \sim 2^\ell} \omega_n v_n \xi'_n \xi''_n \right| \leq C 2^{-r_1 - r_2} \sum_{|n| \sim 2^\ell} |\omega_n| \, |v_n| \tag{3.16}$$

we obtain that

$$\mathbb{E}_\omega \left[(3.11) \right] \leq C \sum_{r_1, r_2 \in \mathbb{Z}_+} \min \left(\sqrt{\ell}, 2^{-r_1 - r_2} 8^\ell \right) \|v\|_3 \leq C \ell^{5/2} \|v\|_3. \tag{3.17}$$

Recalling (1.5), it follows from (3.17) that indeed

$$\mathbb{E}_\omega \left[(3.10) \right] < C 2^{-\varepsilon \ell} \ell \kappa \tag{3.18}$$

which is the desired inequality (2.14).

This proves Theorem 3.

Remarks. (1) From deterministic point of view, the preceding shows that if we fix $\varepsilon, \tau > 0$ and consider a (non-random) potential V s.t.

$$\left\| \{ |n|^\varepsilon V_n | n \in \mathbb{Z}^2 \} \right\|_{3/2} < \kappa(\tau, \varepsilon) \tag{3.19}$$

then $H = \Delta + V$ has a.c. spectrum in $\{ E \in [-4, 4] | \tau < |E| < 4 - \tau \}$.

Thus, also, if

$$\left\| \{ |n|^\varepsilon V_n | n \in \mathbb{Z}^2 \} \right\|_{3/2} < \infty \tag{3.20}$$

then $[-4, 4] \subset ac - Spec H$.

Indeed, following the approach described above, it suffices to verify the inequality

$$(2.13) = \|\rho_{\lambda_2}\mathcal{F}V_\ell\mathcal{F}^{-1}\|_{L^2(\Gamma_{\lambda_1},d\sigma_{\lambda_1})\to L^2(\Gamma_{\lambda_2},d\sigma_{\lambda_2})} < C\kappa 2^{-\varepsilon\ell}$$

which immediately follows from Lemma 2.3 and Hölder's inequality.

Thus if $\mu_i \in L^2(\Gamma_{\lambda_i})$, $\|\frac{d\mu_i}{d\sigma_{\lambda_i}}\|_2 \le 1 (i = 1, 2)$, we have

$$\sum_{|n|\sim 2^\ell} |V_n|\,|\hat{\mu}_1(n)|\,|\hat{\mu}_2(n)| \le \|V_\ell\|_{\frac{3}{2}}\,\|\hat{\mu}_1\|_6\,\|\hat{\mu}_2\|_6 < C\kappa 2^{-\varepsilon\ell}.$$

(2) In view of the Carleson-Sjölin theorem

$$\|\hat{\mu}\|_p \le C_p \left\|\frac{d\mu}{d\sigma}\right\|_p \quad \text{for } p > 4 \text{ and } \mu \in L^p(S^1) \tag{3.21}$$

(S^1=unit circle), one may wonder if (2.13), (2.14) may not be obtained under weaker assumption on V. Clearly, the condition

$$\|\{|n|^\varepsilon v_n\}\|_{\ell^4(\mathbb{Z}^2)} < \infty \tag{3.22}$$

would be natural and optimal.

The validity of (2.13), (2.14) seems to require the stronger condition (1.4), (1.6) however. This may easily be seen as follows. Clearly

$$\|\rho\mathcal{F}V_\ell\mathcal{F}^{-1}\|^2_{L^2(S_1)\to L^2(S_1)}$$

$$= \|\rho\mathcal{F}V_\ell^2\mathcal{F}^{-1}\|_{L^2(S_1)\to L^2(S_1)}$$

$$= \sup_{\mu_i\in L^2(S_1),\|\frac{d\mu_i}{d\sigma}\|_2\le 1} \left|\sum_{|n|\sim 2^\ell} V_n^2\hat{\mu}_1(n)\hat{\mu}_2(n)\right|. \tag{3.23}$$

At this point, the randomness in the potential disappeared.

Denote $N = 2^\ell$ and let

$$\frac{d\mu_1}{d\sigma} = \frac{d\mu_2}{d\sigma} = N^{1/4}\mathcal{X}_\gamma \tag{3.24}$$

where γ denotes an arc in S^1 of size $10^{-3}N^{-1/2}$ say, centered at $(1, 0)$

Thus (3.24) gives the proper normalization in $L^2(S^1)$.

Clearly

$$\hat{\mu}_i(n) \sim N^{-1/4} \text{ for } n \in R = [-N, N] \times [-N^{1/2}, N^{1/2}]$$

$$(3.23) \gtrsim N^{-1/2}\sum_{n\in R} V_n^2$$

forcing an $\ell^3(\mathbb{Z}^2)$-bound on V.

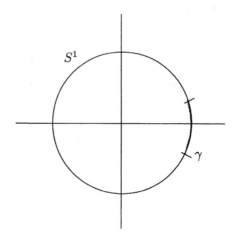

In order to proceed under the weaker assumption (3.22), one would need to eliminate the self-energy loops by a renormalization of V_ω as $\tilde{V}_\omega = V_\omega + W$, with W the (non-random) potential

$$W_n = R_0(E + io). \int V_\omega(n)^2 d\omega.$$

II Construction of Extended States for Lattice Schrödinger Operators on $\mathbb{Z}^d$ ($d \geq 5$) with Slowly Decaying Random Potential

In what follows, we will construct for $d \geq 5$ proper extended states for the random lattice Schrödinger operator on $\mathbb{Z}^d$

$$H_\omega = \Delta + \left(\kappa\omega_n|n|^{-\alpha} + 0(\kappa^2|n|^{-2\alpha})\right)\delta_{nn'}$$

for $\alpha > \frac{1}{3}$. The term $0(\kappa^2|n|^{-2\alpha})\delta_{nn'}$ refers to a deterministic potential arising from suitable renormalizations. Perturbation of the free Laplacian is done at a specific energy, much in the spirit of [S].

The interest of the result lies in the fact that it exhibits a higher dimensional phenomenon, since in 1D, for $\alpha < \frac{1}{2}$, there are a.s. no proper extended states. It is also likely that with additional work, the argument may be carried through for all $\alpha > 0$.

Note. The subsequent numbering only refers to Chapter II.

1 Green's Function Estimate for Certain Deterministic Perturbations

Redefine the Laplacian by subtracting $2d$ from the lattice Laplacian, i.e.

$$-\hat{\Delta}(\xi) = 2d - 2\left(\sum_{j=1}^{d} \cos 2\pi\xi_j\right) = +|\xi|^2 + 0(|\xi|^4). \qquad (1.1)$$

We first prove the following

Lemma 1.2. *Let*

$$H = \Delta + cMd + dMc$$

where

(i) *M is a convolution operator on $\mathbb{Z}^d$ with smooth Fourier-multiplier $\hat{M}(\xi)$ s.t. $\hat{M}(\xi)$ is an even function of $\xi_1, \ldots, \xi_d$ and*

$$\hat{M}(\xi) = |\xi|^2 + 0(|\xi|^4) \qquad (1.3)$$

(ii) *c and d are diagonal operators given by real sequences $(c_n)_{n\in\mathbb{Z}^d}, (d_n)_{n\in\mathbb{Z}^d}$*

$$|c_n|, |d_n| \le \kappa|n|^{-\alpha} \qquad (1.4)$$
$$|c_{n+e_j} - c_n| \lesssim \kappa|n|^{-\alpha-1}$$
$$|d_{n+e_j} - d_n| < \kappa|n|^{-\alpha-1}$$
$$(j = 1, \ldots, d) \qquad (1.5)$$

where $\alpha > 0$ and κ small ($e_j = j^{th}$ unit vector of $\mathbb{Z}^d$).

Then

$$|(H + io)^{-1}(n, n')| < C|n - n'|^{-(d-2)}. \qquad (1.6)$$

Proof. From (1.1), (1.3), we may write

$$M = \Delta M_1$$

where M_1 is a convolution operator with

$$\hat{M}_1(\xi) = 1 + \frac{0(|\xi|^4)}{|\xi|^2 + 0(|\xi|^4)} \qquad (1.7)$$

hence

$$\partial^{(\alpha)}\hat{M}_1(\xi) \in L_\xi^1 \text{ for } |\alpha| < d + 2. \qquad (1.8)$$

Thus

$$|M_1(n, n')| < C\frac{1}{|n - n'|^{d+2-}}. \qquad (1.9)$$

We are replacing next $cMd + dMc$ by $cdM + Mcd$ and evaluate the difference. It follows from (1.5)

$$\left|[cMd + dMc - (cdM + Mcd)](m, n)\right|$$
$$= |c_m d_n + d_m c_n - c_m d_m - c_n d_n| \, |M(m, n)|$$

$$= |c_n - c_m|\, |d_n - d_m|\, |M_{m,n}|$$

$$< C\kappa^2 \frac{|n-m|^2}{(|n|+|m|)^2(|n| \wedge |m|)^{2\alpha}}|M(m,n)|$$

$$\lesssim \kappa^2 \frac{1}{(|n|+|m|)^2(|n| \wedge |m|)^{2\alpha}|n-m|^{d+1}}. \qquad (1.10)$$

Thus

$$H = \Delta + cdM + Mcd + P_1$$
$$= (1+cdM_1)\Delta(1+M_1cd) - cdM_1\Delta M_1cd + P_1 \qquad (1.11)$$

where P_1 satisfies (1.10).

The operator $M_1 \Delta M_1 = M M_1$ is again a convolution operator and clearly

$$\widehat{MM_1}(\xi) = \hat{M}(\xi)\hat{M}_1(\xi)$$

$$= \hat{M}(\xi) + \frac{0(|\xi|^6)}{|\xi|^2 + 0(|\xi|^4)}. \qquad (1.12)$$

Therefore

$$|(MM_1)(n,n')| < \frac{1}{|n-n'|^{d+4-}}. \qquad (1.13)$$

Repeating the preceding with M replaced by MM_1 and c,d replaced by $\frac{cd}{\sqrt{2}}$ (hence α by 2α) we have again

$$\left| cdM M_1cd - \frac{1}{2}c^2d^2 M M_1 - \frac{1}{2}MM_1c^2d^2 \right|$$

$$\lesssim \kappa^4 \frac{1}{(|n|+|m|)^2(|n| \wedge |m|)^{4\alpha}|n-m|^{d+1}} \qquad (1.14)$$

and

$$(1.11) = \left(1 + cdM_1 - \frac{1}{2}c^2d^2 M_1^2\right)\Delta\left(1 + M_1cd - \frac{1}{2}M_1^2c^2d^2\right)$$

$$- \frac{1}{2}c^2d^2 M_1^2 \Delta M_1cd - \frac{1}{2}cdM_1\Delta M_1^2c^2d^2$$

$$+ \frac{1}{4}c^2d^2 M_1^2 \Delta M_1^2c^2d^2 + P_2 \qquad (1.15)$$

with P_2 satisfying (1.10).

Each of the operators $M_1^2\Delta M_1$, $M_1^2\Delta M_1^2$ will still satisfy (1.13) (in fact an even stronger property) and we may repeat the construction.

After s steps, one obtains clearly an operator of the form

$$\left(1 + cdM_1 - \frac{1}{2}c^2d^2 M_1^2 + \cdots\right)\Delta\left(1 + M_1cd - \frac{1}{2}M_1^2c^2d^2 + \cdots\right)$$

$$+ \sum \left[(cd)^k \tilde{M} (cd)^{k'} + (cd)^{k'} \tilde{M} (cd)^k \right] + P \tag{1.16}$$

where

$$k + k' \geq s \tag{1.17}$$

$\tilde{M}$ are convolution operators satisfying

$$|\tilde{M}(n, n')| < \frac{1}{|n - n'|^{d+4-}} \tag{1.18}$$

and

$$|P(n, n')| < \frac{\kappa^2}{(|n| + |n'|)^2 (|n| \wedge |n'|)^{2\alpha} |n - n'|^{d+1}}. \tag{1.19}$$

Thus, from (1.9)

$$\left| \left[(cd)^k \tilde{M} (cd)^{k'} \right] (n, n') \right| \lesssim \kappa^{2s} \frac{1}{|n|^{2k\alpha} |n'|^{2k'\alpha} |n - n'|^{d+3}}$$

$$\lesssim \frac{\kappa^{2s}}{(|n| \wedge |n'|)^{2\alpha s} |n - n'|^{d+3}}. \tag{1.20}$$

Letting $s = s(\alpha)$ be large enough, we may therefore get H in the form

$$H = \left(1 + cdM_1 - \frac{1}{2} c^2 d^2 M_1^2 + \cdots \right) \Delta \left(1 + M_1 cd - \frac{1}{2} M_1^2 c^2 d^2 + \cdots \right) + P' \tag{1.21}$$

where taking (1.19), (1.20) into account the matrix P' satisfies (assuming $\alpha < \frac{1}{2}$)

$$|P'(n, n')| < \frac{\kappa^2}{(|n| + |n'|)^{2+\alpha} |n - n'|^{d+\frac{1}{2}}}. \tag{1.22}$$

Also, by (1.9)

$$Q = cdM_1 - \frac{1}{2} c^2 d^2 M_1^2 + \cdots$$

satisfies in particular

$$|Q(n, n')| < \frac{\kappa^2}{|n - n'|^{d+\frac{19}{10}}} \tag{1.23}$$

and $1 + Q$ is invertible by a Neumann series.

Thus

$$H = (1 + Q) \Delta (1 + Q^*) + P' = (1 + Q)(\Delta + P'')(1 + Q^*) \tag{1.24}$$

where

$$P'' = (1 + Q)^{-1} P' (1 + Q^*)^{-1}. \tag{1.25}$$

Hence, from (1.22), (1.23)

$$|P''(n,n')|$$

$$< C\kappa^2 \sum_{n_1,n_2 \in \mathbb{Z}^d} \frac{1}{|n-n_1|^{d+\frac{19}{10}}} \frac{1}{|n_1-n_2|^{d+\frac{1}{2}}(|n_1|+|n_2|)^{2+\alpha}} \frac{1}{|n_2-n'|^{d+\frac{19}{10}}}$$

$$< c\kappa^2 \frac{1}{|n-n'|^{d+\frac{1}{2}}|n|^{1+\frac{\alpha}{2}}|n'|^{1+\frac{\alpha}{2}}}. \tag{1.26}$$

Replacing Δ by $\Delta + io$, (1.24) implies

$$(H+io)^{-1} = (1+Q^*)^{-1}(\Delta+io+P'')^{-1}(1+Q)^{-1} \tag{1.27}$$

where we expand further

$$(\Delta+io+P'')^{-1} = \left(1+(\Delta+io)^{-1}P''\right)^{-1}(\Delta+io)^{-1}$$
$$= \sum_{s\geq 0} \left[(\Delta+io)^{-1}P''\right]^s (\Delta+io)^{-1}. \tag{1.28}$$

Estimate in (1.28)

$$\left|\left[\left((\Delta+io)^{-1}P''\right)^s(\Delta+io)^{-1}\right](n,n')\right|$$

$$\leq C^s \sum_{n_1,n_2,\ldots,n_{2s}} \frac{1}{|n-n_1|^{d-2}}|P''(n_1,n_2)| \frac{1}{|n_2-n_3|^{d-2}}|P''(n_3,n_4)|$$

$$\cdots \frac{1}{|n_{2s}-n'|^{d-2}}. \tag{1.29}$$

Write

$$\frac{1}{|m-m'|^{d-2}} < C \sum_n \frac{1}{|m-n|^{d-1}|n-m'|^{d-1}}$$

and, from (1.26)

$$\sum_{n_1,n_2} \frac{1}{|m-n_1|^{d-1}}|P''(n_1,n_2)| \frac{1}{|n_2-m'|^{d-1}}$$

$$\leq c\kappa^2 \sum_{n_1,n_2} \frac{1}{|m-n_1|^{d-1}|n_1|^{1+\frac{\alpha}{2}}|n_1-n_2|^{d+\frac{1}{2}}|n_2|^{1+\frac{\alpha}{2}}|n_2-m'|^{d-1}}$$

$$< c\kappa^2 \frac{1}{|m-m'|^{d-1}(|m| \wedge |m'|)^{1+\alpha}}. \tag{1.30}$$

Observe also that

$$\sum_n \frac{1}{|m-n|^{d-1}(|m| \wedge |n|)^{1+\alpha}|n-m'|^{d-1}(|m'| \wedge |n|)^{1+\alpha}}$$

$$< \frac{C}{|m-m'|^{d-1}(|m| \wedge |m'|)^{1+\alpha}}$$

which shows that the estimate (1.30) is preserved under multiplication.

Returning to (1.29), it follows from (1.30) and the preceding that

$$(1.29) < C^s \kappa^{2s} \sum_{m_1,\ldots,m_{s+1}} \frac{1}{|n-m_1|^{d-1}} \left(\frac{1}{|m_1-m_2|^{d-1}(|m_1| \wedge |m_2|)^{1+\alpha}} \right)$$

$$\cdots \left(\frac{1}{|m_s-m_{s+1}|^{d-1}(|m_s| \wedge |m_{s+1}|)^{1+\alpha}} \right) \frac{1}{|m_{s+1}-n'|^{d-1}}$$

$$< C^s \kappa^{2s} \sum_{m_1,m_{s+1}} \frac{1}{|n-m_1|^{d-1}} \frac{1}{|m_1-m_{s+1}|^{d-1}(|m_1| \wedge |m_{s+1}|)^{1+\alpha}}$$

$$\frac{1}{|m_{s+1}-n'|^{d-1}}$$

$$< (C\kappa^2)^s \frac{1}{|n-n'|^{d-2}}. \tag{1.31}$$

Consequently, summing over s, it follows from (1.28) that

$$|(\Delta + io + P'')^{-1}(n,n')| < \frac{2}{|n-n'|^{d-2}} \tag{1.32}$$

and from (1.27), (1.23), that

$$|(H+io)^{-1}(n,n')| < C \sum_{n_1,n_2} \frac{1}{|n-n_1|^{d+\frac{19}{10}}} \frac{1}{|n_1-n_2|^{d-2}} \frac{1}{|n_2-n'|^{d+\frac{19}{10}}}$$

$$< \frac{C}{|n-n'|^{d-2}}.$$

This proves inequality (1.6).

Remarks.

(1) Smoothness condition on $\hat{M}$ may be weakened to

$$\partial^{(\alpha)}\hat{M}(\xi) \in L^1_\xi \text{ for } |\alpha| < d+4. \tag{1.33}$$

(2) The proof of Lemma 1.2 shows that

$$\left|(H+io)^{-1}(n,n')\right| < C|n-n'|^{-(d-2)} \tag{1.34}$$

whenever H has the form

$$H = \Delta + A + P$$

where A is a convex combination of operators $cMd + dMc$ as described in Lemma 1.2 and

$$|P(n,n')| < \frac{\kappa}{|n-n'|^{d+\delta}(|n|+|n'|)^{2+\delta}} \tag{1.35}$$

for some $\delta > 0$ and $\kappa = \kappa(\delta)$.

2 A Probabilistic Estimate

Let for simplicity the random variables $(\omega_n)_{n \in \mathbb{Z}^d}$ be Bernoulli.

Considering a s-tuple $(n_1, \ldots, n_s)$, we say that there is 'cancellation' if

$$\omega_{n_1} \cdots \omega_{n_s} = 1. \tag{2.1}$$

Say that $(n_1, \ldots, n_s)$ is 'admissible' if for any segment $1 \leq s_1 < s_2 \leq s$, the sub-complex $(n_{s_1}, n_{s_1+1}, \ldots, n_{s_2})$ does not cancel.

Use the notation $\sum_{n_1,\ldots,n_s}^{(*)}$ to indicate summation restricted to admissible s-tuples.

The interest of this notion is clear from the following

Lemma 2.2. *For $s \geq 2$*

$$\left\| \sum_{n_1,\ldots,n_s}^{(*)} \omega_{n_1} \cdots \omega_{n_s} a^{(0)}_{n,n_1} a^{(1)}_{n_1 n_2} \cdots a^{(s)}_{n_s,n'} \right\|_{L^2_\omega}$$

$$\leq C_s \left[\sum_{n_1,\ldots,n_s} |a^{(0)}_{n,n_1} \cdots a^{(s)}_{n_s,n'}|^2 \right]^{1/2} . \tag{2.3}$$

Proof. We may clearly assume $a^{(j)}_{m,n} \geq 0$.

Since in the $\sum^*$ summation no $(n_1, \ldots, n_s)$ cancels, there is some index $n_{s'}$ which is not repeated or repeated an odd number of times.

Specifying a subset I of $\{1, \ldots, s\}$ of odd size (at most 2^s possibilities), we consider now s-tuples of the form

$$(\nu^{(1)}, m, \nu^{(2)}, m, \nu^{(3)}, m \cdots)$$

where $m \in \mathbb{Z}^d$ appears on the I-places and $\nu^{(1)}, \nu^{(2)}, \ldots$ are admissible complexes indexed by sub-intervals of $\{1, \ldots, s\}$ determined by I.

Thus, enlarging the $\sum^{(*)}$-sum, (which we may by the positivity assumption), it follows that

$$\left\| \sum_{n_1,\ldots,n_s}^{(*)} \omega_{n_1} \cdots \omega_{n_s} a^{(0)}_{n,n_1} \cdots a^{(s)}_{n_s,n'} \right\|_{L^2_\omega}$$

$$\leq \sum_I \left\| \sum_{m \in \mathbb{Z}^d} \omega_m \left[\sum_{\substack{\nu^{(1)} \\ \| \\ (n_1,\ldots,n_{s_1})}}^{(*)} \omega_{n_1} \cdots \omega_{n_{s_1}} a^{(0)}_{n,n_1} \cdots a^{(s_1)}_{n_{s_1},m} \right] \right.$$

$$\left. \left[\sum_{\substack{\nu^{(2)} \\ \| \\ (n_{s_1+2},\ldots,n_{s_2})}}^{(*)} \omega_{n_{s_1+2}} \cdots \omega_{n_{s_2}} a^{(s_1+1)}_{m,n_{s_1+2}} \cdots a^{(s_2)}_{n_{s_2},m} \right] \cdots \right\|_{L^2_\omega} \tag{2.4}$$

where, for fixed m,

$$m \notin \{n_1, \ldots, n_{s_1}, n_{s_1+2}, \ldots, n_{s_2}, n_{s_2+2}, \ldots\}.$$

We used here that if $(n_1, \ldots, n_s)$ is admissible, then so is $(n_{s_1}, n_{s_1+1}, \ldots, n_{s_2})$ for all $1 \le s_1 \le s_2 \le s$.

Enlargement of the original sum $\sum^{(*)}$ enables thus to get the product structure in (2.4). The number of factors is at least 2.

Thus the preceding and a standard decoupling argument implies

$$(4) \le \sum_I \left[\sum_{m \in \mathbb{Z}^d} \left\| \left[\sum_{\nu^{(1)}}^{(*)}\right] \left[\sum_{\nu^{(2)}}^{(*)}\right] \cdots \right\|_{L_\omega^2}^2 \right]^{1/2}. \tag{2.5}$$

Next, by Hölder's inequality and moment-equivalence, we get

$$\sum_I \left[\sum_{m \in \mathbb{Z}^d} \left\| \sum_{\nu^{(1)}}^{(*)} \right\|_{L_\omega^2}^2 \left\| \sum_{\nu^{(2)}}^{(*)} \right\|_{L_\omega^2}^2 \cdots \right]^{1/2}. \tag{2.6}$$

Preceding by induction on s, we obtain thus

$$\sum_I \left[\sum_{m \in \mathbb{Z}^d} \left[\sum_{n_1, \ldots, n_{s_1}} |a_{n,n_1}^{(0)} \cdots a_{n_{s_1},m}^{(s_1)}|^2 \right] \right.$$
$$\left. \left[\sum_{n_{s_1+2}, \cdots, n_{s_2}} |a_{m,n_{s_1+2}}^{(s_1+1)} \cdots a_{n_{s_2},m}^{(s_2)}|^2 \right] \cdots \right]^{1/2} < (2.3).$$

This proves Lemma 2.2.

Expressions considered in Lemma 2.1 appear when writing out matrix elements of products

$$\left(A^{(0)} V_\omega A^{(1)} V_\omega \cdots A^{(s)} \right)(n, n')$$

where the $A^{(s')}$ are matrices and V_ω a random potential

$$V_\omega(n) = \omega_n v_n.$$

We use the notation

$$\left(A^{(0)} V_\omega A^{(1)} V_\omega \cdots A^{(s)} \right)^{(*)} \tag{2.7}$$

to indicate that, when writing out the matrix product as a sum over multi-indices, we do restrict the sum to the admissible multi indices.

Lemma 2.2 then implies that

$$\mathbb{E}_\omega \left[|(A^{(0)} V_\omega A^{(1)} V_\omega \cdots A^{(s)})^{(*)}(n, n')| \right]$$
$$< C_s \left[\sum_{n_1, \cdots n_s} |v_{n_1}|^2 \cdots |v_{n_s}|^2 |A^{(0)}(n, n_1)|^2 \cdots |A^{(s)}(n_s, n')|^2 \right]^{1/2}. \tag{2.8}$$

3 Green's Function Estimate

Returning to (1.1), let $d \geq 5$ and denote $G_0 = (-\Delta)^{-1}$, i.e.

$$G_0(n, n') = \int \frac{e^{-2\pi i (n-n').\xi}}{-\hat{\Delta}(\xi)} d\xi$$

$$= \int \frac{e^{-2\pi i (n-n')\xi}}{|\xi|^2 + 0(|\xi|^4)} d\xi$$

hence

$$|G_0(n, n')| < C \frac{1}{|n-n'|^{d-2}}. \tag{3.1}$$

Let

$$V_\omega(n) = \omega_n v_n \quad (\omega_n \text{assumed Bernoulli}) \tag{3.2}$$

$$v_n = \kappa |n|^{-\alpha} \tag{3.2'}$$

where we assume

$$\frac{2}{5} < \alpha < \frac{1}{2}. \tag{3.3}$$

(the argument will be developed further in §5 to cover $\alpha > \frac{1}{3}$).

Clearly

$$W_2 = \int V_\omega G_0 V_\omega d\omega \tag{3.4}$$

is the diagonal operator

$$W_2(n) = G_0(0, 0) v_n^2 \tag{3.4'}$$

(observe that $G_0(0,0)$ is real).

Denote by W_4 the operator

$$W_4(n, n') = \begin{cases} v_n^2 v_{n'}^2 G_0(n, n')^3 \text{ for } n \neq n' \\ 0 \text{ otherwise} \end{cases} \tag{3.5}$$

hence from (3.1), (3.2')

$$|W_4(n, n')| < \frac{\kappa^4}{|n|^{2\alpha}|n'|^{2\alpha}|n-n'|^{3(d-2)}}. \tag{3.6}$$

This operator arises from the 4-tuples

Notice that
$$W_4 = cMc + \left(\hat{K}(0) - K(0)\right)c^2 \tag{3.7}$$

where
$$c_n = \frac{\kappa^2}{|n|^{2\alpha}} = v_n^2 \tag{3.8}$$

and M is the convolution operator with symbol
$$\hat{K} - \hat{K}(0), \qquad \hat{K} = \hat{G}_0 * \hat{G}_0 * \hat{G}_0 \tag{3.9}$$

which is an even and symmetric function in $\xi_1, \ldots, \xi_d$ and satisfying
$$\partial_\xi^{(\alpha)} \hat{K} \in L_\xi^1 \quad \text{for} \quad |\alpha| < 3(d-2).$$

In order to meet the condition (1.33), we require thus $3(d-2) \geq d+4$, i.e. $d \geq 5$.

Since from the preceding
$$\hat{K}(\xi) - \hat{K}(0) = .|\xi|^2 + 0(|\xi|^4) \tag{3.10}$$

the operator $W = cMc$ clearly satisfies the conditions of Lemma 1.2.

Thus from (1.6)
$$|(-\Delta + W + io)^{-1}(n, n')| < C|n - n'|^{-(d-2)}. \tag{3.11}$$

We renormalize V_ω as
$$\tilde{V}_\omega = V_\omega + W_2 - \rho v^4 \tag{3.12}$$

denoting
$$\sigma = G_0(0,0) \quad \text{and} \quad \rho = 2\sigma^3 - \hat{K}(0).$$

Thus from (3.7)
$$W_4 = W + (\sigma^3 - \rho)v^4.$$

Consider
$$H = -\Delta + \tilde{V}_\omega \tag{3.13}$$

with Green's function $G = G(z)$.

For notational simplicity, we also denote $G(io)$ by G.

From the resolvent identity, we get
$$G = G_0 - G\tilde{V}G_0 = G_0 - GVG_0 - GW_2G_0 + \rho G v^4 G_0$$

and iterating (using the (*)-notation for admissible complexes - cf. (2.7)) we obtain

$$= G_0 - G_0VG_0 + G\tilde{V}G_0VG_0 - GW_2G_0 + \rho G v^4 G_0$$
$$= G_0 - G_0VG_0 + GVG_0VG_0^{(*)} + \rho G v^4 G_0 + GW_2G_0VG_0 - \underaccent{\sim}{\rho G v^4 G_0VG_0}$$

$$= G_0 - G_0VG_0 + G_0VG_0VG_0^{(*)} - G\tilde{V}(G_0VG_0VG_0)^*$$
$$+ GW_2G_0VG_0 + \rho Gv^4G_0 - \rho Gv^4G_0VG_0$$

$$= G_0 - G_0VG_0 + G_0VG_0VG_0^{(*)} - (GVG_0VG_0VG_0)^* + \sigma GW_2VG_0$$
$$- GW_2(G_0VG_0VG_0)^* + \rho Gv^4G_0 - \rho Gv^4G_0VG_0 + \rho Gv^4(G_0VG_0VG_0)^*$$

$$= G_0 - G_0VG_0 + G_0VG_0VG_0^{(*)} - (G_0VG_0VG_0VG_0)^*$$
$$+ G\tilde{V}(G_0VG_0VG_0VG_0)^* + \sigma G_0W_2VG_0 - \sigma G\tilde{V}G_0W_2VG_0$$
$$- GW_2(G_0VG_0VG_0)^* + \rho Gv^4G_0 - \rho Gv^4G_0VG_0 + \rho Gv^4(G_0VG_0VG_0)^*$$

$$= G_0 - G_0VG_0 + G_0VG_0VG_0^{(*)} - (G_0VG_0VG_0VG_0)^* + \sigma G_0W_2VG_0$$
$$+ (GVG_0VG_0VG_0VG_0)^* - \sigma GW_2(VG_0VG_0)^* - \sigma(GVG_0W_2VG_0)^*$$
$$- \sigma^3 Gv^4G_0 + GW_4G_0 - \sigma G(W_2 - \rho v^4)G_0W_2VG_0$$
$$+ G(W_2 - \rho v^4)(G_0VG_0VG_0VG_0)^* + \rho Gv^4G_0$$
$$- \rho Gv^4G_0VG_0 + \rho Gv^4(G_0VG_0VG_0)^*$$

$$G = G_0 - G_0VG_0 + (G_0VG_0VG_0)^{(*)} - (G_0VG_0VG_0VG_0)^* + \sigma G_0W_2VG_0$$
$$+ (G_0VG_0VG_0VG_0VG_0)^* - \sigma G_0W_2(VG_0VG_0)^* - \sigma(G_0VG_0W_2VG_0)^* \tag{3.14}$$

$$- G\tilde{V}(G_0VG_0VG_0VG_0VG_0)^* + \sigma G\tilde{V}G_0W_2(VG_0VG_0)^*$$

$$+ \sigma G\tilde{V}(G_0VG_0W_2VG_0)^* \tag{3.15}$$

$$+ GWG_0 \tag{3.16}$$

$$- \sigma G(W_2 - \rho v^4)G_0W_2VG_0 + G(W_2 - \rho v^4)(G_0VG_0VG_0VG_0)^*$$

$$- \rho Gv^4G_0VG_0 + \rho Gv^4(G_0VG_0VG_0)^*. \tag{3.15}$$

(We use here the notation $\sim$ for contributions at least of order 5 in V – these terms are not expanded further.)

The next step is to move (3.16) to the left member. We get $G(1 - WG_0)$. Multiply then both sides on the right by $(1 - WG_0)^{-1}$ and observe that $G_0(1 - WG_0)^{-1} = (-\Delta - W)^{-1} \equiv G_0'$, where, by (3.11)

$$|G_0'(n, n')| < C|n - n'|^{-(d-2)}. \tag{3.17}$$

This gives

$$G = A + GB \tag{3.18}$$

with

$$A = G_0' - G_0VG_0' + (G_0VG_0VG_0')^* - (G_0VG_0VG_0VG_0')^*$$

$$+ \sigma G_0 W_2 V G_0' + (G_0 V G_0 V G_0 V G_0 V G_0')^* - \sigma (G_0 W_2 V G_0 V G_0')^*$$
$$- \sigma (G_0 V G_0 W_2 V G_0')^* \qquad (3.19)$$

and

$$B = - \tilde{V} \big((G_0 V G_0 V G_0 V G_0')^* + \sigma (G_0 W_2 V G_0 V G_0')^*$$
$$+ \sigma (G_0 V G_0 W_2 V G_0')^* \big)$$
$$+ (W_2 - \rho v^4) \big[- \sigma G_0 W_2 V G_0' + (G_0 V G_0 V G_0 V G_0')^* \big]$$
$$+ \rho v^4 \big[- G_0 V G_0' + (G_0 V G_0 V G_0')^* \big]. \qquad (3.19')$$

Apply inequality (2.8) to estimate the matrix elements of the (random) matrices A, B given by (3.19), (3.19'). Clearly, with large probability

$$|A(n, n')| < C |n - n'|^{-(d-2)}. \qquad (3.20)$$

We have indeed

$$\mathbb{E}_\omega \big[|(G_0 V G_0 V G_0 V G_0')^* (n, n')|^2 \big]^{1/2}$$
$$\lesssim \Bigg[\sum_{n_1, n_2, n_3, n_4} \frac{1}{|n - n_1|^{2(d-2)}} \frac{\kappa^2}{|n_1|^{2d}} \frac{1}{|n_1 - n_2|^{2(d-2)}} \frac{\kappa^2}{|n_1|^{2\alpha}} \frac{1}{|n_2 - n_3|^{2(d-2)}}$$
$$\frac{\kappa^2}{|n_3|^{2\alpha}} \frac{1}{|n_3 - n_4|^{2(d-2)}} \frac{\kappa^2}{|n_4|^{2\alpha}} \frac{1}{|n_4 - n'|^{2(d-2)}} \Bigg]^{1/2}$$
$$< C \kappa^4 \frac{1}{\min(|n|^{4\alpha}, |n'|^{4\alpha})} \frac{1}{|n - n'|^{d-2}}. \qquad (3.21)$$

Estimate (3.21) also holds for the matrices

$$(G_0 W_2 V G_0 V G_0')^* \text{ and } (G_0 V G_0 V W_2 G_0')^*.$$

Similarly

$$\mathbb{E}_\omega \big[|(G_0 W_2 V G_0')(n, n')| \big] \lesssim \frac{1}{\min(|n|^{3\alpha}, |n'|^{3\alpha}) |n - n'|^{d-2}}$$
$$\mathbb{E}_\omega \big[|(G_0 V G_0 V G_0')^* (n, n')| \big] \lesssim \frac{1}{\min(|n|^{3\alpha}, |n'|^{3\alpha}) |n - n'|^{d-2}}.$$

Consequently

$$\mathbb{E}_\omega \big[|B(n, n')| \big] < C \frac{\kappa^5}{\min(|n|^{5\alpha}, |n'|^{5\alpha}) |n - n'|^{d-2}}. \qquad (3.22)$$

Write from (3.18)

$$G = A(1 - B)^{-1} \qquad (3.23)$$

where, from (3.22) and the assumption

$$\alpha > \frac{5}{2}$$

$1 - B$ may be inverted by a Neumann series

$$(1 - B)^{-1} = 1 - B'$$

with B' satisfying (3.22).

Thus

$$G = A - AB'$$

with A satisfying (3.20) and

$$|(AB')(n, n')| < \sum_{n_1} \frac{1}{|n - n_1|^{d-2} \min(|n'|^{5\alpha}, |n_1|^{5\alpha})|n_1 - n'|^{d-2}}$$

$$\lesssim \frac{1}{|n'|^{5\alpha}|n - n'|^{d-4}} + \frac{1}{|n - n'|^{d-2}}.$$

Thus

$$|G(n, n')| \lesssim \frac{1}{|n'|^{5\alpha}|n - n'|^{d-4}} + \frac{1}{|n - n'|^{d-2}}$$

and by self-adjointness considerations, also

$$|G(n, n')| \lesssim \frac{1}{|n|^{5\alpha}|n - n'|^{d-4}} + \frac{1}{|n - n'|^{d-2}}.$$

Therefore we get in conclusion the estimate

$$|G(n, n')| < \frac{C}{|n - n'|^{d-2}} \tag{3.24}$$

for the Green's function $G(0 + io)$ of $H = -\Delta + \tilde{V}$.

4 Construction of an Extended State

Denote $\hat{\delta}_0 \in \ell^\infty(\mathbb{Z}^d)$ the vector with 1-coordinates thus

$$\hat{\delta}_0(n) = 1 \quad \text{for all } n \in \mathbb{Z}^d. \tag{4.1}$$

Thus

$$\Delta\hat{\delta}_0 = 0 \tag{4.2}$$

and $\hat{\delta}_0$ is an extended state for the free Laplacian Δ.

In order to construct a (proper) extended state for $H = -\Delta + \tilde{V}_\omega$ we will proceed in 2 steps.

First we construct an extended state for the operator $-\Delta - W \equiv H'_0$ with W introduced in (3.11).

We use the construction described in §1, in particular (1.24), which does apply to H_0' (see the discussion in §3)).

Thus

$$-H_0' = (1 + Q)(-\Delta + P'')(1 + Q^*) \tag{4.3}$$

where

$$|Q(n, n')| \lesssim \frac{\kappa^2}{|n - n'|^{d + \frac{19}{10}}} \tag{4.4}$$

and

$$|P''(n, n')| \lesssim \frac{\kappa^2}{|n - n'|^{d + \frac{1}{2}} |n|^{1 + \frac{\alpha}{2}} |n'|^{1 + \frac{\alpha}{2}}} \tag{4.5}$$

(in fact α replaced by 2α, which will be irrelevant); see (1.23), (1.26).

Writing formally

$$\zeta = (1 + Q^*)^{-1}(1 + G_0 P'')^{-1}\hat{\delta}_0 \tag{4.6}$$

it follows from (4.2), (4.3) that

$$H_0'\zeta = 0.$$

We justify (4.6). We claim that ζ is a perturbation of $\hat{\delta}_0$ in $\ell^\infty(\mathbb{Z}^d)$. To see this, it will suffice to show that

$$\|Q^*\|_{\ell^\infty(\mathbb{Z}^d) \to \ell^\infty(\mathbb{Z}^d)} < \kappa \tag{4.7}$$

and

$$\|G_0 P''\|_{\ell^\infty(\mathbb{Z}^d) \to \ell^\infty(\mathbb{Z}^d)} < \kappa \tag{4.8}$$

or equivalently

$$\|Q\|_{\ell^1(\mathbb{Z}^d) \to \ell^1(\mathbb{Z}^d)} < \kappa \tag{4.7'}$$

$$\|(P'')^* G_0\|_{\ell^1 \to \ell^1} < \kappa. \tag{4.8'}$$

Assertion (4.7′) is obvious from (4.4).

To verify (4.8′), estimate for fixed n_0

$$\sum_n \left|\left((P'')^* G_0\right)(n, n_0)\right|$$

$$\leq C\kappa^2 \sum_{n, n_1} \frac{1}{|n - n_1|^{d + \frac{1}{2}} |n|^{1 + \frac{\alpha}{2}} |n_1|^{1 + \frac{\alpha}{2}}} \frac{1}{|n_1 - n_0|^{d-2}}$$

$$< C\kappa^2 \sum_{n_1} \frac{1}{|n_1|^{2+\alpha}} \frac{1}{|n_1 - n_0|^{d-2}} < C\kappa^2,$$

using (4.5).

Therefore

$$\zeta = \hat{\delta}_0 + o(1) \quad (\text{in } \ell^\infty(\mathbb{Z}^d)). \tag{4.9}$$

Next, we construct an extended state for

$$H = -\Delta + \tilde{V} = H_0' + W + \tilde{V}$$

by defining

$$\eta = \zeta - G(W + \tilde{V})\zeta. \tag{4.10}$$

Again, we show that η is a perturbation of ζ in $\ell^\infty(\mathbb{Z}^d)$ (with large probability in ω).

Denoting as in §3 by $G_0' = (H_0')^{-1}$, the resolvent identity implies that

$$G = G_0' - G(W + \tilde{V})G_0' \tag{4.11}$$

hence, recalling (3.18).

$$-G(W + \tilde{V}) = (G - G_0')H_0'$$
$$= (A - G_0')H_0' + GBH_0'. \tag{4.12}$$

Recalling the definition (3.19), (3.19') of A, B, it follows from (4.12) that

$$- G(W + \tilde{V})\zeta =$$
$$- G_0 V\zeta + (G_0 V G_0 V)^*\zeta - (G_0 V G_0 V G_0 V)^*\zeta + \sigma G_0 W_2 V\zeta +$$
$$(G_0 V G_0 V G_0 V G_0 V)^*\zeta - \sigma(G_0 W_2 V G_0 V)^*\zeta - \sigma(G_0 V G_0 W_2 V)^*\zeta \tag{4.13}$$
$$+ G\Theta \tag{4.14}$$

with

$$\Theta = \Theta_\omega =$$
$$- \tilde{V}\big((G_0 V G_0 V G_0 V G_0 V)^*\zeta + \sigma(G_0 W_2 V G_0 V)^*\zeta + \sigma(G_0 V G_0 W_2 V)^*\zeta\big) \tag{4.15}$$
$$+ (W_2 - \sigma W_2^2)[-\sigma G_0 W_2 V\zeta + (G_0 V G_0 V G_0 V)^*\zeta] - \rho v^4 G_0 V\zeta$$
$$+ \rho v^4 (G_0 V G_0 V)^*\zeta. \tag{4.16}$$

The bounds on (4.13), (4.15), (4.16) are probabilistic in ω.

It is important here that ζ does not depend on ω. Thus

$$|(G_0 V\zeta)(n)| = \left| \sum_{n'} G_0(n, n')v_{n'}\omega_{n'} \right|$$

has expectation

$$< C\kappa \left[\sum_{n'} \frac{1}{|n - n'|^{2(d-2)}} |n'|^{-2\alpha} \right]^{1/2} < C\kappa |n|^{-\alpha}. \tag{4.17}$$

The other terms in (4.13) satisfy similar (stronger) estimates. Again we exploit the ( )*-restriction here. Thus with large probability,

$$(4.13) = o(1) \text{ in } \ell^\infty(\mathbb{Z}^d)$$

(this is immediate from (4.17)).

Consider next (4.15). Here the estimates need to be done more carefully. We have

$$\mathbb{E}_\omega \left| \left((G_0 V G_0 V G_0 V G_0 V)^* \zeta \right)(n) \right|$$

$$\leq C\kappa^4 \left(\sum_{n_1,n_2,n_3,n_4} \frac{1}{|n-n_1|^{2(d-2)}|n_1|^{2\alpha}|n_1-n_2|^{2(d-2)}|n_2|^{2\alpha}\cdots|n_4|^{2\alpha}} \right)^{1/2}$$

$$< C\kappa^4|n|^{-4\alpha} \tag{4.18}$$

and similarly for $(G_0 W_2 V G_0 V)^* \zeta$, $(G_0 V G_0 W_2 V)^* \zeta$. Thus, with large probability, we may ensure that the n-coordinate of (4.15) is bounded by

$$C_\varepsilon \kappa^5 |n|^\varepsilon |n|^{-5\alpha} \tag{4.19}$$

(for any $\varepsilon > 0$).

Similarly, the n-coordinate of (4.16) is bounded by

$$C_\varepsilon \kappa^5 |n|^\varepsilon \left[|n|^{-2\alpha}|n|^{-3\alpha} + |n|^{-4\alpha}|n|^{-\alpha} \right]. \tag{4.20}$$

Thus, from (4.19), (4.20) with large probability

$$|\Theta_n| < \kappa |n|^{-5\alpha+} \text{ for all } n \in \mathbb{Z}^d. \tag{4.21}$$

Recalling that $\alpha > \frac{5}{2}$ and the bound (3.24) on the Green's function G, we conclude that (4.14) satisfies

$$|(G\Theta)_n| < C\kappa \left(\sum_{n_1} \frac{1}{|n-n_1|^{d-2}|n_1|^{5\alpha-}} \right) < C\kappa \frac{1}{|n|^{5\alpha-2-}}$$

hence

$$G\Theta = o(1) \text{ in } \ell^\infty(\mathbb{Z}^d).$$

From the preceding, (4.13), (4.14) $= o(1)$ in $\ell^\infty(\mathbb{Z}^d)$ which proves our claim about η. This completes the proof of existence of a (proper) extended state for $H = -\Delta + \tilde{V}_\omega = -\Delta + \kappa\omega_n|n|^{-\alpha} + \kappa^2\sigma|n|^{-2\alpha} + \kappa^4(\hat{K}(0) - 2\sigma^3)|n|^{-4\alpha}, \alpha > \frac{2}{5}$ and κ small enough (with high probability in ω).

5 Relaxing the Condition on α

The purpose of previous analysis was to obtain (proper) extended states for $H_\omega = -\Delta + V_\omega$ with random potential $V_n = |n|^{-\alpha}\omega_n$, for some $\alpha < \frac{1}{2}$. This exhibits thus a higher dimensional phenomenon, in the sense that for $d = 1$, a.s., any extended state of $H_\omega = -\Delta + |n|^{-\alpha}\omega_n, \alpha < \frac{1}{2}$, is in $\ell^2(\mathbb{Z})$. The previous assumption $\alpha > \frac{2}{5}$ may be weakened by continuation of the perturbative expansion (3.14)–(3.16) of the Green's function G in higher powers of V. It is reasonable to expect this type of argument to succeed for any fixed $\alpha > 0$ (with a number of resolvent iterations dependent on α). To achieve this requires further renormalizations (cf. [S], §3) and taking care of certain additional difficulties due to the presence of a potential. Notice that, from the technical side, our approach differs from that in [S] in the sense that we do not rely on the Feynman diagram machinery (but use the estimate from §2 instead).

The technology developed here allows us easily to deal with smaller values of α by carrying out a few more steps. Again, only renormalization by diagonal operators is required. There is one additional idea that will appear in the next iteration. We assume now

$$\frac{1}{3} < \alpha < \frac{1}{2}$$

for which we establish the estimate (3.24) on the Green's function and the existence of a proper extended state for H_ω.

Consider the expansion (3.14)–(3.16) for G. The terms in (3.15) needs to be developed up to order 6, which we mark again as $\sim\!\!\sim$.

Thus

$$
\begin{aligned}
&- G\tilde{V}(G_0 V G_0 V G_0 V G_0)^* \\
={}&- GV(G_0 V G_0 V G_0 V G_0)^* + G(W_2 - \rho v^4)(G_0 V G_0 V G_0 V G_0)^* \\
={}&- G(V G_0 V G_0 V G_0 V G_0)^* - GW_2(G_0 V G_0 V G_0 V G_0)^* \\
&+ \sigma GW_2(V G_0 V G_0 V G_0)^* + \sigma GV W_4 G_0 \\
&- GW_4 G_0 V G_0 + \sigma GW_4 V G_0 + GV D_4 G_0 \\
&+ G(W_2 - \rho v^4)(G_0 V G_0 V G_0 V G_0 V G_0)^*
\end{aligned}
\tag{5.1}
$$

where D_4 denotes the diagonal operator

$$D_n = (W_4 G_0)(n, n) = V_n^2 \left[\sum_{n'} v_{n'}^2 G_0(n - n')^4 \right] \tag{5.2}$$

hence

$$|D_4(n)| < C \frac{\kappa^4}{|n|^{4\alpha}}. \tag{5.2'}$$

Next

$$
\begin{aligned}
&\sigma G\tilde{V} G_0 W_2(V G_0 V G_0)^* \\
={}&\sigma GV G_0 W_2(V G_0 V G_0)^* - \sigma G(W_2 - \rho v^4)G_0 W_2(V G_0 V G_0)^*
\end{aligned}
$$

$$= \sigma G(VG_0W_2VG_0VG_0)^* + \sigma^3 Gv^4G_0VG_0 - \sigma^4 Gv^4VG_0$$
$$- \sigma G(W_2 - \rho v^4)G_0W_2(VG_0VG_0)^* \tag{5.3}$$

and

$$\sigma G\tilde{V}(G_0VG_0W_2VG_0)^*$$
$$= \sigma GV(G_0VG_0W_2VG_0)^* - \sigma G(W_2 - \rho v^4)(G_0VG_0VW_2G_0)^*$$
$$= \sigma G(VG_0VG_0W_2VG_0)^* + \sigma GW_2G_0W_2VG_0 - \sigma^4 Gv^4VG_0$$
$$- \sigma G(W_2 - \rho v^4)(G_0VG_0VW_2G_0)^*. \tag{5.4}$$

Substituting (5.1)–(5.4) in (3.14)–(3.16), it follows

$$G = G_0 - G_0VG_0 + (G_0VG_0VG_0)^* - (G_0VG_0VG_0VG_0)^* + \sigma G_0W_2VG_0$$
$$+ (G_0VG_0VG_0VG_0VG_0)^* - \sigma G_0W_2(VG_0VG_0)^* - \sigma(G_0VG_0W_2VG_0)^*$$
$$- G(VG_0VG_0VG_0VG_0VG_0)^* + \sigma GW_2(VG_0VG_0VG_0)^*$$
$$+ \sigma G(VG_0W_2VG_0VG_0)^* + \sigma G(VG_0VG_0W_2VG_0)^* + \sigma GVW_4G_0$$
$$+ \sigma GW_4VG_0 - G(W_4 + \rho v^4 - \sigma^3 v^4)G_0VG_0 - 2\sigma^4 Gv^4VG_0$$
$$+ GVDG_0 + G(W_2 - \rho v^4)(G_0VG_0VG_0VG_0VG_0)^*$$
$$- \sigma G(W_2 - \rho v^4)G_0W_2(VG_0VG_0)^* - \sigma G(W_2 - \rho v^4)(G_0VG_0W_2VG_0)^*$$
$$+ GWG_0 + \sigma \rho Gv^4G_0W_2VG_0 + \rho Gv^4(G_0VG_0VG_0)^*$$
$$- \rho Gv^4(G_0VG_0VG_0VG_0)^*$$
$$= A + GB + GWG_0 - GWG_0VG_0 \tag{5.5}$$

where

$$A = G_0 - G_0VG_0 + (G_0VG_0VG_0)^* - (G_0VG_0VG_0VG_0)^* + \sigma G_0W_2VG_0$$
$$+ (G_0VG_0VG_0VG_0VG_0)^* - \sigma G_0W_2(VG_0VG_0)^* - \sigma(G_0VG_0W_2VG_0)^*$$
$$- G_0(VG_0VG_0VG_0VG_0VG_0)^* + \sigma G_0W_2(VG_0VG_0VG_0)^*$$
$$+ \sigma G_0(VG_0VW_2G_0VG_0)^* + \sigma G_0(VG_0VG_0W_2VG_0)^* - 2\sigma^4 G_0v^4VG_0$$
$$+ \sigma G_0(VW_4 + W_4V)G_0 + G_0VD_4G_0 \tag{5.6}$$

and

$$B = \tilde{V}G_0(VG_0VG_0VG_0VG_0VG_0)^* - \sigma\tilde{V}G_0W_2(VG_0VG_0VG_0)^*$$
$$- \sigma\tilde{V}(G_0VG_0VW_2G_0VG_0)^* - \sigma\tilde{V}G_0(VG_0VG_0VW_2G_0)^*$$
$$- \sigma\tilde{V}G_0(VW_4 + W_4V)G_0 + 2\sigma^4\tilde{V}G_0v^4VG_0 - \tilde{V}G_0VDG_0$$
$$+ (W_2 - \rho v^4)[(G_0VG_0VG_0VG_0VG_0)^* - \sigma G_0W_2(VG_0VG_0)^*$$
$$- \sigma G_0(VG_0VW_2G_0)^*]$$

$$+ \rho v^4 \left[\sigma G_0 W_2 V G_0 + (G_0 V G_0 V G_0)^* - (G_0 V G_0 V G_0 V G_0)^* \right]. \quad (5.7)$$

Recall from §3 that

$$W = v^2 M v^2$$

where M is a convolution operator satisfying

$$|M(n, n')| < C|n - n'|^{-3(d-2)}$$

and

$$\hat{M}(\xi) = \cdot|\xi|^2 + 0(|\xi|^4).$$

Hence, we may factorize

$$M = M_1 \Delta \quad (5.8)$$

with

$$|M_1(n, n')| < C|n - n'|^{-(d+2)}. \quad (5.9)$$

Write then

$$W_{n,n'} = v_n^2 M(n - n') v_{n'}^2 = v_n^4 M(n - n') + v_n^2 M(n - n')(v_{n'}^2 - v_n^2)$$

hence

$$W = v^4 M_1 \Delta + P \quad (5.10)$$

with

$$|P(n, n')| < C \frac{\kappa^4}{|n|^{2\alpha}} \frac{1}{|n - n'|^{3(d-2)}} \left| \frac{1}{|n'|^{2\alpha}} - \frac{1}{|n|^{2\alpha}} \right|$$
$$< C \kappa^4 |n|^{-4\alpha-1} |n - n'|^{-2(d-2)}. \quad (5.11)$$

From (5.9), (5.10), (5.11)

$$W G_0 = v^4 M_1 + P G_0 \quad (5.12)$$

satisfies

$$|(W G_0)(n, n')| < C \kappa^4 \left(|n|^{-4\alpha} |n-n'|^{-(d+2)} + |n|^{-(1+4\alpha)} |n-n'|^{-(d-2)} \right). \quad (5.13)$$

Write

$$W G_0 V G_0 = v^4 M_1 V G_0 + P G_0 V G_0$$

$$G W G_0 V G_0 = G v^4 M_1 V G_0 + G P G_0 V G_0$$
$$= G_0 v^4 M_1 V G_0 - G \tilde{V} G_0 v^4 M_1 V G_0 + G P G_0 V G_0. \quad (5.14)$$

Rewrite (5.5) using (5.12), (5.14) as

$$G = A' + G B' \quad (5.15)$$

with

$$A' = A - G_0 v^4 M_1 V G_0 \qquad (5.15')$$

and

$$B' = B + v^4 M_1 + P G_0 + \tilde{V} G_0 v^4 M_1 V G_0 - P G_0 V G_0. \qquad (5.15'')$$

We have

$$\mathbb{E}_\omega \left[|G_0 V_\omega G_0(n, n')| \right] < C\kappa \left(|n|^{-\alpha} + |n'|^{-\alpha} \right) |n - n'|^{-(d-2)}$$

and from (5.11)

$$\mathbb{E}_\omega \left[|P G_0 V G_0(n, n')| \right] < C\kappa^5 \sum |n|^{-(1+4\alpha)} |n - n_1|^{-2(d-2)} |n_1 - n'|^{-(d-2)}$$
$$< C\kappa^5 |n|^{-(1+4\alpha)} |n - n'|^{-(d-2)}.$$

Also

$$\mathbb{E}_\omega \left[|(G_0 v^4 M_1 V_\omega G_0)(n, n')| \right]$$
$$< C\kappa \left(\sum_{n_1} |(G_0 v^4 M_1)(n, n_1)|^2 |n_1|^{-2\alpha} |n_1 - n'|^{-2(d-2)} \right)^{\frac{1}{2}}$$
$$< C\kappa^5 \left(\sum_{n_1} |n - n_1|^{-2(d-2)} |n_1|^{-10\alpha} |n_1 - n'|^{-2(d-2)} \right)^{\frac{1}{2}}$$
$$< C\kappa^5 \left(|n|^{-5\alpha} + |n'|^{-5\alpha} \right) |n - n'|^{-(d-2)}$$

hence

$$\mathbb{E}_\omega \left[|(\tilde{V} G_0 v^4 M_1 V G_0)(n, n')| \right] < C\kappa^6 \left(|n|^{-6\alpha} + |n'|^{-6\alpha} \right) |n - n'|^{-(d-2)}.$$

Consequently

$$\mathbb{E}_\omega \left[|A'(n, n')| \right] < C |n - n'|^{-(d-2)} \qquad (5.16)$$

and

$$\mathbb{E}_\omega \left[|B'(n, n')| \right] < C\kappa^6 \left(|n|^{-6\alpha} + |n'|^{-6\alpha} \right) |n - n'|^{-(d-2)}$$
$$+ \kappa^4 |n|^{-4\alpha} |n - n'|^{-(d+2)} + \kappa^5 |n|^{-(1+4\alpha)} |n - n'|^{-(d-2)}$$
$$< C\kappa |n - n'|^{-(d+2)} + \kappa \left(|n|^{-6\alpha} + |n'|^{-6\alpha} \right) |n - n'|^{-(d-2)}.$$
$$(5.17)$$

Thus (5.16), (5.17) correspond to the bounds (3.20), (3.22), except for the first form in (5.17) which is harmless.

Rewriting (5.15) as

$$G = A'(1 - B')^{-1}$$

the assumption $\alpha > \frac{1}{3}$ and estimates (5.16), (5.17) permit then again to establish the bound

$$|G(n, n')| < C|n - n'|^{-(d-2)} \tag{5.18}$$

(as in §3).

To construct the extended state η of $H = -\Delta + \tilde{V}$, proceed as in §4. Thus, recalling the notation

$$H_0' = -\Delta + W, \quad G_0' = (H_0')^{-1}$$

we rewrite (5.5) as

$$
\begin{aligned}
G &= A(1 - WG_0)^{-1} + GB(1 - WG_0)^{-1} - GWG_0VG_0(1 - WG_0)^{-1} \\
&= A(-\Delta)G_0' + GB(-\Delta)G_0' - GWG_0VG_0' \\
&\quad \text{by (5.10)} \\
&= A(-\Delta)G_0' + GB(-\Delta)G_0' - Gv^4 M_1 V G_0' - GPG_0VG_0' \\
&= A(-\Delta)G_0' - G_0 v^4 M_1 V G_0' + G[B(-\Delta)G_0' + \tilde{V}G_0 v^4 M_1 V G_0' - PG_0 V G_0'].
\end{aligned}
\tag{5.19}
$$

From (4.10), (4.11), the extended state η of H is given by

$$\eta = \zeta + (G - G_0')H_0'\zeta = \hat{\delta}_0 + o(1) + (G - G_0')H_0'\zeta \quad \text{in} \quad \ell^\infty(\mathbb{Z}^d).$$

By (5.19)

$$
\begin{aligned}
(G - G_0')H_0'\zeta &= (A - G_0)(-\Delta)\zeta - G_0 v^4 M_1 V \zeta \\
&\quad + G[B(-\Delta) + \tilde{V}G_0 v^4 M_1 V - PG_0 V]\zeta \\
&= o(1) \quad \text{in} \quad \ell^\infty(\mathbb{Z}^d).
\end{aligned}
$$

References

[Bo]   Bourgain, J. (2002): On random Schrödinger operators on $\mathbb{Z}^2$. Discrete Contin. Dyn. Syst., **8(1)**, 1–15

[Ka]   Kato, T. (1976): Perturbation Theory of Linear Operators. Springer Grundlehren, Vol. 132

[K-L-S] Kiselev, A., Last, Y., Simon, B. (1998): Modified Prüfer and EFGP transforms and the spectral analysis of one-dimensional Schrödinger operators. Comm. Math. Phys., **194**, 1–45

[P-T]   Pajor, A., Tomczak-Jaegermann, N. (1986): Subspaces of small codimension of finite dimensional Banach spaces. Proc. AMS, **97**, 637–642

[Si]   Simon, B. (1982): Some Jacobi matrices with decaying potential and dense point spectrum. Comm. Math. Phys., **87**, 253–258

[S]   Spencer, T. Lipschitz tails and localization. Preprint

[St]   Stein, E. (1993): Harmonic Analysis: Real-Variable Methods, Orthogonality, and Oscillatory Integrals. Princeton Math Series, No. 43

On Long-Time Behaviour of Solutions of Linear Schrödinger Equations with Smooth Time-Dependent Potential

J. Bourgain*

Institute for Advanced Study, Princeton, NJ 08540, USA *bourgain@math.ias.edu*

In what follows, the spatial dimension d is assumed $d \geq 3$. We consider equations of the form

$$iu_t + \Delta u + V(x,t)u = 0 \qquad (0.1)$$

where V is bounded and $\sup_t |V(t)|$ compactly supported (or with rapid decay for $|x| \to \infty$).

Further, appropriate smoothness assumptions on V will be made.

The issues considered here are

(i) Decay estimates for $t \to \infty$
(ii) Given $u(0) \in H^s(\mathbb{R}^d)$, $s > 0$, the behaviour of

$$\|u(t)\|_{H^s} \text{ for } t \to \infty.$$

The first part of the paper deals with small potentials (in fact, (i) is only addressed in this context). Results in a similar spirit may have been obtained earlier.

The second and main part of the paper addresses (ii) for large potentials. It turns out that the situation is roughly analogous as in the case of periodic boundary conditions (see [B1], [B2]).

More precisely, assuming V smooth (but, unlike in the case of periodic bc, only smoothness in the x-variable is involved). Then

$$\|u(t)\|_{H^s} \leq C_\varepsilon |t|^\varepsilon \, \|u(0)\|_{H^s} \text{ for all } \varepsilon > 0 \qquad (0.2)$$

and the $|t|^\varepsilon$-factor cannot be removed.

1 Small Potentials

We prove the following

Proposition 1. *Consider the equation*

$$iu_t + \Delta u + V(x,t)u = 0 \qquad (d \geq 3) \qquad (1.0)$$

where V is a complex potential (we do not use self-adjointness here) satisfying

---

* This note is mainly motivated from discussions with I. Rodnianski and W. Schlag and their forthcoming paper [R-S].

(1.1) $\sup_t |V(t)|$ *is compactly supported or with rapid decay for* $|x| \to \infty$

(1.2) $\sup_t \|V(t)\|_2 < \gamma$ *small (depending on assumption (1.1) and (1.3))*

(1.3) $\sup_t \|V\|_\infty < C_1$.

Then

(1.4) *Assuming moreover* $\sup_t \|V(\cdot, t)\|_{H^s} < \gamma$ *for some* $s > \frac{d}{2} - 1$, *the usual* $L^1 - L^\infty$ *decay estimate holds*

$$\|u(t)\|_\infty < C|t|^{-\frac{d}{2}} \|u(0)\|_1 \tag{1.5}$$

(1.6) $\sup_t \|u(t)\|_2 \leq C\|u(0)\|_2$

(1.7) *Assume* V *smooth (with uniform bounds in time). Then, for all* $s \geq 0$

$$\sup_t \|u(t)\|_{H^s} \leq C_s \|u(0)\|_{H^s}.$$

A. Denote $\varphi = u(0)$.

It follows from (1.0) and Duhamel's formula that

$$u(t) = e^{it\Delta}\varphi + i \int_o^t e^{i(t-\tau)\Delta}\big[u(\tau)V(\tau)\big]d\tau$$

where

$$\|e^{it\Delta}\varphi\|_\infty < C|t|^{-\frac{d}{2}} \|\varphi\|_1.$$

Performing a Fourier decomposition in the x-variable, write

$$V = V_0 + \sum_{j \geq 1} V_j \tag{1.8}$$

where

$$\mathrm{supp}\, \mathcal{F}_x V_j \subset B(0, 2^{j+2}) \backslash B(0, 2^j)^{(*)}. \tag{1.9}$$

Estimate for $t > 0$

$$\left\| \int_0^t e^{i(t-\tau)\Delta}\big[u(\tau)V(\tau)\big]d\tau \right\|_\infty \leq \sum_j \left\| \int_0^t e^{i(t-\tau)\Delta}\big[u(\tau)V_j(\tau)\big]d\tau \right\|_\infty.$$

Fix j and denote

$$\delta_j = 2^{-j}.$$

Write

$$\left\| \int_0^t e^{i(t-\tau)\Delta}[u(\tau)V_j(\tau)]d\tau \right\|_\infty$$

_____

$^{(*)}$ $\mathcal{F}_x$ denotes 'Fourier transform' in the x-variable.

$$\leq \left(\int_0^{\delta_j \wedge t} + \int_{\delta_j}^{t-\delta_j} + \int_{t-\delta_j}^t \right) \|e^{i(t-\tau)\Delta}[u(\tau)V_j(\tau)]\|_\infty d\tau$$

$$= (1.10) + (1.11) + (1.12)$$

From the assumption (1.4)

$$\|V_j(t)\|_2 \leq 2^{-js}\|V(t)\|_{H^s} \leq C\gamma 2^{-js} \tag{1.13}$$

$$\left\|\mathcal{F}_x[V_j(t)]\right\|_1 \leq C 2^{j\frac{d}{2}}\|V_j(t)\|_2 < C\gamma 2^{j(\frac{d}{2}-s)}. \tag{1.14}$$

Recall also that

$$\left|\left[e^{it\Delta}(e^{i\zeta \cdot x}\psi)\right](x)\right| = \left|(e^{it\Delta}\psi)(x + 2\zeta t)\right|.$$

Hence

$$\|e^{it\Delta}(e^{i\zeta \cdot x}\psi)\|_\infty = \|e^{it\Delta}\psi\|_\infty$$

$$\|e^{it\Delta}(v\psi)\|_\infty \leq \|\hat{v}\|_1 \|e^{it\Delta}\psi\|_\infty. \tag{1.15}$$

From (1.14), (1.15)

$$(1.10) \leq \delta_j \sup_{0\leq\tau\leq\delta_j\wedge t} \left\|e^{i(t-\tau)\Delta}[u(\tau)V_j(\tau)]\right\|_\infty$$

$$< C\gamma 2^{j(\frac{d}{2}-s-1)} \sup_{0\leq\tau\leq\delta_j\wedge t} \left\|e^{i(t-\tau)\Delta}u(\tau)\right\|_\infty \tag{1.16}$$

$$(1.11) \leq C \int_{\delta_j}^{t-\delta_j} |t - \tau|^{-\frac{d}{2}}\|u(\tau)V_j(\tau)\|_1 d\tau. \tag{1.17}$$

Since (1.1), we may ensure that V_j satisfies also

$$\|V_j(\tau)\|_1 < C\gamma 2^{-js}. \tag{1.18}$$

Hence

$$(1.17) \leq C\gamma 2^{-js} \int_{\delta_j}^{t-\delta_j} |t - \tau|^{-\frac{d}{2}}\|u(\tau)\|_\infty d\tau$$

$$\leq C\gamma 2^{-js}\delta_j^{1-\frac{d}{2}}t^{-\frac{d}{2}} \sup_{0\leq\tau\leq t} \left(\tau^{\frac{d}{2}}\|u(\tau)\|_\infty\right)$$

$$\leq C\gamma 2^{-j(s+1-\frac{d}{2})}t^{-\frac{d}{2}} \sup_{0\leq\tau\leq t} \left(\tau^{\frac{d}{2}}\|u(\tau)\|_\infty\right). \tag{1.19}$$

Again from (1.14), (1.15)

$$(1.12) \leq \delta_j \sup_{t-\delta_j<\tau<t} \left\|e^{i(t-\tau)\Delta}[u(\tau)V_j(\tau)]\right\|_\infty$$

$$< C\gamma 2^{j(\frac{d}{2}-s-1)} \sup_{t-\delta_j<\tau\leq t} \left\|e^{i(t-\tau)\Delta}u(\tau)\right\|_\infty. \tag{1.20}$$

Collecting estimates (1.16), (1.19), (1.20), it follows

$$t^{d/2}\|u(t)\|_\infty \leq C\|\varphi\|_1 + C_s\gamma \sup_{o\leq\tau\leq t} (t^{d/2}\|e^{i(t-\tau)\Delta}u(\tau)\|_\infty)$$
$$+ C_s\gamma \sup_{0\leq\tau\leq t} (\tau^{\frac{d}{2}}\|u(\tau)\|_\infty). \qquad (1.21)$$

Consider the second term in (1.21). From Duhamel's formula, we get also

$$|e^{i(t-\tau)\Delta}u(\tau)| \leq |e^{it\Delta}\varphi| + \left|\int_0^\tau e^{i(t-\tau')\Delta}[u(\tau')V(\tau')]d\tau'\right|$$
$$\leq |e^{it\Delta}\varphi| + \int_0^t \left|e^{i(t-\tau')\Delta}[u(\tau')V(\tau')]\right|d\tau'$$

and the last term may be estimated as before.

Therefore, also

$$\sup_{0\leq\tau\leq t} \left[t^{d/2}\|e^{i(t-\tau)\Delta}u(\tau)\|_\infty\right] \leq (1.21)$$

implying for γ small enough

$$\sup_{0\leq\tau\leq t} t^{d/2}\left\|e^{i(t-\tau)\Delta}u(\tau)\right\|_\infty \leq C\|\varphi\|_1. \qquad (1.22)$$

In particular (1.5) holds.

B. We will need the following

Lemma 2.1. *For $d \geq 2$*

$$\left\|e^{it\Delta}\varphi\right\|_{L^2_t L^2_{x,\text{loc}}} \leq C\|\varphi\|_2. \qquad (2.2)$$

Proof. From the local smoothing inequality

$$\left\|D_x^{1/2}e^{it\Delta}\varphi\right\|_{L^2_t L^2_{x,\text{loc}}} \leq C\|\varphi\|_2. \qquad (2.3)$$

Thus it suffices to prove (2.2) for supp $\hat\varphi \subset B(0,1)$.

Write

$$e^{it\Delta}\varphi(x) = \int \hat\varphi(\xi)e^{i(x\cdot\xi+t|\xi|^2)}d\xi \sim \int_0^1 \left[\int_{S^{d-1}} \hat\varphi(r.\zeta)e^{irx\zeta}r^{d-1}d\zeta\right]e^{itr^2}dr. \qquad (2.4)$$

Make change of variable $s = r^2$. We obtain

$$\int_0^1 \left[\int_{S^{d-1}} \cdots\right]e^{its}\frac{1}{\sqrt{s}}ds.$$

Take L_t^2-norm and apply Parseval. We get the bound

$$\left[\int_0^1 |\cdots|^2 \frac{1}{s} ds\right]^{1/2} = \left[\int_0^1 |\ |^2 \frac{1}{r} dr\right]^{1/2}. \tag{2.5}$$

Estimate pointwise in x

$$\left|\int_{S^{d-1}} \hat{\varphi}(r\zeta) e^{irx\cdot\zeta} r^{d-1} d\zeta\right| \le \left[\int_{S^{d-1}} |\hat{\varphi}(r\zeta)| d\zeta\right] r^{d-1}$$
$$\le \left[\int_{S^{d-1}} |\hat{\varphi}(r\zeta)|^2 d\zeta\right]^{1/2} r^{d-1}.$$

Hence

$$(2.5) \le \left[\int_0^1 \int_{S^{d-1}} |\hat{\varphi}(r\zeta)|^2 \cdot r^{2d-3} d\zeta dr\right]^{1/2}$$
$$\le \left[\int_0^1 \int_{S^{d-1}} |\hat{\varphi}(r\zeta)|^2 r^{d-1} d\rho d\zeta\right]^{1/2} \qquad (d \ge 2)$$
$$= \|\hat{\varphi}\|_2.$$

This proves Lemma 2.1.

We now prove (1.6) of Proposition 1.

Assume $\varphi \in L^2$. From Duhamel's formula and (2.2)

$$\|u\|_{L_t^2 L_x^2(\text{loc})}$$
$$\le \|e^{it\Delta}\varphi\|_{L_t^2 L_x^2(\text{loc})} + \left\{\int dt \left\|\int_0^t e^{i(t-\tau)\Delta}[u(\tau)V(\tau)]d\tau\right\|_{L_x^2(\text{loc})}^2\right\}^{1/2}$$
$$\le C\|\varphi\|_2 + \left\|\int_0^{t-\delta} e^{i(t-\tau)\Delta}[u(\tau)V(\tau)]d\tau\right\|_{L_t^2 L_x^\infty} + \left\|\int_{t-\delta}^t \cdots\right\|_{L_t^2 L_x^2}. \tag{2.6}$$

Second term in (2.6) is bounded by

$$\left\|\int_0^{t-\delta} |t-\tau|^{-\frac{d}{2}} \|u(\tau)V(\tau)\|_1 d\tau\right\|_{L_t^2}$$
$$\le \gamma \left\|\int_0^{t-\delta} |t-\tau|^{-\frac{d}{2}} \|u(\tau)\|_{L_{x,\text{loc}}^2} d\tau\right\|_{L_t^2} \qquad \text{(by (1.2))}$$
$$\le \gamma \left\|\int_0^\infty (\delta + |t-\tau|)^{-\frac{d}{2}} \|u(\tau)\|_{L_{x,\text{loc}}^2} d\tau\right\|_{L_t^2}$$
$$< C_\delta \gamma \|u\|_{L_t^2 L_{x,\text{loc}}^2} \qquad (d \ge 3). \tag{2.7}$$

Third term in (2.6) is bounded by

$$\left\| \int_{t-\delta}^{t} \|u(\tau)V(\tau)\|_{L_x^2}\,d\tau \right\|_{L_t^2} \leq \|V\|_\infty \left\| \int_{t-\delta}^{t} \|u(\tau)\|_{L_{\mathrm{loc}}^2}\,d\tau \right\|_{L_t^2}$$

$$\leq C_1 \delta \|u\|_{L_t^2 L_{x,\mathrm{loc}}^2}. \tag{2.8}$$

From (2.6), (2.7), (2.8), it follows that

$$\|u\|_{L_t^2 L_{x,\mathrm{loc}}^2} \leq C\|\varphi\|_2. \tag{2.9}$$

Next, estimate

$$\|u(t)\|_2 \leq \|\varphi\|_2 + \left\| \int_0^t e^{i(t-\tau)\Delta}[u(\tau)V(\tau)]d\tau \right\|_2$$

and second term using duality by

$$\int_0^t \int |e^{i\tau\Delta}\psi|.|u(\tau)| \ |V(\tau)|dxd\tau$$

$$\leq C_1 \int_0^\infty \|e^{i\tau\Delta}\psi\|_{L_{x,\mathrm{loc}}^2} \|u(\tau)\|_{L_{x,\mathrm{loc}}^2}\,d\tau \qquad (\text{where } \|\psi\|_2 = 1)$$

$$< CC_1\|\psi\|_2\|\varphi\|_2 \qquad (\text{by (2.2) and (2.9)})$$

$$< C\|\varphi\|_2.$$

This proves (1.6).

C. Assume V smooth (a more restricted assumption is easily derived from what follows).

We prove (1.7) of Proposition 1.

Take $s = 1$ (the general case is similar).

Thus

$$iu_t + \Delta u + Vu = 0$$

implying

$$i(D_x u)_t + \Delta(D_x u) + VD_x u + (D_x V)u = 0. \tag{3.1}$$

Hence

$$i \rightarrow \overbrace{\begin{pmatrix} u \\ Du \end{pmatrix}} + \Delta \begin{pmatrix} u \\ Du \end{pmatrix} + \begin{pmatrix} V & 0 \\ DV & V \end{pmatrix}\begin{pmatrix} u \\ Du \end{pmatrix} = 0. \tag{3.2}$$

Since the proof of (1.6) extends to the vector valued case (and does *not* use self-adjointness), (3.2) implies

$$\|u(t)\|_2 + \|D_x u(t)\|_2 \leq C\big(\|u(0)\|_2 + \|D_x u(0)\|_2\big)$$

thus

$$\|u(t)\|_{H^1} \leq C\|u(0)\|_{H^1}.$$

This proves Proposition 1.

2 Large Smooth Potentials

We still assume $d \geq 3$. The following statement establishes a growth estimate for higher Sobolev norms.

Proposition 2. *Consider the equation*

$$iu_t + \Delta u + V(x,t)u = 0$$

where V is real, bounded, and $\sup |V(t)|$ compactly supported (or with rapid decay for $|x| \to \infty$). (No smallness assumption).
   Assume

$$\sup_t \left\| D_x^{(s')} V(t) \right\|_\infty < C_{s'} \text{ for all } s'. \tag{4.1}$$

Then

$$\|u(t)\|_{H^s} \leq C_\varepsilon |t|^\varepsilon \|u(0)\|_{H^s} \text{ for all } \varepsilon > 0. \tag{4.2}$$

Remark. No smoothness assumption in t is made; similar statement under less restrictive assumptions than (4.1) result from the argument below.

Proof. Define

$$\|f\| = \inf_{f=f_1+f_2} (\|f_1\|_2 + \|f_2\|_\infty). \tag{4.3}$$

Assume $\varphi = u(0) \in H^1$.
   We first make an estimate on $\|D_x u(t)\|$.
   From Duhamel's formula

$$\|D_x u(t)\| \leq \|e^{it\Delta} D_x \varphi\|_2 + \left\| \int_0^{t-A} e^{i(t-\tau)\Delta} D_x[u(\tau)V(\tau)] d\tau \right\|_\infty$$

$$+ \left\| \int_{t-A}^t e^{i(t-\tau)\Delta} D_x[u(\tau)V(\tau)] d\tau \right\|_2$$

$$\leq \|\varphi\|_{H^1} + (4.4) + (4.5)$$

$$(4.4) \leq \int_0^{t-A} |t-\tau|^{-\frac{d}{2}} \big[\|D_x u(\tau)V(\tau)\|_1 + \|u(\tau) \cdot D_x V(\tau)\|_1 \big] d\tau$$

$$\leq \int_0^{t-A} |t-\tau|^{-\frac{d}{2}} \big[C\|D_x u(\tau)\|_{L^1_{loc}} + C\|u(\tau)\|_2 \big] d\tau$$

$$\lesssim A^{1-\frac{d}{2}} \sup_{\tau<t} \|D_x u(\tau)\| + C\|\varphi\|_2. \tag{4.6}$$

To estimate (4.5), take $\psi \in L^2, \|\psi\|_2 = 1$ and write

$$\int_{t-A}^t \int \left| D_x^{1/2}[e^{i\tau\Delta}\psi] \right| \left| D_x^{1/2}[u(\tau)V(\tau)] \right| dx d\tau$$

$$\leq A^{1/2} \left\| D_x^{1/2}[e^{i\tau\Delta}\psi] \right\|_{L^2_\tau L^2_{x,loc}} \sup_{\tau \in [t-A,t]} \left\| D_x^{1/2}[u(\tau)V(\tau)] \right\|_2$$

$$\leq CA^{1/2}\|u(0)\|_2^{1/2}\left(\|u(0)\|_2 + \sup_{\tau < t}\|D_x u(\tau)\|\right)^{1/2} \tag{4.7}$$

(using (2.3) and interpolation).

Thus from (4.6), (4.7)

$$\|D_x u(t)\| \leq \|\varphi\|_{H^1} + CA^{1/2}\|\varphi\|_2 + CA^{1-\frac{d}{2}}\sup_{\tau < t}\|D_x u(\tau)\|$$
$$+ CA^{1/2}\|\varphi\|_2^{1/2}\left[\sup_{\tau < t}\|D_x u(\tau)\|\right]^{1/2}. \tag{4.8}$$

This implies by appropriate choice of A that

$$\sup_t \|D_x u(t)\| \leq C\|\varphi\|_{H^1}. \tag{4.9}$$

Similarly, one establishes that

$$\sup_t \|D_x^{(s)} u(t)\| \leq C_s\|\varphi\|_{H^s}. \tag{4.10}$$

Next, estimate

$$\|u(t)\|_{H^s} \leq \|\varphi\|_{H^s} + \int_0^t \left\|[u(\tau)V(\tau)]\right\|_{H^s}d\tau$$
$$\leq \|\varphi\|_{H^s} + t\sup_{\tau < t}\|u(\tau)V(\tau)\|_{H^s}$$
$$\leq \|\varphi\|_{H^s} + tC_s(V)\sup_{\substack{t \\ s' \leq s}}\|D_x^{(s')}u(\tau)\|$$
$$< C_s(V)t\|\varphi\|_{H^s} \tag{4.11}$$

by (4.10).

Thus the linear flow map $S_t : H^s \to H^s$ has a norm bounded by $C_s(V).t$. For $s = 0, S_t$ is unitary.

By interpolation, we get for given $s > 0$ and $s_1 > s$ large

$$\|S_t\|_{H^s \to H^s} \leq \left[C_{s_1}(V)t\right]^{\frac{s}{s_1}}.$$

Hence, since V is assumed smooth

$$\|u(t)\|_{H^s} \leq C_\varepsilon(V)t^\varepsilon\|\varphi\|_{H^s} \text{ for all } \varepsilon > 0.$$

This proves Proposition 2.

3 An Example

The factor T^ε in (4.2) is necessary. To exhibit such growth phenomenon, we need presence of bound states. Thus first consider $-\Delta + v(x) = H_0$ with a bound state φ and next certain time dependent perturbations $H_0 + \varepsilon V(x, t)$. Rather than defining H_0 as above, let

$$H_0 = -\Delta - P \tag{5.0}$$

where P is a real and smooth Fourier multiplier s.t.

$$\operatorname{supp} \hat{P} \subset B(0, 2) \text{ and } \hat{P}(\xi) = |\xi|^2 \text{ for } |\xi| \leq 1. \tag{5.1}$$

Let then $0 \leq \varphi \leq 1$ be a rapidly decaying function such that $\operatorname{supp} \hat{\varphi} \subset B(0, 1)$. Thus from (5.0), (5.1)

$$H_0 \varphi = 0. \tag{5.2}$$

This alternative construction will avoid certain technical difficulties, since the spectral projections related to H_0 are now simply Fourier multipliers.

Let $d = 3$.

(ii) Fix large time T. Let N be a large number and $\xi_0 = N e_1 \in \mathbb{R}^3$.

Let $V = V(x, t)$ be real and satisfying $\operatorname{supp} \mathcal{F}_x V \subset B(\xi_0, 1) \cup B(-\xi_0, 1)$ (to be specified later).

Let $0 \leq \eta \leq 1$ be a bumpfunction with rapid decay for $|x| \to \infty$ such that

$$\operatorname{supp} \hat{\eta} \subset B(0, 1) \tag{5.3}$$

$$V\eta \text{ decays rapidly for } |x| \to \infty. \tag{5.4}$$

Consider the linear Schrödinger equation

$$iu_t - (\Delta + P)u + \varepsilon V \eta u = 0 \text{ for } 0 \leq t \leq T \tag{5.5}$$

with datum

$$u(0) = \varphi.$$

Write

$$u = \varphi + U$$

with U satisfying, by (5.2)

$$\begin{cases} iU_t - (\Delta + P)U + \varepsilon V \eta \varphi + \varepsilon V \eta U = 0 \\ U(0) = 0. \end{cases} \tag{5.6}$$

Denote Q a smooth Fourier multiplier such that

$$\hat{Q}(\xi) = 0 \text{ if } |\xi| < \frac{N}{8} \text{ and } \hat{Q}(\xi) = 1 \text{ for } |\xi| > \frac{N}{4} \tag{5.7}$$

$$|\partial^\alpha \hat{Q}| < C N^{-|\alpha|}. \tag{5.8}$$

(iii) From Duhamel's formula and (5.6)

$$U(T) = i\varepsilon \int_0^T e^{i(T-\tau)H_0}\left[V(\tau)\eta\varphi + V(\tau)\eta U(\tau)\right]d\tau. \tag{5.9}$$

From the local smoothing inequality and (5.1), (5.4), (5.7)

$$\varepsilon\left\|\int_0^T e^{i(T-\tau)H_0}Q\left[V(\tau)\eta U(\tau)\right]d\tau\right\|_{L^2}$$

$$= \varepsilon\left\|\int_0^T e^{i(T-\tau)\Delta}Q\left[V(\tau)\eta U(\tau)\right]d\tau\right\|_{L^2}$$

$$\leq \varepsilon N^{-1/2}\left\|\int_0^T e^{i(T-\tau)\Delta}D_x^{1/2}\left[\eta V(\tau)U(\tau)\right]\right\|_{L^2}$$

$$\leq C\varepsilon N^{-1/2}T^{1/2}\sup_{\tau < T}\|U(\tau)\|_{L^2_{\mathrm{loc}}}. \tag{5.10}$$

We estimate $\sup_{\tau < T}\|U(\tau)\|_{L^2_{\mathrm{loc}}}$.

From (5.9)

$$\|QU(T)\|_{L^2_{\mathrm{loc}}} = \varepsilon\left\|\int_0^T e^{i(T-\tau)\Delta}Q\left[V(\tau)\eta u(\tau)\right]d\tau\right\|_{L^2_{\mathrm{loc}}}$$

$$\leq \varepsilon\left\{\int_0^{T-1}\left\|e^{i(T-\tau)\Delta}Q\left[V(\tau)\eta u(\tau)\right]\right\|_\infty d\tau\right.$$

$$\left. + \int_{T-1}^T\left\|e^{i(T-\tau)\Delta}Q\left[V(\tau)\eta u(\tau)\right]\right\|_2 d\tau\right\}$$

$$\leq C\varepsilon\left\{\int_0^{T-1}|T-\tau|^{-3/2}\|V(\tau)\eta u(\tau)\|_1 d\tau + \sup_{\tau \leq T}\|u(\tau)\|_2\right\}$$

$$\leq C\varepsilon\sup_{\tau \leq T}\|u(\tau)\|_2 < C\varepsilon. \tag{5.11}$$

Also, since $V(\tau)\eta\varphi = Q[V(\tau)\eta\varphi]$

$$(I-Q)U(T) = i\varepsilon\int_0^T e^{i(T-\tau)H_0}(I-Q)\left[V(\tau)\eta(QU(\tau))\right]d\tau$$

$$\|(I-Q)U(T)\|_2 \leq \varepsilon T\sup_{\tau \leq T}\left\|V(\tau)\eta(QU(\tau))\right\|_2$$

$$\leq C\varepsilon T\sup_{\tau \leq T}\|QU(\tau)\|_{L^2_{\mathrm{loc}}}$$

$$< C\varepsilon^2 T \tag{5.12}$$

by (5.11).

Estimating $\|U(\tau)\|_{L^2_{\mathrm{loc}}} \leq \|QU(\tau)\|_{L^2_{\mathrm{loc}}} + \|(I-Q)U(\tau)\|_2$, (5.11), (5.12) imply

$$\sup_{\tau < T}\|U(\tau)\|_{L^2_{\mathrm{loc}}} < C(\varepsilon + \varepsilon^2 T). \tag{5.13}$$

Substitution in (5.10) gives therefore

$$\varepsilon \left\| \int_0^T e^{i(T-\tau)H_0} Q\left[V(\tau)\eta U(\tau)\right] d\tau \right\|_2 < C\varepsilon^2 N^{-1/2} T^{1/2}(1+\varepsilon T). \quad (5.14)$$

Consider next

$$\left\| \int_0^T e^{i(T-\tau)H_0}\left[V(\tau)\eta\varphi\right] d\tau \right\|_2 = \left\| \int_0^T e^{-i\tau\Delta}\left[V(\tau)\eta\varphi\right] d\tau \right\|_2. \quad (5.15)$$

Define

$$\varphi_1(x) = e^{ix\cdot\xi_0}\varphi(x) \quad (5.16)$$

$$\psi = \psi_\omega = \sum_{\substack{j\in\mathbb{Z} \\ \frac{1}{2}NT<j\le NT}} \omega_j e^{-i\frac{j}{N}\Delta}\varphi_1$$

where $\{\omega_j\}_{j\in\mathbb{Z}}$ are independent Bernoulli variables. $\quad (5.17)$

Write

$$\left|(e^{-is\Delta}\varphi_1)(x)\right| = \left| \int e^{ix(\xi_0+\xi)}e^{-i|\xi_0+\xi|^2 s}\hat{\varphi}(\xi)d\xi \right|$$

$$= \left| \int_{|\xi|<1} e^{i[(x-2s\xi_0)\xi - s|\xi|^2]}\hat{\varphi}(\xi)d\xi \right|. \quad (5.18)$$

Since for $|\xi| < 1$

$$\left|\partial_\xi\left[(x-2s\xi_0)\xi - s|\xi|^2\right]\right| > |x - 2s\xi_0| - 2|s|$$

it follows that (for a large constant C)

$$\left|(e^{-is\Delta}\varphi_1)(x)\right| \lesssim |x-2s\xi_0|^{-C} \text{ if } |x - 2s\xi_0| > 10|s|. \quad (5.19)$$

Define
$$V(x,t) = \text{Re}\,(e^{it\Delta}\psi) \text{ or } V(x,t) = \text{Im}\,(e^{it\Delta}\psi). \quad (5.20)$$
It follows from (5.17) that

$$\mathbb{E}_\omega\left[\|\psi_\omega\|_2\right] \sim (NT)^{1/2}\|\varphi_1\|_2$$

and we may thus assume

$$\|\psi\|_2 < C(NT)^{1/2}. \quad (5.21)$$

Next
$$\left|e^{it\Delta}\psi(x)\right| \le \sum_{\substack{j\in\mathbb{Z} \\ j\sim NT}} \left|(e^{i(t-\frac{j}{N})\Delta}\varphi_1)(x)\right|. \quad (5.22)$$

If $|x| < \frac{1}{2}|t - \frac{j}{N}|N$, (5.19) implies that

$$\left|(e^{i(t-\frac{j}{N})\Delta}\varphi_1)(x)\right| \lesssim |Nt - j|^{-C}.$$

If $|x| > \frac{1}{2}|t - \frac{j}{N}|N$, then

$$|\eta(x)| < |Nt - j|^{-C}.$$

Hence

$$\left|(e^{i(t-\frac{j}{N})\Delta}\varphi_1)\eta\right| < (1 + |Nt - j|)^{-C}(1 + |x|)^{-C}. \tag{5.23}$$

Thus from (5.20), (5.22), (5.23)

$$|V(x,t)\eta(x)| < \sum_{j\in\mathbb{Z}}(1 + |Nt - j|)^{-C}(1 + |x|)^{-C} < C(1 + |x|)^{-C}. \tag{5.24}$$

Returning to (5.15), we get by (5.21) and appropriate choice in (5.20)

$$(5.15) = \left\|\int_0^T e^{-i\tau\Delta}[V(\tau)\eta\varphi]d\tau\right\|_2$$

$$> c(NT)^{-1/2}\int_0^T\int|e^{i\tau\Delta}\psi|^2\eta\varphi \tag{5.25}$$

$$= c(NT)^{-1/2}\int_0^T\int\left|\sum_j\omega_j e^{i(\tau-\frac{j}{N})\Delta}\varphi_1\right|^2\eta\varphi.$$

Averaging over ω, we get a lower bound

$$(NT)^{-1/2}\sum_{\substack{j\in\mathbb{Z}\\ \frac{NT}{2}<j<NT}}\int_{\frac{j}{N}}^{\frac{j+1}{N}}d\tau\int dx|e^{i(\tau-\frac{j}{N})\Delta}\varphi_1|^2\eta\varphi. \tag{5.26}$$

It follows from (5.18) that for $|\tau - \frac{j}{N}| < \frac{1}{100N}$,

$$\left|(e^{i(\tau-\frac{j}{N})\Delta}\varphi_1)\right| \approx \left|\varphi\left(\cdot + 2\left(\tau - \frac{j}{N}\right)\xi_0\right)\right| \approx |\varphi| = \varphi.$$

Therefore clearly

$$(5.15), (5.25) > c(NT)^{-1/2}NT\frac{1}{100N}\int\eta\varphi^3 > c\left(\frac{T}{N}\right)^{1/2}. \tag{5.27}$$

Collecting estimates (5.14), (5.27) it follows from (5.9) that

$$\|QU(T)\|_2 > c\varepsilon\left(\frac{T}{N}\right)^{1/2} - C\varepsilon^2\left(\frac{T}{N}\right)^{1/2}(1 + \varepsilon T). \tag{5.28}$$

Choosing $T \sim \varepsilon^{-2}$, (5.28) implies

$$\|Qu(T)\|_2 = \|QU(T)\|_2 > CN^{-1/2}. \tag{5.29}$$

Let s_1 be an arbitrary large (fixed) exponent.

In order for $\varepsilon V \eta$ to satisfy a bound

$$\varepsilon \|V(\tau)\eta\|_{H^{s_1}} \sim \varepsilon N^{s_1} < 1 \tag{5.30}$$

take $\varepsilon = N^{-s_1}$, hence we have

$$T \sim N^{2s_1}$$

and, from (5.29), for $s > \frac{1}{2}$

$$\|u(T)\|_{H^s} > cT^{\frac{2s-1}{4s_1}}. \tag{5.31}$$

The construction described above provides thus arbitrary smooth time-dependent potentials $\varepsilon V(\tau)\eta$ for which (5.5) admits a solution u, supp $\widehat{u(0)} \subset B(0,1)$ and satisfying (5.31) for some (fixed) large time T.

(iv) To obtain a "full counter-example", we will glue constructions as performed above on disjoint time intervals.

Thus define a potential

$$W(x,t) = \sum_{r=1}^{\infty} \varepsilon_r \chi_r(t) V_r(x,t) \eta(x) \tag{5.32}$$

where χ_r are localizing to disjoint time intervals $[\frac{T_r}{2}, T_r]$ (take χ_r smooth), T_r increases rapidly, $\varepsilon_r = T_r^{-1/2}, N_r = T_r^{\frac{1}{4r}}$ and V_r introduced as above.

Thus (letting $s_1 = r$ in the preceding), we may insure

$$\left\|\partial_x^{(\alpha)}(\varepsilon_r V_r \eta)\right\|_{L_x^2} + \left\|\partial_t^{(\beta)}(\varepsilon_r V_r \eta)\right\|_{L_x^2} < 2^{-r} \text{ for } |\alpha|, \beta \leq r.$$

Therefore, we get clearly $\forall \alpha, \beta$

$$\sup_{x,t} \left|\partial_x^{(\alpha)} \partial_t^{(\beta)} W\right| < C_{\alpha\beta}. \tag{5.33}$$

Moreover $\sup_t |W(x,t)|$ decreases rapidly for $|x| \to \infty$, cf. (5.24).

Fix r and consider the initial value problem on $[2T_{r-1}, T_r]$

$$\begin{cases} iu_t - (\Delta + P)u + Wu = 0 \\ u(t = 2T_{r-1}) = \varphi \end{cases}$$

or, equivalently

$$\begin{cases} iu_t - (\Delta + P)u + \varepsilon_r V_r \eta u = 0 \\ u\big(t = \tfrac{1}{2}T_r\big) = \varphi. \end{cases}$$

Then we have by (5.31)

$$\|u(T_r)\|_{H^s} > T_r^{\frac{2s-1}{10r}}. \tag{5.34}$$

(Choose $s > \frac{1}{2}$ a fixed exponent.)

Denote $S(t, t')$ the flow map associated to $i\partial_t - (\Delta + P) + W$. Reformulating the preceding

$$\|S(0, T_r)S(0, 2T_{r-1})^{-1}\varphi\|_{H^s} = \|S(2T_{r-1}, T_r)\varphi\|_{H^s} > T_r^{\frac{2s-1}{10r}}. \tag{5.35}$$

Define

$$\tilde{\varphi} = \sum_{r \geq 2} \sigma_r T_r^{-\frac{2s-1}{20r}} S(0, 2T_{r-1})^{-1}\varphi \tag{5.36}$$

where $\sigma_r = \pm 1$ to be specified.

By crude estimate, we obtain for any $s' \geq 0$ that

$$\|\tilde{\varphi}\|_{H^{s'}} \leq \sum_r T_r^{-\frac{2s-1}{20r}} \|S(0, 2T_{r-1})^{-1}\varphi\|_{H^{s'}} \leq \sum_r T_r^{-\frac{2s-1}{20r}} B(T_{r-1}, s') < C_{s'},$$

for T_r chosen sufficiently rapidly increasing.

Also

$$\|S(0, T_r)\tilde{\varphi}\|_{H^s} > \left\| \sum_{r' < r} \sigma_{r'} T_{r'}^{-\frac{2s-1}{20r'}} S(0, T_r)S(0, 2T_{r'-1})^{-1}\varphi \right.$$
$$\left. + \sigma_r T_r^{-\frac{2s-1}{20r}} S(0, T_r)S(0, 2T_{r-1})^{-1}\varphi \right\|_{H^s}$$
$$- \sum_{r' > r} T_{r'}^{-\frac{2s-1}{20r'}} B(T_{r'-1}, s). \tag{5.37}$$

We may, assuming $\sigma_{r'}, r' < r$, obtained, choose σ_r s.t. the first term in (5.37) is at least $T_r^{-\frac{2s-1}{20r}}$ (5.35) $> T_r^{\frac{2s-1}{20r}}$. Hence

$$\|S(0, T_r)\tilde{\varphi}\|_{H^s} > T_r^{\frac{2s-1}{20r}} - 1.$$

The conclusion is therefore

Proposition 3. *Let $d = 3$ and denote $\gamma(t)$ any increasing function s.t.*

$$\lim_{r \to \infty} \frac{\log \gamma(t)}{\log t} = 0.$$

There is a linear Schrödinger equation IVP

$$iu_t - (\Delta + P)u + W(x,t)u = 0, u(0) = \tilde{\varphi} \in \bigcap_{s'} H^{s'}$$

with P as above, W real and

$$\sup_{x,t} |\partial_x^{(\alpha)} \partial_t^{(\beta)} W| < C_{\alpha\beta} \qquad \forall \alpha, \beta$$

$$\sup_t |W(t)| \text{ with fast decay for } |x| \to \infty$$

and such that for $s > \frac{1}{2}$

$$\varlimsup_{t \to \infty} \frac{\|u(t)\|_{H^s}}{\gamma(t)} = \infty.$$

References

[B1]   Bourgain, J. (1999): Growth of Sobolev norms in linear Schrödinger equations with quasi-periodic potential. Comm. Math. Phys., **204(1)**, 207–247

[B2]   Bourgain, J. (1999): On growth of Sobolev norms in linear Schrödinger equations with smooth time dependent potential. J. Anal. Math., **77**, 315–348

[R-S] Rodnianski, I., Schlag, W. (2001): Time decay for solutions of Schrödinger equations with rough and time-dependent potential. Preprint

On the Isotropy-Constant Problem for "PSI-2"-Bodies

J. Bourgain

Institute for Advanced Study, Princeton, NJ 08540, USA *bourgain@math.ias.edu*

1 Introduction and Statement of the Result

Assume K is a convex symmetric body in $\mathbb{R}^n$, $\mathrm{Vol}_n K \equiv V_n(K) = 1$. Assume further that K is in an isotropic position, i.e.

$$\int_K x_i x_j dx = L_K \delta_{ij} \qquad (1 \leq i,j \leq n).$$

It is known that L_K is bounded from below by a universal constant and, at this point, still an open problem whether L_K admits a universal upperbound (thus independent of K and the dimension n). This problem has several geometric reformulations. To mention one (the "high-dimensional" version of the Busemann–Petty problem): Does every convex symmetric body in $\mathbb{R}^n$, $V_n(K) = 1$, admit a co-dimension-one section $K \cap H$ (H-hyperplane) satisfying

$$V_{n-1}(K \cap H) > c$$

with $c > 0$ a universal constant?

Presently, the best (general) upperbound for L_K is

$$L_K < Cn^{1/4}(\log n) \qquad (1.1)$$

obtained in [Bo]. The present note is a direct outgrowth of the argument in [Bo]. A key ingredient in the proof of (1.1) is indeed inequalities of the form

$$\|\langle x, \xi \rangle\|_{L^{\psi_2}(K)} \leq A \|\langle x, \xi \rangle\|_{L^2(K)} \qquad (1.2)$$

valid for all linear forms $\langle \cdot, \xi \rangle$ considered as functions on K, $V_n(K) = 1$. Here $L^{\psi_q}(K), \psi_q(t) = e^{t^q} - 1$, refers to the usual Orlicz-spaces on K, dx. Recall that in general, there is the weaker inequality

$$\|\langle x, \xi \rangle\|_{L^{\psi_1}(K)} \leq C \|\langle x, \xi \rangle\|_{L^2(K)} \qquad (1.3)$$

with c an absolute constant (see [Bo], which contains also similar results for polynomials of bounded degree).

Definition. *We say that K is a "ψ_2-body" if the linear forms restricted to K satisfy (1.2) for some constant A.*

Theorem.
$$L_K < C(A). \tag{1.4}$$

More precisely

$$L_K < C.A. \log(1 + A). \tag{1.5}$$

Remark. According to a recent result of Barthe and Koldobsky [B-K], the ℓ_n^q-balls $(2 \leq q \leq \infty)$, normalized in measure, are ψ_2-bodies.

In the next section, we prove the Theorem. We will first prove (1.4) and then, with some extra care, (1.5). The argument relies on probabilistic results, such as Talagrand's majorizing measure and its consequences for subgaussian processes and also "standard" facts and methods from the "Geometry of Banach Spaces" for which the reader is referred to [Pi].

Acknowledgement. The author is grateful to V. Milman and especially A. Giannopoulos for comments.

2 Proof of the Theorem

Assume $K, V(K) = 1$, in an isomorphic position and satisfying (1.2) (invariant under affine transformation). Denote L_K by L and let C stand for various constants.

(i) Assumption (1.2) implies the following fact.

Let $\| \ \|$ be any norm on $\mathbb{R}^n$ and $\{g_i | i = 1, \ldots, n\}$ independent normalized Gaussians. Let $\{v_i | i = 1, \ldots, n\}$ be arbitrary vectors in $\mathbb{R}^n$. Then

$$\frac{1}{A.L.} \int_K \left\| \sum_{i=1}^n x_i v_i \right\| dx \leq C \int \left\| \sum_{i=1}^n g_i(\omega) v_i \right\| d\omega. \tag{2.1}$$

Some comments about this inequality. Denote $T = \{t \in \mathbb{R}^n | \ \|t\|_* \leq 1\}$, where $\| \ \|_*$ is the norm dual to $\| \ \|$. Consider the process

$$X_t(x) = \sum_{i=1}^n \frac{x_i}{A.L.} \langle v_i, t \rangle$$

satisfying, by (4.3) and the fact that K is in an isotropic position,

$$\|X_t - X_{t'}\|_{L^{\psi_2}(K)} \leq \left\| \sum_{i=1}^n \frac{x_i}{L} \langle v_i, t - t' \rangle \right\|_{L_K^2} = \left(\sum_{i=1}^n |\langle v_i, t - t' \rangle|^2 \right)^{1/2} \equiv d(t, t'). \tag{2.2}$$

Thus (2.2) means that (X_t) is subgaussian wrt the pseudo-metric d on T. We then combine the majorizing measure theorems of Preston [Pr1],[Pr2] (see also [Fe]) and Talagrand [T] to get

$$\int_K \sup_{t \in T} |X_t| \le C \int \sup_{t \in T} |Y_t| \qquad (2.3)$$

with Y_t the Gaussian process

$$Y_t(\omega) = \sum_{i=1}^{n} g_i(\omega)\langle v_i, t\rangle$$

(see [L-T], Theorem 12.16). Clearly (2.1) is equivalent to (2.3).

(ii) We replace K by

$$K_1 = K \cap \left[|x| < C_2 L\sqrt{n}\right] \qquad (2.4)$$

where C_2 is a sufficiently large constant (see also the remarks in the next section).

In particular

$$\text{Vol}\, K_1 \approx 1$$

and for $|\xi| = 1$

$$\int_{K_1} |\langle x, \xi\rangle|^2 dx \ge \int_K |\langle x, \xi\rangle|^2 dx - \|\langle x, \xi\rangle\|_{L^4(K)}^2 V(K\backslash K_1)^{1/2}$$

$$\ge L^2 - CL^2 V(K\backslash K_1)^{1/2} > \frac{1}{2}L^2 \qquad (2.5)$$

(we do use here the equivalence of all moments for linear functionals on $K - a$ consequence of (1.3)).

Denote by $\|\ \|$ the norm induced on $\mathbb{R}^n$ by K_1 and $\|\ \|_*$ its dual. Thus from (2.4)

$$\|x\| \ge \frac{1}{C_2 L\sqrt{n}} |x|$$
$$\|x\|_* \le C_2 L\sqrt{n}\, |x|. \qquad (2.6)$$

(iii) With the above notations, we prove the following fact

Lemma. *Let E be a subspace of $\mathbb{R}^n$, $\dim E > \frac{n}{2}$ such that*

$$\|x\|_* \ge \rho L\sqrt{n}|x| \text{ for } x \in E. \qquad (2.7)$$

Then, for $0 < \delta < \frac{1}{2}$, there is a subspace F of E satisfying

$$\dim F > (1 - \delta)\dim E \qquad (2.8)$$

and

$$\|x\|_* \ge c\delta^2 \left(\log \frac{1}{\rho}\right)^{-1} \frac{L}{A}\sqrt{n}|x| \text{ for } x \in F. \qquad (2.9)$$

Proof. It follows from (2.6), (2.7) that the Euclidean distance

$$d_{E, \| \ \|_*} \leq C_2 \rho^{-1}.$$

Thus, considering the ℓ-ellipsoid of $E, \| \ \|_*$, we obtain (cf. [Pi] for instance)

$$\int \left\| \sum_{i=1}^{m} \lambda_i g_i(\omega) e_i \right\|_* \lesssim \left(\log \frac{C}{\rho} \right) m \tag{2.10}$$

$$\int \left\| \sum_{i=1}^{m} \lambda_i^{-1} g_i(\omega) e_i \right\|_{(E, \| \ \|_*)^*} < C. \tag{2.11}$$

Here $m = \dim E, (e_i)_{1 \leq i \leq m}$ is an appropriate $0B$ in E and $\lambda_1 \leq \lambda_2 \leq \cdots \leq \lambda_m$.

We first exploit (2.11). Fix $0 < \varepsilon < 1$. From the M_*-lower bound, there is a subspace E_1 of E,

$$\dim E_1 > (1 - \varepsilon)m \tag{2.12}$$

such that

$$\|x\|_* \geq c\varepsilon^{1/2} \sqrt{m} \left(\sum \lambda_i^{-2} x_i^2 \right)^{1/2} \quad \text{for } x \in E_1. \tag{2.13}$$

Thus we use here the "M_*-result" (see again [Pi]) to a subspace of $\mathbb{R}^n, \| \ \|_*$.

Next, we use (2.10) together with (2.1) (which remains obviously true with K replaced by K_1). Thus, letting in (2.1) $\| \ \| = \| \ \|_*$ and $v_i = \lambda_i e_i$, it follows that

$$\int_{K_1} \left\| \sum_{i=1}^{m} \lambda_i \frac{x_i}{L} e_i \right\|_* dx \lesssim \left(\log \frac{C}{\rho} \right) Am$$

$$\frac{L}{2} \sum \lambda_i < \sum \lambda_i \left(\int_{K_1} \frac{x_i^2}{L} dx \right) \lesssim \left(\log \frac{C}{\rho} \right) Am \tag{2.14}$$

where we also use (2.5).

From (2.14), (2.13)

$$\lambda_{(1-\varepsilon)m} \lesssim \frac{A}{\varepsilon L} \left(\log \frac{C}{\rho} \right)$$

$$\|x\|_* \geq c\frac{L}{A} \varepsilon^{3/2} \left(\log \frac{C}{\rho} \right)^{-1} \sqrt{n} \left(\sum_{i \leq (1-\varepsilon)m} x_i^2 \right)^{1/2} \quad \text{for } x \in E_1.$$

Restrict further x to the space $E_2 = E_1 \cap [e_i | i \leq (1 - \varepsilon)m)]$. Then

$$\dim E_2 \geq (1 - 2\varepsilon)m$$

and for $x \in E_2$

$$\|x\|_* \geq c\varepsilon^{3/2} \left(\log \frac{1}{\rho} \right)^{-1} \frac{L}{A} \sqrt{n} |x|. \tag{2.15}$$

Thus let $\varepsilon \sim \delta, F = E_2$ and (2.15) gives (2.9). This proves the Lemma.

(iii) Notice that (2.5) implies in particular that for $|\xi| = 1$

$$\|\xi\|_* \geq \frac{L}{\sqrt{2}}$$

so that for $E_0 = R^n$, (2.7) holds with $\rho = \rho_0 \sim \frac{1}{\sqrt{n}}$. We then perform the "usual" flag construction

$$E_0 \supset E_1 \supset E_2 \supset \cdots \supset E_s \supset E_{s+1} \supset \cdots$$

of subspaces E_s, $\dim E_s = m_s > \frac{1}{2}n$, using the Lemma.

Assume

$$\|x\|_* \geq \rho_s L \sqrt{n} |x| \text{ for } x \in E_s. \tag{2.16}$$

Take $\delta_{s+1} = \left(\log \frac{1}{\rho_s}\right)^{-2}$, so that by (2.8)

$$\dim E_{s+1} = m_{s+1} > \left(1 - \frac{1}{(\log \frac{1}{\rho_s})^2}\right) m_s \tag{2.17}$$

and, by (2.9)

$$\rho_{s+1} \sim \left(\log \frac{1}{\rho_s}\right)^{-5} A^{-1}. \tag{2.18}$$

It follows from (2.17), (2.18) that then, assuming $\rho_{s-1} < A^{-2}$

$$\dim E_s > \left[\prod_{s' < s} \left(1 - \frac{1}{(\log \frac{1}{\rho_{s'}})^2}\right)\right] n$$
$$> \left(1 - \frac{2}{(\log \frac{1}{\rho_{s-1}})^2}\right) n$$

hence

$$m_s > \left(1 - C(A\rho_s)^{1/3}\right) n. \tag{2.19}$$

Regarding volume, we get (cf. [Pi])

$$V(K_1^0) > \left(\frac{c}{n}\right)^n \quad \text{(by reverse Santalo inequality)} \tag{2.20}$$

$$V(K_1^0 \cap E_s) \leq V(K_1^0 \cap E_{s+1}) \cdot V\left(P_{E_{s+1}^\perp \cap E_s}(K_1^0 \cap E_s)\right). \tag{2.21}$$

Since by (2.16)

$$K_1^0 \cap E_s \subset \frac{1}{\rho_s L \sqrt{n}} B \tag{2.22}$$

($B=$ Euclidean ball),
we get

$$V\big(P_{E_s \cap E_{s+1}^\perp}(K_1^0 \cap E_s)\big) < \left(\frac{C}{\rho_s L\sqrt{n}\sqrt{m_s - m_{s+1}}}\right)^{m_s - m_{s+1}}. \tag{2.23}$$

By iteration of (2.21) and (2.22), (2.23)

$$V(K_1^0) \leq V(K_1^0 \cap E_s) \prod_{s' \leq s} V\big(P_{E_{s'-1} \cap E_{s'}^\perp}(K_1^0 \cap E_{s'-1})\big)$$

$$\leq \left(\frac{C}{\rho_s L n}\right)^{m_s} \prod_{s' \leq s} \left(\frac{C}{\rho_{s'-1} L\sqrt{n}\sqrt{m_{s'-1} - m_{s'}}}\right)^{m_{s'-1} - m_{s'}}$$

$$\leq \left(\frac{1}{\rho_s}\right)^{m_s} \left(\frac{C}{Ln}\right)^n \left[\prod_{s' \leq s} \left(\frac{1}{\rho_{s'-1}}\right)^{n(\log \rho_{s'-1}^{-1})^{-2}}\right]$$

$$\left[\prod_{s' \leq s} \left(\frac{n}{m_{s'-1} - m_{s'}}\right)^{\frac{m_{s'-1} - m_{s'}}{2}}\right]$$

$$< \left(\frac{1}{\rho_s}\right)^{m_s} \left(\frac{C}{Ln}\right)^n \tag{2.24}$$

where we also used (2.17),(2.18).

We have chosen s such that $\rho \sim c(A)$, $m_s > \frac{n}{2}$ (cf. (2.19)).

Finally, from (2.20), (2.24)

$$L < C\rho_s^{-\frac{m_s}{n}} < C(A)$$

proving part (1.4) of the Theorem.

(iv) Finally, we prove the Theorem in the more precise form (1.5). The construction in (iii) terminates at s such that $\rho_s \geq A^{-2}$. Inequalities (2.20), (2.24) clearly imply

$$\left(\frac{C}{n}\right)^n < V(K_1^o \cap E_s)C^n \left(\frac{1}{Ln}\right)^{n-m_s}$$
$$V(K_1^o \cap E_s) > c^n n^{-m_s} L^{n-m_s}. \tag{2.25}$$

Take then in the lemma $E = E_s, m = m_s, \rho = \rho_s \geq A^{-2}$.

It follows from (2.11) and Sudakov's inequality that

$$\frac{V(K_1^o \cap E)}{V(B_m)} < C^m \prod_{i=1}^m \frac{\lambda_i}{\sqrt{m}} \tag{2.26}$$

and from (2.25), (2.26)

$$L^{n-m} < C^m \prod \lambda_i$$

hence

$$\frac{1}{m} \sum_{i=1}^{m} \lambda_i > cL^{\frac{n-m}{m}}.$$

Substituting this last inequality in (2.14) then indeed gives

$$L \lesssim L^{\frac{n}{m}} \lesssim A.\log(1 + A).$$

This proves the Theorem.

3 Remarks

(1) A. Giannopoulos [G] kindly pointed out to the author that if K is a ψ_2-body, then K is in fact already contained in a ball of radius $CAL\sqrt{n}$ and hence, by the Theorem, in a ball of radius $CA^2 \log A\sqrt{n}$ (in particular, K° has finite volume ratio). We repeat his argument.

As a consequence of the log-concavity of the section function, we have

$$\|\xi\|_{K^\circ} \leq \|\langle x, \xi \rangle\|_{L^n(K)}.$$

Hence, if K satisfies (1.2)

$$\|\xi\|_{K^\circ} \leq C\sqrt{n}AL|\xi|. \tag{3.1}$$

Thus, if we don't care about the final estimate $C(A)$ in (1.4), replacement of K by K_1 in the proof of the Theorem is unnecessary.

(2) Observe also that we only used the bound

$$\|\langle x, \xi \rangle\|_{L^{\psi_2}(K_1)} \leq AL|\xi| \tag{3.2}$$

with K_1 defined as above. Writing for $|\xi| \leq 1$ and $x \in K_1$

$$|\langle x, \xi \rangle| < C\sqrt{n}L \qquad \text{(by (1.4))}$$

$$\int_K \left\{ \exp c\frac{|\langle x, \xi \rangle|^2}{\sqrt{n}L^2} \right\} dx \leq \int_K \left\{ \exp c'\frac{|\langle x, \xi \rangle|}{L} \right\} dx < C$$

(3.2) therefore holds with $A \sim n^{1/4}$ (without further assumptions).

The general bound (1.1) is then implied by (1.5).

References

[B-K] Barthe, F., Koldobsky A. Extremal slabs in the cube and the Laplace transform. Adv. Math., to appear

[Bo] Bourgain, J. (1991): On the distribution of polynomials on high-dimensional convex sets. Lecture Notes in Math., **1469**, 127–137

[Fe]   Fernique, X. (1975): Régularité des trajectories des fonctions aléatoires gaussiennes. Lecture Notes in Math., **480**, 1–96

[G]   Giannopoulos, A. Private communication

[L-T]   Ledoux, M., Talagrand, M. (1991): Probability in Banach Spaces. Isoperimetry und processes, Ergebnisse der Mathematik and ihrer Grenzgebiete (3), 23, Springer

[Pi]   Pisier, G. (1989): The Volume of Convex Bodies and Banach Space Geometry. Cambridge Tracts in Math., **94**, Cambridge University Press

[Pr1]   Preston, C. (1971): Banach spaces arising from some integral inequalities. Indiana Math. J., **20**, 997–1015

[Pr2]   Preston, C. (1972): Continuity properties of some Gaussian processes. Ann. Math. Statist., **43**, 285–292

[T]   Talagrand, M. (1987): Regularity of Gaussian processes. Acta Math., **159**, 99–149

On the Sum of Intervals

E.D. Gluskin

School of Mathematical Sciences, Tel Aviv University, Tel Aviv 69978, Israel
gluskin@post.tau.ac.il

> *To mark the 70th birthday of my*
> *dear teacher M.Z. Solomyak*

Let $u_1, u_2, \ldots, u_N \in S^{n-1}$ be a sequence of N unit vectors. A body $V = V(u_1, \ldots, u_N)$ is defined as follows:

$$V = \left\{ \lambda_1 u_1 + \lambda_2 u_2 + \cdots + \lambda_N u_N : |\lambda_i| \leq 1, i = 1, \ldots, N \right\}. \tag{1}$$

In geometric language the body V is the Minkowski sum of N intervals with endpoints $\pm u_i$ correspondingly. The main purpose of this note is to investigate the order of the aspherical constant $d(V)$ of the body $V(u_1, \ldots, u_N)$ for an optimal choice of vectors $u_1, \ldots, u_N$. Let us recall that the aspherical constant $d(V)$ of a central symmetric body V is defined as follows

$$d(V) = \inf \left\{ \frac{R}{r} : rD \subset V \subset RD \right\},$$

where $D \subset \mathbb{R}^n$ is the unit Euclidean ball. The optimal value of $d(V(u_1, \ldots, u_N))$, which will be denoted by $d_{n,N}$, describes the rate of approximation of the Euclidean ball by zonotopes. The quantity $d_{n,N}$ was studied by many authors, especially for the case $N/n \to \infty$ (see [FLM], [BeMc], [BLM]). The problem of the precise bounds of $d_{n,N}$ for all N, n was in particular discussed by Milman [M]. To answer his question, we prove the following result.

Proposition 1. *Let n, N be some integers with $n < N$. Then for some universal constant $C > 0$, the following inequality holds:*

$$\frac{1}{C} \min \left\{ \sqrt{n}, 1 + \sqrt{\frac{N}{N-n} \log \frac{N}{N-n}} \right\}$$
$$\leq \inf_{(u_1, \ldots, u_N) \in (S^{n-1})^N} d(V(u_1, \ldots, u_N))$$
$$\leq C \min \left\{ \sqrt{n}, 1 + \sqrt{\frac{N}{N-n} \log \frac{N}{N-n}} \right\}.$$

It is not surprising that the order of $d_{n,N}$ coincides with one of $\sqrt{N}$ times $(N-n)$-Kolmogorov width of an N-dimensional Euclidean ball with respect to an ℓ_∞ metric provided $N \leq 2n$. The close connection between these problems is rather well known, but we do not know any explicit treatment of the relation between both problems. The main part of the paper presents some improvement of the [GG] arguments, which immediately leads to the proof of Theorem 1. Incidentally, a new bound of the Kolmogorov numbers is obtained for some class of linear operators. For the convenience of the reader who isn't familiar with the widths, we mostly avoid using the width theory terminology.

At the end of the paper, we consider the sums of the intervals of the different lengths. It turns out that the same bounds as in Proposition 1 hold true in this general situation (see Proposition 4 and the Remark following it).

We use the following notation: $[a]$ is an integral part of number a, $\binom{m}{n} = \frac{m!}{n!(m-n)!}$ is a binomial coefficient, $[1 : N]$ stands for the set of all integers between 1 and N.

The standard basis of $\mathbb{R}^N$ will be denoted by (e_i). As usual, we do not notate the dimension of the space but use the same symbol (e_i) to denote the bases of the different Euclidean spaces. The coordinates of a vector $x \in \mathbb{R}^N$ will be denoted by x_i or by $x(i)$. S^{N-1} is the unit sphere of $\mathbb{R}^N$, while μ is a normalized Lebesgue measure on it.

For $x \in \mathbb{R}^N$, we define as usual $\|x\|_p = (\sum_{i=1}^N |x_i|^p)^{1/p}$, for $1 \leq p < \infty$ and $\|x\|_\infty = \max_{i=1}^N |x_i|$. ℓ_p^N stands for $\mathbb{R}^N$ occupied with the norm $\|\cdot\|_p$. The unit ball of a Banach space X is denoted by B_X. Let $\mathcal{B}$ be a subset of finite dimensional normed space X and $L \subset X$ be some linear subspace of X. The deviation of $\mathcal{B}$ from L is defined as follows:

$$\rho_X(\mathcal{B}, L) = \sup_{x \in \mathcal{B}} \inf_{y \in L} \|x - y\| .$$

Using the Hahn–Banach theorem, it is not difficult to see that[1]

$$\rho_X(\mathcal{B}, L) = \sup_{f \in L^\perp \cap B_{X^*}} \|f\|_{\mathcal{B}^0} , \tag{2}$$

where $L^\perp = \{f \in X^* : f|_L = 0\}$ is the annihilator of L and the seminorm $\|\cdot\|_{\mathcal{B}^0}$ is defined by

$$\|f\|_{\mathcal{B}^0} = \sup_{x \in \mathcal{B}} |f(x)| .$$

For a linear operator $T : \mathbb{R}^n \to \mathbb{R}^N$ we denote by $\alpha(T)$ and $\beta(T)$ the following quantities:

$$\alpha(T) = \sup_{\xi \in S^{n-1}} \|T\xi\|_1 , \quad \beta(T) = \sup_{\xi \in S^{n-1}} \frac{1}{\|T\xi\|_1} . \tag{3}$$

Certainly $\alpha(T)$ is the norm of the operator T from ℓ_2^n to ℓ_1^N. Note that by duality, one can compute the aspheric constant of a body V, symmetric with respect to the origin, as follows:

---

[1] This duality relation is a very important tool in width theory (see e.g. [I],[ST]).

$$d(V) = \frac{\sup_{\xi \in S^{n-1}} \sup_{x \in V} |\langle x, \xi \rangle|}{\inf_{\xi \in S^{n-1}} \sup_{x \in V} |\langle x, \xi \rangle|} .$$

So, if V is given by (1) and linear operator T is defined by $T^* e_i = u_i$, then

$$d(V) = \alpha(T)\beta(T) . \tag{4}$$

Therefore, Proposition 1 follows immediately from the next three statements.

Lemma 1. *For any integer n and N and for any linear operator $T : \mathbb{R}^n \to \mathbb{R}^N$, the following inequality holds*

$$\alpha(T) \geq \frac{1}{2\sqrt{n}} \sum_{i=1}^{N} \|T^* e_i\|_2 .$$

Lemma 2. *For any integer n and N s.t. $n < N < \frac{6}{5}n$ and for any linear operator $T : \mathbb{R}^n \to \mathbb{R}^N$ s.t. $\|T^* e_i\|_2 = 1$ for $i = 1, \ldots, N$, the following inequality holds*

$$\beta(T) \geq C_1 \left\{ \min \left\{ 1, \sqrt{\frac{1}{N-n} \log \left(\frac{N}{N-n} \right)} + 1 \right\} \right\} ,$$

where C_1 is a universal constant.

Proposition 2. *For any integer n and N, $n < N < 4n/3$, there exists a linear operator $T : \mathbb{R}^n \to \mathbb{R}^N$ s.t. $\|T^* e_i\|_2 = 1$ for $i = 1, \ldots, N$ and*

$$\alpha(T)\beta(T) \leq \min \left\{ \sqrt{n}, C_2 \sqrt{\frac{N}{N-n}} \log \frac{N}{N-n} \right\} ,$$

where C_2 is a universal constant.

Remark. Lemma 1 as well as Proposition 2 are well known to experts. For the convenience of the reader, we present their proofs below.

Proof of Proposition 1. Due to equality (4) and Lemmas 1 and 2, the left-hand inequality holds for $N < 6n/5$. But for $N \geq 6n/5$ it is reduced to the inequality $d(V) \geq 1$, which holds for any V. By (4) and Proposition 2, the right-hand inequality holds for $N < 4n/3$. To complete the proof, it is enough to observe that $d_{n,k} \leq 2d_{n,N}$ for any $k > N$. Indeed this inequality follows immediately from the definitions for $k < 2N$ and we conclude by applying the inequality $d_{n,2N} \leq d_{n,N}$.   □

Proof of Lemma 1. Recall that for any vector $u \in \mathbb{R}^n$ one has (see e.g. (2.15) of [FLM])

$$\int_{S^{n-1}} |<u, \xi>| d\mu(\xi) = \frac{\|u\|_2}{\sqrt{\pi}} \frac{\Gamma(\frac{n}{2})}{\Gamma(\frac{n+1}{2})} \geq \frac{\|u\|_2}{2\sqrt{n}} .$$

It follows that

$$\alpha(T) \;=\; \sup_{\xi \in S^{n-1}} \|T\xi\|_1 \geq \int_{S^{n-1}} \|T\xi\|_1 d\mu(\xi)$$

$$= \sum_{i=1}^{N} \int_{S^n} |\langle \xi, T^* e_i \rangle| d\mu(\xi) \geq \frac{1}{2\sqrt{n}} \sum_{i=1}^{N} \|T^* e_i\|_2 \;.$$

$\square$

For the proof of Lemma 2, we need the following elementary and fairly well-known fact (see e.g. [GG], Lemma 1). For the reader's convenience, we reproduce here its proof.

Lemma 3. *Let K be a subset of some Banach space X. Suppose that for some positive $\varepsilon > 0$, $\kappa > 0$ and some integer k, the set K contains $M > (2 + \kappa)^k$ points $x_1, \ldots, x_M \in K$, s.t. $\|x_i - x_j\| > 4\varepsilon$, for any $i \neq j$, $1 \leq i, j \leq M$, and $\|x_i\| < \kappa\varepsilon$ for any i, $1 \leq i \leq M$. Then for any k-dimensional subspace $L \subset X$ one has*

$$\rho_X(K, L) \geq \varepsilon \;.$$

Proof. On the contrary, suppose that there exists a subspace L with

$$\dim L = k < \frac{\log |M|}{\log(2 + \kappa)} \quad \text{s.t.} \quad \rho_X(K, L) < \varepsilon \;.$$

Then for $i = 1, \ldots, M$ there exists $y_i \in L$ s.t. $\|y_i - x_i\| < \varepsilon$. We have

$$\|y_i\| \leq \|y_i - x_i\| + \|x_i\| < (1 + \kappa)\varepsilon \;.$$

On the other hand, for $i \neq j$

$$\|y_i - y_j\| \geq \|x_i - x_j\| - \|x_i - y_i\| - \|x_j - y_j\| > 2\varepsilon \;.$$

So the union of balls $y_i + \varepsilon(B_X \cap L)$ is disjoint and is contained in the set $(2 + \kappa)\varepsilon(B_X \cap L)$. A comparison of volumes leads to the following inequality

$$M\varepsilon^k = \frac{1}{\text{vol}(B_X \cap L)} \text{vol}\left(\bigcup_{i=1}^{M} (y_i + \varepsilon(B_X \cap L)) \right)$$

$$\leq \frac{1}{\text{vol}(B_X \cap L)} \text{vol}((2 + \kappa)\varepsilon(B_X \cap L)) = (2 + \kappa)^k \varepsilon^k,$$

which contradicts the condition $M > (2 + \kappa)^k$. $\square$

Remark. Certainly the last part of the proof is a well-known Kolmogorov volumetric bound for ε-entropy of the ball ([KT], see also [N]).

Proof of Lemma 2. Let us observe that by (2), one has

$$\beta(T) = \sup_{x \in ImT} \frac{\|T^{-1}x\|_2}{\|x\|_1} = \rho_{\ell_\infty^N}\left(V, (ImT)^\perp\right), \tag{5}$$

where $V = \{f : \|T^*f\|_2 \leq 1\}$ and $(ImT)^\perp$ is the orthogonal complement of the image of operator T. By the parallelogram law, for any set A of integers $A \subset [1 : N]$ one has

$$\frac{1}{2^{|A|}} \sum_{\varepsilon_i = \pm 1} \left\| \sum_{i \in A} \varepsilon_i T^* e_i \right\|_2^2 = \sum_{i \in A} \|T^* e_i\|_2^2 = |A|$$

(the outer sum here is taken over all $2^{|A|}$ possible choices of the signs). Therefore, for any $A \subset [1 : N]$ there exists a vector $x_A \in \mathbb{R}^N$. s.t.

$$|x_A(i)| = 1 \text{ for } i \in A, \quad |x_A(i)| = 0 \text{ for } i \notin A \text{ and } \|T^* x_A\|_2 \leq \sqrt{|A|} \,.$$

The last inequality means that $|A|^{-1/2} x_A \in V$. Set $k = N - n$ and let ℓ be a minimal integer satisfying

$$8^k \leq \left(\frac{N}{3k}\right)^\ell,$$

that is, $\ell = 1 + [3k \log 2 / \log(N/3k)]$. The condition $N < 6n/5$ implies an inequality $k < N/6$. Therefore, $\ell \leq 3k < N/2$. Hence the following inequality holds

$$8^k \leq \left(\frac{N}{\ell}\right)^\ell < \binom{N}{\ell}.$$

It follows that the family of the vectors $|A|^{-1/2} x_A$, where A runs over all subsets of $[1 : N]$ with ℓ elements, satisfies all conditions of Lemma 3 with

$$X = \ell_\infty^n \,, \quad \varepsilon = 1/4\sqrt{\ell} \,, \quad \kappa = 4 \,, \quad k = N - n$$

and $M = \binom{N}{\ell}$. Consequently for any subspace, $L \subset \ell_\infty^N$ of dimension k, one has

$$\rho_{\ell_\infty^N}(V, L) \geq \frac{1}{4\sqrt{\ell}} = \frac{1}{4}\left(\left\lceil \frac{3k \log 2}{\log \frac{N}{3k}} \right\rceil + 1\right)^{-1/2}.$$

In particular, the last inequality holds for $L = (ImT)^\perp$ and we conclude by Eq. (5).   □

In fact the proof of Lemma 2 leads to some result on the Kolmogorov width. To formulate it, let us recall that for a given normed space X, a natural number n and $1 \leq p \leq 2$, the constant $T_p(X, n)$ is the smallest T such that (see e.g. [MS], n. 9)

$$\left(\frac{1}{2^n} \sum_{\varepsilon_i = \pm 1} \left\| \sum_{i=1}^n \varepsilon_i x_i \right\|^2\right)^{1/2} \leq T \left(\sum_{i=1}^n \|x_i\|^p\right)^{1/p}$$

for all $x_1, \ldots, x_n \in X$ (the first sum here is taken over all 2^n possible choices of signs).

Recall that for a given subset K of a normed space Y, the n-Kolmogorov width $d_n(K, Y)$ is defined as follows:

$$d_n(K, Y) = \inf_{\substack{L \subset Y \\ \dim L \leq n}} \rho_X(K, L) \,,$$

where the inf is taken over a collection of all linear suspaces $L \subset Y$ of dimension $\dim L \leq n$. Next for a linear operator S from $\mathbb{R}^d$ to some normed space X we denote by $K_S \subset \mathbb{R}^d$ the following set

$$K_S = \left\{ x \in \mathbb{R}^d : \|Sx\|_X \leq 1 \right\} \,.$$

Proposition 3. *Let S be a linear operator from $\mathbb{R}^d$ to some normed space X such that for at least N ($N \leq d$) indices i one has $\|Se_i\|_X \leq 1$. Then for any $k \leq N/6$ and any p, $1 \leq p \leq 2$, the following inequality holds*

$$d_k(K_S, \ell_\infty^d) \geq \frac{c}{T_p(X, n)} \min \left\{ 1, \left(\frac{\log \frac{N}{k}}{k} \right)^{1/p} \right\} ,$$

where c is some universal constant and $n = 1 + [3k \frac{\log 2}{\log(\frac{N}{3k})}]$.

We omit the proof of this proposition, which repeats that of Lemma 2.

Proof of Proposition 2. It is fairly well known that for $k < N/4$ with some universal constant c_3, the following inequality holds:[2]

$$\mu \left\{ x \in S^{N-1} : \sum_{i=1}^{k} |x_i|^2 > 1/2 \right\} \leq e^{-c_3 N} \,.$$

Next, as usual, $O(N)$ stands for the group of all orthogonal operators in $\mathbb{R}^N$, while ν_N (or just ν) stands for a normalized Haar measure on $O(N)$. For $W \in O(N)$ we denote by $w_j \in \mathbb{R}^N$ the jth column of the matrix W and by (w_{ij}) its entries. Note that when W runs on $O(N)$, any of its rows runs on S^{N-1} and the measure ν on $O(N)$ thus induces the measure μ on S^{N-1}. It follows that $\nu\{W \in O(N)$: for any i, $i = 1, \ldots, N$ $\sum_{j=n+1}^{N} |w_{ij}|^2 \leq 1/2\} > 1 - Ne^{-c_3 N}$. For N large enough the last quantity is bigger than $1/2$. Now, it is known[3] that for some universal constant C_4, the ν-measure $W \in O(N)$ such that

---

[2] This fact follows easily from the concentration measure phenomenon for the sphere S^{N-1} (see (2.6) of [FLM]). Certainly it can be proven by direct computation (see e.g. [A])

[3] This fact is proven in [G] [GG]. See also [Mk] for a simple proof and [GM] for a general discussion of the problem. We also wish to mention here that [G] was greatly influenced by Kashin's work [K] and its dual exposition to Mityagin [Mit].

$$\sup_{(\xi_1,\ldots,\xi_n)\neq 0} \frac{\left\| \sum_{j=1}^{n} \xi_j w_j \right\|_2}{\left\| \sum_{j=1}^{n} \xi_j w_j \right\|_1} \leq \min\left\{ 1, C_4 \sqrt{\frac{\log \frac{N}{N-n}}{N-n}} \right\}, \qquad (6)$$

is exponentially close to 1 and in particular is bigger than $1/2$ for N large enough. So for N large enough there exists an operator $W \in O(N)$ satisfying (6) and s.t. for any i, $i = 1, \ldots, N$,

$$\sum_{j=1}^{n} |w_{ij}|^2 = 1 - \sum_{j=n+1}^{N} |w_{ij}|^2 > \frac{1}{2}. \qquad (7)$$

It is quite easy to construct for any N an operator $W \in O(N)$, satisfying (7) only. For example, if $N = 2k$ is even, one can use the operator given by the matrix

$$\frac{1}{\sqrt{2}} \begin{pmatrix} I_k & -I_k \\ I_k & I_k \end{pmatrix},$$

where I_k is the $k \times k$ unit matrix. It is clear that for bounded N any operator $W \in O(N)$ satisfies (6), probably with a bigger constant. Thus, through a correction of the constant C_4, one gets for any n and $N < 4n/3$ operators, $W \in O(N)$, satisfying both (6) and (7). Now let $S : \mathbb{R}^n \to \mathbb{R}^N$ be defined by $Se_i = w_i, i = 1, \ldots, n$, where W satisfies (6) and (7). Since the vectors w_j are orthonormal, we have $\alpha(S) \leq \sqrt{N}$. On the other hand, the inequality (6) is equivalent to

$$\beta(S) \leq \min\left\{ 1, C_4 \sqrt{\frac{1}{N-n} \log \frac{N}{N-n}} \right\}$$

while (7) means that

$$\frac{1}{\sqrt{2}} \leq \|S^* e_j\|_2 \leq 1 \qquad j = 1, \ldots, N.$$

Therefore operator T defined by

$$T^* e_j = \frac{1}{\|S^* e_j\|_{\ell_2^n}} S^* e_j,$$

gives the desired example.   □

Proposition 4. *Let T be a linear operator from $\mathbb{R}^n$ to $\mathbb{R}^N$ with $\alpha(T) \leq 1$. Then*

$$\beta(T) \geq c \min\left\{ \sqrt{n}, 1 + \sqrt{\frac{N}{N-n} \log \frac{N}{N-n}} \right\},$$

where $c > 0$ is some universal constant.

Proof. Let us denote by $\gamma_2(T)$ the Hilbert–Schmidt norm of the operator T:

$$\gamma_2(T) = \left(\sum_{i=1}^{N} \|T^*e_i\|_2^2 \right)^{1/2} .$$

It is well known that $\gamma_2(T) \leq \alpha(T)$. Indeed, by the parallelogram law,

$$\gamma_2(T) = \left(\frac{1}{2^N} \sum_{\varepsilon_i=\pm 1} \left\| T^* \left(\sum_{i=1}^{N} \varepsilon_i e_i \right) \right\|_2^2 \right)^{1/2} .$$

Then by duality

$$\|T^*x\|_2 \leq \alpha(T)\|x\|_\infty$$

for all $x \in \mathbb{R}^N$ and the inequality $\gamma_2(T) \leq \alpha(T)$ follows. So for T satisfying $\alpha(T) \leq 1$ there exists at least $[N/2]+1$ indices i such that $\|T^*e_i\|_2 \leq \sqrt{2/N}$. Applying Proposition 3 to the operator $S = \sqrt{N/2}\, T^*$, with $d = N$ and $[N/2] + 1$ instead of N, one gets

$$d_{N-n}\left(\{f : \|T^*f\|_2 \leq 1\},\ \ell_\infty^N \right) \geq c\sqrt{N/2}\, \min\left\{ 1,\ \sqrt{\frac{\log(\frac{N}{2(N-n)})}{N-n}} \right\} ,$$

provided that $N - n \leq N/12$. Due to (5), the last inequality implies the desired estimate for $n \geq \frac{11}{12}N$. Now it is enough to use the elementary bound $\alpha(T) \cdot \beta(T) \geq 1$ to complete the proof for $n < \frac{11}{12}N$.   □

Remarks. 1. By (4), Proposition 4 gives the desired bound for the aspherical constant of the sum of N arbitrary length intervals.
2. Using the factorization theorem [Mau] instead of the inequality $\alpha(T) \geq \gamma_2(T)$ permits the proof of Proposition 4 to be reduced to [GG] directly.

Acknowledgement. I would like to take this opportunity to express my gratitude to Vitali Milman for stimulating this work. I would also like to thank the referee for his valuable remarks.

References

[A]   Arstein, S. Proportional concentration phenomena on the sphere. Israel J. of Math., to appear
[BeMc] Betke, V., McMullen, P. (1983): Estimating the sizes of convex bodies for projections. J. London Math. Soc., **27**, 525–538
[BLM]   Bourgain, J., Lindenstrauss, J., Milman, V. (1989): Approximation of zonoids by zonotops. Acta Math., **162**, 73–141
[FLM]   Figiel, T., Lindenstrauss, J., Milman, V. (1977): The dimension of almost spherical sections of convex bodies. Acta Math., **129**, 53–94

[GG]   Garnaev, A.Yu., Gluskin, E.D. (1984): On widths of the Euclidean ball. Dok. Akad. Nauk SSSR, **277**, 1048–1052; English transl. in Soviet Math. Dokl., **30** (1984)

[GM]   Giannopoulos, A.A., Milman, V.D. (1998): Mean width and diameter of proportional sections of a symmetric convex body. J. reine angew. Math., **497**, 113–139

[G]   Gluskin, E.D. (1983): Norms of random matrices and widths of finite-dimensional sets. Mat. Sb. **120(162)**, 180–189; English transl. in Math. USSR Sb., **48** (1984)

[I]   Ismagilov, R.S. (1974): Diameters of sets in normed linear spaces and approximation of functions by trigonometric polynomials. Uspekhi Mat. Nauk, **29**, no. 3(177), 161–178; English transl. in Russian Math. Surveys, **29** (1974)

[K]   Kashin, B.S. (1977): Widths of certain finite-dimensional sets and classes of smooth functions. Izv. Acad. Nauk SSSR Ser. Mat., **41**, 334–351; English transl. in Math. USSR Izv., **11** (1977)

[KT]   Kolmogorov, A.N., Tikhomirov, V.M. (1959): ε-entropy and ε-capacity of sets in functional spaces. Uspekhi Mat. Nauk, **14**, No. 22(86), 3–86; English transl. in Amer. Math. Soc. Trans., (2)17 (1961)

[Mk]   Makovoz, Y. (1988): A simple proof of an inequality in the theory of n-widths. In: B. Seadov et al. (eds.) Constructive Theory of Functions. Sofia, 305–308

[Mau]   Maurey, B. (1974): Théoremes de factorisation pour les opérateurs linéaires a valeurs dans les espaces L_p. Astérisque, No. 11

[M]   Milman, V. (2000): Topics in asymptotic geometric analysis. GAFA 2000, Visions in Mathematics, Special Volume, Part II, 792–815

[MS]   Milman V.D., Schechtman G. (1986): Finite-dimensional normed spaces. Lecture Notes in Mathematics, **1200**, Springer-Verlag, Berlin

[Mit]   Mityagin, B.S. (1977): Random matrices and subspaces. In: B.S. Mityagin (ed.) Geometry of Linear Spaces and Operator Theory. Yaroslav. Gos. Univ., Yaroslavé, 175–202 (Russian)

[N]   Neumann, J. von (1942): Approximative properties of matrices of high finite order. Port. Math., **3**, 1–62

[ST]   Solomiak, M.Z., Tihomirov, V.M. (1967): Geometric characteristics of the imbedding of the classes W_p^α in C. Izv., Vysš. Včebu. Zaved. Matematika, No. 10(65), 76–82 (Russian)

Note on the Geometric-Arithmetic Mean Inequality

E. Gluskin and V. Milman

School of Mathematical Sciences, Tel Aviv University, Tel Aviv 69978, Israel
gluskin@post.tau.ac.il, milman@post.tau.ac.il

In this note, we put together a few observations in the reverse direction in the classical geometric-arithmetic mean inequality which we will study in the form:

$$\sqrt{\frac{1}{n}\sum_1^n \lambda_i^2} \geq \left(\prod_1^n \lambda_i\right)^{1/n}, \qquad \lambda_i > 0. \tag{1}$$

We show that this inequality is, in fact, asymptotic equivalence with very high probability and also in some other sense connected with the linear structure of the vectors $\lambda = (\lambda_i) \in \mathbb{R}^n$. These observations are "standard" from the point of view of the Asymptotic Theory of Normed Spaces but may be useful for purely analytical purposes.

1. Let $x = (x_i)_1^n \in S^{n-1}$, i.e. $\sum_1^n x_i^2 = 1$. Then (1) states

$$\left(\prod_1^n |x_i|\right)^{1/n} \leq \frac{1}{\sqrt{n}}.$$

We equip S^{n-1} with the probability rotation invariant measure $\sigma(x)$.

Proposition 1. Prob$\{x \in S^{n-1} \mid (\prod |x_i|)^{1/n} < \theta/\sqrt{n}\} \leq (C\sqrt{\theta})^n$ *for some absolute constant* $C > 0$. *Say* $C = 1,6$ *suffices. (And therefore, the reverse geometric-arithmetic mean inequality holds for* $x \in S^{n-1}$: $(\prod |x_i|)^{1/n} \geq \theta/\sqrt{n} = \theta\, (\frac{1}{n}\sum_1^n x_i^2)^{1/2}$ *with the probability above* $1 - (C\sqrt{\theta})^n$.)

Proof. Let φ be a positive homogeneous degree α function on $\mathbb{R}^n$, i.e. $\varphi(tx) = t^\alpha\varphi(x)$, $t > 0$. Then for any continuous positive function $f : \mathbb{R} \to \mathbb{R}^+$, one has

$$\int_{\mathbb{R}^n} \varphi(x)f(|x|)dx = \int_0^\infty r^\alpha \cdot r^{n-1}f(r)dr \int_{x \in S^{n-1}} \varphi(x)dx.$$

Apply this formula to the functions $\varphi(x) = (\prod_1^n |x_i|)^p$ and $f(r) = e^{-r^2/2}$. Then $\alpha = np$ and we have

$$\int_{\mathbb{R}^n} \prod |x_i|^p e^{-\sum |x_i|^2/2}dx = \int_{S^{n-1}} \prod |x_i|^p \frac{dx}{\sigma_{n-1}} \cdot \sigma_{n-1} \int_0^\infty r^{n+np-1}e^{-r^2/2}dr, \tag{2}$$

where σ_{n-1} denotes the Lebesgue measure of S^{n-1}.

The last integral in (2) as well as the left hand one are easily expressed in terms of Γ-functions. So, we see that

$$\int_{S_{n-1}} \left(\prod |x_i| \right)^p d\sigma(x) = \frac{2}{\sigma_{n-1}} \cdot \frac{\Gamma(\frac{p+1}{2})^n}{\Gamma(\frac{n+np}{2})} .$$

In particular, taking $p = -1/2$ and recalling that $\sigma_{n-1} = \frac{2\pi^{n/2}}{\Gamma(\frac{n}{2})}$ we obtain

$$\mathbb{E}_{S^{n-1}} \left(\prod |x_i|^{-1/2} \right) = \int_{S^{n-1}} \prod |x_i|^{-1/2} d\sigma(x) = \frac{\Gamma(\frac{n}{2})}{\Gamma(\frac{n}{4})} \cdot \frac{\Gamma(\frac{1}{4})^n}{\pi^{n/2}} \leq (Cn^{1/4})^n$$

for some absolute constant $C > 0$ (one may take $C = \frac{\Gamma(\frac{1}{4})}{\sqrt{\pi}e^{1/4}} \sim 1,593$). Therefore, probability

$$P\left\{ x \in S^{n-1} \mid \prod |x_i|^{-1/2} > \left(\frac{\sqrt{n}}{\theta}\right)^{n/2} \right\} = P\left\{ x \in S^{n-1} \mid \prod |x_i|^{1/n} < \frac{\theta}{\sqrt{n}} \right\}$$
$$\leq (C\sqrt{\theta})^n ,$$

by the Chebyshev inequality.

2. Since the first investigations on Dvorezky's Theorem (see [M] and the notion of spectrum there), it has become a common fact in Asymptotic Theory (see [FLM], [MS], [K]) that if some given functional on $\mathbb{R}^n$ has a sharp concentration then there also exists a subspace of proportional dimension θn, $0 < \theta < 1$, such that the restriction of this functional on its unit sphere is almost a constant.

However, interestingly and obviously, it is not so for the geometric mean. Indeed, if a subspace $L \subset \mathbb{R}^n$ has dimension $\dim L \geq 2$ then there is $x \in L \cap S^{n-1}$ such that at least one of its coordinates x_i is zero and $\prod_i^n |x_i| = 0$.

To avoid this obstacle one can consider a slightly different functional

$$\varphi(x) = \max_{1 \leq k \leq n} \sqrt{k} \left(\prod_1^k x_i^* \right)^{1/k} , \tag{3}$$

where as usual $(x_i^*)_{i=1}^n$ is a non-increasing rearrangement of the sequence $(|x_i|)_{i=1}^n$, i.e., say, $x_1^* = \max |x_i|$. Note that inequality (1) means that

$$\varphi(x) \leq \|x\|_{\ell_2} = \sqrt{\sum |x_i|^2} ,$$

and we are studying its reverse.

The functional φ is known to be equivalent to weak ℓ_2-norm

$$\|x\|_{\ell_{2,\infty}} := \max_k \frac{1}{\sqrt{k}} \sum_{i=1}^{k} x_i^* ,$$

and more precisely

$$\varphi(x) \le \|x\|_{\ell_{2,\infty}} \le 2\varphi(x) . \tag{4}$$

Indeed, the left side follows by the geometric-arithmetic mean inequality. Also $x_k^* \le \frac{\varphi(x)}{\sqrt{k}}$ for $k = 1, \ldots n$, and therefore

$$\sum_{1}^{k} x_i^* \le \varphi(x) \sum_{1}^{k} \frac{1}{\sqrt{i}} \le 2\varphi(x)\sqrt{k} ,$$

which implies the right side.

Certainly $\ell_{2,\infty}$ norm is very close to ℓ_2 norm. Particularly, for any vector for which ℓ_1 and ℓ_2 norms are equivalent the norms $\ell_{2,\infty}$ and ℓ_2 are also equivalent. The well known Kashin [K] theorem claims that also for some θn-dimensional subspaces ℓ_1^n and ℓ_2^n norms are equivalent (for an almost isometric corresponding fact, see [FLM] and for the right behavior of parameters when θ approaches 1, see [GG]). More precisely:
for any θ, $0 < \theta < 1$, there exists a subspace $L \subset \mathbb{R}^n$ of dimension $k = [\theta n]$ such that

$$\frac{1}{\sqrt{n}}\|x\|_{\ell_1^n} \le \|x\|_{\ell_2^n} \le \frac{C(\theta)}{\sqrt{n}}\|x\|_{\ell_1^n} \tag{5}$$

for any $x \in L$. The function $C(\theta)$ above depends only on θ and it is known ([GG]) that $C(\theta) \sim \sqrt{\frac{1}{1-\theta} \log \frac{1}{1-\theta}}$ when θ approaches 1. Also, (5) is satisfied for an exponentially close to 1 measure of k-dimensional subspaces (with probability above $1 - e^{-ck}$ for a universal number $c > 0$).

From (4) it follows that for such subspaces L and any $x \in L$ also

$$\varphi(x) \le \|x\|_{\ell_2^n} \le 2C(\theta)\varphi(x) .$$

In fact, the functional φ in the last inequalities may be changed to a smaller one which is more useful. Introduce for $m = [n/4C(\theta)^2]$

$$\widetilde{\varphi}(x) = \sqrt{m}\left(\prod_{1}^{m} x_i^*\right)^{1/m} .$$

Then for any $x \in \mathbb{R}^n$ one has

$$\|x\|_{\ell_1^n} \le (n - m)x_m^* + \sqrt{m}\|x\|_{\ell_2^n} .$$

Now, whenever (5) is satisfied the following inequality also holds

$$\|x\|_{\ell_2^n} \le C(\theta)\sqrt{n}x_m^* + \tfrac{1}{2}\|x\|_{\ell_2^n}$$

and consequently

$$\widetilde{\varphi}(x) \leq \|x\|_{\ell_2^n} \leq 2C(\theta)\sqrt{n}x_m^* \leq 4C^2(\theta)\widetilde{\varphi}(x) \ .$$

Summarizing the information we collected above we have

Proposition 2. *Let $f(x) = -\frac{1}{x}\log x$. For some universal constant $c > 0$, any integer n and any θ, $0 < \theta < 1$, $k = [\theta n]$, a random k-dimensional subspace $L \subset \mathbb{R}^n$ with high probability satisfies: for any $x \in L$*

$$\sqrt{m}\left(\prod_1^m x_i^*\right)^{1/m} \geq \frac{c}{f(1-\theta)}\|x\|_{\ell_2^n} \ ,$$

where $m = [cn/f(1-\theta)]$.

Kashin also proved that for a random orthogonal matrix $u \in O(n)$ with probability above $1 - e^{-cn}$ the following inequality holds:

$$\|x\|_{\ell_2^n} \leq \frac{c}{\sqrt{n}}\left(\|x\|_{\ell_1^n} + \|ux\|_{\ell_1^n}\right) \ .$$

Exactly as before one gets a similar corollary in our case.

Corollary. *There are universal constants $c_i > 0$, $i = 1, 2, 3$ such that for $k = [c_1 n]$ for a random operator $u \in O(n)$ with probability above $1 - e^{-c_2 n}$ and any $x \in \mathbb{R}^n$ either*

$$\sqrt{k}\left(\prod_1^k x_i^*\right)^{1/k} \geq c_3\|x\|_{\ell_2}$$

or this inequality is satisfied for the vector $y = ux$.

3. Note also that for any $C > 1$ the set of positive vectors

$$G_{1;n}(C) = \left\{\overline{x} \in (\mathbb{R}^n)^+ \mid \frac{1}{n}\sum_1^n x_i \leq C\left(\prod_1^n x_i\right)^{1/n}\right\}$$

is a convex cone. (It is a trivial consequence of the geometric-arithmetic mean inequality.) Convex sets

$$G_{1;n}(C) \cap \left\{\overline{x} \mid \sum_1^n x_i = 1\right\}$$

are interesting objects to study.

There are other related convex sets. Let $\{x_i > 0\}_1^n$ and $E_j = 1/\binom{n}{j}$ $\sum_{1 \leq i_1 < \ldots < i_j \leq n} x_{i_1} x_{i_2} \cdots x_{i_j}$, $1 \leq j \leq n$, be the normalized elementary symmetric functions. Then the classical inequalities of Maclaurin's state, for $j > i$, are

$$E_i^{1/i} \geq E_j^{1/j} \ .$$

Then, for $i = 1$ and any $C > 1$, sets

$$\left\{ \overline{x} \in (\mathbb{R}^n)^+ \mid E_1 \leq CE_j^{1/j} \right\} := G_{j;n}(C)$$

are convex cones, and again, it means that convex sets

$$G_{j;n}(C) \cap \left\{ \overline{x} \mid \sum_1^n x_i = 1 \right\}$$

describe reverse Maclaurin's inequalities.

To prove this fact one should use the Lopez-Marcus [LM] inequalities: for any $\overline{x}$ and $\overline{y} \in (\mathbb{R}^n)^+$ and any j, $1 \leq j \leq n$,

$$E_j \left(\frac{\overline{x} + \overline{y}}{2} \right)^{1/j} \geq \frac{E_j(\overline{x})^{1/j} + E_j(\overline{y})^{1/j}}{2} .$$

Indeed, let $\overline{x}, \overline{y} \subset G_{j;n}(C)$. Then

$$E_j \left(\frac{\overline{x} + \overline{y}}{2} \right)^{1/j} \geq \frac{E_j(\overline{x})^{1/j} + E_j(\overline{y})^{1/j}}{2} \geq \frac{1}{C} \frac{E_1(\overline{x}) + E_1(\overline{y})}{2} = \frac{1}{C} E_1 \left(\frac{\overline{x} + \overline{y}}{2} \right) .$$

References

[GG]   Garnaev, A.Yu., Gluskin, E.D. (1984): The widths of a Euclidean ball. (Russian) Dokl. Akad. Nauk SSSR, **277(5)**, 1048–1052

[FLM]   Figiel, T., Lindenstrauss, J., Milman, V.D. (1977): The dimension of almost spherical sections of convex bodies. Acta Math., **139(1-2)**, 53–94

[K]   Kashin, B.S. (1977): Sections of some finite-dimensional sets and classes of smooth functions. Izv. Akad. Nauk SSSR, Ser. Mat., **41**, 334–351

[LM]   Marcus, M., Lopes, L. (1956): Inequalities for symmetric functions and Hermitian matrices. Canad. J. Math., **8**, 524–531

[M]   Milman, V.D. (1971): A new proof of A. Dvoretzky's theorem on cross-sections of convex bodies. (Russian) Funkcional. Anal. i Priložen., **5(4)**, 28–37 (English translation in: Functional Analysis and Applications)

[MS]   Milman, V.D., Schechtman, G. (1986): Asymptotic theory of finite-dimensional normed spaces. With an appendix by M. Gromov. Lecture Notes in Mathematics, **1200**, Springer-Verlag, Berlin

Supremum of a Process in Terms of Trees

Olivier Guédon[1] and Artem Zvavitch[2]

[1] Equipe d'Analyse, Université Paris 6, 4 Place Jussieu, Case 186, 75005 Paris, France *guedon@ccr.jussieu.fr*
[2] Mathematics Department, Mathematical Sciences Bldg, University of Missouri, Columbia, MO 65211, USA *zvavitch@math.missouri.edu*

Summary. In this paper we study the quantity $\mathbb{E}\sup_{t\in T} X_t$, where X_t is some random process. In the case of the Gaussian process, there is a natural sub-metric d defined on T. We find an upper bound in terms of labelled-covering trees of (T, d) and a lower bound in terms of packing trees (this uses the knowledge of packing numbers of subsets of T). The two quantities are proved to be equivalent via a general result concerning packing trees and labelled-covering trees of a metric space. Instead of using the majorizing measure theory, all the results involve the language of entropy numbers. Part of the results can be extended to some more general processes which satisfy some concentration inequality.

1 Introduction

Let (T, d) be a compact metric space and for all $t \in T$, X_t be a collection of random variables such that $\mathbb{E}X_t = 0$. The aim of this paper is to present a different approach to the theory of majorizing measures. To avoid the problem of measurability of $\sup_{t\in T} X_t$, we take, as usual, the following definition:

$$\mathbb{E}\sup_{t\in T} X_t = \sup\left\{\mathbb{E}\sup_{t\in T_f} X_t,\ T_f \text{ finite subset of } T\right\}. \qquad (*)$$

It allows us to assume without loss of generality that (T, d) is in fact a finite metric space, which will make the presentation of the statements clearer. It means that in a general compact metric space (T, d), we take a very fine net on the set T to approach the quantity $\mathbb{E}\sup_{t\in T} X_t$. We want to present a new way to provide an estimate of this quantity where $(X_t)_{t\in T}$ is in particular a Gaussian process. In this case, there is a natural sub-metric d defined on T by $d(s, t)^2 = \mathbb{E}|X_s - X_t|^2$ and of course, by taking a quotient, we can assume that d is a metric on T. We recall a result of Talagrand in terms of the majorizing measure

Theorem [T1]. *If T is a finite set, $(X_t)_{t\in T}$ is a Gaussian process with the natural sub-metric d associated, then, up to universal constants, $\mathbb{E}\sup_{t\in T} X_t$ is similar to the quantity*

$$\inf_{t\in T} \sup \int_0^\infty \sqrt{\log\frac{1}{\mu(B(t,\varepsilon))}}d\varepsilon,$$

where the infimum is taken over all probability measures on T, and $B(t,\varepsilon) = \{s \in T, d(s,t) \leq \varepsilon\}$.

One of the biggest problems is to provide a uniform approach for construction of a "good" measure.

The main tools of this paper are "packing" and "labelled-covering" trees. The idea to use these objects comes from works of Talagrand [T3], [T5]. In [T3] he defined the notion of an s-tree and he shows, using the majorizing measures technique, that an s-tree provides an estimate for $\mathbb{E}\sup_{t \in T} X_t$. This point of view has been very fruitful in the study of embeddings of subspace of L_p into ℓ_p^n for $0 < p < 1$ [Z]. Here we would like to present a geometrical method for providing bounds for the supremum of a process which satisfies a concentration type inequality, where instead of measures, we will consider special families of sets of our metric space T. The main idea is to present a straightforward technique which like the theorems of Dudley and Sudakov involves the language of entropy numbers.

There are two different sections in this paper. First, we present the notions of packing and labelled-covering trees and define how to measure the size of such trees. The main result of this part is a general comparison of these two quantities. The second part is devoted to the study of upper and lower bounds of (∗) when the process satisfies a concentration type inequality. We obtain an improvement of Dudley's result which gives directly, iterating this result, an upper bound in function of the size of labelled-covering trees of the compact metric space (T, d). For the lower bound, an additional hypothesis is a Sudakov type minoration of $\mathbb{E}\sup(X_{t_1}, \ldots, X_{t_N})$ for well separated points in T. In this part, we consider for simplicity a particular case of the Gaussian process but the spirit of this idea allows generalization when the process satisfies other types of concentrations and other Sudakov type minorations [L], [T2]. We obtain an expression in terms of the size of packing trees and combining this with the result of the first part, it shows that in the Gaussian (or Euclidean) case, all these quantities are similar up to universal constants.

2 Trees of Sets

Consider a finite metric space (T, d).

Recall that a tree of subsets of T is a finite collection $\mathcal{F}$ of subsets of T with the property that for all $A, B \in \mathcal{F}$, either $A \cap B = \emptyset$, or $A \subset B$, or $B \subset A$. We say that B is a *son* of A if $B \subset A$, $B \neq A$ and

$$C \in \mathcal{F}, \quad B \subset C \subset A \quad \Longrightarrow \quad C = B \text{ or } C = A.$$

We assume that $\mathcal{A}_1$ consists of one single set (this is the root of $\mathcal{F}$) and that for each $k \in \mathbb{N}^*$, $\mathcal{A}_{k+1}$ is a finite collection of subsets of T such that each of them is a son of a set in $\mathcal{A}_k$. A *branch* of $\mathcal{F}$ is a sequence $A_1 \supset A_2 \supset \ldots$

such that A_{k+1} is a son of A_k. A branch is *maximal* if it is not contained in a longer branch. To each $A \in \mathcal{F}$ we denote by $N(A)$ the number of sons of A. Let $B_1, \ldots, B_{N(A)}$ be sons of A. We denote by ℓ_A a one-to-one map

$$\ell_A : \{B_1, \ldots, B_{N(A)}\} \to \{1, \ldots, N(A)\}.$$

Consider some fixed number $r \geq 120$. A tree $\mathcal{F}$ is called a *packing tree* if to each $A \in \mathcal{F}$, we can associate an integer $n(A) \in \mathbb{Z}$ such that

1) for all sons B of A, $\mathrm{diam}(B) \leq 2r^{-n(A)}$,
2) if B and B' are two distinct sons of A then $d(B, B') \geq 30r^{-n(A)}$.

We define the *size* $\gamma_p(\mathcal{F}, d)$ *of a packing tree* $\mathcal{F}$ to be the infimum over all possible maximal branches of

$$\sum_{k \geq 1} r^{-n(A_k)} \sqrt{\log\left(N(A_k)\right)}.$$

A tree $\mathcal{F}$ is called a *labelled-covering* tree if

1) for any $t \in T$ there is a maximal branch $A_1 \supset A_2 \supset \ldots$ such that $t = \bigcap A_k$,
2) to each $A \in \mathcal{F}$ is associated a labelled function ℓ_A (which numerates each son of A) and an integer $n(A) \in \mathbb{Z}$ such that $\mathrm{radius}(A) \leq r^{-n(A)}$ (we allow $n(A) = +\infty$ when the set A is a single point).

Finally we define the *size* $\gamma_c(\mathcal{F}, d)$ *of a labelled-covering tree* $\mathcal{F}$ as the supremum over all possible maximal branches of

$$\sum_{k \geq 1} r^{-n(A_k)} \sqrt{\log\left(e\ell_{A_k}(A_{k+1})\right)}.$$

We denote by $Cov(T, d)$ (respectively, $Pac(T, d)$) the set of all labelled-covering (respectively, packing) trees in T. The first theorem shows a connection between the definitions of size of packing trees and of labelled-covering trees.

Theorem 1. *There exists a constant $C > 1$ such that for any finite metric space (T, d)*

$$\inf_{\mathcal{F} \in Cov(T,d)} \gamma_c(\mathcal{F}, d) \leq C \sup_{\mathcal{F} \in Pac(T,d)} \gamma_p(\mathcal{F}, d).$$

To prove it, we will use the following theorem due to Talagrand.

Theorem [T5]. *Consider a finite metric space (T, d) and the largest $i \in \mathbb{Z}$ such that $\mathrm{radius}(T) \leq r^{-i}$. Assume that for $j \geq i$ there are functions $\phi_j : T \to \mathbb{R}^+$ with the following property:*

For any point s of T, any integer $j \geq i$ and $N \geq 1$, if $t_1, \ldots t_N$ are N points in $B(s, r^{-j})$ such that

$$d(t_{l'}, t_l) \geq r^{-j-1}, \text{ for any } l, l' \leq N, l \neq l',$$

then we have

$$\phi_j(s) \geq \alpha r^{-j} \sqrt{\log N} + \min_{l \leq N} \phi_{j+2}(t_l). \tag{1}$$

Assume also that $(\phi_j)_{j \geq i}$ is a decreasing sequence of functions. Then

$$\inf_{\mathcal{F} \in Cov(T,d)} \gamma_c(\mathcal{F}, d) \leq \frac{5}{\alpha} \sup_{t \in T} \phi_i(t).$$

For completeness of the paper, we reproduce here a proof of this result which is almost the proof of Proposition 4.3 of [T5].

Proof. Our goal is to construct a labelled-covering tree $\mathcal{F}$ such that

$$\sum_{k \geq 1} r^{-n(A_k)} \sqrt{\log \left(e \ell_{A_k}(A_{k+1}) \right)} \leq C \sup_{t \in T} \phi_i(t),$$

for any branch $\{A_1 \supset \ldots \supset A_k \supset \ldots\}$ in $\mathcal{F}$.

We will inductively construct our covering tree.

First step: $k = 1$.
The first step consists of taking $A_1 = T$, $n(A_1) = n(T) = i$ and we define $a_1(A_1) \in A_1$ such that

$$A_1 \subset B\big(a_1(A_1), r^{-i}\big).$$

Iterative step: from k to $k+1$.
Assume that we have constructed the kth level A_k of the tree $\mathcal{F}$ (which is a covering of the set T) such that

1) $T = A_k^1 \cup \ldots \cup A_k^d$,
2) for each set A_k of this covering, either A_k is a single point or there exists $a_k(A_k) \in A_k$ such that $A_k \subset B(a_k(A_k), r^{-n(A_k)})$ with the biggest possible integer $n(A_k)$.

If all the sets of this covering consist of single points then the construction is finished (and this situation will appear because T is a finite set). Now we show how to partition any given element A_k of this covering. If A_k is a single point then $n(A_k) = +\infty$ and $A_1 \supset \ldots \supset A_k$ is a maximal branch so we have nothing to do. Assume now that A_k is not a single point. We pick $t_1 \in A_k$ such that

$$\phi_{n(A_k)+2}(t_1) = \max \left\{ \phi_{n(A_k)+2}(t); t \in A_k \right\}.$$

Then the first son of A_k is

$$B_1 = A_k \cap B(t_1, r^{-n(A_k)-1})$$

and $a_{k+1}(B_1) = t_1$. We define $n(B_1)$ as the biggest integer such that $B_1 \subset B(t_1, r^{-n(B_1)})$. To construct B_2 we repeat this procedure, replacing A_k by $A_k \setminus B_1$. This set is not empty because $r > 2$, A_k is not a single point and by the maximum condition on $n(A_k)$.

Finally we have constructed points $t_1, \ldots, t_N$ $(N \geq 2)$ and sons $B_1, \ldots, B_N$ such that for any $m \in \{1, \ldots, N\}$,

$$t_m \in A_k \setminus \bigcup_{l < m} B(t_l, r^{-n(A_k)-1})$$

and

$$\phi_{n(A_k)+2}(t_m) = \max \left\{ \phi_{n(A_k)+2}(t); t \in A_k \setminus \bigcup_{l < m} B(t_l, r^{-n(A_k)-1}) \right\}.$$

It is clear (by construction) that $B_1, \ldots, B_N$ are sons of A_k, form a covering of A_k and that $n(B_m) \geq n(A_k) + 1$. Also by construction $d(t_l, t_{l'}) \geq r^{-n(A_k)-1}$, $d(a_k(A_k), t_m) \leq r^{-n(A_k)}$ and taking $j = n(A_k)$, we obtain by definition of our functions ϕ_j that for any $m \in \{1, \ldots, N\}$,

$$\phi_{n(A_k)}(a_k(A)) \geq \alpha r^{-n(A_k)} \sqrt{\log m} + \min_{l \leq m} \phi_{n(A_k)+2}(t_l).$$

We labelled the sons by setting $\ell_A(B_m) = m$ so

$$\phi_{n(A_k)}(a_k(A)) \geq \alpha r^{-n(A_k)} \sqrt{\log \ell_A(B_m)} + \min_{l \leq m} \phi_{n(A_k)+2}(t_l).$$

By construction of the points $\{t_l\}$, if $l < l'$,

$$\phi_{n(A_k)+2}(t_l) \geq \phi_{n(A_k)+2}(t_{l'}),$$

so we get

$$\min_{l \leq m} \phi_{n(A_k)+2}(t_l) \geq \phi_{n(A_k)+2}(t_m).$$

At this stage, for each set A_k of our starting covering of T, we have constructed a labelled function ℓ_{A_k}, sons who form a covering of A_k such that for all sons A_{k+1} of A_k, $n(A_{k+1}) \geq n(A_k) + 1$, and point $a_{k+1}(A_{k+1})$ such that

$$\phi_{n(A_k)}(a_k(A_k)) \geq \alpha r^{-n(A_k)} \sqrt{\log \ell_{A_k}(A_{k+1})} + \phi_{n(A_k)+2}(a_{k+1}(A_{k+1})).$$

Next we observe that of course, for all sons A_{k+2} of A_{k+1}, $a_{k+2}(A_{k+2}) \in A_{k+1}$ so by construction of $a_{k+1}(A_{k+1})$,

$$\phi_{n(A_k)+2}(a_{k+1}(A_{k+1})) \geq \phi_{n(A_k)+2}(a_{k+2}(A_{k+2})).$$

But $n(A_{k+2}) \geq n(A_{k+1}) + 1 \geq n(A_k) + 2$ (by construction) and as $(\phi_j)_{j \geq i}$ is a decreasing sequence of functions,

$$\phi_{n(A_k)+2}\big(a_{k+2}(A_{k+2})\big) \geq \phi_{n(A_{k+2})}\big(a_{k+2}(A_{k+2})\big),$$

and finally, for all branches $A_k \supset A_{k+1} \supset A_{k+2}$,

$$\phi_{n(A_k)}\big(a_k(A_k)\big) \geq \alpha r^{-n(A_k)} \sqrt{\log \ell_{A_k}(A_{k+1})} + \phi_{n(A_{k+2})}\big(a_{k+2}(A_{k+2})\big).$$

Conclusion.

If we sum up the last inequality for $k \geq 1$, we get

$$\phi_{n(A_1)}\big(a_1(A_1)\big) + \phi_{n(A_2)}\big(a_2(A_2)\big) \geq \alpha \sum_{k \geq 1} r^{-n(A_k)} \sqrt{\log \ell_{A_k}(A_{k+1})}$$

which gives (because the sequence $(\phi_j)_{j \geq i}$ is decreasing), for all branches $A_1 \supset \ldots \supset A_k \supset \ldots$ of the labelled-covering tree $\mathcal{F}$

$$\alpha \sum_{k \geq 1} r^{-n(A_k)} \sqrt{\log \ell_{A_k}(A_{k+1})} \leq 2 \sup_{t \in T} \phi_i(t).$$

Now call

$$S_1 = \sup_{\text{maximal branch}} \sum_{k \geq 1} r^{-n(A_k)} \sqrt{\log \ell_{A_k}(A_{k+1})}$$

and

$$S_2 = \sup_{\text{maximal branch}} \sum_{k \geq 1} r^{-n(A_k)} \sqrt{\log e \ell_{A_k}(A_{k+1})}.$$

It is clear that $S_1 \geq r^{-n(A_1)} \sqrt{\log 2}$. By construction, for all sons A_{k+1} of A_k, $n(A_{k+1}) \geq n(A_k) + 1$ then for all maximal branch $A_1 \supset \ldots \supset A_k \supset \ldots$ of the labelled-covering tree $\mathcal{F}$,

$$\sum_{k \geq 1} r^{-n(A_k)} \sqrt{\log e \ell_{A_k}(A_{k+1})} \leq \sum_{k \geq 1} r^{-n(A_k)} \left(1 + \sqrt{\log \ell_{A_k}(A_{k+1})}\right)$$

$$\leq S_1 + \frac{r-1}{r} r^{-n(A_1)} \leq \frac{5}{2} S_1$$

because r is large enough. It proves that for this tree $\mathcal{F}$,

$$S_2 \leq 5 S_1 / 2 \leq 5 \sup_{t \in T} \phi_i(t) / \alpha.$$

$\square$

Proof (of Theorem 1). Let i be the largest integer such that radius$(T) \leq r^{-i}$. For a set $A \subset T$, let $\gamma_p(A) = \sup_{\mathcal{F} \in Pac(A,d)} \gamma_p(\mathcal{F}, d)$, and for all integers $j \geq i$, define the function $\phi_j : T \to \mathbb{R}^+$ by

$$\forall s \in T, \quad \phi_j(s) = \gamma_p\big(B(s, 2r^{-j})\big).$$

The sequence $(\phi_j)_{j\geq i}$ is decreasing and, by definition of $i \in \mathbb{Z}$,

$$\sup_{t\in T} \phi_i(t) = \gamma_p(T).$$

To prove Theorem 1, we need to check assumption (1) of the previous theorem. Fix some $j \geq i$ and $s \in T$. Let $t_1, \ldots, t_N$ be points in $B(s, r^{-j})$ with $d(t_l, t_{l'}) \geq r^{-j-1}$, then

$$\phi_{j+2}(t_l) = \gamma_p\big(B(t_l, 2r^{-j-2})\big) \qquad \text{and} \qquad B(t_l, 2r^{-j-2}) \subset B(s, 2r^{-j})$$

and

$$d\big(B(t_l, 2r^{-j-2}), B(t_{l'}, 2r^{-j-2})\big) \geq r^{-j-1} - 4r^{-j-2} \geq \frac{1}{4}r^{-j-1}.$$

Consider in $B(s, 2r^{-j})$ a two level packing tree whose first level is $B(s, 2r^{-j})$ and whose second level consists of

$$\big\{B_l = B(t_l, 2r^{-j-2})\big\}_{l\leq N}.$$

Take $n(B(s, 2r^{-j})) = j+2$ then for each son B, B' of $B(s, 2r^{-j})$, $\operatorname{diam}(B) \leq 2r^{-n(B(s,2r^{-j}))}$ and

$$d(B, B') \geq \frac{1}{4}r^{-j-1} \geq 30\, r^{-n(B(s,2r^{-j}))} = \frac{30}{r}r^{-j-1}$$

because r is large enough ($r = 120$). By definition of the size of packing trees,

$$\gamma_p\big(B(s, 2r^{-j})\big) \geq r^{-j-2}\sqrt{\log N} + \min_{l\leq N}\gamma_p\big(B(t_l, 2r^{-j-2})\big),$$

or

$$\phi_j(s) \geq \frac{1}{r^2}r^{-j}\sqrt{\log N} + \min_{l\leq N}\phi_{j+2}(t_l),$$

so we can apply the previous theorem with $\alpha = 1/r^2$.   □

3 Application to Random Processes

Let (T, d) be a finite metric space, and for all $t \in T$, X_t be a collection of random variables such that $\mathbb{E}X_t = 0$. In this part, we show how the quantities defined in the above sections are related to the study of $\mathbb{E}\sup_{t\in T} X_t$.

We will say that the process $(X_t)_{t\in T}$ satisfies a concentration inequality (H) if there exists $c > 0$ such that

$$\begin{cases} \text{for all subsets } A \subset T, \text{ for all } t_0 \in T, \\ \text{if } Y_{A,t_0} = \sup_{t\in A}(X_t - X_{t_0}) \text{ and } \sigma = \sup_{t\in A} d(t, t_0) \text{ then} \\ \forall u \geq 0, \mathbb{P}\big(|Y_{A,t_0} - \mathbb{E}Y_{A,t_0}| \geq u\big) \leq 2\exp\left(-c\left(\frac{u}{\sigma}\right)^2\right). \end{cases}$$

Remark. This hypothesis (H) implies a deviation inequality: for all $(s,t) \in T$,

$$\mathbb{P}(|X_s - X_t| \geq u) \leq 2 \exp\left(-c\left(\frac{u}{d(s,t)}\right)^2\right).$$

Indeed, choose $s = t_0$ and $A = \{t\}$ then $\sigma = d(s,t)$ and $Y_{A,t_0} = X_t - X_{t_0}$ which gives the result. Maurey and Pisier ([P] Theorem 4.7) have proved that (H) is satisfied for the Gaussian process (with $c = 2/\pi^2$) and Talagrand [T4] proved it for the Bernoulli process. We don't know if it is true for a general subgaussian process, i.e. a process which satisfies only a deviation inequality as above.

3.1 Relation with the Size of Covering Trees

When the process $(X_t)_{t \in T}$ satisfies such a concentration inequality, we obtain an upper bound of $\mathbb{E}\sup_{t \in T} X_t$ in terms of the size of labelled-covering trees of T with respect to the metric d. The next result is an improvement of Lemma 3.4.4 in [Fe] which was the usual Dudley's upper bound.

Theorem 2. *If the process $(X_t)_{t \in T}$ satisfies a concentration inequality (H), there exists a constant $C_1 > 0$ (depending only on the constant c in (H)) such that for all $N \in \mathbb{N}^*$, for all subsets $A_1, \ldots, A_N$ of T, and $A = A_1 \cup \ldots \cup A_N$, we have*

$$\mathbb{E}\sup_{t \in A} X_t \leq \sup_{1 \leq \ell \leq N}\left(C_1 \operatorname{diam}A\sqrt{\log e\ell} + \mathbb{E}\sup_{t \in A_\ell} X_t\right).$$

Proof. Let $t_0 \in A$ then $\mathbb{E}\sup_{t \in A} X_t = \mathbb{E}\sup_{t \in A}(X_t - X_{t_0})$. For all $\ell \in \{1, \ldots, N\}$, let $Y_\ell = \sup_{t \in A_\ell}(X_t - X_{t_0})$ then

$$\mathbb{E}\sup_{t \in A} X_t = \mathbb{E}\sup_{1 \leq \ell \leq N} Y_\ell.$$

Let S be defined by

$$S = \sup_{1 \leq \ell \leq N}\left(c_1 \operatorname{diam}(A)\sqrt{\log e\ell} + \mathbb{E}\sup_{t \in A_\ell} X_t\right),$$

where c_1 will be defined later in accordance with the constant $c > 0$ in the hypothesis (H).
As $\sup_{1 \leq \ell \leq N} Y_\ell$ is a non-negative random variable,

$$\mathbb{E}\sup_{1 \leq \ell \leq N} Y_\ell = \int_0^{+\infty} \mathbb{P}\big(\exists \ell \in \{1 \ldots, N\}, Y_\ell > u\big)du$$

$$\leq K + \int_K^{+\infty} \mathbb{P}\big(\exists \ell \in \{1 \ldots, N\}, Y_\ell > u\big)du.$$

By definition of S, for all $\ell \in \{1, \ldots, N\}$,

$$S \geq \left(c_1 \operatorname{diam}(A)\sqrt{\log e\ell} + \mathbb{E}Y_\ell\right),$$

so by choosing $K = S$, we obtain

$$\mathbb{E}\sup_{1 \leq \ell \leq N} Y_\ell \leq S + \int_S^{+\infty} \mathbb{P}\left(\exists \ell, Y_\ell - \mathbb{E}Y_\ell > u - S + c_1 \operatorname{diam}(A)\sqrt{\log e\ell}\right) du$$

$$\leq S + \sum_{\ell=1}^N \int_0^{+\infty} \mathbb{P}\left(Y_\ell - \mathbb{E}Y_\ell > u + c_1 \operatorname{diam}(A)\sqrt{\log e\ell}\right) du.$$

To conclude, we know that for all $t \in A_\ell$, $d(t, t_0) \leq \operatorname{diam}(A)$ and by the concentration inequality (H), we have

$$\mathbb{E}\sup_{1 \leq \ell \leq N} Y_\ell \leq S + 2 \sum_{\ell=1}^{+\infty} \int_0^{+\infty} \exp\left(-c\left(\frac{u}{\operatorname{diam}(A)} + c_1\sqrt{\log e\ell}\right)^2\right) du$$

$$\leq S + \sqrt{\frac{\pi}{c}} \operatorname{diam}(A) \sum_{\ell=1}^{+\infty} \exp\left(-cc_1^2 \log(e\ell)\right)$$

$$\leq S + \frac{1}{e^2}\sqrt{\frac{\pi}{c}} \operatorname{diam}(A) \sum_{\ell=1}^{+\infty} \frac{1}{\ell^2},$$

choosing c_1 such that $cc_1^2 = 2$. Because $\log e\ell \geq 1$, we have proved the theorem with

$$C_1 = \frac{1}{\sqrt{c}}\left(\sqrt{2} + \frac{\pi^{3/2}}{6e^2}\right).$$

$\square$

Now, it is very easy to deduce the following result.

Corollary 3. *If the process $(X_t)_{t \in T}$ satisfies a concentration inequality (H) then there exists a constant $C > 1$ (depending only on the constant c in (H)) such that*

$$\mathbb{E}\sup_{t \in T} X_t \leq C \inf_{\mathcal{F} \in Cov(T,d)} \gamma_c(\mathcal{F}, d).$$

Proof. Let $\mathcal{F}$ be a labelled-covering tree of T with respect to the metric d. Then by Theorem 2, we deduce that

$$\mathbb{E}\sup_{t \in T} X_t \leq \sup_{A_i \text{ sons of } T} \left(C_1 \operatorname{diam}T \sqrt{\log e\ell_T(A_i)} + \mathbb{E}\sup_{t \in A_i} X_t\right).$$

Now iterating this procedure over a particular son that realizes this maximum (it is finite because T is finite and note also that by the hypothesis on a labelled-covering tree, the last term of the sum will be $\mathbb{E}X_{t_i} = 0$ because the last sons must be a single point), we deduce that

$$\mathbb{E}\sup_{t \in T} X_t \leq C_1 \sup_{\text{maximal branch}} \sum_{k \geq 1} 2\, r^{-n(A_k)} \sqrt{\log e\ell_{A_k}(A_{k+1})}.$$

This is true for all labelled-covering trees so it gives exactly the stated result.
$\square$

3.2 Relation with the Size of Packing Trees

To study a lower bound of $\mathbb{E}\sup_{t\in T}X_t$, we would like to start with the following theorem due to Talagrand [T5], which will lead us to the idea of how to bound $\mathbb{E}\sup_{t\in T}X_t$, where $(X_t)_{t\in T}$ is a Gaussian process, using packing trees.

Theorem [T5]. *Consider a Gaussian process* $(X_t)_{t\in T}$, d *the natural submetric associated and sets* $\{B_l\}_{l\leq N}$ *with* $N\geq 2$. *Assume that* $d(B_l,B_{l'})\geq 15\,u$ *for all integers* $l,l'\leq N, l\neq l'$ *and* $\mathrm{diam}(B_l)\leq u$. *Consider* $A=\bigcup_{l\leq N}B_l$, *then*

$$\mathbb{E}\sup_{t\in A}X_t\geq C\,u\sqrt{\log N}+\min_{l\leq N}\mathbb{E}\sup_{t\in B_l}X_t,$$

where $C=\pi/\sqrt{2}>2$.

Proof. The proof of this theorem is based on the following two classical lemmas.

Lemma. *Under the assumptions of the previous theorem, let* $t_l\in B_l$ *and* $Y_\ell=\sup_{t\in B_\ell}(X_t-X_{t_l})$ *then*

$$\mathbb{E}\sup_{\ell\in\{1,\dots,N\}}|Y_\ell-\mathbb{E}Y_\ell|\leq u\,\frac{\pi}{\sqrt{2}}\sqrt{\log eN}.$$

Remark. This result was also used to obtain the classical Dudley upper bound in terms of entropy numbers [Fe] but is weaker than Theorem 2. As $\sup_{1\leq\ell\leq N}|Y_\ell-\mathbb{E}Y_\ell|$ is a non-negative random variable,

$$\mathbb{E}\sup_{1\leq\ell\leq N}|Y_\ell-\mathbb{E}Y_\ell|=\int_0^{+\infty}\mathbb{P}\big(\exists\,\ell\in\{1\dots,N\},|Y_\ell-\mathbb{E}Y_\ell|>t\big)dt$$

$$\leq K+\sum_{\ell=1}^N\int_K^{+\infty}\mathbb{P}\big(|Y_\ell-\mathbb{E}Y_\ell|>t\big)dt$$

$$\leq K+2\sum_{\ell=1}^N\int_K^{+\infty}\exp\left(-c\Big(\frac{t}{u}\Big)^2\right)dt,$$

by the concentration inequality (H) and because $\mathrm{diam}B_l\leq u$. The result follows choosing $K=u/\sqrt{c}\sqrt{\log N}$ (and recall that in this case, we could take $c=2/\pi^2$). $\quad\square$

The next result is a Sudakov type inequality. There are many methods to obtain this kind of inequality. For the Gaussian case, we could see it as an application of Slepian's lemma but there is another method which can be generalized to other processes in the paper of Talagrand [T2] and in the paper of Latała [L].

Lemma. *If $t_1, \ldots, t_N$ $(N \geq 2)$ are well separated points in T, i.e. assume that there exists $u > 0$ such that for all $l \neq l'$, $d(t_l, t_{l'}) \geq 15\,u$, then*

$$\mathbb{E} \sup_{i \in \{1, \ldots, N\}} X_{t_i} \geq u\,\pi\sqrt{2\log eN}.$$

Proof. Let $g_1, \ldots, g_N$ be i.i.d. random normal Gaussian variables and define the process $Y_1, \ldots, Y_N$ by $Y_i = \frac{15}{\sqrt 2}\,u\,g_i$. Then it is clear that for all $l \neq l'$,

$$\mathbb{E}|X_{t_l} - X_{t_{l'}}|^2 = d(t_l, t_{l'})^2 \geq \mathbb{E}|Y_l - Y_{l'}|^2.$$

As

$$\mathbb{E}\sup(g_1, \ldots, g_N) \geq \sqrt{\frac{\log N}{\pi \log 2}} \geq \sqrt{\frac{\log eN}{\pi \log 2e}}$$

for $N \geq 2$ (see for example formula 1.7.1 in [Fe]), the result follows easily by an application of Slepian's comparison property.   □

Combining these two lemmas, it is very easy to finish the proof of the previous theorem.

$$\begin{aligned}
\mathbb{E}\sup_{t \in A} X_t &= \mathbb{E}\sup_{l \leq N}(Y_l - \mathbb{E}Y_l) + \mathbb{E}Y_l + X_{t_l} \\
&\geq \min_{l \leq N}\mathbb{E}Y_l + \mathbb{E}\sup(X_{t_1}, \ldots, X_{t_N}) - \mathbb{E}\sup_{l \leq N}|Y_l - \mathbb{E}Y_l| \\
&\geq \frac{\pi}{\sqrt 2}u\sqrt{\log N} + \min_{l \leq N}\mathbb{E}\sup_{t \in B_l} X_t.
\end{aligned}$$

□

Using this theorem we deduce the following corollary.

Corollary 4. *There is a universal constant $C > 0$ such that, if $(X_t)_{t \in T}$ is a Gaussian process and d the natural sub-metric associated, then*

$$C \sup_{\mathcal{F} \in Pac(T,d)} \gamma_p(\mathcal{F}, d) \leq \mathbb{E}\sup_{t \in T} X_t.$$

Proof. Let $\mathcal{F}$ be a packing tree of T with respect to the metric d. For any element A of this packing tree, let $B_1, \ldots, B_{N(A)}$ be the sons of $A \in \mathcal{F}$. Then we use the previous lemma with $u = 2r^{-n(A)}$ (because $d(B_l, B_{l'}) \geq 15u$ and $\mathrm{diam}(B_l) \leq 2r^{-n(A)} \leq u$) to get

$$\mathbb{E}\sup_{t \in A} X_t \geq Cr^{-n(A)}\sqrt{\log N(A)} + \min_{l \leq N(A)}\mathbb{E}\sup_{t \in B_l} X_t.$$

Now iterate this formula over a particular son which realizes this minimum to deduce that

$$\mathbb{E}\sup_{t \in T} X_t \geq C \inf_{\text{maximal branch}} \sum_{k \geq 1} r^{-n(A_k)}\sqrt{\log\left(N(A_k)\right)}.$$

This is true for all packing trees and it finishes the proof.   □

To conclude this part, we just want to state the result we can deduce from Theorem 1, Corollary 3 and Corollary 4 in the case of the Gaussian process.

Theorem 5. *Let T be a finite set, $(X_t)_{t \in T}$ a Gaussian process with $\mathbb{E}X_t = 0$ and d the natural sub-metric associated, then, up to universal constants, the three quantities*

$$\mathbb{E}\sup_{t \in T} X_t, \quad \inf_{\mathcal{F} \in Cov(T,d)} \gamma_c(\mathcal{F}, d) \quad \text{and} \quad \sup_{\mathcal{F} \in Pac(T,d)} \gamma_p(\mathcal{F}, d)$$

are similar.

References

[D]   Dudley, R.M. (1967): The sizes of compact subsets of Hilbert space and continuity of Gaussian processes. J. Funct. Anal., **1**, 290–330

[Fe]   Fernique, X. (1997): Fonctions aléatoires gaussiennes, vecteurs aléatoires gaussiens. Publications du Centre de Recherches Mathématiques, Montréal

[L]   Latała, R. (1997): Sudakov minoration principle and supremum of some processes. Geom. Funct. Anal., **7**, 936–953

[P]   Pisier, G. (1989): The Volume of Convex Bodies and Banach Space Geometry. Cambridge Tracts in Math., **94**, Cambridge University Press, Cambridge

[T1]   Talagrand, M. (1987): Regularity of Gaussian processes. Acta Math., **159**, 99–149

[T2]   Talagrand, M. (1994): The supremum of some canonical processes. American Journal of Mathematics, **116**, 283–325

[T3]   Talagrand, M. (1995): Embedding subspaces of L_p in L_p^N. Geometric Aspects of Functional Analysis (Israel, 1992-1994), Oper. Theory Adv. Appl., **77**, Birkhaüser, Basel, 311–325

[T4]   Talagrand, M. (1995): Concentration of measure and isoperimetric inequalities in product spaces. Publications Mathématiques de l'I.H.E.S., **81**, 73–205

[T5]   Talagrand, M. (1996): Majorizing measures: the generic chaining. Annals of Probability, **24(3)**, 1049–1103

[Z]   Zvavitch, A. (2000): More on embedding subspaces of L_p into ℓ_p^N, $0 < p < 1$. GAFA Seminar 1996-2000, Lecture Notes in Math., **1745**, 269–281

Point Preimages under Ball Non-Collapsing Mappings[*]

Olga Maleva

Department of Mathematics, The Weizmann Institute of Science, Rehovot 76100, Israel *maleva@wisdom.weizmann.ac.il*

Summary. We study three classes of Lipschitz mappings of the plane: Lipschitz quotient mappings, ball non-collapsing mappings and locally ball non-collapsing mappings. For each class, we estimate the maximum cardinality of point preimage in terms of the ratio of two characteristic constants of the mapping. For Lipschitz quotients and for Lipschitz locally BNC mappings, we provide a complete scale of such estimates, while for the intermediate class of BNC mappings the answer is not complete yet.

1. Let X and Y be metric spaces. The class of Lipschitz mappings $f\colon X \to Y$ is defined by the condition: $f(B_r(x)) \subset B_{Lr}(f(x))$ for all points x of X and all positive r (by $B_r(x)$ we denote an open ball of radius r, centered at x). Here L is a constant depending on the mapping f but not on the point x; the infimum of all possible such L is called the Lipschitz constant of f.

In a similar way, co-Lipschitz mappings $f\colon X \to Y$ are defined by the condition $f(B_r(x)) \supset B_{cr}(f(x))$, where the positive constant c is independent of x and r; the supremum of all such c is called the co-Lipschitz constant of the mapping f. (In some fundamental papers, e.g. [JLPS], the co-Lipschitz constant of the mapping is defined as infimum over all c', such that $f(B_r(x)) \supset B_{r/c'}(f(x))$.)

By definition, a Lipschitz quotient mapping is a mapping that satisfies both of the above conditions, i.e. is L-Lipschitz and c-co-Lipschitz for some constants $0 < c \le L < \infty$.

The recently developed theory of Lipschitz quotient mappings between Banach spaces raised many interesting questions about the properties of these mappings. Here we are interested in the case when X and Y are finite dimensional Banach spaces.

The paper [JLPS] contains far-reaching results for Lipschitz quotient mappings $f\colon \mathbb{R}^2 \to \mathbb{R}^2$. In particular, it is proved there that the preimage of each point under such an f is finite. The question whether the same is true for Lipschitz quotients $f\colon \mathbb{R}^n \to \mathbb{R}^n$ for $n \ge 3$ is still open, although the following result concerning this was obtained in [M]: There is a $\rho_n < 1$ such that if the ratio of co-Lipschitz and Lipschitz constants of such a mapping is greater than ρ_n, then the mapping is one-to-one. It was also proved in [M] that the cardinality of the preimage of a point under a Lipschitz quotient mapping

---

[*] Supported by the Israel Science Foundation.

of the plane does not exceed the ratio between its Lipschitz constant L and co-Lipschitz constant c with respect to the Euclidean norm.

In section 2 of the present paper, we generalize this result to the case of arbitrary norm. One important situation is when the ratio c/L is greater than $1/2$, then the mapping is a homeomorphism. In section 3, we discuss the question whether the bound $c/L \leq 1/\max_x \#f^{-1}(x)$ is tight.

In section 4, we study so-called ball non-collapsing (BNC) mappings. We say that a mapping $f : X \to Y$ is C-ball non-collapsing, if for any $x \in X$ and $r > 0$ one has

$$f\big(B_r(x)\big) \supset B_{Cr}(y) \qquad (*)$$

for some $y \in Y$. This property generalizes co-Lipschitzness. We will say that a mapping is C locally BNC, if for any $x \in X$ there exists $\varepsilon = \varepsilon(x) > 0$ such that $(*)$ holds for all $r \leq \varepsilon$.

Note that ball non-collapsing mappings can be very far from being co-Lipschitz: e.g., the mapping $F(x,y) = (x, |y|)$ from $\mathbb{R}^2$ to itself is $1/2$ BNC, but is not co-Lipschitz (its image is not the whole plane).

The local ball non-collapsing property does not imply in general the global property, as demonstrated by another plane-folding example: $F_1(x,y) = (x, |y - [y + \frac{1}{2}]|)$, where $[t]$ stands for the integer part of t. This mapping is locally $1/2$ ball non-collapsing, but is not globally ball non-collapsing for any constant.

However, it turns out that in particular cases, the local BNC property may even imply co-Lipschitzness, though with smaller constant: it is easy to show (see Lemma 4, section 4 that if the Lipschitz constant of a Lipschitz, locally BNC mapping f is less than twice the BNC constant, then f is a Lipschitz quotient mapping. For the mappings of the plane this immediately yields finiteness of point preimages. But we obtain a stronger result. In Theorem 2 we show that such a mapping f is a bi-Lipschitz homeomorphism, that is, the preimage of each point consists of one point. On the other hand, the above example of locally BNC mapping $F_1(x,y)$ shows that as soon as the ratio of constants is less than or equal to one half, the locally BNC mapping may have infinite point preimages.

The idea of folding the plane infinitely many times has to be modified in order to construct an example of a Lipschitz globally BNC mapping of the plane with infinite point preimage. In section 5 we discuss the modified construction, but it yields the BNC constant less than (and arbitrarily close to) one third of the Lipschitz constant. Thus, we do not know exactly how large the point preimages in the global BNC case can be, when the ratio of constants is in the interval $[1/3, 1/2]$.

2.   This section is devoted to Lipschitz quotient mappings. We would like to prove the following theorem, which is a generalization of a similar result in [M] to the case of arbitrary norm.

Theorem 1. *If $f\colon (\mathbb{R}^2, \|\cdot\|) \to (\mathbb{R}^2, \|\cdot\|)$ is an L-Lipschitz and c-co-Lipschitz mapping with respect to any norm $\|\cdot\|$ and*

$$\max_{x \in \mathbb{R}^2} \# f^{-1}(x) = n,$$

then $c/L \leq 1/n$.

Proof. The proof will follow the same scheme as the proof of [M, Theorem 2]. We will only explain the details needed for the argument to work in case of arbitrary norm. We consider the decomposition $f = P \circ h$, where $h\colon \mathbb{R}^2 \to \mathbb{R}^2$ is a homeomorphism and $P(z)$ is a polynomial of one complex variable (see [JLPS]). Clearly, $\deg P = \max_{x \in \mathbb{R}^2} \# f^{-1}(x) = n$. We may also assume that $f(0) = 0$ and $L = \operatorname{Lip}(f) = 1$.

Assume $c > 1/n$, then there exists $\varepsilon > 0$ such that $c_1 = c(1 - \varepsilon) > 1/n$.

We omit the proof of the following lemma, since it would in fact repeat the proof of [M, Lemma 1]:

Lemma 1. *There exists an R such that for any x with $\|x\| \geq R$ one has $\|f(x)\| \geq c_1 \|x\|$.* □

Let us show that for large enough r the index of the image $f(\partial B_r^{\|\cdot\|}(0))$ around zero is equal to n.

Lemma 2. *There exists $d > 1$ such that for any $\rho > d$*

$$\operatorname{Ind}_0 f\big(\partial B_\rho^{\|\cdot\|}(0)\big) = \operatorname{Ind}_0 P\Big(h\big(\partial B_\rho^{\|\cdot\|}(0)\big)\Big) = n.$$

Proof. Denote the Euclidean norm of $x \in \mathbb{R}^2$ by $|x|$. By [M, Lemma 3] there exists such σ that $\operatorname{Ind}_0 f(\partial B_\sigma^{|\cdot|}(0)) = n$, and all preimages of zero under f lie in $B_\sigma^{|\cdot|}(0)$. Take d such that $\|x\| \geq d$ implies $|x| \geq \sigma$, and let $\rho \geq d$. Since the set $B_\rho^{\|\cdot\|}(0) \setminus B_\sigma^{|\cdot|}(0)$ does not contain preimages of zero, one has

$$\operatorname{Ind}_0 f\big(\partial B_\rho^{\|\cdot\|}(0)\big) = \operatorname{Ind}_0 f\big(\partial B_\sigma^{|\cdot|}(0)\big) = n.$$

□

The last lemma in the proof of Theorem 1 is rather obvious in the Euclidean case, but needs some technical work in the case of arbitrary norm and the corresponding Hausdorff measure. By the k-dimensional Hausdorff measure of a Borel set A we mean

$$\mathcal{H}_k(A) = \sup_{\delta > 0} \inf \left\{ \sum_{j=1}^\infty (\operatorname{diam} C_j)^k \mid A \subset \bigcup_{j=1}^\infty C_j, \operatorname{diam} C_j \leq \delta \right\}$$

(cf. [F, 2.8.15]). The diameter in this definition is with respect to the metric given by the norm $\|\cdot\|$. Note that $\mathcal{H}_k$ is so normalized that the 1-Hausdorff measure of a segment $[x, y]$ is equal to $\|x - y\|$.

Lemma 3. *If $\Gamma\colon [0,1] \to \mathbb{R}^2$ is a closed curve with $\|\Gamma(t)\| \geq r$ for all $t \in [0,1]$ and $\mathrm{Ind}_0\,\Gamma = n$, then the length of Γ in the sense of the 1-dimensional Hausdorff measure $\mathcal{H}_1$ is at least $n\mathcal{H}_1(\partial B_r(0))$.*

Proof. In order to prove Lemma 3, it suffices to prove it in the case $n = 1$, since a closed curve of index n can be split into n closed curves of index 1.

Note first that there exist convex polygons inscribed in the sphere $\partial B_r(0)$ with perimeter arbitrarily close to $\mathcal{H}_1(\partial B_r(0))$.

Indeed, fix positive ε and take $\delta > 0$ such that for any covering of $\partial B_r(0)$ by balls of diameters less than δ, the sum of the diameters is at least $\mathcal{H}_1(\partial B_r(0)) - \varepsilon$. Consider the family of all balls with centers on $\partial B_r(0)$ and diameters less than δ. By the Besicovitch Covering Theorem (see [F, 2.8.15]) there exists a countable subfamily of disjoint balls $\{B_i\}$, which covers almost all of $\partial B_r(0)$. Since the remaining part of $\partial B_r(0)$ is of $\mathcal{H}_1$ measure zero, it can be covered by a collection of balls with diameters less than δ and sum of diameters less than ε. Therefore, $\sum_i \mathrm{diam}(B_i) \geq \mathcal{H}_1(\partial B_r(0)) - 2\varepsilon$.

Choose m such that $\sum_{i \leq m} \mathrm{diam}(B_i) \geq \mathcal{H}_1(\partial B_r(0)) - 3\varepsilon$. The perimeter of the convex polygon whose vertices are the centers of $B_1, \ldots, B_m$ is then at least $\mathcal{H}_1(\partial B_r(0)) - 3\varepsilon$, since the balls are disjoint.

Thus it is enough to consider a convex polygon γ inside the ball $B_r(0)$, and to prove that $\mathcal{H}_1(\Gamma) \geq \mathcal{H}_1(\gamma)$.

Let us note that the $\mathcal{H}_1$-length of a planar curve is at least the $\|\cdot\|$-distance between its endpoints. This can be shown by replacing the curve by a broken line of nearly the same $\mathcal{H}_1$-length (which may be achieved by a procedure similar to inscribing a polygon in a sphere as above) and using the triangle inequality. Therefore, if we replace an arc of a curve by a straight line segment, we do not make the curve longer (this is similar to the case of Euclidean norm, except that in some norms a curve may have length equal to the distance between its endpoints even if it is not a straight line).

Successively replacing arcs of the curve Γ by straight line segments containing sides of the polygon γ, we do not increase the $\mathcal{H}_1$-length, and in a finite number of steps will replace Γ by γ. $\quad\square$

To conclude the proof of Theorem 1, note that 1-Lipschitz mappings do not increase the Hausdorff measure. Therefore the $\mathcal{H}_1$-length of $\Gamma = f(\partial B_\rho(0))$ cannot exceed $\mathcal{H}_1(\partial B_\rho(0))$. On the other hand, if ρ is sufficiently large, then by Lemma 2, $\mathrm{Ind}_0\,\Gamma = n$, and by Lemma 1, $\|y\| \geq c_1\rho$ for any $y \in \Gamma$. So by Lemma 3 the $\mathcal{H}_1$-length of Γ is at least $nc_1\mathcal{H}_1(\partial B_\rho(0))$. Since $nc_1 > 1$, this is a contradiction which finishes the proof of the theorem. $\quad\square$

3. Having proved such a theorem, one would like to know if the $1/n$ bounds are precise. In the case of Euclidean norm the mappings $\phi_n(re^{i\theta}) = re^{ni\theta}$ have the ratio of constants equal to $1/n$ and maximum cardinality of a point preimage equal to n. Unfortunately, this does not immediately generalize to the case of arbitrary norm.

We are able to construct examples of such mappings in the situation when the unit ball is a regular polygon (or, of course, its affine equivalent). The ℓ_∞ norm is then a particular case of this. The idea of construction is as follows. Let V_0 be a vertex of the unit sphere $S = \{x \colon \|x\| = 1\}$. If x is a point on S, let $\arg_{\|\cdot\|}(x)$ be the length of the arc of S between V_0 and x in the counterclockwise direction, measured by the Hausdorff measure $\mathcal{H}_1$ corresponding to the metric defined by the norm $\|\cdot\|$. We define $\psi_n(rx) = ry$, where $r \geq 0$, and y is such a point on S that $\arg_{\|\cdot\|}(y) = n \arg_{\|\cdot\|}(x)$. One easily checks that the Lipschitz constant of ψ_n is equal to n. To check that the co-Lipschitz constant is equal to 1, one may consider a local inverse of ψ_n (see Lemma 5 below) and satisfy oneself that this inverse does not increase the $\|\cdot\|$-distance.

We do not know of such examples for other norms, so despite the feeling that the converse of the theorem holds for any norm (that is, there exist mappings with maximum of n point preimages and the ratio of constants equal to $1/n$), this question remains open.

4.   Now we would like to switch from Lipschitz quotient mappings to more general locally BNC mappings of $\mathbb{R}^2$ with the distance defined by an arbitrary norm $\|\cdot\|$. Our next goal will be to obtain a result which links the maximum cardinality of a point preimage to the ratio of the BNC constant C and the Lipschitz constant L of the mapping. This result, which is Theorem 2 below, deals only with the case $C/L > 1/2$. Recall that if $C/L \leq 1/2$, point preimages can be infinite (an example is given in Section 1). However, we know this only for Lipschitz, locally BNC mappings of the plane. See the next section for a discussion of the case $C/L \leq 1/2$ for Lipschitz, globally BNC mappings of $\mathbb{R}^2$.

We start with a simple lemma for BNC mappings between metric spaces.

Lemma 4. *If a mapping f between two normed spaces X and Y is L-Lipschitz and is locally C-BNC with $C/L > 1/2$ then f is $c = (2C - L)$ co-Lipschitz.*

Proof. Consider any point x and radius $R \leq \varepsilon(x)$, where $\varepsilon(x)$ is from the definition $(*)$ of local BNC property of the mapping f. There exists a point y such that $B_{CR}(y) \subset fB_R(x) \subset B_{LR}(f(x))$. Then the distance $\mathrm{dist}(y, f(x))$ does not exceed $(L - C)R < CR$. Now since $B_{CR - \mathrm{dist}(y, f(x))}(f(x))$ is contained in $B_{CR}(y)$, we conclude that the mapping f is locally $C - (L - C) = (2C - L)$ co-Lipschitz. This implies that f is globally $(2C - L)$ co-Lipschitz. For a proof that local co-Lipschitzness at every point implies global co-Lipschitzness see, for example, [C, Section 4]. $\square$

We proved in Theorem 1 that for an L-Lipschitz and c-co-Lipschitz mapping from the plane to itself, the cardinality of a point preimage is not greater than L/c. We thus have a

Corollary. *If $f \colon \mathbb{R}^2 \to \mathbb{R}^2$ is L-Lipschitz and C locally BNC with $C/L > 1/2$ then*

$$\max_{x \in \mathbb{R}^2} \# f^{-1}(x) \leq \tfrac{L}{2C-L}.$$

The bound on the right blows up when C/L is larger than but close to $1/2$. Our aim now is to improve the bound to the best possible one, that is, to prove that a C locally BNC and L-Lipschitz mapping with $C/L > 1/2$ is in fact a homeomorphism, i.e. the preimage of each point is a single point.

We will need several lemmas.

Lemma 5 (Local invertibility of a Lipschitz quotient mapping). *Let $f \colon \mathbb{R}^2 \to \mathbb{R}^2$ be a Lipschitz quotient mapping. There exists a finite subset $\mathcal{A}$ of $\mathbb{R}^2$ such that if Ω is a connected simply connected open domain which does not intersect with $\mathcal{A}$, then for any point x such that $y = f(x) \in \Omega$ there exists a mapping $\phi = \phi_{x,y} \colon \Omega \to \mathbb{R}^2$ which satisfies $\phi(y) = x$ and $f \circ \phi = \mathrm{Id}_\Omega$. This mapping ϕ is open and is locally $1/c$-Lipschitz, where c is the co-Lipschitz constant of f.*

Proof. By [JLPS] any such f is a composition $P \circ h$ of a polynomial P with a homeomorphism h. Let $\mathcal{A}$ be the finite set $\{P(z) \mid P'(z) = 0\}$. If Ω is a connected simply connected open domain which does not intersect with $\mathcal{A}$, then the polynomial P has a unique inverse, which is an analytic function p defined on Ω such that $p(y) = h(x)$. Define $\phi = h^{-1} \circ p$. It is clear that $\phi(y) = x$ and $f \circ \phi = \mathrm{Id}_\Omega$.

Since ϕ is a composition of a homeomorphism h^{-1} and an analytic function p, whose derivative $p'(\omega) = \frac{1}{P'(p(\omega))}$ is nonzero, we conclude that ϕ is open.

Suppose $\omega \in \Omega$ and $r > 0$ is so small that $B_{cr}(\omega) \subset \Omega$ and $B_r(\phi(\omega)) \subset \phi(\Omega)$. Then co-Lipschitzness of f implies that $\phi B_{cr}(\omega) \subset B_r(\phi(\omega))$, so ϕ is locally c^{-1}-Lipschitz, where c is the co-Lipschitz constant of f. □

Lemma 6. *Assume that a mapping f between two finite dimensional normed spaces X and Y is C locally BNC and is differentiable at a point a. Then for any $\epsilon > 0$ there exists $r = r(\epsilon, a)$ such that $fB_\rho(a) \supset B_{(C-\epsilon)\rho}(f(a))$ for $\rho \leq r$.*

Proof. Let $d_a f$ be the differential of f at a, so that $f(a+h) = f(a) + (d_a f)h + o(h)$. We will show now that $(d_a f)B_1(0) \supset B_C(0)$. Then for every $\epsilon > 0$ one can find r such that $\|o(h)\| < \epsilon \|h\|$ for $\|h\| \leq r$. It follows that for $\rho \leq r$ the image $fB_\rho(a)$ contains the ball centered at $f(a)$ of radius $C\rho - \epsilon\rho = \rho(C - \epsilon)$.

Assume $C_1 = \min_{\|x\|=1} \|d_a f(x)\| < C$. Then $(d_a f)B_1(0) \not\supset B_{C_1(1+\epsilon)}(0)$ for every $\epsilon > 0$ (thus, in particular, $(d_a f)B_1(0) \not\supset B_C(0)$).

It follows that $(d_a f)B_1(0) \not\supset B_{C_1(1+\epsilon)}(x)$ for any $x \in (d_a f)B_1(0)$ and $\epsilon > 0$. Indeed, assuming $(d_a f)B_1(0) \supset B_R(x)$ one gets $(d_a f)B_1(0) \supset -B_R(x) = B_R(-x)$ and thus

$$(d_a f)B_1(0) \supset \mathrm{conv}\big(B_R(x), B_R(-x)\big) \supset B_R(0).$$

Take r such that $\|o(h)\| < \frac{C-C_1}{2}\|h\|$ for $\|h\| \leq r$. Then for any $\rho \leq r$ one has

$$fB_\rho(a) \subset \Sigma = f(a) + \rho(d_a f)B_1(0) + B_{\rho\frac{C-C_1}{2}}(0).$$

The latter does not contain a ball of radius greater than $\frac{C+C_1}{2}\rho$ (the proof of this uses that $(d_a f)B_1(0)$ is convex), and in particular we conclude that Σ (and therefore $fB_\rho(a)$) does not contain a ball of radius $C\rho$, in contradiction to the local C-BNC property of f. $\square$

In what follows we will assume that $f(0) = 0$.

The next key lemma is an analogue of Lemma 1 for Lipschitz quotient mappings, but in the case of BNC mappings the proof becomes technically more complicated.

Lemma 7. *If a mapping $f : \mathbb{R}^2 \to \mathbb{R}^2$ is L-Lipschitz and is locally C-BNC with $C/L > 1/2$ and $f(0) = 0$, then for any $C' < C$ there exists $R > 0$ such that $\|f(x)\| \geq C'\|x\|$ for any $\|x\| \geq R$. Consequently, $fB_r(0) \supset B_{C'r}(0)$ for all $r \geq R$.*

Proof. Assume $L = 1$, set $M = 1 + \max_{f(z)=0}\|z\|$ and consider $R = 4M/(C - C')$. Assume that there exists a point x_0 such that $\|x_0\| = r \geq R$ and $\|f(x_0)\| < C'r$. There exists $\varepsilon > 0$ such that for all $y \in U(x_0, \varepsilon) = \{y : \|y\| = \|x_0\|$ and $\|y - x_0\| < \varepsilon\}$ one has $\|f(y)\| < C'r$.

Note that there exists $x_1 \in U(x_0, \varepsilon)$ and $\varepsilon' > 0$ such that $U(x_1, \varepsilon') \subset U(x_0, \varepsilon)$ and

$$\Omega = \cup_{y \in U(x_1, \varepsilon')}\big(0, 2f(y)\big)$$

is such a domain as was described in Lemma 5 (i.e., Ω does not contain $P(z)$ such that $P'(z) = 0$). Here $(0, a)$ is the straight line interval between 0 and a in $\mathbb{R}^2$. Let $\phi = \phi_{x_1, f(x_1)} : \Omega \to \mathbb{R}^2$ be the mapping from Lemma 5. Note that $\phi(\Omega)$, being open, contains an open neighbourhood of x_1, so there exists $\varepsilon_1 : 0 < \varepsilon_1 < \varepsilon'$, such that $U(x_1, \varepsilon_1) \subset \phi(\Omega)$. Then $\phi f(y) = y$ for any $y \in U(x_1, \varepsilon_1)$, since $\phi|_\Omega$ is a 1-1 mapping.

Since ϕ is locally Lipschitz, and is defined in an open cone, $\phi(0)$ is also well-defined.

In what follows, we are going to use both the Lebesgue measure $\mathcal{L}_k$ and the Hausdorff measure $\mathcal{H}_k$ for $k = 1, 2$. Recall that in $\mathbb{R}^k$ the measure $\mathcal{L}_k$ coincides with $\mathcal{H}_k$ on Borel sets. But the measure $\mathcal{H}_k$ is defined also in spaces of dimension different from k; if ψ is a Lipschitz mapping and A is such a set that $\mathcal{H}_k(A) = 0$, then $\mathcal{H}_k(\psi(A)) = 0$. In particular, if A is a Borel set in $\mathbb{R}^k$ such that $\mathcal{L}_k(A) = 0$, and $\psi : \mathbb{R}^k \to \mathbb{R}^k$ is Lipschitz, then $\mathcal{L}_k(\psi(A)) = 0$.

We know that f is $\mathcal{L}_2$-almost everywhere differentiable on $\phi(\Omega)$. Let $\mathcal{D} = \{t \in \phi(\Omega) \mid f$ is differentiable at $t\}$. Since $\mathcal{H}_2(\phi(\Omega) \setminus \mathcal{D}) = 0$ and f is Lipschitz, we conclude that the set $\Omega \setminus f(\mathcal{D})$ is also of $\mathcal{L}_2$ measure zero. Then by Fubini's theorem there exists a point y in $U(x_1, \varepsilon_1)$, such that almost every point of the interval $(0, 2f(y))$ with respect to $\mathcal{L}_1$ measure is in $f(\mathcal{D})$. Now consider the restriction of ϕ onto the segment $[0, f(y)]$. This restriction is a Lipschitz mapping from $[0, f(y)]$ to $\mathbb{R}^2$; therefore $\mathcal{H}_1$-almost every point of the curve $\gamma = \phi([0, f(y)])$ is in $\mathcal{D}$, that is f is $\mathcal{H}_1$-almost everywhere differentiable on γ. Let $\mathcal{B} = \mathcal{D} \cap \gamma$ be the set of points on γ where f is differentiable.

Since $\frac{C+C'}{2} < C$, by Lemma 6 for each differentiability point $z \in \mathcal{B}$ there exists $r_z > 0$ such that $fB_\rho(z) \supset B_{\rho(C+C')/2}(f(z))$ for any $\rho \leq r_z$.

Let $\mathcal{H}_1(\gamma)$ be the 1-Hausdorff measure of γ. There exists $\tau > 0$ such that if almost all of γ is covered by balls of diameter at most τ, then the sum of diameters of the balls is at least $\mathcal{H}_1(\gamma) - \frac{M}{2}$ (we defined M in the beginning of the proof). Without loss of generality we may assume that $\tau < M/2$.

Consider $\mathcal{F} = \{B_\rho(z) \mid z \in \mathcal{B}, \ \rho \leq \min\{r_z, \tau/2\}\}$. By the Besicovitch Covering Theorem (see [F, 2.8.15]) there exists a countable disjoint subcollection $\mathcal{F}_0$ of $\mathcal{F}$, which covers almost all of $\mathcal{B}$, therefore almost all of γ, with respect to the measure $\mathcal{H}_1$. Then

$$\sum_{B \in \mathcal{F}_0} \operatorname{diam} B \geq \mathcal{H}_1(\gamma) - \frac{M}{2}.$$

On the other hand the f-image of each ball $B \in \mathcal{F}_0$ contains a ball with center on $[0, f(y)]$ and of radius $r(B)\frac{C+C'}{2}$. Note that $\mathcal{F}_1 = \{B_{\rho(C+C')/2}(f(z)) \mid B_\rho(z) \in \mathcal{F}_0\}$ is a family of nonintersecting balls with centers on the interval $[0, f(y)]$, therefore

$$\frac{C+C'}{2} \sum_{B \in \mathcal{F}_0} \operatorname{diam} B = \sum_{B \in \mathcal{F}_1} \operatorname{diam} B \leq \|f(y)\| + \tau \frac{C+C'}{2}.$$

Thus

$$\|f(y)\| \geq \left(\mathcal{H}_1(\gamma) - \frac{M}{2}\right)\frac{C+C'}{2} - \tau\frac{C+C'}{2} \geq (\mathcal{H}_1(\gamma) - M)\frac{C+C'}{2}.$$

Note also that $\mathcal{H}_1(\gamma) \geq \|y\| - \|\phi(0)\| \geq r - M$ (see the explanation in the proof of Lemma 3), so $\|f(y)\| \geq (r - 2M)\frac{C+C'}{2}$. But we assumed that $\|f(y)\| < C'r$, so one gets

$$C'r > \frac{C+C'}{2}r - 2M,$$

or, equivalently, $2M > \frac{C-C'}{2}r$, which contradicts $r \geq R = \frac{4M}{C-C'}$. $\quad\square$

Theorem 2. *Let $\mathbb{R}^2$ be equipped with an arbitrary norm $\|\cdot\|$. If $f: \mathbb{R}^2 \to \mathbb{R}^2$ is an L-Lipschitz and C locally ball non-collapsing mapping with $C/L > 1/2$, then*

$$\#f^{-1}(x) = 1$$

for any point $x \in \mathbb{R}^2$.

Proof. By Lemma 4, such a mapping f is a Lipschitz quotient mapping. Let $n = \max_{x \in \mathbb{R}^2} \# f^{-1}(x)$. We may assume $f(0) = 0$.

Fix any C', such that $L/2 < C' < C$. Then by Lemma 7 there exists R such that $\|f(x)\| \geq C'\|x\|$ for all $\|x\| \geq R$. By Lemma 2, there exists $r > R$ such that $|\operatorname{Ind}_0 f(\partial B_r(0))| = n$.

Then by Lemma 4 the $\mathcal{H}_1$-length of $f(\partial B_r(0))$ is at least $nC'\mathcal{H}_1(\partial B_r(0))$, which is strictly greater than $\frac{nL}{2}\mathcal{H}_1(\partial B_r(0))$. But since f is L-Lipschitz, the length of $f(\partial B_r(0))$ is at most $L\mathcal{H}_1(\partial B_r(0))$. Hence $\frac{nL}{2} < L$, therefore $n = 1$. This finishes the proof of the theorem. $\square$

5.   The last question we would like to discuss here is what happens when a globally BNC mapping has a ratio of constants less than or equal to 1/2. The plane folding example, $F(x, y) = (x, |y|)$, where $C/L = 1/2$, shows that such a mapping neither has to be co-Lipschitz, nor is necessarily 1-1. However, the mapping in this example has point preimages of finite maximum cardinality 2.

On the other hand a mapping with ratio C/L less than 1/3 may have infinite point preimages. An example to this end is the following. For an interval $I = [a, b]$ in $\mathbb{R}^1$ define the "hat function" $h_I(x)$ by $\frac{b-a}{2} - |x - \frac{a+b}{2}|$. Now let the mapping $\zeta_A : \mathbb{R}^1 \to \mathbb{R}^1$, where $A > 1$, be defined by

$$\zeta_A(x) = \begin{cases} x, & \text{if } x \leq 0, \\ (-1)^k h_{[A^{-k}, A^{-k+1}]}(x), & \text{if } A^{-k} \leq x \leq A^{-k+1}, \ k \text{ a positive integer,} \\ x - 1, & \text{if } x > 1. \end{cases}$$

Obviously, ζ_A is a 1-Lipschitz function. One can check that ζ_A is BNC with constant $C = \frac{1-A^{-2}}{3-A^{-2}}$. Then the function $f(x, y) = (x, \zeta_A(y))$ is a Lipschitz and BNC mapping of the plane, with infinite point preimages, and the ratio of constants less than but arbitrarily close to 1/3 (at least with respect to a norm $\|\cdot\|$ for which $\|(x, y)\| = \|(x, -y)\|$).

Note that a point preimage under a Lipschitz BNC mapping may even be uncountable. For example, if

$$E = [0, 1] \setminus \bigcup_{k, n \geq 0} \left(\frac{3k + 1}{3^n}, \frac{3k + 2}{3^n} \right)$$

is a Cantor set on $[0, 1]$, the mapping $g(x) = \operatorname{dist}(x, E)$ is 1-Lipschitz and is globally BNC, whose zeros set is E.

We also have a proof that in 1-dimensional space the bound of 1/3 cannot be improved (that is, if a Lipschitz and BNC mapping has infinite point preimages, then the ratio of constants C/L is strictly less than 1/3). Thus, we have no definite results concerning point preimages under Lipschitz globally BNC mappings of the plane whose ratio of constants is between 1/3 and 1/2.

Let us summarize the results concerning the estimates of the maximum cardinality of the preimage of a point under the three classes of Lipschitz mappings of the plane. Let L be the Lipschitz constant of a mapping. If a mapping is Lipschitz quotient with co-Lipschitz constant c, the preimage of a point consists of at most L/c points. If a mapping is (globally) BNC with BNC constant C, then in the case $C/L > 1/2$ a point preimage is a single point, in the case $C/L < 1/3$ it can be infinite, and in the case $1/3 \le C/L \le 1/2$ we have no definite answer. And if a mapping is locally BNC with BNC constant C, the complete answer is as follows. If $C/L > 1/2$, a point preimage is a single point, and in the case $C/L \le 1/2$ a point preimage can be infinite.

Acknowledgement. I thank Professor Gideon Schechtman for his continuing support throughout my research in this field.

References

[C]   Csörnyei, M. (2001): Can one squash the space into the plane without squashing? Geom. Funct. Anal., **11(5)**, 933–952

[JLPS] Johnson, W.B., Lindenstrauss, J., Preiss, D., Schechtman, G. (2000): Uniform quotient mappings of the plane. Michigan Math. J., **47**, 15–31

[M]   Maleva, O. Lipschitz quotient mappings with good ratio of constants. Mathematika, to appear

[F]   Federer, H. (1996): Geometric Measure Theory. Springer

Some Remarks on a Lemma of Ran Raz

Vitali Milman and Roy Wagner

School of Mathematical Sciences, Tel Aviv University, Tel Aviv 69978, Israel
milman@post.tau.ac.il, pasolini@post.tau.ac.il

In this note we will review a Lemma published by Ran Raz in [R], and suggest improvements and extensions. Raz' Lemma compares the measure of a set on the sphere to the measure of its section with a random subspace. Essentially, it is a sampling argument. It shows that, in some sense, we can simultaneously sample a function on the entire sphere and in a random subspace.

In the first section we will discuss some preliminary ideas, which underlie the lemma and our interest in it. We will view a random subspace as the span of random points, without discussing the sampling *inside* the subspace. We will demonstrate how substantial results follow from this elementary approach. In the second section we will review the original proof of Raz' Lemma, analyse it, and improve the result. In the final section we will extend the Lemma to other settings.

1 Random Points Span a Random Subspace, and What It Has to Do with Medians, Spectra and Concentration

1.1 An Appetiser

The starting point of our discussion, and an important ingredient in Raz' Lemma, is the following simple observation regarding random subspaces: to choose a random k-dimensional subspace in $\mathbb{R}^n$ is nothing more than to scatter k random points on the sphere.

Indeed, the Lebesgue measure on the sphere is the unique normalised Haar measure invariant under rotations. In other words, fix $U \in O(n)$; if $y_1, \ldots, y_n$ are independent and uniform, then so are $U(y_1), \ldots, U(y_n)$. Therefore span$\{y_1, \ldots, y_n\}$ has the same distribution as span$\{U(y_1), \ldots, U(y_n)\} = U(\text{span}\{y_1, \ldots, y_n\})$. So the distribution of the span of k random points on the sphere is the unique rotation invariant distribution on the Grassmanian $G_{n,k}$.

This observation alone has surprising strength. Take a continuous function f on S^{n-1} with median $\mathcal{M}$ (namely both $\mu(f(x) > \mathcal{M})$ and $\mu(f(x) < \mathcal{M})$ do not exceed $1/2$). The probability that k independent uniformly distributed points be on the same side of the median is at most 2^{-k+1}. The probability that their span is on one side of the median is even smaller. Since their span is uniform in the Grassmanian, we find that those elements of $G_{n,k}$, which do not intersect the median, measure less than 2^{-k+1}.

Playing with parameters, one can see that in a random orthogonal decomposition of R^n into $\log n$-dimensional subspaces, each component will intersect the median, and thus produce $n/\log n$ orthogonal points on the median. Using the additional information from Raz' Lemma (to be introduced below), we can obtain the same result for an orthogonal decomposition into 130-dimensional subspaces, yielding $n/130$ orthogonal points on the median. These facts are brought only to illustrate possible application avenues for the ideas we promote. We omit the details, because the above results are not optimal for this specific instance. Indeed, as G. Schechtmann observed, if we orthogonally decompose the space into 2-dimensional subspaces, each component is 50% likely to intersect the median. In expectation, therefore, half of the components intersect the median, and yield $n/4$ orthogonal points on the median.

There is a small notable advantage to the non-optimal arguments over the final argument sketched in the last paragraph — the first arguments are "high probability" arguments, and can therefore be used in conjunction with other "high probability" restrictions. Finally, in [YY] topological considerations provide any real continuous function on S^{n-1} with n orthogonal equivalued points. In our discussion we secure less points, but prescribe them on the median.

1.2 From Spectrum to Concentration

It has by now become a standard turn of narrative to begin with concentration of measure and culminate with existence of spectrum. One proves first, using an isoperimetric inequality, that a Lipschitz function is close to its median everywhere but for a small-measure exceptional set. Then one deduces that the function can have only small oscillations on random subspaces, which makes it close to a constant. In our jargon we will say it has *spectrum* (cf. [M]).

The ideas of the previous section allow to reverse the plot. We will first invoke some 20-year-old unpublished discussions between M. Gromov and the first named author in order to obtain spectrum. We will then go from there to prove concentration. One protagonist finds itself missing from our new storyline; isoperimetry will play no role. Instead, we will only require the classical cap volume estimate:

$$\mu_{S^n}(x_1 \geq \epsilon) \leq \sqrt{\frac{\pi}{8}} e^{-\frac{1}{2}\epsilon^2(n-2)} . \tag{1}$$

A large portion of our account will take place on the Stieffel manifold $W_{n,2} = \{(x,u) \mid x \in S^{n-1}, u \in S^{n-2}(T_x)\}$ equipped with the local product topology. An element of this manifold is a pair of normalised orthogonal vectors (T_x is the tangent plane at x).

Note that the action of $O(n)$ on S^{n-1} naturally extends to a transitive action on $W_{n,2}$, inducing the measure

$$\mu_{W_{n,2}}(A) = \mu_{O(n)}\left(\{T \mid T(\underline{e}_1, \underline{e}_2) \in A\}\right) ,$$

where $\underline{e}_1$ and $\underline{e}_2$ are two orthogonal unit vectors used as reference. As the Stiefel manifold is metrically equivalent to joining two copies of $D^{n-1} \times S^{n-2}$, it carries a δ-net of cardinality at no more than $(\frac{C}{\delta})^{2n}$.

Let F be a 1-Lipschitz function on the sphere. We will view ∇F, the gradient of F, as a function on $W_{n,2}$ with the action $\nabla F(x,u) = \nabla_x F \cdot u$. Since we want F to have a gradient, and since we want to induce from the behaviour of the gradient on a net to its behaviour on the entire space, we must replace F for a while with a smoother look-alike. We will introduce

$$f(x) = \operatorname{Ave}_{y \in B_\eta(x) \cap S^{n-1}} F(y) ,$$

where $B_\eta(x)$ is the ball of radius η around x (η will be selected later). Since F is 1-Lipschitz, this new function is differentiable with

1. $|\nabla f(x,u)| \le 1$ for all $(x,u) \in W_{n,2}$,
2. the Lipschitz constant of $\nabla f(\cdot, u)$ is smaller than $\frac{C}{\eta}$, and
3. $|f(x) - F(x)| \le \eta$ for all $x \in S^{n-1}$.

Let us now agree that C and C' represent universal constants, which do not retain the same value throughout the text, and commence arguing.

Step 1. The action of ∇f on $W_{n,2}$ is small everywhere but for a small-measure exceptional set.

Proof. Fix $x \in S^{n-1}$. The linear functional $\nabla f(x, \cdot)$ on the tangent space T_x has a 1–codimensional kernel. If $u \in S(T_x)$ is ϵ–approximated by a $v \in \ker(\nabla f(x, \cdot))$, then

$$|\nabla f(x,u)| \le |\nabla f(x, u - v)| + |\nabla f(x,v)| \le \|u - v\| \le \epsilon .$$

According to (1), the measure of directions u with the above property is at least $1 - e^{-C\epsilon^2 n}$. As this holds for every $x \in S^{n-1}$, we conclude

$$\mu_{W_{n,2}}\{(x,u) \mid |\nabla f(x,u)| \le \epsilon\} \ge 1 - e^{-C\epsilon^2 n} . \tag{2}$$

□

The information gathered in the first step above allows us to simultaneously map any $e^{C\epsilon^2 n}$ elements of $W_{n,2}$ to elements with small ∇f using a single orthogonal transformation. Indeed,

$$\mu_{O(n)}\left(\{T \mid |\nabla f(T(x,u))| \le \epsilon\}\right) = \mu_{W_{n,2}}\left(\{(x,u) \mid |\nabla f(x,u)| \le \epsilon\}\right)$$
$$\ge 1 - e^{-C\epsilon^2 n}$$

and therefore

$$\mu_{O(n)}\left(\left\{T \mid |\nabla f(T(x_i, u_i))| \le \epsilon, 1 \le i < e^{C\epsilon^2 n}\right\}\right) > 0 .$$

As usual, if we take $e^{C'\epsilon^2 n}$ elements, with $C' < C$, only an exponentially small measure of operators in $O(n)$ will fail to do the job.

Step 2. F has small oscillations on random subspaces.

Proof. As we observed, $(\frac{C'}{\delta})^{2k}$ points suffice for a δ-net on $W_{k,2}(\mathbb{R}^k)$. Assume that $(\frac{C'}{\delta})^{2k} < e^{C\epsilon^2 n}$. By the above, this net can be orthogonally mapped to points where ∇f is bounded by ϵ (as a function on $W_{n,2}$). The new points still form a δ-net in $W_{k,2}(V^k)$ for some $V_k \in G_{n,k}$. We can now establish that ∇f is small on the entire $W_{k,2}(V^k)$.

Indeed, any $(x, u) \in W_{k,2}(V^k)$ has a δ–neighbour (x', u') in the net, so we get

$$|\nabla f(x,u)| \le |\nabla f(x,u) - \nabla f(x,u')| + |\nabla f(x,u') - \nabla f(x',u')| + |\nabla f(x',u')|$$
$$\le \delta + \frac{C}{\eta} \cdot \delta + \epsilon .$$

The combination of f's small directional derivatives with the sphere's bounded diameter secures small oscillations for f. These small oscillations easily transfer to the original F.

Take $x_1, x_2 \in V^k$, and $\eta \le 1$. We get

$$|F(x_0) - F(x_1)| \le \max_{(x,u)\in W_{k,2}(V^k)} |\nabla f(x,u)| \cdot \|x_0 - x_1\| + 2\eta$$
$$\le \frac{C}{\eta} \cdot \delta + \epsilon + 2\eta .$$

We now set $\epsilon \le 1$, $\eta = \epsilon$, $\delta = \epsilon^2$, and $k = \lfloor C\epsilon^2 n/\log\frac{1}{\epsilon^2}\rfloor$, and obtain the following:

Statement. With probability at least $1 - e^{-C\epsilon^2 n}$, the 1-Lipschitz function F oscillates by no more than ϵ on random $\lfloor C\epsilon^2 n/\log\frac{1}{\epsilon^2}\rfloor$-dimensional subspaces. $\square$

Step 3. The measure of points, where F is ϵ-close to its median, is at least $1 - e^{-C\epsilon^2 n}$ (as long as $\epsilon > C'\frac{\log n}{\sqrt{n}}$).

Proof. From the previous section we know that a k-dimensional subspace intersects the median of F with probability at least $1 - 2^{-k+1}$. Combining this information with the even greater odds of k-dimensional spectrum, we find that only an e^{-Ck} proportion of k-dimensional subspaces allow F to diverge more than ϵ from its median (note that this argument requires $k \ge 2$, but this is supported by the restriction on ϵ).

This estimate readily carries over to at most e^{-Ck} proportion of points on the sphere, where F can diverge by ϵ from its median; we are close to our goal, but not quite there.

In order to get an improved estimate to the proportion of points where F diverges by at least ϵ from its median, let's recycle again the idea that a random k-dimensional subspace is the span of k random points.

$P_{V \in G_{n,k}}(F$ is ϵ-far from its median on all points of $V)$

$= P_{\{x_i\}_{i=1}^k \in (S^{n-1})^k}(F$ is ϵ-far from its median on span$\{x_1, \ldots, x_k\})$

$\leq P_{\{x_i\}_{i=1}^k \in (S^{n-1})^k}(F$ is ϵ-far from its median on $\{x_1, \ldots, x_k\}) \leq (e^{-C'k})^k$.

We again confront this estimate with the estimate for k-dimensional spectrum in the Statement above. Since $k^2 > C\epsilon^2 n$ for the stated values of ϵ and k, we conclude that at least $1 - e^{-C\epsilon^2 n}$ proportion of k-dimensional subspaces, and hence at least such a proportion of points of the sphere, are sent by F ϵ-close to the median. By repeating the above argument, we can improve slightly the restriction on ϵ, and bring the enumerator towards $\sqrt{\log(n)}$. $\square$

While we demonstrated this line of thought only for the sphere, it can be easily imagined in other contexts. Complementing the well-known "concentration implies spectrum" principle, we should encourage a general "spectrum implies concentration" ideology. Raz' Lemma, as discussed below, is one of the components of this ideology.

2 Raz' Lemma

2.1 Raz' Original Argument

Up to this point we only used the simple fact that independent random points sample random subspaces. We will now see that the same random points, even though they needn't be independent with respect to the subspace they span, are still tame enough to provide useful information regarding their span.

Raz' Lemma, introduced in [R], is a special deviation inequality on the Grassmanian. Let μ be the normalised Haar measure on the unit sphere, and ν be the Haar probability measure on $G_{n,k}$. If we restrict μ to subspaces according to the formula $\mu|_V(A) = \mu_V(A \cap V)$, where μ_V is the normalised Haar measure on V, we'll find that

$$\mu(A) = \int_{V \in G_{n,k}} \mu|_V(A) \, d\nu.$$

We may now ask how well $\mu|_V(A)$ is concentrated as a function of V on $G_{n,k}$. It turns out, that even though $\mu|_V(A)$ needn't even be continuous, we still get a concentration inequality similar to that of Lipschitz functions.

Theorem 1 (Raz' Lemma). *Let* $C \subseteq S^{n-1}$ *and denote* $\mu(C) = c$. *Then:*

$$\nu\big(|\mu|_V(C) - c| \geq \varepsilon\big) \leq \frac{4e^{-\frac{\varepsilon^2 k}{2}}}{\varepsilon},$$

where ν *and* $\mu|_V$ *are as above.*

Remark 1. In the following text we will use P for the product of k Haar probability measures on k copies of S^{n-1}. Recall that we have established that a sequence uniform in P spans a subspace uniform in ν. Therefore, where an event in $(S^{n-1})^k$ depends only on the span of the sequence of vectors, the probability P becomes synonymous with ν. In fact, where no harm is expected, we intend to generally confuse subspaces and their spanning sequences.

Proof. Let $y_1, \ldots, y_k$ be independent μ-uniform variables. We already established that $\text{span}\{y_1, \ldots, y_k\}$ is a uniform element of the unique rotation invariant measure on $G_{n,k}$. The behaviour of the y_i's inside their span is more delicate. If $y_1, \ldots, y_k$ were independent inside their span, they would simultaneously sample both A and $A \cap V$ quite well, and thus secure an easy proof of Raz' Lemma. However, restricting a measure to a zero measure subset depends on further desired properties, and our needs preclude independence inside the span. We will motivate our choice of restricted measure and justify its properties in the next section. For now, just assume that y_i is Haar-uniform in $\text{span}\{y_1, \ldots, y_k\}$, and nothing more.

Let's start with the upper tail estimate. We will study the event

$$B = \left\{ (y_i)_{i=1}^k \mid \mu|_{\text{span}\{y_i\}_{i=1}^k}(C) \geq c + 2\varepsilon \right\}.$$

We use the elementary inequality:

$$P(B) \leq \frac{P(A)}{P(A|B)},$$

into which we substitute

$$A = \left\{ (y_i)_{i=1}^k \mid \frac{\sum_{i=1}^k 1_C(y_i)}{k} - c \geq \varepsilon \right\}.$$

$P(A)$ is estimated from above by $e^{-2\varepsilon^2 k}$ (computer scientists call this sampling estimate a Chernoff-type bound, but it goes back to Kolmogorov, see [L], section 18.1). In order to estimate $P(A|B)$, let's take a subspace V where B holds, and consider first

$$P\big(A \mid \text{span}\{y_i\}_{i=1}^k = V \big).$$

Chebyshev's inequality bounds this from below by:

$$\frac{\mathbb{E}\left(\frac{\sum 1_C(y_i)}{k} \mid \text{span}\{y_i\}_{i=1}^k = V \right) - (c+\varepsilon)}{\max\left(\frac{\sum 1_C(y_i)}{k} \mid \text{span}\{y_i\}_{i=1}^k = V \right) - (c+\varepsilon)}.$$

Since each y_i is uniform in $\text{span}\{y_i\}_{i=1}^k$, and regardless of their conditional dependence, this equals

$$\frac{\mu|_V(C) - (c + \varepsilon)}{1 - (c + \varepsilon)} .$$

Since V was chosen such that $\mu|_V(C) \geq c + 2\varepsilon$, we get

$$P\big(A \mid \text{span}\{y_i\}_{i=1}^k = V\big) \geq \varepsilon ,$$

and deduce

$$P(A|B) \geq \varepsilon. \tag{3}$$

This last transition is, from the point of view of measure theory, the most delicate point of the proof. It would follow, if, for example, the global measure were the average of the restricted measures. This property is the key to our choice of restriction of the measure to a subspace. We will defer the details to the next section, so that we can now directly conclude

$$P(B) \leq \frac{P(A)}{P(A|B)} \leq \frac{e^{-2\varepsilon^2 k}}{\varepsilon} .$$

If we rescale ε, and consider the symmetric lower tail estimate, we find the theorem proved. $\square$

We would like to point out that Raz' Lemma is not optimal in the case of fixed ε and k, and c tending to 0 (or, symmetrically, to 1). Obviously $P_V(|\mu|_V(C) - c| > \varepsilon)$ should tend to zero as c tends to zero. Nevertheless, the bound we have does not emulate this property.

The reason for this lies in the sampling estimate we quote. While $P\big(\frac{\sum_{i=1}^k 1_C(y_i)}{k} - c > \varepsilon\big)$ tends to zero with c, the bound $e^{-2\varepsilon^2 k}$ does not. To adapt Raz' Lemma to such a marginal situation, one must replace the sampling estimate in the proof. Fortunately, one of the advantages of the proof is its remarkable modularity and resilience to variations.

2.2 Conditional (In)dependence

The restriction of a measure to a zero measure subset requires additional structure in order to make sense. For example, while volume in $\mathbb{R}^n$ is uniquely determined, the definition of surface area depends on Euclidean structure.

For the definition of the restricted measure $P\big(A \mid \text{span}\{y_i\}_{i=1}^k = V\big)$ in the proof (to which we will refer in short as $P(A|V)$) we shall take

$$\int_{x_1 \in S^{n-1}} \cdots \int_{x_k \in S^{n-1}} \int_{T \in O(n, \text{span}\{x_1, \dots, x_k\} \to V)} \chi_A\big(T(x_1), \dots, T(x_k)\big) ,$$

where the integral is calculated with respect to normalised Haar measures, and $O(n, U \to V)$ stands for the collection of unitary operators in $O(n)$, which map the subspace U to the subspace V. The definition aims at validating the deduction of (3), which would follow from the equality

$$\int_{V \in G_{n,k}} P(A|V) d\nu = P(A).\tag{4}$$

Indeed,

$$\int_{V \in G_{n,k}} \int_{(x_i)_{i=1}^k \in (S^{n-1})^k} \int_{T \in O(n, \text{span}\{x_1,...,x_k\} \to V)} \chi_A\big(T(x_1),\ldots,T(x_k)\big)$$

$$= \int_{(x_i)_{i=1}^k \in (S^{n-1})^k} \int_{V \in G_{n,k}} \int_{T \in O(n, \text{span}\{x_1,...,x_k\} \to V)} \chi_A\big(T(x_1),\ldots,T(x_k)\big)$$

$$= \int_{(x_i)_{i=1}^k \in (S^{n-1})^k} \int_{T \in O(n)} \chi_A\big(T(x_1),\ldots,T(x_k)\big)$$

$$= \int_{T \in O(n)} \int_{(x_i)_{i=1}^k \in (S^{n-1})^k} \chi_A\big(T(x_1),\ldots,T(x_k)\big)$$

$$= \int_{T \in O(n)} \int_{(x_i)_{i=1}^k \in (S^{n-1})^k} \chi_A(x_1,\ldots,x_k)$$

$$= \int_{(x_i)_{i=1}^k \in (S^{n-1})^k} \chi_A(x_1,\ldots,x_k) = P(A).$$

The transition between the second and third lines follows from the uniqueness of the normalised Haar measure on $O(n)$.

Our definition also conforms to our earlier statement that the marginal distribution of a single coordinate with respect to $P(\cdot|V)$ is simply the Haar probability measure on the sphere. Indeed, substitute into the definition a set of the form $A \times S^{n-1} \times \cdots S^{n-1}$, follow a reasoning similar to the above, and you will simply get $\mu|_V(A)$.

Finally, the restricted measure $P(\cdot|V)$ is no longer the product of k Haar measures on $S(V)$. Indeed, let A be the set of all closely clustered k-tuples in $(S^{n-1})^k$. The global measure $P(A)$ will be strictly smaller than the product measure of A in $(S(V))^k$ for all subspaces V (by well-known concentration of measure estimates). Therefore, if $P(\cdot|V)$ were the product of k Haar measures, we would be in violation of the just-proven equality (4).

The loss of independence when restricting to a subspace is responsible for the 'bad' denominator in Raz' estimate. In the next section we will present a trick, which allows one to erase the denominator altogether.

2.3 Getting Rid of the Denominator

We will use here the 'tensorisability' of the exponential estimate of $P(A)$ to suppress the linear factor arising from the estimate of $P(A|B)$.

Fix $k < n$ and arbitrary m. Consider $y_1, \ldots y_{mk}$ independent Haar-uniform variables in S^{n-1}. Now the sequence $\{\text{span}\{y_i\}_{i=(j-1)k+1}^{jk}\}_{j=1}^m$ is Haar-uniform and independent in $G_{n,k}$, and for every $1 \le \ell \le k$ the variable $y_{(j-1)k+\ell}$ is Haar-uniform in $\text{span}\{y_i\}_{i=(j-1)k+1}^{jk}$ (we make no claim as to their conditional independence here).

We will consider the event:

$$A = \left\{ (y_i)_{i=1}^{mk} \ \middle| \ \frac{\sum_{i=1}^{mk} 1_C(y_i)}{mk} - c > \varepsilon \right\}$$

and

$$B = B_1 \cap B_2 \cap \cdots \cap B_m ,$$

where

$$B_j = \left\{ (y_i)_{i=1}^{mk} \ \middle| \ \mu|_{\mathrm{span}\{y_i\}_{i=(j-1)k+1}^{jk}}(C) \geq c + 2\varepsilon \right\} .$$

Just as before $P(A) \leq e^{-2\varepsilon^2 mk}$. Furthermore, by Chebyshev

$$P(A \mid \forall j : \mathrm{span}\{y_i\}_{i=(j-1)k+1}^{jk} = V_j)$$

$$\geq \frac{\mathbb{E}\left(\frac{\sum 1_C(y_i)}{mk} \ \middle| \ \forall j : \mathrm{span}\{y_i\}_{i=(j-1)k+1}^{jk} = V_j \right) - (c+\varepsilon)}{\max\left(\frac{\sum 1_C(y_i)}{mk} \ \middle| \ \forall j : \mathrm{span}\{y_i\}_{i=(j-1)k+1}^{jk} = V_j \right) - (c+\varepsilon)}$$

$$\geq \frac{\frac{1}{m}\sum_j \mu|_{V_j}(C) - (c+\varepsilon)}{1 - (c+\varepsilon)} .$$

As before, we conclude that $P(A|B) \geq \varepsilon$.

Since the original y_i's are independent, so are the events B_i, And we get $P(B) = P(B_1) \cdot \ldots \cdot P(B_m) = P(B_1)^m$. It follows that

$$P(B_1)^m \leq \frac{P(A)}{P(A|B)} \leq \frac{e^{-2\varepsilon^2 km}}{\varepsilon} .$$

Letting m go to infinity, we find:

$$P(B_1) \leq e^{-2\varepsilon^2 k} .$$

Again, rescaling and repeating for the lower tail, we conclude with

Theorem 2 (Improved Raz' Lemma). *In the same setting as Theorem 1,*

$$\nu\left(\left| \mu \right|_V (C) - c \right| \geq \varepsilon) \leq 2e^{-\frac{\varepsilon^2 k}{2}} .$$

3 Extensions

3.1 From S^{n-1} to $G_{n,m}$

In this section we will transport the result from S^{n-1} (which, for the purpose of this discussion, works like $G_{n,1}$) to $G_{n,m}$. The same can be repeated for various related homogeneous manifolds.

Theorem 3. *Let $C \subseteq G_{n,m}$ and $\mu(C) = c$ (where μ is the normalised Haar-measure on $G_{n,m}$). Choose k such that $km \leq n$, then*

$$P_{V \in G_{n,km}} \left(\left| \mu(C \mid G_{km,m}(V)) - c \right| > \varepsilon \right) \leq 2e^{-\frac{\varepsilon^2 k}{2}} .$$

Proof. Let $V_1, \ldots, V_k$ be independent Haar-uniform variables in $G_{n,m}$. The span$\{V_i\}_{i=1}^{k}$ is uniform in $G_{n,mk}$ and each V_i is uniform in $G_{mk,m}(\text{span}\{V_i\}_{i=1}^{k})$.
Let

$$A = \left\{ (V_i)_{i=1}^{k} \mid \frac{\sum_{i=1}^{k} 1_C(V_i)}{k} - c \geq \varepsilon \right\}$$

and

$$B = \left\{ (V_i)_{i=1}^{k} \mid \mu\big|_{G_{km,m}(\text{span}\{V_i\}_{i=1}^{k})}(C) \geq c + 2\varepsilon \right\} .$$

The remainder of the argument works just as before. By a sampling estimate

$$P(A) \leq e^{-2\varepsilon^2 k} .$$

By Chebyshev's inequality

$$P\big(A \mid \text{span}\{V_i\}_{i=1}^{k} = V\big) \geq \frac{\mu\big|_{G_{km,m}(V)}(C) - (c + \varepsilon)}{1 - (c + \varepsilon)} ,$$

and so $P(A|B) \geq \varepsilon$. The same trick as in section 2.3 can write off $P(A|B)$, and we obtain the desired result. $\square$

3.2 From Indicators to General Functions

The purpose of this section is to replace the function 1_C in the previous sections by any other function. For this we require the following inequality by Kolmogorov (as quoted in [L], section 18.1).

Theorem 4 (Kolmogorov's Inequality). *Let f be a function on a probability space with $\mathbb{E}f = M$, $\|f - M\|_2 = s$ and $\|f - M\|_\infty = b$. Let $(y_i)_{i=1}^{k}$ be independent random variables on the domain of f.*
   1. *If $\varepsilon \leq \frac{s^2}{b}$ then:*

$$P\left(\left| \frac{\sum_{i=1}^{k} f(y_i)}{k} - M \right| \geq \varepsilon \right) \leq 2e^{-\frac{1}{4}(\frac{\varepsilon}{s})^2 k} .$$

   2. *If $\varepsilon \geq \frac{s^2}{b}$ then:*

$$P\left(\left| \frac{\sum_{i=1}^{k} f(y_i)}{k} - M \right| \geq \varepsilon \right) \leq 2e^{-\frac{1}{4}(\frac{\varepsilon}{b})k} .$$

Kolmogorov's inequality allows to extend Raz' technique to general functions.

Theorem 5. *Let* $f : S^{n-1} \to \mathbb{R}$, *and* M, s, *and* b *as in the previous theorem.*
   1. *If* $\varepsilon \leq \frac{s^2}{b}$ *then:*

$$P_{V \in G_{n,k}}\left(\left|\mathbb{E}|_V (f) - M\right| > \varepsilon\right) \leq 2 \cdot e^{-\frac{1}{16}(\frac{\varepsilon}{s})^2 k} .$$

   2. *If* $\varepsilon \geq \frac{s^2}{b}$ *then:*

$$P_{V \in G_{n,k}}\left(\left|\mathbb{E}|_V (f) - M\right| > \varepsilon\right) \leq 2 \cdot e^{-\frac{1}{8}(\frac{\varepsilon}{b})k} .$$

Note that if we substitute characteristic functions into f, we reproduce the original statement for sets (up to constants).

Proof. Substitute into Raz' argument the events:

$$A = \left\{ (y_i)_{i=1}^k \,\middle|\, \frac{\sum_{i=1}^k f(y_i)}{k} - M \geq \varepsilon \right\} \quad \text{and}$$

$$B = \left\{ (y_i)_{i=1}^k \,\middle|\, \mathbb{E}|_V (f) \geq M + 2\varepsilon \right\} .$$

Estimate $P(A)$ by Kolmogorov's bound, and use Chebyshev's inequality as above:

$$P\left(A \mid \mathrm{span}\{y_i\}_i = V\right) \geq \frac{\mathbb{E}|_V (f) - (M + \varepsilon)}{(b + M) - (M + \varepsilon)} ,$$

and so

$$P(A|B) \geq \frac{\varepsilon}{b}.$$

Using the trick from section 2.3 to write off $P(A|B)$, we obtain the promised result. □

References

[L]   Loève, M. (1993): Probability Theory, 3rd edition. Van Nostrand Reinholt, New York

[M]   Milman, V. (1988): The heritage of P. Lévy in geometrical functional analysis. Astérisque, **157–158**, 273–301

[MS]   Milman, V.D., Schechtman, G. (1986): Asymptotic theory of finite dimensional normed spaces. Lecture Notes in Mathematics, **1200**, Springer-Verlag, Berlin

[R]   Raz, R. (1999): Exponential separation of quantum and classical communication complexity. Proceedings of the 31st Annual ACM Symposium on Theory of Computing (STOC'99) ACM Press, Atlanta, Georgia, 358–367

[YY]   Yamabe, H., Yajubô, Z. (1950): On the continuous function defined on the sphere. Osaka Math. Jour., **2**, 19–22

On the Maximal Perimeter of a Convex Set in $\mathbb{R}^n$ with Respect to a Gaussian Measure

Fedor Nazarov

Department of Mathematics, Michigan State University, East Lansing, MI 48824-1027, USA *fedja@math.msu.edu*

Introduction

Let A be an $n \times n$ positive definite symmetric matrix and let

$$d\gamma_A(y) = \varphi_A(x)\,dx = (2\pi)^{-\frac{n}{2}}\sqrt{\det A}\, e^{-\frac{\langle Ax, x\rangle}{2}}\,dx$$

be the corresponding Gaussian measure. Let

$$\Gamma(A) = \sup\left\{\frac{\gamma_A(Q_h \setminus Q)}{h} : Q \subset \mathbb{R}^n \text{ is convex}, h > 0\right\}$$

where Q_h denotes the set of all points in $\mathbb{R}^n$ whose distance from Q does not exceed h.

Since, for convex Q, one has $Q_{h'+h''} \setminus Q = [(Q_{h'})_{h''} \setminus Q_{h'}] \cup [Q_{h'} \setminus Q]$, the definition of $\Gamma(A)$ can be rewritten as

$$\Gamma(A) = \sup\left\{\limsup_{h\to 0+} \frac{\gamma_A(Q_h \setminus Q)}{h} : Q \subset \mathbb{R}^n \text{ is convex}\right\}$$

$$= \sup\left\{\int_{\partial Q} \varphi_A(y)\,d\sigma(y) : Q \subset \mathbb{R}^n \text{ is convex}\right\}$$

where $d\sigma(y)$ is the standard surface measure in $\mathbb{R}^n$.

Making the change of variable $x \to Bx$ where B is the (positive definite) square root of A, the last expression can be rewritten as

$$\sup\left\{\int_{\partial Q} \varphi(y)|B\nu_y|\,d\sigma(y) : Q \subset \mathbb{R}^n \text{ is convex}\right\}$$

where $\varphi(y) = (2\pi)^{-\frac{n}{2}} e^{-\frac{|y|^2}{2}}$ is the density of the standard Gaussian measure $d\gamma$ in $\mathbb{R}^n$ and ν_y is the unit normal vector to the boundary ∂Q of the body Q at the point $y \in \partial Q$.

Recall that the Hilbert-Schmidt norm $\|A\|_{\text{H-S}}$ of a positive definite symmetric matrix A is defined as the square root of the sum of squares of all entries of A or, which is the same, as the square root of the sum of squares of the eigenvalues of A. The aim of this paper is to prove the following

Theorem. *There exist absolute constants $0 < c < C < +\infty$ such that*

$$c\sqrt{\|A\|_{\text{H-S}}} \leqslant \Gamma(A) \leqslant C\sqrt{\|A\|_{\text{H-S}}}.$$

A few words should, probably, be said about the history of the question. To the best of my knowledge, it was S. Kwapien who first pointed out that it would be desirable to have good estimates for $\Gamma(I_n)$ (i.e., for the maximal perimeter of a convex body with respect to the standard Gaussian measure). The only progress that has been made was due to K. Ball who in 1993 proved the inequality $\Gamma(I_n) \leqslant 4n^{\frac{1}{4}}$ for all $n \geqslant 1$ and observed that a cube in $\mathbb{R}^n$ may have its Gaussian perimeter as large as $\sqrt{\log n}$ (see [B]). Many people seemed to believe that the logarithmic order of growth must be the correct one and that it is the upper bound that needs to be improved. If it were the case, it would open a road to essentially improving some constants in various "convex probability" theorems (see [Be1],[Be2] for a nice example). Alas, as it turned out, K. Ball's estimate is sharp.

As to the proof of the theorem, I cannot shake the feeling that there should exist some simple and elegant way leading to the result. Unfortunately, what I can present is a pretty boring and technical computation. So I encourage the reader to stop reading the paper here and to (try to) prove the theorem by himself.

The Case $A = I_n$

We shall be primarily interested in the behavior of $\Gamma(I_n)$ for large n. Our first goal will be to prove the asymptotic upper bound

$$\limsup_{n \to \infty} \frac{\Gamma(I_n)}{n^{\frac{1}{4}}} \leqslant \pi^{-\frac{1}{4}} < 0.76\,,$$

which, with some extra twist, can be improved to

$$\limsup_{n \to \infty} \frac{\Gamma(I_n)}{n^{\frac{1}{4}}} \leqslant (2\pi)^{-\frac{1}{4}} < 0.64\,.$$

While this result is essentially equivalent to that of K. Ball, our proof will use different ideas and yield more information about the possible shapes of convex bodies with large Gaussian perimeter.

As to the estimates from below, we shall show that

$$\liminf_{n \to \infty} \frac{\Gamma(I_n)}{n^{\frac{1}{4}}} \geqslant e^{-\frac{5}{4}} > 0.28\,.$$

First of all, note that in the definition of $\Gamma(I_n)$ we may restrict ourselves to convex bodies Q containing the origin. One of the most natural ways to estimate the integral $\int_{\partial Q} \varphi(y)\,d\sigma(y)$ is to introduce some "polar coordinate system" $x = X(y,t)$ in $\mathbb{R}^n$ with $y \in \partial Q$, $t \geqslant 0$. Then we can write

$$1= \int_{\mathbb{R}^n} \varphi(x)\, dx = \int_{\partial Q} \left[\int_0^\infty \varphi(X(y,t)) D(y,t)\, dt \right] d\sigma(y) = \int_{\partial Q} \varphi(y) \xi(y)\, d\sigma(y)$$

$$(*)$$

where $D(y,t)$ stands for the determinant of the differential $\frac{\partial X(y,t)}{\partial y\, \partial t}$ of the mapping $\partial Q \times (0,+\infty) \ni (y,t) \to X(y,t) \in \mathbb{R}^n$ and

$$\xi(y) = \varphi(y)^{-1} \int_0^\infty \varphi(X(y,t)) D(y,t)\, dt.$$

This yields the estimate

$$\int_{\partial Q} \varphi(y)\, d\sigma(y) \leqslant \frac{1}{\min_{\partial Q} \xi}.$$

There are two natural polar coordinate systems associated with a convex body Q containing the origin. The first one is given by the mapping $X_1(y,t) = ty$. Then

$$D_1(y,t) = t^{n-1} |y| \alpha(y)$$

where $\alpha(y)$ is the cosine of the angle between the "radial vector" y and the unit outer normal vector ν_y to the surface ∂Q at the point y. So, in this case, we have

$$\xi_1(y) = e^{\frac{|y|^2}{2}} \left[\int_0^\infty |y| t^{n-1} e^{-\frac{t^2 |y|^2}{2}}\, dt \right] \alpha(y)$$

$$= |y|^{-(n-1)} e^{\frac{|y|^2}{2}} \left[\int_0^\infty t^{n-1} e^{-\frac{t^2}{2}}\, dt \right] \alpha(y).$$

It is not hard to see that the function $f(t) := t^{n-1} e^{-\frac{t^2}{2}}$ is nice enough for the application of the Laplace asymptotic formula. Since it attains its maximum at $t_0 = \sqrt{n-1}$ and since $\frac{d^2}{dt^2} \log f(t_0) = -2$, we get

$$\int_0^\infty f(t)\, dt = \left[\sqrt{\pi} + o(1) \right] f(t_0).$$

Observing that $\frac{d^2}{dt^2} \log f(t_0) \leqslant -1$ for all $t > 0$, we get

$$f(t) \leqslant f(t_0) e^{-\frac{(t-t_0)^2}{2}} \qquad \text{for all } t > 0.$$

Bringing these estimates together, we conclude that

$$\xi_1(y) \geqslant e^{\frac{(|y| - \sqrt{n-1})^2}{2}} \left[\sqrt{\pi} + o(1) \right] \alpha(y).$$

Unfortunately, as one can easily see, $\alpha(y)$ can be very close to 0 at some points, so we cannot get an estimate for the Gaussian perimeter of an *arbitrary* convex body Q using ξ_1 alone. Nevertheless, let us mention here that if we know in advance that Q contains a ball of radius $R > 0$ centered at the origin, we may use the elementary inequality $\alpha(y) \geqslant \frac{R}{|y|}$ and conclude (after some not very hard computations) that

$$\min_{\partial Q} \xi_1 \geqslant [1 + o(1)] \frac{\sqrt{\pi}\,R}{\sqrt{n}}.$$

Thus, if R is much greater than $n^{\frac{1}{4}}$, the Gaussian perimeter of Q is much less than $n^{\frac{1}{4}}$. It is interesting to compare this observation with the construction of the convex body Q with large perimeter below: the body we shall construct will have the ball of radius $n^{\frac{1}{4}}$ as its inscribed ball!

Let us now consider the second natural "polar coordinate system" associated with Q, which is given by the mapping $X_2(y, t) = y + t\nu_y$. The reader may object that it is a coordinate system in $\mathbb{R}^n \setminus Q$, not in $\mathbb{R}^n$, but this makes things only better because now we can write $1 - \gamma(Q)$ instead of 1 on the left hand side of the inequality (∗) (it is this improvement that, exploited carefully, yields the extra factor of $2^{-\frac{1}{4}}$). It is not hard to check that $X_2(y, t)$ is an expanding map in the sense that $|X_2(y', t') - X_2(y'', t'')|^2 \geqslant |y' - y''|^2 + (t' - t'')^2$ and, therefore, $D_2(y, t) \geqslant 1$ for all $y \in \partial Q$, $t > 0$. This results in the inequality

$$\xi_2(y) \geqslant \int_0^\infty e^{-t|y|\alpha(y)} e^{-\frac{t^2}{2}}\,dt \geqslant \frac{1}{|y|\alpha(y) + 1}.$$

This expression can also be small, but only if $\alpha(y)$ is *large*. Thus, it seems to be a good idea to bring these two estimates together and to write

$$\int_{\partial Q} \varphi(y)\, \Xi(y)\, d\sigma(y) \leqslant 2$$

where

$$\Xi(y) = \xi_1(y) + \xi_2(y) \geqslant [1 + o(1)] \cdot \left\{ e^{\frac{(|y| - \sqrt{n-1})^2}{2}} \sqrt{\pi}\,\alpha(y) + \frac{1}{|y|\alpha(y) + 1} \right\}.$$

It is a simple exercise in elementary analysis now to show that the minimum of the right hand side over all possible values of $|y|$ and $\alpha(y)$ is $[2 + o(1)]\pi^{\frac{1}{4}} n^{-\frac{1}{4}}$ attained at $|y| \approx \sqrt{n-1}$, $\alpha(y) \approx (\pi n)^{-\frac{1}{4}}$.

Note that if $|y|$ or $\alpha(y)$ deviate much from these values ($|y|$ on the additive and $\alpha(y)$ on the multiplicative scale), the corresponding value of $\Xi(y)$ is much greater than $n^{-\frac{1}{4}}$. Thus, if a convex body Q with the Gaussian perimeter comparable to $n^{\frac{1}{4}}$ exists at all, a noticeable part of its boundary (in the sense of angular measure) should lie in the constant size neighborhood of the sphere S of radius $\sqrt{n}$ centered of the origin, $\alpha(y)$ being comparable to $n^{-\frac{1}{4}}$ on that part of the boundary. At first glance, this seems unfeasible because what it means is that the boundary of Q should simultaneously be very close to the sphere S and very transversal (almost orthogonal!) to it. Actually, it leaves one with essentially one possible choice of the body Q for which it is impossible to "do something" to essentially improve the upper bound: the regular polyhedron with inscribed radius of $n^{\frac{1}{4}}$ and circumscribed radius of $\sqrt{n}$ (there is no such deterministic thing, to be exact, but there is a good random substitute).

The fastest way to get the estimate $\Gamma(I_n) \geqslant \text{const}\, n^{\frac{1}{4}}$ seems to be the following. Observe, first of all, that the polar coordinate system $X_1(y, t)$ can be used to obtain the inequality

$$\int_{\partial Q} \varphi(y)\, d\sigma(y) \geqslant \int_{(\partial Q)'} \varphi(y)\, d\sigma(y) \geqslant \text{const}\, n^{\frac{1}{4}} \gamma(\mathcal{K}_Q)$$

where

$$(\partial Q)' = \left\{ y \in \partial Q : \big| |y| - \sqrt{n-1} \big| \leqslant 1,\ \tfrac{1}{2} n^{-\frac{1}{4}} \leqslant \alpha(y) \leqslant 2 n^{-\frac{1}{4}} \right\}$$

and $\mathcal{K}_Q = \{ ty : y \in (\partial Q)', t \geqslant 0 \}$ is the cone generated by $(\partial Q)'$. Let now H be a hyperplane tangent to the ball of radius $n^{\frac{1}{4}}$ centered at the origin. Let S be the (smaller) spherical cap cut off from the sphere $\mathbb{S}$ of radius $\sqrt{n}$ centered at the origin by the hyperplane H, let $\widetilde{H} = \{ y \in H : \sqrt{n} - 1 \leqslant |y| \leqslant \sqrt{n} \}$, and let $\widetilde{S}$ be the radial projection of $\widetilde{H}$ to the sphere $\mathbb{S}$. Now, instead of one hyperplane H, take N independent random hyperplanes H_j and consider the convex body Q that is the intersection of the corresponding half-spaces. A point $y \in \widetilde{H}_j$ belongs to $(\partial Q)'$ unless it is cut off by one of the other hyperplanes H_k. Note that if a point $y \in \widetilde{H}_j$ is cut off by a hyperplane H_k, then its radial projection to the sphere $\mathbb{S}$ belongs to S_k. Thus,

$$\gamma(\mathcal{K}_Q) \geqslant \sum_{j=1}^{N} \lambda\left(\widetilde{S}_j \setminus \bigcup_{k: k \neq j} S_j \right)$$

where $d\lambda$ is the normalized (by the condition $\lambda(\mathbb{S}) = 1$) angular measure on $\mathbb{S}$. Since the random hyperplanes H_j are chosen independently, the expectation of the right hand side equals $N(1 - \lambda(S))^{N-1} \lambda(\widetilde{S})$. Observing that, for large n, $\lambda(S)$ is small compared to 1 and choosing $N \approx \lambda(S)^{-1}$, we get the estimate

$$\gamma(\mathcal{K}) \geqslant \text{const}\, \frac{\lambda(\widetilde{S})}{\lambda(S)}.$$

A routine computation shows that the ratio $\frac{\lambda(\widetilde{S})}{\lambda(S)}$ stays bounded away from 0 as $n \to \infty$, finishing the proof. While this (sketch of a) proof is missing a few technical details, I included it in the hope that it might give the reader a clearer picture of how the example was constructed than the completely formal reasoning below aimed at obtaining the largest possible coefficient in front of $n^{\frac{1}{4}}$ rather than at making the geometry transparent.

The formal construction runs as follows. Consider N (a large integer to be chosen later) independent random vectors x_j equidistributed over the unit sphere in $\mathbb{R}^{n+1}$ (this 1 is added just to avoid indexing φ) and define the (random) polyhedron

$$Q := \{x \in \mathbb{R}^{n+1} : \langle x, x_j \rangle \leqslant \rho\}.$$

In other words, Q is the intersection of N random half-spaces bounded by hyperplanes H_j whose distance from the origin is ρ. The expectation of the Gaussian perimeter of Q equals

$$N \frac{1}{\sqrt{2\pi}} e^{-\frac{\rho^2}{2}} \int_{\mathbb{R}^n} \varphi(y) \left(1 - p(|y|)\right)^{N-1} dy$$

where $p(r)$ is the probability that a fixed point whose distance from the origin equals $\sqrt{r^2 + \rho^2}$ is separated from the origin by one random hyperplane H_j. It is easy to compute $p(r)$ explicitly: it equals

$$\left[\int_{-\sqrt{r^2+\rho^2}}^{\sqrt{r^2+\rho^2}} \left(1 - \frac{t^2}{r^2 + \rho^2}\right)^{\frac{n-1}{2}} dt \right]^{-1} \int_{\rho}^{\sqrt{r^2+\rho^2}} \left(1 - \frac{t^2}{r^2 + \rho^2}\right)^{\frac{n-1}{2}} dt.$$

This is quite a cumbersome expression so let us try to find a good asymptotics for it when $\rho = e^{O(1)} n^{\frac{1}{4}}$ and $r = \sqrt{n-1} + w$, $|w| < O(1)$. The first integral then becomes a typical exercise example for the Laplace asymptotic formula and we get it equal to $\sqrt{2\pi} + o(1)$. Using the inequality $(1 - a) \leqslant e^{-\frac{a^2}{2}} e^{-a}$ ($a > 0$), we can estimate the second integral by

$$\int_{\rho}^{\infty} \exp\left\{ -\frac{n-1}{4(r^2 + \rho^2)^2} t^4 \right\} \exp\left\{ -\frac{n-1}{r^2 + \rho^2} \frac{t^2}{2} \right\} dt$$

$$\leqslant \exp\left\{ -\frac{n-1}{4(r^2 + \rho^2)^2} \rho^4 \right\} \int_{\rho}^{\infty} \exp\left\{ -\frac{n-1}{r^2 + \rho^2} \frac{t^2}{2} \right\} dt.$$

The first factor is asymptotically equivalent to $e^{-\frac{\rho^4}{4n}}$ in the ranges of ρ and r we are interested in. To estimate the second factor, let us observe that, for every $a > 0$,

$$\int_{\rho}^{\infty} e^{-a\frac{t^2}{2}} dt \leqslant \int_{\rho}^{\infty} e^{-a\frac{\rho^2}{2}} e^{-a\rho(t-\rho)} dt = \frac{1}{a\rho} e^{-a\frac{\rho^2}{2}}.$$

Observe also that under our restrictions for r and ρ, we have

$$\frac{n-1}{r^2 + \rho^2} = 1 - \frac{2w}{\sqrt{n}} - \frac{\rho^2}{n} + o(n^{-\frac{1}{2}}).$$

Bringing all the estimates together, we arrive at the inequality

$$p(r) \leqslant [1+o(1)]\frac{1}{\sqrt{2\pi}}\frac{1}{\rho}\exp\left\{\frac{\rho^4}{4n}\right\}\exp\left\{\frac{w\rho^2}{\sqrt{n}}\right\}e^{-\frac{\rho^2}{2}} =: L(n,\rho)\exp\left\{\frac{w\rho^2}{\sqrt{n}}\right\}$$

(the reader should treat $o(1)$ in the definition of $L(n,\rho)$ as some hard to compute but definite quantity that depends on n only and tends to 0 as $n \to \infty$). Since we expect the main part of the integral $\int_{\mathbb{R}^n} \varphi(y)(1-p(|y|))^{N-1}dy$ to come from the points y with $|y| \approx \sqrt{n-1}$, which correspond to the values of w close to 0, let us look at what happens if we replace the last factor in our estimate for $p(r)$ by its value at $w=0$, which is just 1. Then $p(r)$ would not depend on r at all and, taking into account that $\int_{\mathbb{R}^n} \varphi(y)\,dy = 1$, we would get the quantity

$$\frac{1}{\sqrt{2\pi}}e^{-\frac{\rho^2}{2}}N\left[1 - L(n,\rho)\right]^{N-1}$$

to maximize. Optimizing first with respect to N (note that $L(n,\rho) \to 0$ as $n \to \infty$), we see that we should take N satisfying the inequality $N \leqslant L(n,\rho)^{-1} \leqslant N+1$, which results in the value of the maximum being

$$[1+o(1)]e^{-1}\rho\exp\left\{-\frac{\rho^4}{4n}\right\}.$$

Optimizing with respect to ρ, we see that the best choice would be $\rho = n^{\frac{1}{4}}$ which would yield the desired asymptotic lower bound $e^{-\frac{5}{4}}n^{\frac{1}{4}}$ for $\Gamma(I_n)$. Now let us use these values of N and ρ and make an accurate estimate of the integral

$$\int_{\mathbb{R}^n} \varphi(y)\left(1-p(|y|)\right)^{N-1}dy \geqslant c\int_{-W}^{W} f(\sqrt{n-1}+w)\left(1-L(n,\rho)\exp\left\{\frac{w\rho^2}{\sqrt{n}}\right\}\right)^{N-1}$$

where $f(t) = t^{n-1}e^{-\frac{t^2}{2}}$ as before, $c = \left(\int_0^\infty f(t)\,dt\right)^{-1}$, and W is some big positive number. Note again that the product $c\,f(\sqrt{n-1}+w) = [1+o(1)]\frac{1}{\sqrt{\pi}}e^{-w^2}$ for fixed w and $n \to \infty$. Also, for fixed w and $n \to \infty$, the second factor in the integral is asymptotically equivalent to $\exp\{-e^w\}$ (recall that $\rho = n^{\frac{1}{4}}$ and, therefore, $\frac{\rho^2}{\sqrt{n}} = 1$). Thus, we obtain the estimate

$$\Gamma_n \geqslant [1+o(1)]e^{-\frac{1}{4}}n^{\frac{1}{4}}\frac{1}{\sqrt{\pi}}\int_{-W}^{W}\exp\{-e^w\}e^{-w^2}\,dw.$$

The integral on the right looks scary, but, since everything except the factor $\exp\{-e^w\}$ is symmetric, we can replace it by

$$\int_{-W}^{W}\frac{\exp\{-e^w\}+\exp\{-e^{-w}\}}{2}e^{-w^2}\,dw.$$

Using the elementary inequality

$$\frac{\exp\{-a\} + \exp\{-\frac{1}{a}\}}{2} \geqslant \frac{1}{e} \quad \text{for all } a > 0,$$

we conclude that

$$\Gamma_n \geqslant [1 + o(1)] e^{-\frac{5}{4}} n^{\frac{1}{4}} \frac{1}{\sqrt{\pi}} \int_{-W}^{W} e^{-w^2} \, dw.$$

It remains to note that $\frac{1}{\sqrt{\pi}} \int_{-W}^{W} e^{-w^2} \, dw$ can be made arbitrarily close to 1 by choosing W large enough.

The General Case

Let us start with two simple reductions. First of all, observe that the estimate we want to prove is homogeneous with respect to A, so, without loss of generality, we may assume that $\text{Tr } A = 1$.

Since the problem is rotation invariant, we may assume that both A and B are diagonal matrices. We shall primarily deal with B, so let us denote the diagonal entries of B by $b_1, \ldots, b_n$ (our normalization condition $\text{Tr } A = 1$ means that $\sum_j b_j^2 = 1$). Denote

$$\mathcal{D} := \sqrt[4]{\sum_j b_j^4}.$$

Note that $0 < \mathcal{D} \leqslant 1$, so $\mathcal{D}^2 \leqslant \mathcal{D}$ and so forth.

Proof of the Estimate $\Gamma(A) \leqslant C\sqrt{\|A\|_{\text{H-S}}}$

We shall follow the idea of K. Ball and use the Cauchy integral formula. Let us recall how it works. Suppose you have two functions $F, G : \mathbb{R}^n \to [0, +\infty)$, a nonnegative homogeneous of degree 1 function Ψ in $\mathbb{R}^n$, and a random unit vector $z_\omega \in \mathbb{R}^n$. Suppose that you can show that for every point $y \in \mathbb{R}^n$ and for every vector $\nu \in \mathbb{R}^n$,

$$\mathcal{E}_\omega \left[|\langle \nu, z_\omega \rangle| \int_{\mathbb{R}} G(y - t z_\omega) dt \right] \geqslant \kappa \, F(y) \Psi(\nu)$$

with some constant $\kappa > 0$, where $\mathcal{E}_\omega$ denotes the expectation with respect to z_ω. Then for any convex body $Q \subset \mathbb{R}^n$,

$$\int_{\partial Q} F(y) \Psi(\nu_y) \, d\sigma(y) \leqslant 2\kappa^{-1} \int_{\mathbb{R}^n} G(x) \, dx.$$

To make this general formula applicable to our special case, we have to choose $F(y) = \varphi(y)$ and $\Psi(\nu) = |B\nu|$. Unfortunately, there is no clearly forced choice of z_ω and G. To choose z_ω, let us observe that our task is to make the "typical value" of $|\langle \nu, z_\omega \rangle|$ approximately equal to $|B\nu|$. The standard way to achieve this is to take $z_\omega = BZ_\omega$ with $Z_\omega = \sum_j \varepsilon_j(\omega) e_j$ where e_j is the orthonormal basis in $\mathbb{R}^n$ in which B is diagonal and $\varepsilon_j(\omega)$ $(\omega \in \Omega)$ are independent random variables taking values ± 1 with probability $\frac{1}{2}$ each. Note that our normalization condition $\sum_j b_j^2 = 1$ guarantees that z_ω is always a unit vector in $\mathbb{R}^n$. The hardest part is finding an appropriate function G. We shall search for $G(x)$ in the form

$$G(x) = \varphi(x)\Xi(x)$$

where $\Xi(x)$ is some relatively tame function: after all, the integral of a function over a random line containing a fixed point is equal to the value at the point times something "not-so-important" (at least, I do not know a better way to evaluate it with *no a priori information*). If $\Xi(x)$ changes slower than $\varphi(x)$, then we may expect the main part of the integral $\int_{\mathbb{R}} G(y - tz_\omega)\, dt$ to come from the points t that lie in a small neighbourhood of $t_0 = \langle y, z_\omega \rangle$, which is the point where the function $t \to \varphi(y - tz_\omega)$ attains its maximum. To make this statement precise, let us observe that

$$\varphi\big(y - (t_0 + \tau)z_\omega\big) = \exp\left\{-\frac{\tau^2}{2}\right\}\exp\left\{\frac{\langle y, z_\omega \rangle^2}{2}\right\}\varphi(y)$$

$$\geqslant \frac{1}{\sqrt{e}}\exp\left\{\frac{\langle y, z_\omega \rangle^2}{2}\right\}\varphi(y)$$

when $|\tau| \leqslant 1$. If our function $\Xi(x)$ satisfies the condition

$$\max\left\{\Xi(x - \tau z_\omega), \Xi(x + \tau z_\omega)\right\} \geqslant \frac{1}{2}\Xi(x) \quad \text{for all } x \in \mathbb{R}^n,\ \omega \in \Omega,\ |\tau| \leqslant 1$$

(which we shall call "weak convexity" condition), then we may estimate the integral from below by $\frac{1}{2\sqrt{e}}\varphi(y)\,\Xi(y - \langle y, z_\omega \rangle z_\omega) \exp\{\langle y, z_\omega \rangle^2/2\}$. Then our "only" task will be to prove the inequality

$$\mathcal{E}_\omega\left[|\langle \nu, z_\omega \rangle| \cdot \Xi\big(y - \langle y, z_\omega \rangle z_\omega\big) \cdot \exp\left\{\frac{\langle y, z_\omega \rangle^2}{2}\right\}\right] \geqslant \kappa|B\nu|. \qquad (**)$$

Let us make a second "natural leap of faith" and assume that Ξ changes so slowly that $\Xi(y - \langle y, z_\omega \rangle z_\omega) \approx \Xi(y)$. Then we can just *compute* the expectation of the product of other two factors and *define* $\Xi(y)$ to be the factor that makes the desired inequality almost an identity (we should pray that after that the loop will close and we shall not have to make a second iteration). Thus, our first task will be to compute the quantity

$$\mathcal{E}_\omega\left[|\langle \nu, z_\omega \rangle| \cdot \exp\left\{\frac{\langle y, z_\omega \rangle^2}{2}\right\}\right] = \mathcal{E}_\omega\left[|\langle B\nu, Z_\omega \rangle| \cdot \exp\left\{\frac{\langle By, Z_\omega \rangle^2}{2}\right\}\right].$$

Since $B\nu$ and By are just two arbitrary vectors in $\mathbb{R}^n$, let us introduce some one-letter notation for them. Let, say, $B\nu = v$ and $By = u$. As usual, we shall write v_j and u_j for the coordinates of v and u in the basis e_j. The unpleasant thing we shall have to face on this way is that even $\mathcal{E}_\omega[\exp\{\langle u, Z_\omega\rangle^2/2\}]$ (forget about the factor $|\langle v, Z_\omega\rangle|$!) is not easy to compute when $|u| > 1$. Fortunately, if we assume in addition that $\Xi(\beta x) \geqslant \Xi(x)$ for all $x \in \mathbb{R}^n$, $\beta \geqslant 1$, then the left hand side of (**) will satisfy a similar inequality with respect to y and, thereby, (**) will hold for all points y with $|By| > 1$ as soon as it holds for all y with $|By| = 1$.

We shall use the formula

$$\exp\left\{\frac{\langle u, Z_\omega\rangle^2}{2}\right\} = \frac{1}{\sqrt{2\pi}} \int_{-\infty}^{\infty} e^{-\frac{t^2}{2}} \exp\{t\langle u, Z_\omega\rangle\}\, dt$$

and write

$$\mathcal{E}_\omega\left[\exp\left\{\frac{\langle u, Z_\omega\rangle^2}{2}\right\}\right] = \frac{1}{\sqrt{2\pi}} \int_{-\infty}^{\infty} e^{-\frac{t^2}{2}} \mathcal{E}_\omega\left[\exp\{t\langle u, Z_\omega\rangle\}\right]\, dt$$

$$= \frac{1}{\sqrt{2\pi}} \int_{-\infty}^{\infty} e^{-\frac{t^2}{2}} \prod_j \left(\frac{e^{-tu_j} + e^{tu_j}}{2}\right)\, dt.$$

Using the elementary inequality $\frac{e^{-s}+e^s}{2} \leqslant e^{\frac{s^2}{2}}\left[1 + \frac{s^4}{26}\right]^{-1}$, we can estimate the last integral from above by

$$\frac{1}{\sqrt{2\pi}} \int_{-\infty}^{\infty} \exp\left\{-\frac{(1-|u|^2)t^2}{2}\right\} \left[1 + \frac{(t\|u\|_4)^4}{26}\right]^{-1}\, dt$$

where $\|u\|_4 := \sqrt[4]{\sum_j u_j^4}$. If $|u| \leqslant 1$, we can say that, from the L^1 point of view, the integrand is hardly distinguishable from the characteristic function of the interval $|t| \leqslant \Delta(u)$ where $\Delta(u) = 1/\max\{\sqrt{1-|u|^2}, \|u\|_4\}$, so the last integral should be, roughly speaking, $\Delta(u)$. A reasonably accurate computation yields the upper bound $\min\{1/\sqrt{1-|u|^2}, 3/\|u\|_4\} \leqslant 3\Delta(u)$. To estimate $\mathcal{E}_\omega[\exp\{\langle u, Z_\omega\rangle^2/2\}]$ from below, we shall use another elementary inequality $\frac{e^{-s}+e^s}{2} \geqslant e^{s^2/2}e^{-s^4/8}$. It yields the lower bound

$$\mathcal{E}_\omega\left[\exp\left\{\frac{\langle u, Z_\omega\rangle^2}{2}\right\}\right] \geqslant \frac{1}{\sqrt{2\pi}} \int_{-\infty}^{\infty} \exp\left\{-\frac{(1-|u|^2)t^2}{2}\right\} \exp\left\{-\frac{(t\|u\|_4)^4}{8}\right\}\, dt$$

$$\geqslant \Delta(u)\frac{1}{\sqrt{2\pi}} \int_{-\infty}^{\infty} e^{-\frac{t^2}{2}} e^{-\frac{t^4}{8}}\, dt \geqslant \frac{1}{2}\Delta(u).$$

Let us now turn to the estimates for the expectation $\mathcal{E}_\omega[|\langle v, Z_\omega\rangle| \exp\{\langle u, Z_\omega\rangle^2/2\}]$. To this end, we shall first estimate $\mathcal{E}_\omega[\exp\{is\langle v, Z_\omega\rangle\} \exp\{\langle u, Z_\omega\rangle^2/2\}]$ where, as usual, $i = \sqrt{-1}$. Again, write

$$\mathcal{E}_\omega\left[\exp\{is\langle v, Z_\omega\rangle\}\exp\left\{\frac{\langle u, Z_\omega\rangle^2}{2}\right\}\right]$$

$$= \frac{1}{\sqrt{2\pi}}\int_{-\infty}^{\infty} e^{-\frac{t^2}{2}}\mathcal{E}_\omega\left[\exp\{is\langle v, Z_\omega\rangle + t\langle u, Z_\omega\rangle\}\right]\,dt$$

$$= \frac{1}{\sqrt{2\pi}}\int_{-\infty}^{\infty} e^{-\frac{t^2}{2}}\prod_j\left(\frac{e^{-(tu_j+isv_j)}+e^{(tu_j+isv_j)}}{2}\right)\,dt.$$

Now note that for all $\alpha, \beta \in \mathbb{R}$, one has

$$\left|\frac{e^{-(\alpha+i\beta)}+e^{(\alpha+i\beta)}}{2}\right| \leqslant \frac{e^{-\alpha}+e^\alpha}{2}$$

and this trivial estimate (the triangle inequality) can be improved to

$$\left|\frac{e^{-(\alpha+i\beta)}+e^{(\alpha+i\beta)}}{2}\right| \leqslant e^{-\delta\beta^2}\left(\frac{e^{-\alpha}+e^\alpha}{2}\right)$$

with some absolute $\delta \in (0,1)$ for $|\alpha|, |\beta| \leqslant 1$. Therefore,

$$\prod_j\left|\frac{e^{-(tu_j+isv_j)}+e^{(tu_j+isv_j)}}{2}\right| \leqslant \prod_j\left(\frac{e^{-tu_j}+e^{tu_j}}{2}\right)$$

for all $t, s \in \mathbb{R}$ and

$$\prod_j\left|\frac{e^{-(tu_j+isv_j)}+e^{(tu_j+isv_j)}}{2}\right| \leqslant e^{-\delta s^2|v|^2}\prod_j\left(\frac{e^{-tu_j}+e^{tu_j}}{2}\right)$$

if $|t| \leqslant \|u\|_\infty^{-1}$ and $|s| \leqslant \|v\|_\infty^{-1}$. Since $\|u\|_\infty^{-1} \geqslant \|u\|_4^{-1} \geqslant \Delta(u)$, we can write

$$\mathcal{E}_\omega\left[\exp\left\{\frac{\langle u, Z_\omega\rangle^2}{2}\right\}\right] - \left|\mathcal{E}_\omega\left[\exp\{is\langle v, Z_\omega\rangle\}\exp\left\{\frac{\langle u, Z_\omega\rangle^2}{2}\right\}\right]\right|$$

$$\geqslant \left(1-e^{-\delta s^2|v|^2}\right)\frac{1}{\sqrt{2\pi}}\int_{-\Delta(u)}^{\Delta(u)} e^{-\frac{t^2}{2}}\prod_j\left(\frac{e^{-tu_j}+e^{tu_j}}{2}\right)\,dt$$

$$\geqslant \left(1-e^{-\delta s^2|v|^2}\right)\Delta(u)\frac{1}{\sqrt{2\pi}}\int_{-1}^{1} e^{-\frac{t^2}{2}}e^{-\frac{t^4}{8}}\,dt \geqslant \frac{1}{2}\left(1-e^{-\delta s^2|v|^2}\right)\Delta(u).$$

On the other hand, the trivial inequality $|\alpha - \alpha e^{i\beta}| \leqslant 2\alpha\min\{|\beta|, 1\}$ $(\alpha > 0,\ \beta \in \mathbb{R})$ yields

$$\left|\mathcal{E}_\omega\left[\exp\left\{\frac{\langle u, Z_\omega\rangle^2}{2}\right\}\right] - \mathcal{E}_\omega\left[\exp\{is\langle v, Z_\omega\rangle\}\exp\left\{\frac{\langle u, Z_\omega\rangle^2}{2}\right\}\right]\right|$$

$$\leqslant 2\mathcal{E}_\omega\left[\min\{|s\langle v, Z_\omega\rangle|, 1\}\exp\left\{\frac{\langle u, Z_\omega\rangle^2}{2}\right\}\right].$$

Bringing these estimates together and taking $s = |v|^{-1}$, we obtain

$$\mathcal{E}_\omega \left[\min\{|\langle v, Z_\omega \rangle|, |v|\} \exp\left\{ \frac{\langle u, Z_\omega \rangle^2}{2} \right\} \right] \geqslant \frac{1}{4}(1 - e^{-\delta})\Delta(u)|v| = 2\eta \Delta(u)|v|$$

where $\eta := \frac{1}{8}(1 - e^{-\delta}) > 0$ is an absolute constant. Obviously, the expectation $\mathcal{E}_\omega[|\langle v, Z_\omega \rangle| \exp\{\langle u, Z_\omega \rangle^2/2\}]$ can be only greater.

This brings us to the idea to take $\Xi(y) = \Delta(By)^{-1} = \max\{\|By\|_4, \sqrt{1 - |By|^2}\}$. This formula makes little sense for $|By| > 1$, so, to be formally correct, we shall distinguish two cases: $|By|^2 \geqslant 1 - \|By\|_4^2$ and $|By|^2 < 1 - \|By\|_4^2$. We shall separate them completely and even construct two different functions Ξ_1 and Ξ_2 serving the first and the second case correspondingly. Note that all points y for which $|By| \geqslant 1$ are covered by the first case, so we need the condition that Ξ be non-decreasing along each ray starting at the origin only for Ξ_1. Let us start with

Case 1: $1 - \|By\|_4^2 \leqslant |By|^2 \leqslant 1$.
In this case the natural candidate for Ξ_1 is $\Xi_1(y) = \|By\|_4$. We have no problem with the "weak convexity" condition because Ξ_1 is even strongly convex. Also, it obviously satisfies $\Xi_1(\beta y) \geqslant \Xi_1(y)$ for every $\beta \geqslant 1$. The only thing we should take care about is the assumption $\Xi_1(y - \langle y, z_\omega \rangle z_\omega) \approx \Xi_1(y)$. What we would formally need here is $\Xi_1(y - \langle y, z_\omega \rangle z_\omega) \geqslant \zeta \Xi_1(y)$ with some absolute $0 < \zeta \leqslant 1$. Unfortunately, it is futile to hope for such an estimate for all $y \in \mathbb{R}^n$ and $\omega \in \Omega$ because it can easily happen that y is collinear with some z_ω and then we shall get $\Xi_1(y - \langle y, z_\omega \rangle z_\omega) = 0$. To exclude this trivial problem, let us bound Ξ from below by some constant. Since our aim is to control the integral of Ξ_1 with respect to the Gaussian measure in $\mathbb{R}^n$, we may just take the maximum of Ξ_1 and its average value with respect to the Gaussian measure $d\gamma(x) = \varphi(x)\,dx$, which is almost the same as $\|\Xi_1\|_{L^4(\mathbb{R}^n, d\gamma)} = \sqrt[4]{3}\sqrt[4]{\sum_j b_j^4} = \sqrt[4]{3}\,\mathcal{D}$. This leads to the revised definition

$$\Xi_1(y) = \max\{\|By\|_4, \mathcal{D}\}$$

(note that this revised function Ξ_1 is still convex and non-decreasing along each ray starting at the origin). The condition $\Xi_1(y - \langle y, z_\omega \rangle z_\omega) \geqslant \zeta \Xi_1(y)$ is then trivially satisfied with $\zeta = 1$ if $\|By\|_4 \leqslant \mathcal{D}$. Assume that $\|By\|_4 \geqslant \mathcal{D}$. Then

$$\Xi_1(y - \langle y, z_\omega \rangle z_\omega) \geqslant \|By - \langle y, z_\omega \rangle Bz_\omega\|_4 \geqslant \|By\|_4 - |\langle By, Z_\omega \rangle| \cdot \|B^2 Z_\omega\|_4$$
$$= \Xi_1(y) - |\langle By, Z_\omega \rangle| \sqrt[4]{\sum_j b_j^8} \geqslant \Xi_1(y) - \mathcal{D}^2 |\langle By, Z_\omega \rangle|.$$

Uniting this estimate with the trivial lower bound $\Xi_1(y - \langle y, z_\omega \rangle z_\omega) \geqslant \mathcal{D}$, we can write

$$\Xi_1(y - \langle y, z_\omega \rangle z_\omega) \geqslant \frac{1}{1 + \mathcal{D}|\langle By, Z_\omega \rangle|} \Xi_1(y)$$

(we used here the elementary estimate $\max\{\alpha - \beta \mathcal{D}, \mathcal{D}\} \geqslant \frac{\alpha}{1+\beta}$). Therefore we shall be able to prove $(**)$ if we demonstrate that under the conditions $\sqrt{1 - |u|^2} \leqslant \|u\|_4$ and $\|u\|_4 \geqslant \mathcal{D}$ (where, as before, $u = By$), the main part of the expectation $\mathcal{E}_\omega[\min\{|\langle v, Z_\omega\rangle|, |v|\}\exp\{\langle u, Z_\omega\rangle^2/2\}]$ comes from those $\omega \in \Omega$ for which $|\langle v, Z_\omega\rangle|$ is not much greater than $\mathcal{D}^{-1}$. To this end, we shall have to prove some "tail estimate" for $\mathcal{E}_\omega[\exp\{\langle u, Z_\omega\rangle^2/2\}]$. Using the inequality

$$\langle u, Z_\omega\rangle^2 \exp\left\{\frac{\langle u, Z_\omega\rangle^2}{2}\right\} \leqslant \frac{1}{\sqrt{2\pi}}\int_{-\infty}^{\infty} t^2 e^{-\frac{t^2}{2}}\exp\{t\langle u, Z_\omega\rangle\}\,dt,$$

we get

$$\mathcal{E}_\omega\left[\langle u, Z_\omega\rangle^2 \exp\left\{\frac{\langle u, Z_\omega\rangle^2}{2}\right\}\right] \leqslant \frac{1}{\sqrt{2\pi}}\int_{-\infty}^{\infty} t^2 e^{-\frac{t^2}{2}}\mathcal{E}_\omega\left[\exp\{t\langle u, Z_\omega\rangle\}\right]\,dt$$

$$= \frac{1}{\sqrt{2\pi}}\int_{-\infty}^{\infty} t^2 e^{-\frac{t^2}{2}}\prod_j\left(\frac{e^{-tu_j} + e^{tu_j}}{2}\right)\,dt$$

$$\leqslant \frac{1}{\sqrt{2\pi}}\int_{-\infty}^{\infty} t^2\left[1 + \frac{(t\|u\|_4)^4}{26}\right]^{-1}\,dt$$

$$= \frac{2}{3}\pi^{\frac{1}{2}}26^{\frac{3}{4}}\|u\|_4^{-3} \leqslant 16\|u\|_4^{-3}.$$

This results in the tail estimate

$$\mathcal{E}_\omega\left[\chi_{\{|\langle u, Z_\omega\rangle| > \beta\|u\|_4^{-1}\}}\exp\left\{\frac{\langle u, Z_\omega\rangle^2}{2}\right\}\right]$$

$$\leqslant \beta^{-2}\|u\|_4^2\mathcal{E}_\omega\left[\langle u, Z_\omega\rangle^2\exp\left\{\frac{\langle u, Z_\omega\rangle^2}{2}\right\}\right] \leqslant 16\beta^{-2}\|u\|_4^{-1}.$$

Choosing $\beta := \frac{4}{\sqrt{\eta}}$ and recalling that $\|u\|_4 \geqslant \mathcal{D}$, we finally get ($u = By$, $v = B\nu$)

$$\mathcal{E}_\omega\left[|\langle v, Z_\omega\rangle| \cdot \Xi_1(y - \langle y, z_\omega\rangle z_\omega) \cdot \exp\left\{\frac{\langle u, Z_\omega\rangle^2}{2}\right\}\right]$$

$$\geqslant \mathcal{E}_\omega\left[\min\{|\langle v, Z_\omega\rangle|, |v|\} \cdot \Xi_1(y - \langle y, z_\omega\rangle z_\omega) \cdot \exp\left\{\frac{\langle u, Z_\omega\rangle^2}{2}\right\}\right]$$

$$\geqslant \frac{1}{1+\beta}\Xi_1(y) \cdot \mathcal{E}_\omega\left[\chi_{\{|\langle u, Z_\omega\rangle| \leqslant \beta\mathcal{D}^{-1}\}} \cdot \min\{|\langle v, Z_\omega\rangle|, |v|\} \cdot \exp\left\{\frac{\langle u, Z_\omega\rangle^2}{2}\right\}\right]$$

$$\geqslant \frac{1}{1+\beta}\Xi_1(y) \cdot \mathcal{E}_\omega\left[\chi_{\{|\langle u, Z_\omega\rangle| \leqslant \beta\|u\|_4^{-1}\}} \cdot \min\{|\langle v, Z_\omega\rangle|, |v|\} \cdot \exp\left\{\frac{\langle u, Z_\omega\rangle^2}{2}\right\}\right]$$

$$\geqslant \frac{\Xi_1(y)}{1+\beta}\left(\mathcal{E}_\omega\left[\min\{|\langle v, Z_\omega\rangle|, |v|\} \cdot \exp\left\{\frac{\langle u, Z_\omega\rangle^2}{2}\right\}\right]\right)$$

$$-|v| \cdot \mathcal{E}_\omega \left[\chi_{\{|\langle u, Z_\omega \rangle| > \beta \|u\|_4^{-1}\}} \exp \left\{ \frac{\langle u, Z_\omega \rangle^2}{2} \right\} \right] \right)$$

$$\geqslant \frac{\varXi_1(y)}{1+\beta} \left(2\eta \|u\|_4^{-1} |v| - 16\beta^{-2} \|u\|_4^{-1} |v| \right) \geqslant \frac{\eta}{1+\beta} \varXi_1(y) \|u\|_4^{-1} |v| = \frac{\eta}{1+\beta} |v|.$$

This finishes Case 1.

Case 2: $|By|^2 \leqslant 1 - \|By\|_4^2.$

To make the long story short, the function $\varXi_2$ that we shall use for this case is

$$\varXi_2(y) = \max \left\{ \sqrt{(1 - |By|^2)_+}, \mathcal{D} \right\}.$$

It is easy to see that

$$\|\varXi_2\|_{L^4(\mathbb{R}^n, d\gamma)}^4 \leqslant \mathcal{D}^4 + \int_{\mathbb{R}^n} (1 - |Bx|^2)^2 \, d\gamma(x) = 3\mathcal{D}^4.$$

To prove "weak convexity", let us observe that

$$1 - |B(x + \tau z_\omega)|^2 = 1 - |Bx|^2 - 2\tau \langle Bx, Bz_\omega \rangle - \tau^2 |Bz_\omega|^2 \geqslant 1 - |Bx|^2 - \mathcal{D}^4$$

if $|\tau| \leqslant 1$ and the sign of τ is opposite to that of $\langle Bx, Bz_\omega \rangle$. Therefore,

$$\sqrt{(1 - |Bx + \tau z_\omega|^2)_+} \geqslant \sqrt{(1 - |Bx|^2)_+ - \mathcal{D}^2} \geqslant \sqrt{(1 - |Bx|^2)_+} - \mathcal{D}$$

for such τ, which is enough to establish the weak convexity property for $\varXi_2$. Now let us turn to the inequality $\varXi_2(y - \langle y, z_\omega \rangle z_\omega) \geqslant \zeta \varXi_2(y)$. Again, it is trivial if $\sqrt{1 - |By|^2} \leqslant \mathcal{D}$. For other y, write

$$1 - |B(y - \langle y, z_\omega \rangle z_\omega)|^2 = 1 - |By|^2 + 2\langle y, z_\omega \rangle \langle By, Bz_\omega \rangle - \langle y, z_\omega \rangle^2 |Bz_\omega|^2.$$

It will suffice to show that the main contribution to the mathematical expectation $\mathcal{E}_\omega[\min\{|\langle v, Z_\omega \rangle|, |v|\} \exp\{\langle u, Z_\omega \rangle^2/2\}]$ is made by those $\omega \in \Omega$ for which

$$2\langle y, z_\omega \rangle \langle By, Bz_\omega \rangle - \langle y, z_\omega \rangle^2 |Bz_\omega|^2 \geqslant -K^2 \mathcal{D}^2$$

where K is some absolute constant (for such ω, one has $\varXi_2(y - \langle y, z_\omega \rangle z_\omega) \geqslant \frac{\varXi_2(y)}{1+K}$). Since $|Bz_\omega|^2 = \mathcal{D}^4$, we can use the tail estimate

$$\mathcal{E}_\omega \left[\chi_{\left\{|\langle u, Z_\omega \rangle| > \beta \frac{1}{\sqrt{1-|u|^2}}\right\}} \exp \left\{ \frac{\langle u, Z_\omega \rangle^2}{2} \right\} \right]$$

$$\leqslant \beta^{-2} (1 - |u|^2) \mathcal{E}_\omega \left[\langle u, Z_\omega \rangle^2 \exp \left\{ \frac{\langle u, Z_\omega \rangle^2}{2} \right\} \right] \leqslant \beta^{-2} \frac{1}{\sqrt{1 - |u|^2}}$$

(which is proved in exactly the same way as the tail estimate in Case 1) to restrict ourselves to $\omega \in \Omega$ satisfying $|\langle y, z_\omega \rangle| \leqslant \beta \mathcal{D}^{-1}$. This allows to bound the subtrahend in the difference $2\langle y, z_\omega \rangle \langle By, Bz_\omega \rangle - \langle y, z_\omega \rangle^2 |Bz_\omega|^2$ by $\beta^2 \mathcal{D}^2$. To bound the minuend from below, we shall use the following

Correlation Inequality

Let $u, w \in \mathbb{R}^n$ satisfy $|u| < 1$, $\langle u, w \rangle \geq 0$. Then

$$\mathcal{E}_\omega \left[\chi_{\left\{ \langle u, Z_\omega \rangle \langle w, Z_\omega \rangle < -\beta \frac{|w|}{\sqrt{1 - |u|^2}} \right\}} \exp \left\{ \frac{\langle u, Z_\omega \rangle^2}{2} \right\} \right] \leq \frac{2}{\sqrt{3}} e^{-\beta} \frac{1}{\sqrt{1 - |u|^2}}.$$

Let us first show that this correlation inequality implies the desired bound for the minuend $2\langle y, z_\omega \rangle \langle By, Bz_\omega \rangle$. Indeed, write

$$\langle y, z_\omega \rangle \langle By, Bz_\omega \rangle = \langle By, Z_\omega \rangle \langle B^3 y, Z_\omega \rangle = \langle u, Z_\omega \rangle \langle B^2 u, Z_\omega \rangle$$

where, as always, $u = By$. Observe that $\langle u, B^2 u \rangle = |Bu|^2 \geq 0$. Therefore, according to the correlation inequality, we may restrict ourselves to $\omega \in \Omega$ satisfying $\langle u, Z_\omega \rangle \langle B^2 u, Z_\omega \rangle \geq -\beta \frac{1}{\sqrt{1 - |u|^2}} |B^2 u|$ where $\beta > 0$ is chosen so large that $\beta^{-2} + \frac{2}{\sqrt{3}} e^{-\beta} \leq \eta$. Now observe that

$$|B^2 u| = \sqrt{\sum_j b_j^4 u_j^2} \leq \sqrt{\sqrt{\sum_j b_j^8} \sqrt{\sum_j u_j^4}} \leq \mathcal{D}^2 \|u\|_4.$$

Thus

$$2\langle u, Z_\omega \rangle \langle B^2 u, Z_\omega \rangle \geq -2\beta \frac{\|u\|_4}{\sqrt{1 - |u|^2}} \mathcal{D}^2 \geq -2\beta \mathcal{D}^2$$

due to our assumption $|u|^2 < 1 - \|u\|_4^2$. To prove the correlation inequality, just take

$$\tilde{u} := u - \frac{\sqrt{1 - |u|^2}}{2|w|} w$$

and observe that $|\tilde{u}|^2 \leq |u|^2 + \frac{1}{4}(1 - |u|^2)$, so $\sqrt{1 - |\tilde{u}|^2} \geq \frac{\sqrt{3}}{2} \sqrt{1 - |u|^2}$. Now we have

$$\mathcal{E}_\omega \left[\chi_{\left\{ \langle u, Z_\omega \rangle \langle w, Z_\omega \rangle < -\beta \frac{|w|}{\sqrt{1 - |u|^2}} \right\}} \exp \left\{ \frac{\langle u, Z_\omega \rangle^2}{2} \right\} \right] \leq e^{-\beta} \mathcal{E}_\omega \left[\exp \left\{ \frac{\langle \tilde{u}, Z_\omega \rangle^2}{2} \right\} \right]$$

because

$$\langle \tilde{u}, Z_\omega \rangle^2 \geq \langle u, Z_\omega \rangle^2 - 2 \frac{\sqrt{1 - |u|^2}}{2|w|} \langle u, Z_\omega \rangle \langle w, Z_\omega \rangle \geq \langle u, Z_\omega \rangle^2 + \beta$$

under the condition $\langle u, Z_\omega \rangle \langle w, Z_\omega \rangle < -\beta \frac{|w|}{\sqrt{1 - |u|^2}}$. It remains to recall that

$$\mathcal{E}_\omega \left[\exp \left\{ \frac{\langle \tilde{u}, Z_\omega \rangle^2}{2} \right\} \right] \leq \frac{1}{\sqrt{1 - |\tilde{u}|^2}} \leq \frac{2}{\sqrt{3}} \frac{1}{\sqrt{1 - |u|^2}}.$$

The upper bound for $\Gamma(A)$ is now completely proved.

Proof of the Estimate $\Gamma(A) \geqslant c\sqrt{\|A\|_{\text{H-S}}}$

Let $\varrho > 0$. Consider the family of random polyhedrons

$$Q(\varrho, N; \omega) := \left\{ x \in \mathbb{R}^n \ : \ \left|\langle x, x^{[k]} \rangle\right| \leqslant \varrho \quad \text{for all } k = 1, \ldots, N \right\}$$

where $x^{[k]} = BZ_\omega^{[k]}$ and $Z_\omega^{[k]}$ $(k \geqslant 1)$ is a sequence of independent random vectors equidistributed with $Z_\omega = \sum_j \varepsilon_j(\omega) e_j$. Let us observe that $|B\nu_y|$ identically equals $\sqrt{\sum_j b_j^4} = \mathcal{D}^2$ on $\partial Q(\varrho; \omega)$. Thus, the inequality $\Gamma(A) \geqslant c\sqrt{\|A\|_{\text{H-S}}} = c\mathcal{D}$ will be proved if we show that at least one polyhedron $Q(\varrho, N; \omega)$ has the Gaussian perimeter of $c\mathcal{D}^{-1}$ or greater.

I tried to use as few non-trivial statements about Bernoulli random variables in this note as possible but I still had to employ the following

Pinnelis Tail Lemma. *Let $u \in \mathbb{R}^n$, $\beta \geqslant 0$. Then*

$$\mathcal{P}_\omega\{\langle u, Z_\omega \rangle \geqslant \beta|u|\} \leqslant K \frac{1}{\sqrt{2\pi}} \int_\beta^\infty e^{-\frac{t^2}{2}} \, dt \leqslant K \frac{1}{1+\beta} e^{-\frac{\beta^2}{2}}$$

where K is some universal constant. Informally speaking, this means that Bernoulli tails do not exceed Gaussian tails.

The simplest and most elegant proof of the Pinnelis Tail Lemma belongs to Sergei Bobkov, who observed that the function

$$\Phi(\beta) = \frac{1}{\sqrt{2\pi}} \int_\beta^\infty e^{-\frac{t^2}{2}} \, dt$$

satisfies the inequality

$$\Phi\left(\frac{\beta - a}{\sqrt{1 - a^2}}\right) + \Phi\left(\frac{\beta + a}{\sqrt{1 - a^2}}\right) \leqslant 2\Phi(\beta)$$

for all $\beta \geqslant \sqrt{3}$, $0 \leqslant a < 1$ (to prove it, just differentiate the left hand side with respect to a and check that the derivative is never positive), which allows to prove the lemma by induction with $K = \frac{1}{2\Phi(\sqrt{3})} < 13$.

We shall show that the "average perimeter" of $Q(\varrho, N; \omega)$ is large. To formalize this, choose some nice continuous non-negative decreasing L^1-function $p : [0, +\infty) \to \mathbb{R}$ (which will serve as the weight with which we shall average with respect to ϱ) and some small $h > 0$.

Note that for each $\varrho > 0$, $N \geqslant 1$, and $\omega \in \Omega$,

$$\gamma\big(Q(\varrho + h, N; \omega)\big) - \gamma\big(Q(\varrho, N; \omega)\big) \leqslant h\Upsilon$$

where Υ is the supremum of all perimeters of our polyhedra with respect to the standard Gaussian measure. Therefore

$$\mathcal{E}_\omega\big[\gamma(Q(\varrho + h, N; \omega))\big] - \mathcal{E}_\omega\big[\gamma(Q(\varrho, N; \omega))\big] \leqslant h\Upsilon.$$

On the other hand,

$$\mathcal{E}_\omega\big[\gamma(Q(\varrho; \omega))\big] = \int_{\mathbb{R}^n} \Big(1 - \mathcal{P}_\omega\{|\langle Bx, Z_\omega\rangle| > \varrho\}\Big)^{N(\varrho)} d\gamma(x).$$

Now take $\varrho_\ell = \ell h$ ($\ell = 1, 2, \dots$), choose some integer-valued positive increasing function $N(\varrho)$, and consider the sumtegral

$$\sum_{\ell=1}^{\infty} p(\varrho_\ell) \int_{\mathbb{R}^n} \Big[(1 - \mathcal{P}_\omega \{|\langle Bx, Z_\omega\rangle| > \varrho_{\ell+1}\})^{N(\varrho_\ell)}$$
$$- (1 - \mathcal{P}_\omega \{|\langle Bx, Z_\omega\rangle| > \varrho_\ell\})^{N(\varrho_\ell)}\Big] d\gamma(x).$$

On one hand, this sumtegral does not exceed $\sum_{\ell=1}^{\infty} p(\varrho_\ell) h\Upsilon \leqslant \Upsilon \int_0^\infty p(\varrho) d\varrho$. On the other hand, since

$$(1 - \alpha)^M - (1 - \beta)^M \geqslant e^{-1} M(\beta - \alpha) \qquad \text{whenever } \alpha \leqslant \beta \leqslant \frac{1}{M},$$

we can change the order of summation and integration (the sumtegrand is nonnegative) and estimate our sumtegral from below by

$$e^{-1} \int_S \Big\{ \sum_{\ell=1}^{\infty} p(\varrho_\ell) N(\varrho_\ell) \Big[\mathcal{P}_\omega \{|\langle Bx, Z_\omega\rangle| > \varrho_\ell\}$$
$$- \mathcal{P}_\omega \{|\langle Bx, Z_\omega\rangle| > \varrho_{\ell+1}\}\Big] \Big\} d\gamma(x)$$

where $S \subset \mathbb{R}^n$ is the set of all points x for which $\mathcal{P}_\omega \{|\langle Bx, Z_\omega\rangle| > \varrho\} \leqslant N(\varrho)^{-1}$ for all $\varrho > 0$. For each fixed $x \in S$, the integrand converges to the mathematical expectation $\mathcal{E}_\omega \big[p(|\langle Bx, Z_\omega\rangle|) N(|\langle Bx, Z_\omega\rangle|)\big]$ as $h \to 0^+$. Therefore, the lower limit of the sumtegral is at least

$$e^{-1} \gamma(S) \mathcal{E}_\omega \Big[p(|\langle Bx, Z_\omega\rangle|) N(|\langle Bx, Z_\omega\rangle|)\Big]$$

as $h \to 0^+$. Comparing the upper and the lower bound, we get the inequality

$$\Upsilon \int_0^\infty p(\varrho) d\varrho \geqslant e^{-1} \gamma(S) \mathcal{E}_\omega \Big[p(|\langle Bx, Z_\omega\rangle|) N(|\langle Bx, Z_\omega\rangle|)\Big].$$

Our aim will be to choose the function $N(\rho)$ sufficiently small to make the set S large on one hand and sufficiently large to make the right hand side much larger than $\int_0^\infty p(\varrho) d\varrho$ on the other hand. Note that the demand that $N(\varrho)$ assume only integer values can be dropped because, given any non-negative function $N(\varrho)$, we can always replace it by the function $\widetilde{N}(\varrho)$ that takes value

1 if $0 \leqslant N(\varrho) \leqslant 2$ and value k if $k < N(\varrho) \leqslant k+1$, $k = 2, 3, \ldots$. This will not reduce the set S and will reduce the mathematical expectation on the right not more than twice.

Now recall that

$$\int_{\mathbb{R}^n} \|Bx\|_4^4 \, d\gamma(x) = 3D^4 \quad \text{and} \quad \int_{\mathbb{R}^n} \left(1 - |Bx|^2\right)^2 \, d\gamma(x) = 2D^4.$$

Therefore, for at least one quarter (with respect to $d\gamma$) of the points $x \in \mathbb{R}^n$, one has both

$$\|Bx\|_4 \leqslant 2D \quad \text{and} \quad |1 - |Bx|^2| \leqslant 2D^2$$

(the measures of the exceptional sets do not exceed $\frac{3}{16}$ and $\frac{1}{2}$ correspondingly).

Now we can use Pinnelis Tail Lemma and observe that for such points x,

$$\mathcal{P}_\omega\left\{|\langle Bx, Z_\omega\rangle| > \varrho\right\} \leqslant 2K \left(1 + \frac{\varrho}{\sqrt{1+2D^2}}\right)^{-1} \exp\left\{-\frac{1}{1+2D^2}\frac{\varrho^2}{2}\right\}.$$

This leads to the choice

$$N(\varrho) := \frac{1}{2K}\left(1 + \frac{\varrho}{\sqrt{1+2D^2}}\right) \exp\left\{\frac{1}{1+2D^2}\frac{\varrho^2}{2}\right\}.$$

Let us now choose the weight p. Since the only mathematical expectations we can easily compute are those of slight perturbations of exponential functions, it seems reasonable to try

$$p(\varrho) := \exp\left\{-\frac{D^2}{3}\frac{\varrho^2}{2}\right\}.$$

With such a choice, we have

$$\mathcal{E}_\omega\left[p(|\langle Bx, Z_\omega\rangle|)\, N\left(|\langle Bx, Z_\omega\rangle|\right)\right] \geqslant \frac{1}{2K}\mathcal{E}_\omega\left[(1 + |\langle u, Z_\omega\rangle|)\exp\left\{\frac{\langle u, Z_\omega\rangle^2}{2}\right\}\right]$$

where

$$u := \sqrt{\frac{1}{1+2D^2} - \frac{D^2}{3}}\, Bx.$$

Note that

$$\|u\|_4 \leqslant \|Bx\|_4 \leqslant 2D \quad \text{and} \quad |u|^2 \geqslant \left(\frac{1}{1+2D^2} - \frac{D^2}{3}\right)(1 - 2D^2) \geqslant 1 - 5D^2.$$

Using the inequality

$$(1 + |\langle u, Z_\omega\rangle|)\exp\left\{\frac{\langle u, Z_\omega\rangle^2}{2}\right\} \geqslant \frac{1}{\sqrt{2\pi}}\int_{-\infty}^{\infty} |t|e^{-\frac{t^2}{2}}\exp\{t\langle u, Z_\omega\rangle\}\, dt,$$

we conclude that

$$\mathcal{E}_\omega \left[(1 + |\langle u, Z_\omega \rangle|) \exp\left\{ \frac{\langle u, Z_\omega \rangle^2}{2} \right\} \right] \geqslant \mathcal{D}^{-2} \cdot \frac{1}{\sqrt{2\pi}} \int_{-\infty}^{\infty} |t| e^{-\frac{5t^2}{2}} e^{-2t^4} \, dt.$$

Since $\int_0^\infty p(\varrho) \, d\varrho = \sqrt{\frac{3\pi}{2}} \mathcal{D}^{-1}$, the desired bound $\Upsilon \geqslant c\mathcal{D}^{-1}$ follows.
The theorem is thus completely proved.

References

[B]   Ball, K. (1993): The reverse isoperimetric problem for Gaussian measure. Discrete Comput. Geom., **10(4)**, 411–420

[Be1]   Bentkus, V. (1986): Dependence of the Berry–Esseen bound on the dimension. Lithuanian Maqth. J., **26(2)**, 205–210

[Be2]   Bentkus, V. (2001): On the dependence of the Berry–Esseen bound on dimension. Max Planck Instutute Preprint series 2001 (53), Bonn. Journal Statistical Planning Inference, to appear

On p-Pseudostable Random Variables, Rosenthal Spaces and l_p^n Ball Slicing[*]

Krzysztof Oleszkiewicz

Institute of Mathematics, Warsaw University, Banacha 2, 02-097 Warszawa, Poland *koles@mimuw.edu.pl*

Summary. We introduce the class of p-pseudostable random variables and investigate some of their properties. Short notes concerning embedding Rosenthal-type spaces into $L_q(0,1)$ and hyperplane sections of the unit ball of l_p^n are added.

Notation. Throughout this paper the symbol $\sim$ denotes equality of distributions; $X, G, X_1, G_1, X_2, G_2, \ldots$ are independent symmetric random variables with $X_n \sim X$ and $G_n \sim G \sim \mathcal{N}(0,1)$ for $n = 1, 2, \ldots$ A Fourier transform of an integrable function $f : R \longrightarrow R$ is defined by $\hat{f}(t) = \int_R e^{itx} f(x) dx$, so that if f is even and continuous, and also $\hat{f}$ is integrable then $(\hat{f})^\wedge = 2\pi f(x)$.

Introduction. A characteristic function of a symmetric p-stable distribution ($0 < p \leq 2$) is of the form $\varphi(t) = e^{-c|t|^p}$ for some $c > 0$. It is well known that no random variable has a characteristic function of the form $\varphi(t) = e^{-c|t|^p}$ for $p > 2, c > 0$. Indeed, assume that $p > 2$ and $\varphi(t) = Ee^{itX} = e^{-c|t|^p}$ for all real t. Then $\varphi''(0) = 0$ and therefore $EX^2 = 0$ implying that $X = 0$ a.s.

Also if we turn to the standard characterization of symmetric p-stable distributions by

$$aX_1 + bX_2 \sim (|a|^p + |b|^p)^{1/p} X$$

we see that $aX_1 + bX_2$ would have a greater second moment than $(|a|^p + |b|^p)^{1/p} X$ for $p > 2$ and $ab \neq 0$ if $EX^2 < \infty$. And we would like to have $EX^2 < \infty$ because the classical p-stable has an absolute q-th moment finite for all $q \in (0, p)$.

However, one can hope that the "overdose" of variance could be extracted in some easy to control way, for example, in the form of the independent Gaussian summand.

Definition 1. *For $p > 2$ we call X a symmetric p-pseudostable random variable (a p-pseudostable) if X is not Gaussian (meaning also that X is not identically zero) and if for any real a and b there exists some real number $v(a,b)$ such that*

$$aX_1 + bX_2 \sim (|a|^p + |b|^p)^{1/p} X + v(a,b)G.$$

We will say that a p-pseudostable X is pure if X is Gaussian-free (i.e. it cannot be expressed as the sum of two independent random variables one of which is a nondegenerate Gaussian).

---

[*] Research partially supported by KBN Grant 2 P03A 043 15.

The following theorem is the main result of this note.

Theorem 1. *All p-pseudostables have finite second moments. If X is a p-pseudostable then it has a characteristic function*

$$\varphi(t) = e^{-c|t|^p - \alpha t^2}$$

for some positive constants c and α. Conversely, if a random variable X has a characteristic function of the above form then X is p-pseudostable. For $p \in (2,4] \cup \bigcup_{k=2}^{\infty} [4k-2, 4k]$ there are no p-pseudostables. For $p \in \bigcup_{k=1}^{\infty} (4k, 4k+2)$ there exist pure p-pseudostables and any p-pseudostable is either pure or it can be expressed as a sum of a pure p-pseudostable and an independent Gaussian summand. Moreover, for fixed p all pure p-pseudostables are dilations of the pure p-pseudostable X with $EX^2 = 1$, whose distribution is uniquely determined. If X is a p-pseudostable then it has a continuous density g (with respect to the Lebesgue measure on R) and the limit $\lim_{t \to \infty} t^{p+1} g(t)$ exists and it is finite and strictly positive. Moreover, g has a zero point if and only if X is pure.

Proof. Assume that X is a p-pseudostable. Let $\varphi(t) = Ee^{itX}$ be the characteristic function of X. Certainly φ is an even and real-valued function on R since X is symmetric. As $\lim_{t \to 0} \varphi(t) = 1$ there is some $t_0 > 0$ such that $\varphi(t) > 0$ for $t \in [0, t_0]$. Let

$$A = \inf_{t \in [2^{-1/p} t_0, t_0]} \varphi(t)^{t^{-p}}.$$

$A \in (0, 1]$ because φ is continuous. There exists some real v such that

$$2^{-1/p} X_1 + 2^{-1/p} X_2 \sim X + vG.$$

Therefore

$$\varphi(2^{-1/p} t) = \varphi(t)^{1/2} e^{-v^2 t^2/4}$$

for any $t \in [0, t_0]$. By iteration we arrive at

$$\varphi(2^{-n/p} t) = \varphi(t)^{2^{-n}} \exp\left(-\frac{1}{2} v^2 (2^{-n/p} t)^2 \cdot \sum_{k=1}^{n} (2^{\frac{2-p}{p}})^k \right)$$

$$\geq \left(\varphi(t)^{t^{-p}} \right)^{(2^{-n/p} t)^p} e^{-\frac{1}{2} v^2 (2^{-n/p} t)^2 \cdot \left(2^{\frac{p-2}{p}} - 1\right)^{-1}}$$

for any positive integer n. For any $s \in (0, t_0]$ there exists $t \in [2^{-1/p} t_0, t_0]$ and a positive integer n such that $s = 2^{-n/p} t$, so that we have

$$\varphi(s) \geq A^{s^p} e^{-Bs^2},$$

where $B = \frac{1}{2} v^2 \cdot (2^{\frac{p-2}{p}} - 1)^{-1}$. Hence $\limsup_{s \to 0} \frac{1 - \varphi(s)}{s^2} < \infty$ and therefore $EX^2 < \infty$. Comparing second moments of $aX_1 + bX_2$ and $(|a|^p + |b|^p)^{1/p} X + v(a, b)G$ we arrive at

$$|v(a,b)| = \sqrt{a^2 + b^2 - (|a|^p + |b|^p)^{2/p}} \sqrt{EX^2}.$$

From the equality

$$\varphi(at)\varphi(bt) = \varphi\big((|a|^p + |b|^p)^{1/p}t\big)e^{-v(a,b)^2 t^2/2}$$

which holds for any real a, b and t we deduce that the continuous function $\psi(t) = \varphi(t^{1/p})e^{EX^2 t^{2/p}/2}$ satisfies $\psi(x)\psi(y) = \psi(x+y)$ for any $x, y > 0$ and $\psi(0) = 1$. Therefore $\psi(t) = e^{-ct}$ for some constant c and all $t \geq 0$. Hence

$$\varphi(t) = e^{-\frac{EX^2}{2}t^2 - c|t|^p}$$

for $t \in R$. We know that $c \geq 0$ because φ is bounded and the case $c = 0$ is excluded since we assume X is not Gaussian. Of course, if some random variable X has a characteristic function of the form $\varphi(t) = e^{-\alpha t^2 - c|t|^p}$ then it satisfies the functional equation which is equivalent to the fact that X is a p-pseudostable with $EX^2 = 2\alpha$. Assume now that $p \in (4k-2, 4k)$ for some integer $k \geq 1$. The function $\varphi(t) = e^{-\alpha t^2 - c|t|^p}$ is $4k-2$ times differentiable and therefore $EX^{4k-2} < \infty$ and $\varphi^{(4k-2)}(t) = -EX^{4k-2}e^{itX}$. Hence

$$\varphi^{(4k-2)}(t) \geq -EX^{4k-2} = \varphi^{(4k-2)}(0).$$

We know that

$$f(x) = x^{-2k}\left(e^{-x} - \sum_{l=0}^{2k-1} \frac{(-1)^l x^l}{l!}\right)$$

is an analytic function. Now, as

$$\varphi(t) = \sum_{l=0}^{2k-1} (-1)^l \frac{(\alpha t^2 + c|t|^p)^l}{l!} + (\alpha t^2 + c|t|^p)^{2k} f(\alpha t^2 + c|t|^p),$$

differentiating each summand separately $4k-2$ times, we see that growth of $\varphi^{(4k-2)}(t)$ at the neighbourhood of zero is determined by the second summand ($l = 1$). In the $(4k-2)$-th derivatives of all other summands there appear either constant summands or the summands with powers of $|t|$ higher than $|t|^{p-4k+2}$. Therefore

$$\varphi^{(4k-2)}(0) - \varphi^{(4k-2)}(t) = cp(p-1)(p-2)\ldots(p-4k+3)|t|^{p-4k+2} + o\big(|t|^{p-4k+2}\big)$$

which contradicts the fact that $\varphi^{(4k-2)}$ has the global minimum at $t = 0$. We have proved that there are no p-pseudostables for $p \in \bigcup_{k=1}^{\infty}(4k-2, 4k)$.

For even $p \geq 4$ we use another argument – Marcinkiewicz' theorem (Th. 2bis of [M]) stating that if $\varphi(t) = e^{W(t)}$ is a characteristic function of some probability distribution and W is a polynomial then $\deg W \leq 2$. In the case of p divisible by 4 we can give a straightforward argument. Since $\varphi(t) = e^{-\alpha t^2 - ct^p}$ belongs to $C^\infty(R)$ all moments of X are finite and $\varphi^{(2l)}(0) =$

$(-1)^l EX^{2l}$ for $l = 0, 1, 2, \ldots$ We know that $\varphi(z) = e^{-\alpha z^2 - c z^p}$ is an entire analytic function and therefore

$$\varphi(z) = \sum_{l=0}^{\infty} \frac{\varphi^{(2l)}(0)}{(2l)!} z^{2l}.$$

Hence

$$e^{\alpha t^2 - ct^p} = \varphi(it) = \sum_{l=0}^{\infty} \frac{(-1)^l \varphi^{(2l)}(0)}{(2l)!} t^{2l} = \sum_{l=0}^{\infty} \frac{EX^{2l}}{(2l)!} t^{2l}.$$

For real t we can use the Fubini theorem because $X^{2l} t^{2l}$ is nonnegative, arriving at

$$Ee^{tX} = E\frac{e^{tX} + e^{-tX}}{2} = E\sum_{l=0}^{\infty} \frac{(tX)^{2l}}{(2l)!} = e^{\alpha t^2 - ct^p}.$$

On the other hand

$$Ee^{tX} \geq P(tX \geq 0) \geq 1/2,$$

so that for t large enough we obtain a contradiction. The case $p = 4k + 2$ $(p > 2)$ is a bit harder. Since the characteristic function $\varphi(t) = e^{-\alpha t^2 - ct^p}$ extends to an entire analytic function we have

$$\limsup_{l \to \infty} \sqrt[2l]{EX^{2l}/(2l)!} = \limsup_{l \to \infty} \sqrt[2l]{|\varphi^{(2l)}(0)|/(2l)!} = 0,$$

so that $z \mapsto Ee^{zX} = E\frac{e^{zX} + e^{-zX}}{2} = \sum_{l=0}^{\infty} \frac{EX^{2l}}{(2l)!} z^{2l}$ is an entire function, too. By the identity principle we get $Ee^{zX} = \varphi(-iz) = e^{\alpha z^2 + cz^p}$ for all complex z. Therefore for $z_p = \cos\frac{2\pi}{p} + i\sin\frac{2\pi}{p}$ and $t > 0$ we have

$$Ee^{t(Re\, z_p)X} = e^{\alpha t^2 \cos^2 \frac{2\pi}{p} + ct^p \cos^p \frac{2\pi}{p}}$$

and

$$E\, Re\, e^{tz_p X} = Re\, Ee^{tz_p X} = e^{\alpha t^2 \cos\frac{\pi}{p} + ct^p} \cos\left(\alpha t^2 \sin\frac{\pi}{p}\right).$$

Hence for $t_n = \sqrt{\frac{2\pi n}{\alpha \sin \frac{\pi}{p}}}$ and integer n great enough there would be

$$Ee^{t_n(Re\, z_p)X} < E\, Re\, e^{t_n z_p X}$$

in contradiction to the fact that $e^{Re\, u} = |e^u| \geq Re\, e^u$ for all complex u.

Assume now that $p \in (4k, 4k+2)$ for some natural $k \geq 1$. Let $f_p(t) = e^{-|t|^p}$ and $F_p = \hat{f}_p$. We will need several lemmas. The first of them is well known (cf. [PS], Part Three, Chapter 4, Problem 154) and covers a wider range of the parameter p. We give its proof for the sake of completeness – later we will use a more refined version of this argument.

Lemma 1. *If $p > 2$ then $F_p : R \longrightarrow R$ is a continuous even function integrable with respect to the Lebesgue measure and such that $\int_R F_p(x)dx = 2\pi$ and*

$$\lim_{x \to \infty} x^{p+1} F_p(x) = -2\Gamma(p+1)\cos\left(\frac{p+1}{2}\pi\right) = 2\Gamma(p+1)\sin\left(\frac{p\pi}{2}\right).$$

Proof. Only the last assertion of the lemma needs proof as it implies the integrability of F_p and therefore $\hat{F}_p = (\hat{f}_p)^\wedge = 2\pi f_p$; in particular

$$\int_R F_p(x)dx = \hat{F}_p(0) = 2\pi f_p(0) = 2\pi.$$

Let

$$h_p(t) = e^{-|t|^p} + |t|^p e^{-|t|} + |t|^{p+1} e^{-|t|} = 1 - \frac{1}{2}|t|^{p+2} + \dots$$

The function h_p is at least $\lceil p \rceil + 1$ times differentiable (to see it expand the exponential terms into power series and note that due to fast enough convergence one can differentiate the series term by term) and all derivatives of h_p up to order $\lceil p \rceil + 1$ are integrable (use the Leibniz rule). Hence by the Riemann-Lebesgue theorem

$$\lim_{x \to \infty} \left| x^{\lceil p \rceil + 1} \hat{h}_p(x) \right| = \lim_{x \to \infty} \left| (h_p^{(\lceil p \rceil + 1)})^\wedge(x) \right| = 0,$$

so that $\hat{h}_p(x) = o(x^{-p-1})$ for $x \to \infty$. Note now that

$$f_p(t) = h_p(t) - |t|^p e^{-|t|} - |t|^{p+1} e^{-|t|}$$

and therefore

$$F_p(x) = \hat{h}_p(x) - \int_R e^{itx}|t|^p e^{-|t|}dt - \int_R e^{itx}|t|^{p+1} e^{-|t|}dt.$$

Hence our assertion immediately follows from the following lemma. $\square$

Lemma 2. *For $p > 0$ we have*

$$\lim_{x \to \infty} x^{p+1} \int_R e^{itx}|t|^p e^{-|t|}dt = 2\Gamma(p+1)\cos\left(\frac{p+1}{2}\pi\right).$$

Proof. We will prove a little more, namely that

$$I_1 = \int_R e^{itx}|t|^p e^{-|t|}dt = \frac{2\Gamma(p+1)}{(1+x^2)^{\frac{p+1}{2}}}\cos\left(\frac{p+1}{2}\pi - (p+1)\arcsin\frac{1}{\sqrt{1+x^2}}\right)$$

which follows from $I_1 = 2\,\mathrm{Re}\,I_2$, where

$$I_2 = \int_0^\infty e^{itx}t^p e^{-t}dt = \left(\cos\left(\frac{p+1}{2}\pi\right) + i\sin\left(\frac{p+1}{2}\pi\right)\right) \cdot \frac{\Gamma(p+1)}{(x+i)^{p+1}}.$$

Here values of $(x + i)^{p+1}$ are taken from the main branch of z^{p+1}. This is a simple transformation of the formula for the characteristic function of the gamma distribution which can be found, for example, in [F]. A simple argument goes as follows: I_2 is an analytic function of x $(Im\,x > -1)$ and the formula for I_2 can be easily checked for x being a purely imaginary complex number; therefore by the identity principle the formula must be valid for all real x. $\square$

Lemma 3. *Let $p \in (4k, 4k + 2)$ for some natural $k \geq 1$. For $\sigma > 0$ let $H_\sigma(x) = EF_p(x + \sigma G)$. Then*

$$\lim_{x \to \infty} x^{p+1} H_\sigma(x) = 2\Gamma(p+1)\sin\left(\frac{p\pi}{2}\right)$$

and there exists $\sigma > 0$ such that $H_\sigma(x) \geq 0$ for all $x \in R$. Moreover, there exists $y_p > 0$ such that $H_\sigma(x) > 0$ for all $x > y_p$ and $\sigma > 0$.

Proof. Let $u(x) = x^{p+1} F_p(x)$. By Lemma 1 u is continuous and

$$\lim_{x \to \infty} u(x) = -2\Gamma(p+1)\cos\left(\frac{p+1}{2}\pi\right).$$

Therefore u and F_p are bounded on R. Let us recall the well known estimate:

$$P(G > x) \leq e^{-x^2} Ee^{xG} = e^{-x^2/2}$$

for $x \geq 0$. Splitting

$$x^{p+1} H_\sigma(x) = E\left(\frac{x}{x+\sigma G}\right)^{p+1} u(x + \sigma G)1_{|G|\leq \frac{x}{2\sigma}} + x^{p+1}EF_p(x + \sigma G)1_{|G|>\frac{x}{2\sigma}}$$

we obtain the first assertion of the lemma by the Lebesgue theorem on majorized convergence – the first term tends to $2\Gamma(p+1)\sin(\frac{p\pi}{2})$ and the second one converges to zero as $x \to \infty$ because its absolute value is bounded by $x^{p+1}\|F_p\|_\infty e^{-\frac{x^2}{8\sigma^2}}$. The second assertion is more delicate. From Lemma 1 it follows that there exist some $x_p > 1000$ and $\varepsilon_p > 0$ such that $\int_{-x_p}^{x_p} F_p(x)dx > \pi$ and $F_p(x) > \frac{\varepsilon_p}{x^{p+1}}$ if $|x| > x_p$. Because of the symmetry we can restrict our considerations to the case $x > 0$. Let $F_p^+ = \max(F_p, 0)$, $F_p^- = \max(-F_p, 0)$ and let $A_p = \int_{-x_p}^{x_p} F_p^+(x)dx$, $B_p = \int_{-x_p}^{x_p} F_p^-(x)dx$; therefore $A_p > B_p > 0$. Choose $y_p > 2x_p$ and such that the following two facts hold simultaneously:

$$\varepsilon_p e^{\frac{1}{8}x^{1/2}} > 4B_p x^{p+\frac{1}{4}}$$

for all $x \geq y_p$ and

$$A_p > B_p \cdot e^{2x_p y_p^{-1/2} + x_p^2 y_p^{-3/2}}.$$

First we will prove that $H_\sigma(x) \geq 0$ for all $x \geq y_p$ and $\sigma > 0$. Let us consider two cases.

Case 1: $\sigma \geq x^{3/4}$.

Then

$$\sqrt{2\pi}\sigma H_\sigma(x) = \int_R F_p(t)e^{-\frac{(x-t)^2}{2\sigma^2}}\,dt \geq \int_{-x_p}^{x_p} F_p(t)e^{-\frac{(x-t)^2}{2\sigma^2}}\,dt$$

$$= e^{-\frac{x^2}{2\sigma^2}}\int_{-x_p}^{x_p} F_p(t)e^{(xt-\frac{t^2}{2})\sigma^{-2}}\,dt.$$

For $|t| < x_p$ we have

$$\left|xt - \frac{t^2}{2}\right|\sigma^{-2} \leq \left(x_p x + \frac{x_p^2}{2}\right)x^{-3/2} = x_p x^{-1/2} + \frac{1}{2}x_p^2 x^{-3/2}$$

$$\leq x_p y_p^{-1/2} + \frac{1}{2}x_p^2 y_p^{-3/2}.$$

From the way in which we chose y_p it follows that

$$\int_{-x_p}^{x_p} F_p(t)e^{(xt-\frac{t^2}{2})\sigma^{-2}}\,dt$$

$$\geq A_p e^{-x_p y_p^{-1/2} - \frac{1}{2}x_p^2 y_p^{-3/2}} - B_p e^{x_p y_p^{-1/2} + \frac{1}{2}x_p^2 y_p^{-3/2}} > 0$$

which ends the proof of Case 1.

Case 2: $\sigma \leq x^{3/4}$.

Then (recall that $x \geq y_p \geq 2x_p \geq 2000$)

$$\sqrt{2\pi}\sigma H_\sigma(x) = \int_R F_p(t)e^{-\frac{(x-t)^2}{2\sigma^2}}\,dt \geq \int_{x/2}^x \frac{\varepsilon_p}{t^{p+1}}e^{-\frac{(x-t)^2}{2\sigma^2}}\,dt - B_p e^{-\frac{(x-x_p)^2}{2\sigma^2}}$$

$$\geq \varepsilon_p x^{-p-1}\int_{x/2}^x e^{-\frac{(x-t)^2}{2\sigma^2}}\,dt - B_p e^{-\frac{x^2}{8\sigma^2}}$$

$$= \varepsilon_p x^{-p-1}\sigma \int_0^{\frac{x}{2\sigma}} e^{-t^2/2}\,dt - B_p e^{-\frac{x^2}{8\sigma^2}}$$

$$\geq \varepsilon_p x^{-p-1}\sigma \int_0^{\frac{x^{1/4}}{2}} e^{-t^2/2}\,dt - B_p e^{-\frac{x^2}{8\sigma^2}}$$

$$\geq \varepsilon_p x^{-p-1}\sigma \int_0^3 e^{-t^2/2}\,dt - B_p e^{-\frac{x^2}{8\sigma^2}}$$

$$\geq \varepsilon_p x^{-p-1}\sigma - B_p e^{-\frac{x^2}{8\sigma^2}}.$$

Define the function $\Psi_x(w) = \varepsilon_p e^{\frac{1}{8}w^2 x^{1/2}} - B_p w x^{p+\frac{1}{4}}$. Note that $\Psi_x(1) = \varepsilon_p e^{\frac{1}{8}x^{1/2}} - B_p x^{p+1/4} > 0$ because of the way in which we chose y_p and similarly $\Psi_x'(1) = 1/4\varepsilon_p x^{1/2}e^{1/8x^{1/2}} - B_p x^{p+1/4} \geq \frac{1}{4}\varepsilon_p e^{1/8x^{1/2}} - B_p x^{p+1/4} > 0$. Since Ψ_x is convex on $[1,\infty)$ it means that $\Psi_x(w) > 0$ for all $w \geq 1$. Hence putting

$w = x^{3/4}/\sigma \geq 1$ we arrive at $\varepsilon_p e^{\frac{x^2}{8\sigma^2}} > B_p x^{p+1}/\sigma$ which ends the proof of Case 2.

To finish the proof of Lemma 3 we show that for $\sigma = \sqrt{(2x_p y_p)/(\ln \frac{A_p}{B_p})}$ there is $H_\sigma(x) \geq 0$ for $x \in [0, y_p]$ and therefore for all real x. Indeed,

$$\sqrt{2\pi}\sigma H_\sigma(x) \geq \int_{-x_p}^{x_p} F_p(t) e^{-\frac{(x-t)^2}{2\sigma^2}} dt$$

$$\geq A_p e^{-\frac{(x+x_p)^2}{2\sigma^2}} - B_p e^{-\frac{(x-x_p)^2}{2\sigma^2}}$$

$$= e^{-\frac{(x+x_p)^2}{2\sigma^2}} (A_p - B_p e^{2x_p y_p/\sigma^2}) = 0.$$

This completes the proof. $\square$

Note that $EF_p(x + \sigma G) = (F_p * \gamma_\sigma)(x)$ where $\gamma_\sigma(x) = \frac{1}{\sqrt{2\pi}\sigma} e^{-\frac{x^2}{2\sigma^2}}$. Lemma 1 and Lemma 3 imply that F_p and $F_p * \gamma_\sigma$ are integrable functions and therefore

$$(F_p * \gamma_\sigma)^\wedge(t) = \hat{F}_p(t) \cdot \hat{\gamma}_\sigma(t) = (\hat{f}_p)^\wedge(t) \cdot e^{-\frac{\sigma^2 t^2}{2}} = 2\pi e^{-|t|^p - \frac{\sigma^2 t^2}{2}}.$$

Hence if $H_\sigma \geq 0$ then $g = 1/(2\pi)H_\sigma$ is the density of a symmetric probability measure on R with a characteristic function $\varphi(t) = e^{-|t|^p - \frac{\sigma^2 t^2}{2}}$. This proves that p-pseudostables do exist for $p \in \bigcup_{k=1}^\infty (4k, 4k+2)$. Note that the arguments used can be easily adapted to prove that there are no p-pseudostables for $p \in \bigcup_{k=1}^\infty (4k-2, 4k)$. For $\sigma' > \sigma$ we have

$$H_{\sigma'}(x) = EF_p\left(x + \sigma G_1 + \sqrt{\sigma'^2 - \sigma^2}G_2\right) = EH_\sigma\left(x + \sqrt{\sigma'^2 - \sigma^2}G\right).$$

Therefore $H_\sigma \geq 0$ implies $H_{\sigma'} > 0$ for $\sigma' > \sigma$. Let $\sigma_p = \inf\{\sigma > 0 : H_\sigma \geq 0\}$. As $t \mapsto e^{-|t|^p - \frac{1}{2}\sigma^2 t^2}$ is a continuous function and it is a pointwise limit of a sequence of characteristic functions $e^{-|t|^p - \frac{1}{2}(\sigma_p + \frac{1}{n})^2 t^2}$ we deduce that it is also a characteristic function and therefore $H_{\sigma_p} \geq 0$. We know that $\frac{1}{2\pi}H_{\sigma_p}$ is the density of a symmetric random variable X. We will prove that X is Gaussian-free. Indeed, assume that X can be expressed as a sum of two independent random variables one of which is a non-degenerate Gaussian. Without loss of generality we can assume that the Gaussian summand is symmetric (transferring its mean to the other summand). Then the other summand would have a characteristic function of the form $e^{-|t|^p - \frac{1}{2}\sigma^2 t^2}$ for some $\sigma < \sigma_p$ yielding $H_\sigma \geq 0$ which contradicts the minimality of σ_p. Hence X is a pure pseudostable and any p-pseudostable with the characteristic function of the form $e^{-|t|^p - \frac{1}{2}\sigma^2 t^2}$ for $\sigma > \sigma_p$ can be expressed as a sum of a pure p-pseudostable having the same distribution as X and an independent $\mathcal{N}(0, \sqrt{\sigma^2 - \sigma_p^2})$ summand. After obvious rescaling the same holds for a random variable with characteristic function $e^{-c|t|^p - \alpha t^2}$ — it is a pure p-pseudostable if $\sqrt{2\alpha}c^{-1/p} = \sigma_p$

and it is a non-pure p-pseudostable if $\sqrt{2a}c^{-1/p} > \sigma_p$. Note that $\sigma_p > 0$ since there are no p-stables for $p > 2$. The above considerations imply that densities of all non-pure p-pseudostables are strictly positive. Now we will prove that the continuous density of a pure p-pseudostable has some zero point. Indeed, we know that functions $H_{\frac{n}{n+1}\sigma_p}$ have some zero points for $n = 1, 2, \ldots$ because they are continuous. Denote by z_n some zero point of $H_{\frac{n}{n+1}\sigma_p}$ (the choice is arbitrary). From Lemma 3 it follows that $|z_n| \leq y_p$ for $n = 1, 2, \ldots$ and therefore we can choose a subsequence z_{n_l} convergent to some point z. Note that F_p is a Lipschitz function as

$$|F_p'(x)| = \left| \int_R it e^{itx - |t|^p} dt \right| \leq \int_R |t| e^{-|t|^p} dt$$
$$= \Gamma\left(1 + \frac{2}{p}\right) \leq \sup_{u \in [1,2]} \Gamma(u) = L < \infty.$$

Therefore all functions H_σ ($\sigma \geq 0$) are also Lipschitz with a Lipschitz constant L. Hence

$$\left| H_{\frac{n_l}{n_l+1}\sigma_p}(z) \right| = \left| H_{\frac{n_l}{n_l+1}\sigma_p}(z) - H_{\frac{n_l}{n_l+1}\sigma_p}(z_{n_l}) \right| \leq L|z - z_{n_l}| \underset{l \to \infty}{\longrightarrow} 0.$$

We also know that for $\sigma' > \sigma$ there is

$$\sup_{x \in R} |H_{\sigma'}(x) - H_\sigma(x)| = \sup_{x \in R} \left| EH_\sigma\left(x + \sqrt{\sigma'^2 - \sigma^2} G\right) - H_\sigma(x) \right|$$
$$\leq L \cdot E|G| \cdot \sqrt{\sigma'^2 - \sigma^2}$$

and therefore $H_{\frac{n_l}{n_l+1}\sigma_p}(z) \to H_{\sigma_p}(z)$ as $l \to \infty$. Hence $H_{\sigma_p}(z) = 0$ and the proof is finished. We have proved Theorem 1.   □

Proposition 1. *The function $p \mapsto \sigma_p$ (where $\sigma_p = \inf\{\sigma > 0 : H_\sigma \geq 0\}$) is continuous on $\bigcup_{k=1}^{\infty}(4k, 4k + 2)$ with*

$$\lim_{p \to 4k+} \sigma_p = \lim_{p \to (4k+2)-} \sigma_p = \infty.$$

Proof. Recall that $f_p(t) = e^{-|t|^p}$ and $F_p = \hat{f}_p$.
    We will need the following lemmas.

Lemma 4. *Let $[p_1, p_2] \subset (4k, 4k + 2)$ for some integer $k \geq 1$. Then there exist x_0 and ε (depending on p_1 and p_2 only) such that $F_p(x) \geq \frac{\varepsilon}{x^{p_2+1}}$ for any $x > x_0$ and $p \in [p_1, p_2]$.*

Proof (sketch). We will follow the approach used in the proof of Lemma 1. However, now we need more precision. For $p \in [p_1, p_2]$ let

$$w_p(t) = e^{-|t|^p} + \sum_{l=0}^{3} \frac{1}{l!} |t|^{p+l} e^{-|t|}.$$

In the neighbourhood of zero $w_p(t) = 1 - \frac{1}{24}|t|^{p+4} + o(|t|^{p+4})$ since $p > 4$. Hence $w_p \in C^{4k+4}(R)$. Using the Leibniz rule for $|t| > 1$ and differentiating the series expansion term by term for $|t| < 1$ we prove that $\sup_{p\in[p_1,p_2]} \|w_p^{(4k+4)}\|_1 < \infty$ and therefore

$$\sup_{p\in[p_1,p_2]} \sup_{x\in R} |x|^{4k+4}|\hat{w}_p(x)| < \infty.$$

Hence there exists some positive constant C depending on p_1 and p_2 only such that $|\hat{w}_p(x)| < \frac{C}{x^{p+2}}$ for all $x > 1$ and $p \in [p_1, p_2]$. According to Lemma 2

$$F_p(x) = \hat{w}_p(x) - \sum_{l=1}^{4} \frac{2\Gamma(p+l)}{(l-1)!(1+x^2)^{\frac{k+l}{2}}} \cos\left(\frac{p+l}{2}\pi - (p+l)\arcsin\frac{1}{\sqrt{1+x^2}}\right).$$

Let $D = 1 + \tan^{-1}(\frac{p_1-4k}{2p_1+2}\pi)$. One easily checks that

$$\inf_{p\in[p_1,p_2]} \inf_{x\geq D} -\cos\left(\frac{p+1}{2}\pi - (p+1)\arcsin\frac{1}{\sqrt{1+x^2}}\right) > 0.$$

Therefore by some elementary estimates we prove that there exists some positive constant M such that for all $x > D$ and all $p \in [p_1, p_2]$ there is $F_p(x) \geq \frac{M}{x^{p+1}} - \frac{1}{Mx^{p+2}}$. The assertion of the lemma easily follows. We omit long but elementary calculations. $\square$

Lemma 5. *Let $[p_1, p_2] \subset (4k, 4k+2)$ for some integer $k \geq 1$. Then there exists some number $y > 0$ (depending on p_1 and p_2 only) such that for all $x > y$, $\sigma > 0$ and $p \in [p_1, p_2]$ there is $H_\sigma(x) = EF_p(x + \sigma G) > 0$.*

Proof. The proof follows closely the proof of Lemma 3. Note that in view of Lemma 4 the main problem remaining is how to deal uniformly (for $p \in [p_1, p_2]$) with x_p, A_p and B_p. First one needs to prove that there exists some $s > 0$ such that for all $p \in [p_1, p_2]$ and all $z \geq s$ there is

$$\int_{-z}^{z} F_p(x)dx > \pi.$$

Then for $y_0 = \max(x_0, s, 2000)$, being the "uniform version" of x_p, put $A_p = \int_{-y_0}^{y_0} F_p^+(x)dx$ and $B_p = \int_{-y_0}^{y_0} F_p^-(x)dx$. One needs to show that $\inf_{p\in[p_1,p_2]} (A_p/B_p) > 1$ and $\sup_{p\in[p_1,p_2]} B_p < \infty$ to complete the proof. The last two assertions are equivalent since $A_p - B_p \in [\pi, 2\pi]$ and they are equivalent to the fact that

$$\sup_{p\in[p_1,p_2]} \int_{-y_0}^{y_0} |F_p(x)|dx < \infty.$$

Step 1: We prove that such s exists.

By the approach used in the proof of Lemma 4 one proves also that there exists $D > 0$ (depending on p_1 and p_2 only) such that $F_p(x) \leq \frac{D}{x^{p_1+1}}$ for any $x > D$ and $p \in [p_1, p_2]$. The proof of this fact is simpler than the proof of Lemma 4 so that we leave it to the reader. Recall that for any p there is $\int_R F_p(x)dx = 2\pi$ and therefore for $z > D$ we have

$$\int_{-z}^{z} F_p(x)dx > 2\pi - 2D \int_z^\infty \frac{dx}{x^{p_1+1}} = 2\pi - \frac{2D}{p_1 z^{p_1}}.$$

Hence there exists $s > 0$ such that for any $z \geq s$ and $p \in [p_1, p_2]$ there is $\int_{-z}^{z} F_p(x)dx > \pi$.

Step 2: We prove that $\sup_{p \in [p_1,p_2]} \int_{-y_0}^{y_0} |F_p(x)|dx < \infty$.

Note that by the Hölder inequality and the Plancherel identity we have

$$\int_{-y_0}^{y_0} |F_p(x)|dx \leq \sqrt{2y_0} \left(\int_{-y_0}^{y_0} |F_p(x)|^2 dx \right)^{1/2} \leq \sqrt{2y_0} \left(\int_R |\hat{f}_p(x)|^2 dx \right)^{1/2}$$

$$= \sqrt{2y_0} \left(2\pi \int_R f_p(t)^2 dt \right)^{1/2} = \sqrt{8\pi y_0} \left(\int_0^\infty e^{-2t^p} dt \right)^{1/2}$$

and the last expression is uniformly bounded for $p \in [p_1, p_2]$. The proof of Lemma 5 is finished.   □

Let $\varphi_p(t) = e^{-|t|^p - \frac{1}{2}\sigma_p^2 t^2}$. Lemma 5 immediately implies that for $x > y$ there is $\hat{\varphi}_p(x) > 0$. Note also that the family $(\hat{\varphi}_p)_{p \in [p_1,p_2]}$ is uniformly Lipschitz. Indeed,

$$|(\hat{\varphi}_p)'(x)| \leq \int_R |t| e^{-|t|^p - \frac{1}{2}\sigma_p^2 t^2} dt \leq 2 \int_0^\infty t e^{-t^p} dt$$

$$= \Gamma\left(1 + \frac{2}{p}\right) \leq \sup_{u \in [1,2]} \Gamma(u) < \infty.$$

Now we are in a position to prove Proposition 1.

Let $p, q_1, q_2, \ldots \in (p_1, p_2)$ with $\lim_{n \to \infty} q_n = p$. Choose a subsequence (q_{n_l}) such that

$$\lim_{l \to \infty} \sigma_{q_{n_l}} = \liminf_{n \to \infty} \sigma_{q_n} = \sigma_{\inf}.$$

Then

$$\varphi_{q_{n_l}} \xrightarrow[l \to \infty]{L_1} e^{-|t|^p - \frac{1}{2}\sigma_{\inf}^2 t^2} = \varphi_{\inf}(t)$$

by the Lebesgue majorized convergence theorem. Therefore $\hat{\varphi}_{q_{n_l}}$ tends uniformly to $\hat{\varphi}_{\inf}$ as $l \to \infty$. Hence $\hat{\varphi}_{\inf} \geq 0$. Assume that $\sigma_{\inf} < \sigma_p$. Then

$$\hat{\varphi}_p(x) = E\hat{\varphi}_{\inf}\left(x + \sqrt{\sigma_p^2 - \sigma_{\inf}^2}G\right) > 0$$

in contradiction to the fact that $\frac{1}{2\pi}\hat{\varphi}_p$ as the continuous density of a pure p-pseudostable must have some zero point. Hence $\sigma_{\inf} \geq \sigma_p$. Now choose a subsequence (q_{n_l}) such that

$$\lim_{l\to\infty} \sigma_{q_{n_l}} = \limsup_{n\to\infty} \sigma_{q_n} = \sigma_{\sup}$$

(a priori it is possible that $\sigma_{\sup} = \infty$). Then, again by the Lebesgue majorized convergence theorem we have

$$\varphi_{q_{n_l}}(t) \xrightarrow[l\to\infty]{L_1} e^{-|t|^p - \frac{1}{2}\sigma_{\sup}^2 t^2} = \varphi_{\sup}(t)$$

(with $\varphi_{\sup} = 0$ if $\sigma_{\sup} = \infty$) and therefore $\hat{\varphi}_{q_{n_l}}$ tends uniformly to $\hat{\varphi}_{\sup}$ as $l \to \infty$. Note that $\frac{1}{2\pi}\hat{\varphi}_{q_{n_l}}$ as the continuous density of a pure q_{n_l}-pseudostable must have some zero point in $[-y, y]$ (since by Lemma 5 it has no zero points outside this interval). As $\hat{\varphi}_{q_{n_l}}$ are uniformly Lipschitz we deduce (the same argument appeared at the very end of the proof of Theorem 1) that $\hat{\varphi}_{\sup}$ also has some zero point. But $\sigma_{\sup} > \sigma_p$ would imply

$$\hat{\varphi}_{\sup}(x) = E\hat{\varphi}_p\left(x + \sqrt{\sigma_{\sup}^2 - \sigma_p^2}\, G\right) > 0$$

since $\frac{1}{2\pi}\hat{\varphi}_p$ is the density of a pure p-pseudostable. The obtained contradiction proves that $\sigma_{\sup} \leq \sigma_p$. We have proved that

$$\liminf_{n\to\infty} \sigma_{q_n} \geq \sigma_p \geq \limsup_{n\to\infty} \sigma_{q_n}$$

and therefore $\sigma_{q_n} \to \sigma_p$ as $n \to \infty$, so that $p \mapsto \sigma_p$ is continuous on (p_1, p_2). Choosing p_1 and p_2 arbitrarily close to $4k$ and $4k+2$ respectively we prove that $p \mapsto \sigma_p$ is continuous on $(4k, 4k+2)$. It remains to investigate the boundary behavior. Assume that there exists a sequence $(p_n) \subset (4k, 4k+2)$ convergent to $4k$ and such that the sequence σ_{p_n} is bounded from above. Then there would exist a subsequence (p_{n_l}) such that $\sigma_{p_{n_l}} \to \sigma$ as $l \to \infty$ for some $\sigma \geq 0$ and therefore $\varphi_{p_{n_l}}(t)$ would tend pointwise to $e^{-t^{4k} - \frac{1}{2}\sigma^2 t^2}$. Hence $e^{-t^{4k} - \frac{1}{2}\sigma^2 t^2}$ as a continuous function being the pointwise limit of the sequence of the characteristic functions would also be a characteristic function in contradiction to the fact that there are no $4k$-pseudostables. Hence $\lim_{p\to 4k+} \sigma_p = \infty$. In a similar way one proves that $\lim_{p\to(4k-2)-} \sigma_p = \infty$. The proof of Proposition 1 is finished. $\quad\square$

Remark 1. Let X be a p-pseudostable with a characteristic function $\varphi(t) = e^{-c|t|^p - \frac{1}{2}t^2}$. Then for all positive integers $l < p$ we have $EX^l = EG^l$.

Proof. It suffices to prove that $\varphi^{(l)}(0)$ does not depend on c for positive integers $l < p$ and this is an easy consequence of the Leibniz rule and the fact that $\frac{d^m}{dt^m}(e^{-c|t|^p}) = 0$ for $t = 0$ and positive integer $m < p$. $\quad\square$

K. M. Ball pointed out that by the so-called "moment method" (in his review article [D] Diaconis traces it back to Chebyshev's proof of the Central Limit Theorem and presents some of its applications) Remark 1 immediately yields the following corollary.

Corollary 1. *If X is a p-pseudostable then*

$$\sup_{t \in R} \left| P\left(X < t\sqrt{EX^2}\right) - P(G < t) \right| \le \sqrt{\frac{\pi}{2(p-1)}}.$$

Proof. $X/\sqrt{EX^2}$ has a characteristic function of the form $\varphi(t) = e^{-c|t|^p - \frac{1}{2}t^2}$. Now it suffices to apply Theorem 2 of [D] and Remark 1 to $X/\sqrt{EX^2}$.   □

Corollary 2. *Under the notation of Proposition 1*

$$\lim_{k \to \infty} \left(\inf_{p \in (4k, 4k+2)} \sigma_p \right) = \infty.$$

Proof. Assume that there exists a sequence (p_n) tending to infinity with $p_n \in \bigcup_{k=1}^{\infty} (4k, 4k+2)$ and $\sup_n \sigma_{p_n} < \infty$. Let Z_n be the pure p_n-pseudostable with $EZ_n^2 = 1$, therefore it has the characteristic function $\psi_n(t) = e^{-\sigma_{p_n}^{-p_n} |t|^{p_n} - \frac{1}{2}t^2}$. Corollary 1 implies that Z_n tends in distribution to G as $n \to \infty$ and therefore $\psi_n(t) \to e^{-t^2/2}$ pointwise, meaning that $\sigma_{p_n}^{-p_n} |t|^{p_n} \to 0$ as $n \to \infty$ for any real t. Taking $t = 2\sup_n \sigma_{p_n}$ we obtain the contradiction which ends the proof. □

Remark 2. If X is a pure p-pseudostable then

$$\lim_{q \to p^-} (p - q)^{1/p} \|X\|_q = \kappa_p \|X\|_2,$$

where $\kappa_p = (\frac{2}{\pi} \Gamma(p+1) \sin(\frac{p\pi}{2}))^{1/p} \sigma_p^{-1}$.

Proof. By Lemma 3 we can precisely describe the limit behavior of the density of X. The assertion follows by some elementary calculation.   □

One of the classical applications of p-stable random variables is the linear isometric embedding of l_p^n space into $L_q(0,1)$ for $0 < q < p \le 2$. The main idea (a so-called representation theorem, cf. [L]) comes from P. Levy, at least for finite n although the application to Banach spaces appeared much later. For embedding l_p^∞ some more effort is needed (see [K] and [BDCK]). Of course there is no linear isomorphic embedding of l_p^∞ into $L_q(0,1)$ for $2 \le q < p \le \infty$ since $L_q(0,1)$ has cotype q and l_p^∞ does not have cotype q. However, using p-pseudostables instead of p-stables we can transfer the ideas to obtain some other results.

We will need the following simple lemma.

Lemma 6. *Let $X, X_1, X_2, \ldots$ be i.i.d. p-pseudostables with $EX^2 = 1$. Then for any $a = (a_1, a_2, \ldots) \in l_2^\infty$ the series $\sum_{n=1}^\infty a_n X_n$ is convergent a.s. and*

$$\sum_{n=1}^\infty a_n X_n \sim \|a\|_p X + \sqrt{\|a\|_2^2 - \|a\|_p^2}\, G.$$

Proof. By the Kolmogorov three series theorem one easily checks that $\sum_{n=1}^\infty a_n X_n$ converges a.s.; therefore the series converges also in distribution to the same limit. To finish the proof it suffices to show that the characteristic functions of $\sum_{n=1}^N a_n X_n$ tend pointwise to the characteristic function of $\|a\|_p X + \sqrt{\|a\|_2^2 - \|a\|_p^2}\, G$ as $N \to \infty$, which is trivial.   □

Before we pass to embedding results let us transfer the cotype argument to the pseudostable setting to obtain some bounds on κ_p and σ_p.

Lemma 7. *For $q \geq 2$ and any real a, b there is*

$$\sum_{l=0}^{\lfloor q/2 \rfloor - 1} \binom{q}{2l} |a|^{q-2l} b^{2l} + \frac{1}{2} \binom{q}{2\lfloor q/2 \rfloor} |a|^{q - 2\lfloor q/2 \rfloor} b^{2\lfloor q/2 \rfloor}$$

$$\leq \frac{|a+b|^q + |a-b|^q}{2}$$

$$\leq \sum_{l=0}^{\lfloor q/2 \rfloor} \binom{q}{2l} |a|^{q-2l} b^{2l} + |b|^q.$$

Proof. Treat the three expressions, which we compare as functions of the parameter b and note that their derivatives up to order $2\lfloor q/2 \rfloor$ agree at $b = 0$, so that it suffices to prove the inequalities for $2\lfloor q/2 \rfloor$-th derivatives, i.e.

$$\frac{|a|^\beta}{2} \leq \frac{|a+b|^\beta + |a-b|^\beta}{2} \leq |a|^\beta + |b|^\beta,$$

where $\beta = q - 2\lfloor q/2 \rfloor \in [0, 2]$, which are elementary and well-known to be true.   □

Lemma 8. *Let X be a p-pseudostable with $EX^2 = 1$. Then*

$$(2 - 2^{\frac{q}{p}}) E|X|^q \leq 2^{\frac{q}{2}+1} E|G|^q$$

for any $q \in [2, p)$.

Proof. Recall that X_1 and X_2 denote independent copies of X. From an elementary inequality $\frac{|a+b|^q + |a-b|^q}{2} \geq |a|^q + |b|^q$ holding for $q \geq 2$ and any real a, b we deduce that

$$E|X_1 + X_2|^q = E\frac{|X_1 + X_2|^q + |X_1 - X_2|^q}{2} \geq E|X_1|^q + E|X_2|^q = 2E|X|^q.$$

On the other hand there is $X_1 + X_2 \sim 2^{1/p}X + \sqrt{2 - 2^{2/p}}\, G$ so that

$$E|X_1 + X_2|^q = E\left|2^{1/p}X + \sqrt{2 - 2^{2/p}}\, G\right|^q$$

$$= E\frac{\left|\sqrt{2 - 2^{2/p}}\, G + 2^{1/p}X\right|^q + \left|\sqrt{2 - 2^{2/p}}\, G - 2^{1/p}X\right|^q}{2}$$

$$\leq \sum_{l=0}^{\lfloor q/2 \rfloor} \binom{q}{2l} E\left|\sqrt{2 - 2^{2/p}}\, G\right|^{q-2l} E(2^{1/p}X)^{2l} + E|2^{1/p}X|^q$$

$$= \sum_{l=0}^{\lfloor q/2 \rfloor} \binom{q}{2l} E\left|\sqrt{2 - 2^{2/p}}\, G_1\right|^{q-2l} E|2^{1/p}G_2|^{2l} + E|2^{1/p}X|^q$$

$$\leq E\left(\left|\sqrt{2 - 2^{2/p}}\, G_1 + 2^{1/p}G_2\right|^q + \left|\sqrt{2 - 2^{2/p}}\, G_1 - 2^{1/p}G_2\right|^q\right)$$

$$\quad + 2^{q/p}E|X|^q$$

$$= 2E|\sqrt{2}G|^q + 2^{q/p}E|X|^q$$

$$= 2^{\frac{q}{2}+1}E|G|^q + 2^{\frac{q}{p}}E|X|^q,$$

where we used Lemma 7, Remark 1 and again Lemma 7. Putting both inequalities together we finish the proof. $\qquad\square$

Proposition 2. *For any* $p \in \bigcup_{k=1}^{\infty}(4k, 4k+2)$ *there is*

$$\kappa_p \leq 2\left(\frac{p\Gamma(\frac{p+1}{2})}{\sqrt{\pi}\ln 2}\right)^{1/p}$$

and for any $\theta \in (0,1)$ *there is*

$$\liminf_{k \to \infty} \left(\inf_{p \in (4k+\theta, 4k+2-\theta)} p^{-1/2}\sigma_p\right) > 0.$$

Proof. The first assertion follows from Lemma 8 and Remark 2 as

$$\lim_{q \to p^-} \frac{(2 - 2^{q/p})^{1/q}}{(p-q)^{1/p}} = \left(\frac{2\ln 2}{p}\right)^{1/p}$$

and

$$\|G\|_p = \sqrt{2}\left(\frac{\Gamma(\frac{p+1}{2})}{\sqrt{\pi}}\right)^{1/p}.$$

The second assertion follows immediately from the first one – note that $\inf_k \inf_{p \in (4k+\theta, 4k+2-\theta)} \sin(\frac{p\pi}{2}) > 0$ for any $\theta \in (0,1)$ and use the fact that $\lim_{u \to \infty} \frac{\Gamma(u)^{1/u}}{u} = 1/e$. This ends the proof. $\qquad\square$

Lemma 9. *If* X *is a* p-pseudostable and $EX^2 = 1$ then $4\|X\|_q \geq \|G\|_q$ for any $q \in [1, p)$.

Proof. As $X_1 + X_2 \sim 2^{1/p} X + \sqrt{2 - 2^{2/p}}\, G$ we have

$$2\|X\|_q = \|X_1\|_q + \|X_2\|_q \geq \|X_1 + X_2\|_q$$
$$= \left\|2^{1/p} X + \sqrt{2 - 2^{2/p}}\, G\right\|_q \geq \sqrt{2 - 2^{2/p}}\, \|G\|_q \geq \frac{1}{2}\|G\|_p$$

since $p > 4$.   □

Remark 3. In the above we have used the well known fact – if Y and Z are independent mean-zero random variables then

$$\max(\|Y\|_q, \|Z\|_q) \leq \|Y + Z\|_q \leq \|Y\|_q + \|Z\|_q$$

for any $q \geq 1$. The first inequality follows by the Jensen inequality:

$$\|Y + Z\|_q^q = E_Y(E_Z|Y + Z|^q) \geq E_Y|Y + EZ|^q = E|Y|^q = \|Y\|_q^q$$

and by the same argument $\|Y + Z\|_q \geq \|Z\|_q$.

Proposition 3. *Under the assumptions of Lemma 6 there is*

$$\frac{1}{32}\left(\|X\|_q\|a\|_p + \|G\|_q\|a\|_2\right) \leq \|\sum_{n=1}^{\infty} a_n X_n\|_q \leq \|X\|_q\|a\|_p + \|G\|_q\|a\|_2$$

for any $q \in [1, p)$.

Proof. From Lemma 6 and Remark 3 we deduce that

$$\frac{1}{2}\left(\|a\|_p\|X\|_q + \sqrt{\|a\|_2^2 - \|a\|_p^2}\,\|G\|_q\right) \leq \|\sum_{n=1}^{\infty} a_n X_n\|_q$$

$$\leq \|a\|_p\|X\|_q + \sqrt{\|a\|_2^2 - \|a\|_p^2}\,\|G\|_q.$$

The second inequality of Proposition 3 follows immediately. To prove the first one we consider two cases.

Case 1: $\|a\|_p^2 \leq \frac{1}{2}\|a\|_2^2$

Then $\sqrt{\|a\|_2^2 - \|a\|_p^2} \geq \frac{1}{2}\|a\|_2$ and therefore $\|\sum_{n=1}^{\infty} a_n X_n\|_q \geq \frac{1}{32}(\|a\|_p\|X\|_q + \|a\|_2\|G\|_q)$.

Case 2: $\|a\|_p^2 \geq \frac{1}{2}\|a\|_2^2$

Then $\|a\|_p \geq \frac{1}{2}\|a\|_2$ and by Lemma 9 we get

$$\|a\|_p\|X\|_q \geq \frac{1}{2}\|a\|_2\|X\|_q \geq \frac{1}{8}\|a\|_2\|G\|_q$$

and therefore

$$\|a\|_p\|X\|_q + \sqrt{\|a\|_2^2 - \|a\|_p^2}\,\|G\|_q \geq \|a\|_p\|X\|_q \geq \frac{1}{16}\|a\|_p\|X\|_q + \frac{1}{16}\|a\|_2\|G\|_q,$$

yielding $\|\sum_{n=1}^{\infty} a_n X_n\|_q \geq \frac{1}{32}(\|a\|_p\|X\|_q + \|a\|_2\|G\|_q)$ which ends the proof.
□

It is well known that any sequence of independent random variables can be realized on the probability space $(0,1)$ equipped with the Lebesgue measure and the σ-field of Borel sets (an easy way to see it is to produce the i.i.d. sequence of random variables uniformly distributed on $[0,1]$ out of Rademacher functions and then to express given probability distributions as images of the uniform ones).

Corollary 3. *Let $p \in \bigcup_{k=1}^{\infty}(4k, 4k+2)$ and $q \in [1,p)$. On a (Rosenthal type) linear space of square summable sequences equipped with the norm*

$$\|a\| = \|X\|_q \|a\|_p + \|G\|_q \|a\|_2$$

let a linear $L_q(0,1)$-valued operator T be given by

$$Ta = \sum_{n=1}^{\infty} a_n X_n,$$

where $X_1, X_2, \ldots$ are i.i.d. p-pseudostables defined on the probability space $((0,1), \lambda_1, \mathcal{B}(0,1))$ and such that $EX_n^2 = 1$. Then

$$\frac{1}{32}\|a\| \leq \|Ta\|_q \leq \|a\|$$

for any $a \in l_2^{\infty}$.

Proof. It is an immediate consequence of Proposition 3. Note that $X_1, X_2, \ldots$ indeed belong to L_q since $E|X|^q < \infty$ for $q \in (0,p)$ – it can be deduced from Theorem 1 or directly from Definition 1, by a simple modification of Step 1 of [KPS].   □

The main disadvantage of Corollary 3 is that we do not have precise information on the possible values of $\frac{\|X\|_q}{\|G\|_q\|X\|_2}$ for X being a p-pseudostable (note that Corollary 3 holds true also for p-pseudostables which are not pure). In fact it seems most interesting when q is close to p. Note that for any even natural number $q < p$ we have

$$\|Ta\|_q = \|G\|_q\|a\|_2 = \left((q-1)!!\right)^{1/q}\|a\|_2$$

as a simple consequence of Remark 1, so that in this case we get an embedding similar to the classical isometric embedding of l_2^{∞} into L_q using Gaussian random variables. If X is a pure p-pseudostable then Remark 2 and Proposition 2 yield that $\lim_{q \to p^-} \frac{\|X\|_q}{\|G\|_q\|X\|_2}(p-q)^{1/p}$ can be bounded from above by some universal constant (not depending on p.) However some lower bound would be much more useful since we are interested in the situation when the l_p-norm summand is as little perturbed by the l_2-summand as possible. We will see in a moment that the range of q's covered by Corollary 3 is far from the best possible for isomorphic embeddings of l_p^{∞} into L_q. Finally, also the condition $p \in \bigcup_{k=1}^{\infty}(4k, 4k+2)$ seems restrictive. To obtain better results on embedding Rosenthal spaces into L_q we will use the following fact due to Hitczenko, Montgomery-Smith and the present author.

Theorem 2. ([HMSO]) *There exist universal positive constants A and B such that for any $p \geq 2$, any natural n and independent symmetric random variables $Y_1, Y_2, \ldots, Y_n$ having logarithmically convex tails (i.e. such that $t \mapsto \ln P(|Y_i| > t)$ is a convex function on R_+) and finite p-th moment, the inequalities*

$$A\left(\left(\sum_{i=1}^{n} E|Y_i|^p\right)^{1/p} + \sqrt{p}\left(\sum_{i=1}^{n} EY_i^2\right)^{1/2}\right)$$

$$\leq \left\|\sum_{i=1}^{n} Y_i\right\|_p \leq B\left(\left(\sum_{i=1}^{n} E|Y_i|^p\right)^{1/p} + \sqrt{p}\left(\sum_{i=1}^{n} EY_i^2\right)^{1/2}\right)$$

hold true.

Corollary 4. *Let $p \geq 2$. If $Y, Y_1, Y_2, \ldots$ are i.i.d. symmetric random variables with logarithmically convex tails defined on the probability space $((0,1), \lambda_1, \mathcal{B}(0,1))$ and such that $EY^2 = 1$ and $E|Y|^p < \infty$ then the linear operator S defined on a linear space of square summable sequences equipped with the norm*

$$\|a\| = \|Y\|_p \|a\|_p + \sqrt{p}\|a\|_2$$

by the formula $Ta = \sum_{n=1}^{\infty} a_n Y_n$ satisfies

$$A\|a\| \leq \|Ta\|_p \leq B\|a\|,$$

where A and B are some universal positive constants.

Proof. The proof follows closely the proof of Lemma 6. To prove that the series $\sum_{n=1}^{\infty} a_n X_n$ is convergent also in L_p note that the Cauchy condition is satisfied (it follows easily by Theorem 2 or by some general theory). $\square$

The above corollary gives a good embedding into $L_p(0,1)$. Before we turn to the embeddings into $L_q(0,1)$ for $q \in [2, p)$ let us determine the possible values of the parameter $s = \frac{\sqrt{p}\|Y\|_2}{\|Y\|_p}$ (of course we are not interested in the case $s > 1$ since then the Banach-Mazur distance from $\|\cdot\|$ to the Euclidean norm is not greater than 2).

Lemma 10. *Let $p > 2$. If $s \in (0, \sqrt{2p}\Gamma(p+1)^{-1/p})$ then there exists a symmetric random variable Y with logarithmically convex tails and all moments finite such that $\frac{\sqrt{p}\|Y\|_2}{\|Y\|_p} = s$.*

Proof. Let Y be a symmetric random variable with $P(|Y| > t) = e^{-t^{1/\theta}}$ for all $t > 0$, where $\theta \geq 1$ is some constant. Y has then the so-called Weibull distribution and it has logarithmically convex tails and all moments finite. Let

$$h(\theta) = \frac{\sqrt{p}\|Y\|_2}{\|Y\|_p} = \sqrt{p\Gamma(2\theta+1)}\Gamma(\theta p+1)^{-1/p}.$$

Since $h(\theta) \to 0$ as $\theta \to \infty$ and h is continuous it takes on all values from the interval $(0, h(1))$ which ends the proof. $\quad\square$

Remark 4. The estimate of Lemma 10 cannot be improved since for any symmetric random variable Y with logarithmically convex tails there is

$$\frac{\|Y\|_p}{\|Y\|_2} \geq \frac{\Gamma(p+1)^{1/p}}{\sqrt{2}}$$

for $p > 2$ (see [HMSO] for the proof).

It seems that there is a gap in our method of isomorphic embedding Rosenthal spaces into L_q for $s \in (q^{-1/2}, 1)$ but in a while we will see that this gap can be easily filled.

Lemma 11. *Let $p > q > 2$ and let $s \geq (\frac{1}{\sqrt{2}})^{\frac{q(p-2)}{p-q}}$. Then for any $a \in l_2^\infty$ there is*

$$\|a\|_p + s\|a\|_2 \leq \|a\|_q + s\|a\|_2 \leq 3(\|a\|_p + s\|a\|_2).$$

Proof. The first inequality is trivial. To prove the second one it suffices to show that $\|a\|_q \leq 2(\|a\|_p + s\|a\|_2)$. Note that by the Hölder inequality

$$\|a\|_q \leq \|a\|_p^{1-\beta}\|a\|_2^\beta,$$

where $\beta = \frac{2(p-q)}{q(p-2)} \in (0, 1)$. Let $t = \|a\|_p/\|a\|_2 \in (0, 1]$ (the case $a = 0$ is trivial). We are to prove that

$$t^{1-\beta} \leq 2(t+s)$$

so it suffices to show that

$$\inf_{t\in(0,1]} t^\beta + st^{\beta-1} \geq \frac{1}{2}.$$

The infimum is attained at $t = \frac{1-\beta}{\beta}s$ or at $t = 1$. The second case is trivial and the first one leads to checking whether $\beta^{-\beta}(1-\beta)^{-(1-\beta)}s^\beta \geq \frac{1}{2}$. Note that $\inf_{u\in(0,1)} u^{-u} = 1$; therefore it suffices to prove that

$$s \geq 2^{1/\beta} = \left(\frac{1}{\sqrt{2}}\right)^{\frac{q(p-2)}{p-q}}$$

which was our assumption. $\quad\square$

Corollary 5. *Let $p > 2$ and $s \in (0,1)$. If $q \in (2,p)$ is such that $s \geq (1/\sqrt{2})^{\frac{q(p-2)}{p-q}}$ then the Rosenthal space with a norm $\|a\| = \|a\|_p + s\|a\|_2$ is in the Banach-Mazur distance not greater than a certain universal constant (not depending on p, s and q) from some linear subspace of $L_q(0,1)$.*

Proof. If $s \in (0, \sqrt{2p}\Gamma(p+1)^{-1/p}$ then the corollary is a direct consequence of Corollary 4, Lemma 10 and Lemma 11. If $s > \sqrt{2p}\Gamma(p+1)^{-1/p}$ then in Corollary 4 replace $Y, Y_1, Y_2, \ldots$ by an i.i.d. sequence $Z, Z_1, Z_2, \ldots$ with $Z \sim \mathcal{E}+cG$, where $\mathcal{E}$ and G are independent, $\mathcal{E}$ is the Weibull distribution with $\theta = 1$ (i.e. a symmetric exponential distribution) and $c > 0$. Note that then, in view of Remark 3, $\|\sum_{n=1}^{\infty} a_n Z_n\|_p$ is up to some universal multiplicative constant equal to

$$\left\| \sum_{n=1}^{\infty} a_n \mathcal{E}_n \right\|_p + c\left\| \sum_{n=1}^{\infty} a_n G_n \right\|_p$$

or

$$\|\mathcal{E}\|_p \|a\|_p + (c + \|\mathcal{E}\|_2)\|G\|_p.$$

Choosing appropriately large c we can represent any value from the interval $(\sqrt{2p}\Gamma(p+1)^{-1/p}, 1)$ as $\frac{c+\|\mathcal{E}\|_2\|G\|_p}{\|\mathcal{E}\|_p}$ up to a universal multiplicative constant since $\sqrt{2p}\Gamma(p+1)^{-1/p}\|\mathcal{E}\|_p/\|G\|_p$ is uniformly bounded away from zero for $p > 2$. Now use Lemma 11 to finish the proof. $\qquad\square$

Remark 5. Of course the constant $1/\sqrt{2}$ in Corollary 5 can be replaced by some other constant $1/C$ with an appropriate change of the bound on the Banach-Mazur distance (the bound on the Banach-Mazur distance grows approximately like C^2 and in a moment we will see that it cannot be essentially improved).

Remark 6. The estimate of Corollary 5 is close to optimal. Let $s \in (0,1)$ and $p > 2$. If the Rosenthal space with a norm $\|a\| = \|a\|_p + s\|a\|_2$ is in the Banach-Mazur distance less than C from some linear subspace of $L_q(0,1)$ for $q \in (2,p)$ then $s \geq (1/C')^{\frac{q(p-2)}{p-q}}$, where $C' > 1$ is some constant depending on C only.

Proof. If $s \geq \frac{2(p-q)}{p(q-2)}$ then $s \geq \frac{2(p-q)}{q(p-2)}$ since $p(q-2) \leq q(p-2)$. Therefore $s \geq (1/2)^{\frac{q(p-2)}{p-q}}$ because $2^{-u} \leq 2/u$ for $u > 0$. Hence we can restrict ourselves to the case $s < \frac{2(p-q)}{p(q-2)}$. Put $n = \lfloor (\frac{2(p-q)}{p(q-2)s})^{\frac{2p}{p-2}} \rfloor \geq 1$. By a standard cotype argument we get

$$C^q E\left\| \sum_{l=1}^{n} r_l v_l \right\|^q \geq \sum_{l=1}^{n} \|v_l\|^q,$$

where $r_1, r_2, \ldots, r_n$ are independent symmetric Bernoulli random variables $(P(r_l = 1) = P(r_l = -1) = 1/2)$ and $v_1, v_2, \ldots, v_n$ are vectors of the Rosenthal space. Taking $v_l = e_l$ (the l-th versor) we arrive at

$$C(n^{1/p} + sn^{1/2}) \geq (1+s)n^{1/q} \geq n^{1/q}.$$

Therefore

$$C \geq \frac{1}{n^{\frac{1}{p}-\frac{1}{q}} + sn^{\frac{1}{2}-\frac{1}{q}}}.$$

By some elementary calculation one checks that the supremum of the expression $\frac{1}{n^{1/p-1/q}+sn^{1/2-1/q}}$ over $n > 0$ is attained at $n_{\max} = (\frac{2(p-q)}{p(q-2)s})^{\frac{2p}{p-2}}$ and is equal to $\beta^{\beta}(1-\beta)^{1-\beta}s^{-\beta}$, where $\beta = \frac{2(p-q)}{q(p-2)}$. Note that $n_{\max} \geq n = \lfloor n_{\max} \rfloor \geq \frac{n_{\max}}{2}$, since we assumed that $s < \frac{2(p-q)}{p(q-2)}$. Therefore

$$C \geq 2^{\frac{1}{p}-\frac{1}{q}} \frac{1}{n_{\max}^{\frac{1}{p}-\frac{1}{q}} + sn_{\max}^{\frac{1}{2}-\frac{1}{q}}} \geq \frac{1}{2}\beta^{\beta}(1-\beta)^{1-\beta}s^{-\beta}$$

and hence

$$s \geq \left(\frac{1}{\sqrt{6C}}\right)^{\frac{q(p-2)}{p-q}},$$

since $u^u \geq e^{-1/e} \geq 1/\sqrt{3}$ for $u \in (0,1)$. Taking $C' = \sqrt{6C}$ we finish the proof.   □

Proposition 4. *There exist universal positive constants A and B such that for any $p > q > 2$ and $s \in (0,1)$ the Rosenthal space with the norm given by $\|a\| = \|a\|_p + s\|a\|_2$ is in the Banach-Mazur distance not greater than $As^{-\frac{2(p-q)}{q(p-2)}}$ from some linear subspace of $L_q(0,1)$ spanned by independent random variables and it is in the Banach-Mazur distance greater than $Bs^{-\frac{2(p-q)}{q(p-2)}}$ from all linear subspaces of $L_q(0,1)$.*

Proof. It is a simple consequence of Remark 5 and Remark 6.   □

Remark 7. Despite the similarities between Corollary 3 and Corollary 4 for any $p \in \bigcup_{k=1}^{\infty}(4k, 4k+2)$ there exists a non-pure p-pseudostable which does not have logarithmically convex tails. Also pure p-pseudostables do not have logarithmically convex tails. The author does not know whether there exists any p-pseudostable with logarithmically convex tail.

Proof. Note that for any $p \in \bigcup_{k=1}^{\infty}(4k, 4k+2)$ one can choose a sequence of non-pure p-pseudostables tending in distribution to some Gaussian random variable. If all of them had logarithmically convex tails then the limit distribution would also have logarithmically convex tails which is not the case. The contradiction ends the proof. Pure p-pseudostables cannot have logarithmically convex tails since by Theorem 1 their continuous densities have zero points.   □

In the end of the paper let us turn to the sections of the unit ball of l_p^n. There are many interesting results concerning this subject due to Ball, Hadwiger, Hensley, Koldobsky, Meyer, Pajor, Vaaler and others (see [BN] for

more references and related results on projections). The result of Ball ([B])
states that among hyperplane sections of the unit cube in R^n the central
section orthogonal to $(1, 1, 0, 0, 0, \ldots, 0)$ has the greatest $(n - 1)$-dimensional
Lebesgue measure. It was of interest whether the same direction of a hyperplane maximizes the $(n - 1)$-dimensional Lebesgue measure of the section of
the unit ball of l_p^n. The answer is negative, at least for small enough values of
p. Perhaps it is positive for p large enough – this problem remains open. In
the recent paper of Barthe and Naor ([BN]) a similar observation was made
for the projections of the unit ball of l_p^n for $p \in (1, 2)$ which in some sense is a
dual problem (although there cannot be any formal duality since the "phase
transition" in [BN] appears for $p = 4/3$ whereas in our considerations there
is no "phase transition" for $p = 4$).

Proposition 5. *Let $A(p, n)$ denote the $(n-1)$-dimensional Lebesgue measure
of the central section of the unit ball of l_p^n with the hyperplane orthogonal to
$(1, 1, \ldots, 1)$ and let $B(p, n)$ denote the $(n - 1)$-dimensional Lebesgue measure
of the central section of the unit ball of l_p^n with the hyperplane orthogonal to
$(1, 1, 0, 0, \ldots, 0)$. Then*

$$\lim_{n \to \infty} \left(\frac{A(p, n)}{B(p, n)} \right)^2 = \frac{\Gamma(1/p)^3 2^{2/p}}{\pi p^2 \Gamma(3/p)}$$

which is greater than 1, for example, for $p = 24$.

Proof. Some computer calculation suggests that the limit is greater than 1
for $p \in (2, p_0)$ and it is less than 1 for $p \in (p_0, \infty)$, where p_0 is some number
close to 26. It is clear that the limit is less than 1 for p large enough since it
tends to $3/\pi$ as $p \to \infty$.

Let $Z_1, Z_2, \ldots, Z_n$ be i.i.d. random variables with the density $g_{p(x)} =
c_p e^{-|x|^p}$, where $c_p = \frac{1}{2\Gamma(1 + \frac{1}{p})}$. To prove the formula for the limit recall the
well known fact that the $(n - 1)$-dimensional Lebesgue measure of the section
of the unit ball of l_p^n with the hyperplane orthogonal to the unit vector $a \in R^n$
is proportional to the value of continuous density g of $a_1 Z_1 + a_2 Z_2 + \ldots + a_n Z_n$
at zero and the proportionality constant depends on p and n only (and it does
not depend on the choice of a). Therefore

$$\frac{A(p, n)}{B(p, n)} = g_{\frac{z_1 + z_2 + \ldots + z_n}{\sqrt{n}}}(0) / g_{\frac{z_1 + z_2}{\sqrt{2}}}(0).$$

Note that

$$g_{\frac{x_1 + x_2}{\sqrt{2}}}(0) = \int_R \left(\sqrt{2} g_p(\sqrt{2} t) \right)^2 dt = 2^{-\frac{1}{2} - \frac{1}{p}} / \Gamma\left(1 + \frac{1}{p}\right).$$

Since $EX^2 = \frac{2 c_p \Gamma(3/p)}{p}$, by the Central Limit Theorem we get

$$\lim_{n \to \infty} g_{\frac{z_1+z_2+\ldots+z_n}{\sqrt{n}}}(0) = \sqrt{\frac{\Gamma(1/p)}{2\pi\Gamma(3/p)}}.$$

Here we used a version of the CLT for i.i.d. random variables stating that the integrability of the characteristic function (φ_Z is integrable due to Lemma 1) implies the uniform convergence of the densities to the normal density, cf. [F] for details. This ends the proof. $\square$

Acknowledgement. I would like to thank Professors Keith Ball, Alexander Koldobsky and Stanisław Kwapień for their comments. Also, it is a pleasure for me to acknowledge the kind hospitality of the Pacific Institute for the Mathematical Sciences in Vancouver, Canada where part of the research was done.

References

[B]   Ball, K.M. (1986): Cube slicing in R^n. Proc. Amer. Math. Soc., **97**, 465–473

[BN]   Barthe, F., Naor, A.: Hyperplane projections of the unit ball of l_p^n. Discrete Comput. Geom., to appear

[BDCK]   Bretagnolle, J., Dacunha-Castelle, D., Krivine, J.L. (1966): Lois stables et espaces L_p. Ann. Inst. H. Poincaré Prob. Stat., **2**, 231–259

[D]   Diaconis, P. (1987): Application of the method of moments in probability and statistics. Proc. Symp. Appl. Math., **37**, 125–142

[F]   Feller, W. (1971): An Introduction to Probability Theory and Its Applications, Vol. II, 2nd Ed. John Wiley & Sons, Inc., New York-London-Sydney

[HMSO]   Hitczenko, P., Montgomery-Smith, S.J., Oleszkiewicz, K. (1997): Moment inequalities for sums of certain independent symmetric random variables. Studia Math., **123**, 15–42

[K]   Kadec, M.I. (1958): On linear dimension of the spaces L_p. Uspekhi Mat. Nauk., **13**, 95–98 (in Russian)

[KPS]   Kwapień, S., Pycia, M., Schachermayer, W. (1996): A proof of a conjecture of Bobkov and Houdré. Electron. Comm. Probab., **1**, 7–10

[L]   Levy, P. (1925): Calcul des Probabilites. Gauthier-Villars, Paris

[M]   Marcinkiewicz, J. (1939): Sur une propriété de la loi de Gauss. Math. Z., **44**, 612–618

[PS]   Pólya, G., Szegö, G. (1972): Problems and Theorems in Analysis, Vol. I (translation from German). Springer-Verlag, Berlin-Heidelberg-New York

Ψ_2-Estimates for Linear Functionals on Zonoids

G. Paouris

Department of Mathematics, University of Crete, Iraklion, Greece
paouris@math.uch.gr

Summary. Let K be a convex body in $\mathbb{R}^n$ with centre of mass at the origin and volume $|K| = 1$. We prove that if $K \subseteq \alpha\sqrt{n}B_2^n$ where B_2^n is the Euclidean unit ball, then there exists $\theta \in S^{n-1}$ such that

$$\|\langle \cdot, \theta \rangle\|_{L_{\psi_2}(K)} \le c\alpha \|\langle \cdot, \theta \rangle\|_{L_1(K)}, \tag{$*$}$$

where $c > 0$ is an absolute constant. In other words, "every body with small diameter has ψ_2-directions". This criterion applies to the class of zonoids. In the opposite direction, we show that if an isotropic convex body K of volume 1 satisfies $(*)$ for every direction $\theta \in S^{n-1}$, then $K \subseteq C\alpha^2\sqrt{n}\log nB_2^n$, where $C > 0$ is an absolute constant.

1 Introduction

We shall work in $\mathbb{R}^n$ which is equipped with a Euclidean structure $\langle \cdot, \cdot \rangle$. The Euclidean norm $\langle x, x \rangle^{1/2}$ is denoted by $|\cdot|$. We write B_2^n for the Euclidean unit ball, S^{n-1} for the unit sphere, and σ for the rotationally invariant probability measure on S^{n-1}.

Throughout this note we assume that K is a convex body in $\mathbb{R}^n$ with volume $|K| = 1$ and centre of mass at the origin. Given $\alpha \in [1, 2]$, the Orlicz norm $\|f\|_{\psi_\alpha}$ of a bounded measurable function $f : K \to \mathbb{R}$ is defined by

$$\|f\|_{\psi_\alpha} = \inf\left\{t > 0 : \int_K \exp\left(\left(\frac{|f(x)|}{t}\right)^\alpha\right) dx \le 2\right\}. \tag{1.1}$$

It is not hard to check that

$$\|f\|_{\psi_\alpha} \simeq \sup\left\{\frac{\|f\|_p}{p^{1/\alpha}} : p \ge 1\right\}. \tag{1.2}$$

Let $y \ne 0$ in $\mathbb{R}^n$. We say that K satisfies a ψ_α-*estimate with constant* b_α *in the direction of* y if

$$\|\langle \cdot, y \rangle\|_{\psi_\alpha} \le b_\alpha \|\langle \cdot, y \rangle\|_1. \tag{1.3}$$

We say that K is a ψ_α-*body with constant* b_α if (1.3) holds for every $y \ne 0$.

It is easy to see that if K satisfies a ψ_α-estimate in the direction of y and if $T \in SL(n)$, then $T(K)$ satisfies a ψ_α-estimate (with the same constant) in the direction of $T^*(y)$. It follows that $T(K)$ is a ψ_α-body if K is a ψ_α-body. By Borell's lemma (see [MiS], Appendix III), every convex body K is a ψ_1-body with constant $b_1 = c$, where $c > 0$ is an absolute constant.

Estimates of this form are related to the hyperplane problem for convex bodies. Recall that a convex body K of volume 1 with centre of mass at the origin is called isotropic if there exists a constant $L_K > 0$ such that

$$\int_K \langle x, \theta \rangle^2 dx = L_K^2 \tag{1.4}$$

for all $\theta \in S^{n-1}$. Every convex body K with centre of mass at the origin has an isotropic image under $GL(n)$ which is uniquely determined up to orthogonal transformations (for more information on the isotropic position, see [MiP]). It follows that the isotropic constant L_K is an invariant for the class $\{T(K) : T \in GL(n)\}$. The hyperplane problem asks if every convex body of volume 1 has a hyperplane section through its centre of mass with "area" greater than an absolute constant. An affirmative answer to this question is equivalent to the following statement: there exists an absolute constant $C > 0$ such that $L_K \leq C$ for every isotropic convex body K.

Bourgain [Bou] has proved that $L_K \leq c\sqrt[4]{n} \log n$ for every origin symmetric isotropic convex body K in $\mathbb{R}^n$ (the same estimate holds true for non-symmetric convex bodies as well; see [D2] and [P]). Bourgain's argument shows that if K is a ψ_2-body with constant b_2, then $L_K \leq cb_2 \log n$ where $c > 0$ is an absolute constant. Examples of ψ_2-bodies are given by the ball and the cube in $\mathbb{R}^n$.

Alesker [A] has proved that the Euclidean norm satisfies a ψ_2-estimate: there exists an absolute constant $C > 0$ such that

$$\int_K \exp\left(\frac{|x|^2}{C^2 I_2^2}\right) dx \leq 2 \tag{1.5}$$

for every isotropic convex body K in $\mathbb{R}^n$, where $I_2^2 = \int_K |x|^2 dx$.

It is not clear if every isotropic convex body satisfies a good ψ_2-estimate for most directions $\theta \in S^{n-1}$; for a related conjecture, see [AnBP]. On the other hand, to the best of our knowledge, even the existence of some good ψ_2-direction has not been verified in full generality. This would correspond to a sharpening of Alesker's result.

Bobkov and Nazarov [BoN] have recently proved that every 1-unconditional and isotropic convex body satisfies a ψ_2-estimate with constant c in the direction $y = (1, 1, \ldots, 1)$, where $c > 0$ is an absolute constant. The purpose of this note is to establish an analogous fact for zonoids.

Theorem 1.1. *There exists an absolute constant $C > 0$ with the following property: For every zonoid Z in $\mathbb{R}^n$ with volume $|Z| = 1$, there exists $\theta \in S^{n-1}$ such that*

$$\left(\int_Z |\langle x, \theta \rangle|^p dx\right)^{1/p} \leq C\sqrt{p} \int_Z |\langle x, \theta \rangle| dx$$

for every $p \geq 1$.

The proof of Theorem 1.1 is presented in Section 2. The argument shows that the same is true for every convex body in $\mathbb{R}^n$ which has a linear image of volume 1 with diameter of the order of $\sqrt{n}$ (we call these "bodies with small diameter"). In Section 3 we show that zonoids belong to this class.

In the opposite direction, we show that every ψ_2-isotropic convex body has small diameter. More precisely, in Section 4 we prove the following.

Theorem 1.2. *Let K be an isotropic convex body in $\mathbb{R}^n$. Assume that K is a ψ_2-body with constant b_2. Then,*

$$K \subseteq Cb_2^2\sqrt{n}\log nB_2^n,$$

where $C > 0$ is an absolute constant.

The letters c, c_1, c_2, c' etc. denote absolute positive constants, which may change from line to line. Wherever we write $a \simeq b$, this means that there exist absolute constants $c_1, c_2 > 0$ such that $c_1 a \leq b \leq c_2 a$. We refer the reader to the books [MiS], [Pi] and [S] for standard facts that we use in the sequel. We thank the referee for suggestions that improved the presentation and some estimates.

2 Bodies with Small Diameter

We say that a convex body K in $\mathbb{R}^n$ with centre of mass at the origin has "small diameter" if $|K| = 1$ and $K \subseteq \alpha\sqrt{n}B_2^n$, where α is "well bounded". Note that a convex body has a linear image with small diameter if and only if its polar body has bounded volume ratio. Our purpose is to show that bodies with small diameter have "good" ψ_2-directions.

Our first lemma follows by a simple computation.

Lemma 2.1. *For every $p \geq 1$ and every $x \in \mathbb{R}^n$,*

$$\left(\int_{S^{n-1}} |\langle x, \theta\rangle|^p \sigma(d\theta)\right)^{1/p} \simeq \frac{\sqrt{p}}{\sqrt{p+n}}|x|. \tag{2.1}$$

Proof. Observe that

$$\int_{B_2^n} |\langle x, y\rangle|^p dy = |B_2^n|\frac{n}{n+p}\int_{S^{n-1}} |\langle x, \theta\rangle|^p \sigma(d\theta).$$

On the other hand,

$$\int_{B_2^n} |\langle x, y\rangle|^p dy = |x|^p \int_{B_2^n} |\langle e_1, y\rangle|^p dy$$

$$= 2|B_2^{n-1}| \cdot |x|^p \int_0^1 t^p(1-t^2)^{(n-1)/2} dt$$

$$= |B_2^{n-1}| \cdot |x|^p \frac{\Gamma\left(\frac{p+1}{2}\right)\Gamma\left(\frac{n+1}{2}\right)}{\Gamma\left(\frac{p+n+2}{2}\right)}.$$

Since $|B_2^k| = \pi^{k/2}/\Gamma\left(\frac{k+2}{2}\right)$, we get

$$\int_{S^{n-1}} |\langle x, \theta \rangle|^p \sigma(d\theta) = \frac{1}{\sqrt{\pi}} \frac{n+p}{n} \frac{\Gamma\left(\frac{p+1}{2}\right)\Gamma\left(\frac{n+2}{2}\right)}{\Gamma\left(\frac{p+n+2}{2}\right)} |x|^p.$$

The result follows from Stirling's formula.   □

Lemma 2.2. *Let K be a convex body in $\mathbb{R}^n$ with volume $|K| = 1$ and centre of mass at the origin. Then,*

$$\sigma\left(\theta \in S^{n-1} : \int_K |\langle x, \theta \rangle|\, dx \geq c_1\right) \geq 1 - 2^{-n},$$

where $c_1 > 0$ is an absolute constant.

Proof. The Binet ellipsoid E of K is defined by

$$\|\theta\|_E^2 = \int_K \langle x, \theta \rangle^2 dx = \langle M_K \theta, \theta \rangle,$$

where $M_K = \left(\int_K x_i x_j dx\right)$ is the matrix of inertia of K (see [MiP]). It is easily checked that $\det M_K = \det M_{TK}$ for every $T \in SL(n)$, and this implies that

$$\int_{S^{n-1}} \|\theta\|_E^{-n} \sigma(d\theta) = \frac{|E|}{|B_2^n|} = (\det M_K)^{-1/2} = L_K^{-n}.$$

Then, Markov's inequality shows that

$$\sigma\left(\theta \in S^{n-1} : \|\theta\|_E \geq L_K/2\right) \geq 1 - \frac{1}{2^n}.$$

Since $L_K \geq c$ and $\|\langle \cdot, \theta \rangle\|_1 \simeq \|\langle \cdot, \theta \rangle\|_2$ (see [MiP]), the result follows.   □

Lemma 2.3. *Let K be a convex body in $\mathbb{R}^n$ with volume $|K| = 1$ and centre of mass at the origin. Assume that $K \subseteq \alpha\sqrt{n}B_2^n$. Then,*

$$\int_{S^{n-1}} \int_K \exp\left(\frac{|\langle x, \theta \rangle|}{c_2 \alpha}\right)^2 dx\sigma(d\theta) \leq 2,$$

where $c_2 > 0$ is an absolute constant.

Proof. For every $s > 0$ we have

$$\int_{S^{n-1}} \int_K \exp\left(\frac{|\langle x, \theta \rangle|}{s}\right)^2 dx\sigma(d\theta) = 1 + \sum_{k=1}^{\infty} \frac{1}{k!s^{2k}} \int_K \int_{S^{n-1}} |\langle x, \theta \rangle|^{2k} \sigma(d\theta) dx.$$

From Lemma 2.1 we see that this is bounded by

$$1 + \sum_{k=1}^{\infty} \frac{1}{k!s^{2k}} \left(\frac{c \cdot 2k}{2k+n}\right)^k \int_K |x|^{2k} dx \leq 1 + \sum_{k=1}^{\infty} \left(\frac{c'\alpha}{s}\right)^{2k},$$

where $c, c' > 0$ are absolute constants. We conclude the proof taking $s = c_2\alpha$ where $c_2 = 2c'$.   □

An application of Markov's inequality gives the following.

Corollary 2.1. *Let K be a convex body in $\mathbb{R}^n$ with volume $|K| = 1$ and centre of mass at the origin. Assume that $K \subseteq \alpha\sqrt{n}B_2^n$. Then, for every $A > 2$ we have*

$$\sigma\left(\theta \in S^{n-1} : \int_K \exp\left(\frac{|\langle x, \theta \rangle|}{c_2\alpha}\right)^2 dx < A\right) > 1 - \frac{2}{A},$$

where $c_2 > 0$ is the constant from Lemma 2.3. $\square$

Theorem 2.1. *Let K be a convex body in $\mathbb{R}^n$ with volume $|K| = 1$ and centre of mass at the origin. Assume that $K \subseteq \alpha\sqrt{n}B_2^n$. There exists $\theta \in S^{n-1}$ such that*

$$\left(\int_K |\langle x, \theta \rangle|^p dx\right)^{1/p} \leq C\alpha\sqrt{p} \int_K |\langle x, \theta \rangle| dx$$

for every $p > 1$, where $C > 0$ is an absolute constant.

Proof. Choose $A = 4$. Using the inequality $e^z > z^k/k!$ $(z > 0)$, Lemma 2.2 and Corollary 2.1 we see that with probability greater than $\frac{1}{2} - \frac{1}{2^n}$ a direction $\theta \in S^{n-1}$ satisfies

$$\int_K |\langle x, \theta \rangle| dx \geq c_1 \quad \text{and} \quad \int_K \exp\left(\frac{|\langle x, \theta \rangle|}{c_2\alpha}\right)^2 dx < 4.$$

It follows that

$$\int_K |\langle x, \theta \rangle|^{2k} dx \leq 4k!(c_2\alpha)^{2k}$$

for every $k \geq 1$, and hence

$$\left(\int_K |\langle x, \theta \rangle|^{2k} dx\right)^{\frac{1}{2k}} \leq c\alpha\sqrt{2k} \leq \frac{c}{c_1}\alpha\sqrt{2k} \int_K |\langle x, \theta \rangle| dx.$$

This is the statement of the theorem for $p = 2k$. The general case follows easily. $\square$

Remarks. (a) Bourgain's argument in [Bou] shows that L_K is bounded by a power of $\log n$ for every convex body K in $\mathbb{R}^n$ if the following statement holds true: If an isotropic convex body W in $\mathbb{R}^n$ is contained in the centered Euclidean ball of radius $\alpha\sqrt{n}L_W$, then W is a ψ_2-body with constant $O(\alpha^s)$. Lemma 2.3 shows that, under the same assumptions, "half" of the directions are ψ_2-directions for W, with constant $c\alpha$.

(b) It can be also easily proved that convex bodies with small diameter have large hyperplane sections (this can be verified in several other ways, but the argument below gives some estimate on the distribution of the volume of their $(n-1)$-dimensional sections).

Proposition 2.1. *Let K be a convex body in $\mathbb{R}^n$ with volume $|K| = 1$ and centre of mass at the origin. Assume that $K \subseteq \alpha\sqrt{n}B_2^n$. Then, for every $t > 0$ we have*

$$\sigma\left(\theta \in S^{n-1} : |K \cap \theta^\perp| \geq \frac{c_3}{t\alpha}\right) \geq 1 - 2e^{-t^2},$$

where $c_3 > 0$ is an absolute constant.

Proof. Applying Jensen's inequality to Lemma 2.3, we get

$$\int_{S^{n-1}} \exp\left(\left(\frac{\int_K |\langle x, \theta\rangle|\, dx}{c_2\alpha}\right)^2\right) \sigma(d\theta) \leq 2.$$

Markov's inequality shows that

$$\sigma\left(\theta \in S^{n-1} : \int_K |\langle x, \theta\rangle|\, dx \geq c_2\alpha t\right) \leq 2e^{-t^2}$$

for every $t > 0$. On the other hand, it is a well-known fact (see [MiP] for the symmetric case) that if K has volume 1 and centre of mass at the origin, then

$$\int_K |\langle x, \theta\rangle|\, dx \simeq \frac{1}{|K \cap \theta^\perp|} \tag{2.2}$$

for every $\theta \in S^{n-1}$. This completes the proof.   □

3 Positions of Zonoids

We first introduce some notation and recall basic facts about zonoids. The support function of a convex body K is defined by $h_K(y) = \max_{x \in K} \langle x, y\rangle$ for all $y \neq 0$. The mean width of K is given by

$$w(K) = 2 \int_{S^{n-1}} h_K(u)\sigma(du).$$

We say that K has minimal mean width if $w(K) \leq w(TK)$ for every $T \in SL(n)$.

Recall also the definition of the area measure σ_K of a convex body K: for every Borel $V \subseteq S^{n-1}$ we have

$$\sigma_K(V) = \nu\left(\{x \in \mathrm{bd}(K) : \text{the outer normal to } K \text{ at } x \text{ is in } V\}\right),$$

where ν is the $(n-1)$-dimensional surface measure on K. It is clear that $\sigma_K(S^{n-1}) = A(K)$, the surface area of K. We say that K has minimal surface area if $A(K) \leq A(TK)$ for every $T \in SL(n)$.

A zonoid is a limit of Minkowski sums of line segments in the Hausdorff metric. Equivalently, a symmetric convex body Z is a zonoid if and only if its polar body is the unit ball of an n-dimensional subspace of an L_1 space; i.e. if there exists a positive measure μ (the supporting measure of Z) on S^{n-1} such that

$$\|x\|_{Z^\circ} = \frac{1}{2} \int_{S^{n-1}} |\langle x, y \rangle| \mu(dy). \tag{3.1}$$

The class of zonoids coincides with the class of projection bodies. Recall that the projection body ΠK of a convex body K is the symmetric convex body whose support function is defined by

$$h_{\Pi K}(\theta) = |P_\theta(K)|, \quad \theta \in S^{n-1}, \tag{3.2}$$

where $P_\theta(K)$ is the orthogonal projection of K onto $\theta^\perp$. From the integral representation

$$|P_\theta(K)| = \frac{1}{2} \int_{S^{n-1}} |\langle u, \theta \rangle| \, d\sigma_K(u) \tag{3.3}$$

which is easily verified in the case of a polytope and extends to any convex body K by approximation, it follows that the projection body of K is a zonoid whose supporting measure is σ_K. Moreover, if we denote by $\mathcal{C}_n$ the class of symmetric convex bodies and by $\mathcal{Z}$ the class of zonoids, Aleksandrov's uniqueness theorem shows that the Minkowski map $\Pi : \mathcal{C}_n \to \mathcal{Z}$ with $K \mapsto \Pi K$, is injective. Note also that $\mathcal{Z}$ is invariant under invertible linear transformations (in fact, $\Pi(TK) = (T^{-1})^*(\Pi K)$ for every $T \in SL(n)$) and closed in the Hausdorff metric. For more information on zonoids, see [S] and [BouL].

We shall see that three natural positions of a zonoid have small diameter in the sense of Section 2. The proof makes use of the isotropic description of such positions which allows the use of the Brascamp-Lieb inequality.

1. Lewis position: A result of Lewis [L] (see also [B]) shows that every zonotope Z has a linear image Z_1 (the "Lewis position" of Z) with the following property: there exist unit vectors $u_1, \ldots, u_m$ and positive real numbers $c_1, \ldots, c_m$ such that

$$h_{Z_1}(x) = \sum_{j=1}^m c_j |\langle x, u_j \rangle|$$

and

$$I = \sum_{j=1}^m c_j u_j \otimes u_j,$$

where I denotes the identity operator in $\mathbb{R}^n$. Using the Brascamp-Lieb inequality, Ball proved in [B] that, under these conditions,

$$|Z_1^\circ| \leq \frac{2^n}{n!} \quad \text{and} \quad B_2^n \subseteq \sqrt{n} Z_1^\circ.$$

The reverse Santaló inequality for zonoids (see [R] and [GoMR]) implies that

$$|Z_1| \geq 2^n \quad \text{and} \quad Z_1 \subseteq \sqrt{n}B_2^n. \tag{3.4}$$

This shows that

$$\text{diam}(Z_1) \leq \sqrt{n}|Z_1|^{1/n}. \tag{A}$$

2. Lowner position: Assume that B_2^n is the ellipsoid of minimal volume containing a zonoid Z_2. Let Z_1 be the Lewis position of Z_2. Then,

$$\frac{|B_2^n|}{|Z_2|} \leq \frac{|\sqrt{n}B_2^n|}{|Z_1|}. \tag{3.5}$$

Now, (3.5) and (3.4) show that

$$\text{diam}(Z_2) \leq 2 \leq |Z_1|^{1/n} \leq \sqrt{n}|Z_2|^{1/n}. \tag{B}$$

3. Minimal mean width position: Assume that $Z_3 = \Pi K$ is a zonoid of volume 1 which has minimal mean width. The results of [GM1] and [GMR] show that the area measure σ_K is isotropic, i.e.

$$\int_{S^{n-1}} \langle u, \theta \rangle^2 d\sigma_K(u) = \frac{A(K)}{n} \tag{3.6}$$

for every $\theta \in S^{n-1}$, where $A(K)$ is the surface area of K. Moreover, a result of Petty [Pe] shows that K has minimal surface area. Now, an application of the Cauchy-Schwarz inequality and (3.6) show that

$$h_{Z_3}(\theta) = \frac{1}{2} \int_{S^{n-1}} |\langle \theta, u \rangle| d\sigma_K(u) \leq \frac{A(K)}{2\sqrt{n}}$$

for every $\theta \in S^{n-1}$. We will use the following fact from [GP]:

Lemma 3.1. *If K has minimal surface area, then*

$$A(K) \leq n|\Pi K|^{1/n}.$$

It follows that $h_{Z_3}(\theta) \leq \sqrt{n}/2$ for every $\theta \in S^{n-1}$. In other words,

$$\text{diam}(Z_3) \leq \sqrt{n}|Z_3|^{1/n}. \tag{C}$$

The preceding discussion shows that zonoids have positions with small diameter. More precisely, we have the following statement.

Theorem 3.1. *Let Z be a zonoid in Lewis or Lowner or minimal mean width position. Then,*

$$\text{diam}(Z) \leq \sqrt{n}|Z|^{1/n}. \quad \square$$

It follows that the results of Section 2 apply to the class of zonoids: every zonoid has ψ_2-directions in the sense of Theorem 1.1.

Remark. We do not know if isotropic zonoids have small diameter. One can check that their mean width is bounded by $c\sqrt{n}$ (it is of the smallest possible order).

4 Isotropic ψ_2-Bodies have Small Diameter

The purpose of this last section is to show that a convex body is a ψ_2-body only if its isotropic position has small diameter. More precisely, we prove the following.

Theorem 4.1. *Let K be an isotropic convex body in $\mathbb{R}^n$. Assume that K is a ψ_2-body with constant b_2. Then,*

$$K \subseteq Cb_2^2 \sqrt{n} \log n B_2^n,$$

where $C > 0$ is an absolute constant.

The proof will follow from two simple lemmas. The idea for the first one comes from [GM2].

Lemma 4.1. *Let K be a convex body in $\mathbb{R}^n$ with volume 1 and centre of mass at the origin. Then, for every $\theta \in S^{n-1}$,*

$$\int_K |\langle x, \theta \rangle|^p dx \geq \frac{\Gamma(p+1)\Gamma(n)}{2e\Gamma(p+n+1)} \max\left\{h_K^p(\theta), h_K^p(-\theta)\right\}.$$

Proof. Consider the function $f_\theta(t) = |K \cap (\theta^\perp + t\theta)|$. Brunn's principle implies that $f_\theta^{1/(n-1)}$ is concave. It follows that

$$f_\theta(t) \geq \left(1 - \frac{t}{h_K(\theta)}\right)^{n-1} f_\theta(0)$$

for all $t \in [0, h_K(\theta)]$. Therefore,

$$
\begin{aligned}
\int_K |\langle x, \theta \rangle|^p dx &= \int_0^{h_K(\theta)} t^p f_\theta(t)dt + \int_0^{h_K(-\theta)} t^p f_{-\theta}(t)dt \\
&\geq \int_0^{h_K(\theta)} t^p \left(1 - \frac{t}{h_K(\theta)}\right)^{n-1} f_\theta(0)dt \\
&\quad + \int_0^{h_K(-\theta)} t^p \left(1 - \frac{t}{h_K(-\theta)}\right)^{n-1} f_\theta(0)dt \\
&= f_\theta(0)\left(h_K^{p+1}(\theta) + h_K^{p+1}(-\theta)\right)\int_0^1 s^p(1-s)^{n-1}ds \\
&= \frac{\Gamma(p+1)\Gamma(n)}{\Gamma(p+n+1)} f_\theta(0)\left(h_K^{p+1}(\theta) + h_K^{p+1}(-\theta)\right) \\
&\geq \frac{\Gamma(p+1)\Gamma(n)}{2\Gamma(p+n+1)} f_\theta(0)\left(h_K(\theta) + h_K(-\theta)\right) \\
&\quad \cdot \max\left\{h_K^p(\theta), h_K^p(-\theta)\right\}.
\end{aligned}
$$

Since K has its centre of mass at the origin, we have $\|f_\theta\|_\infty \leq e f_\theta(0)$ (see [MM]), and hence

$$1 = |K| = \int_{-h_K(-\theta)}^{h_K(\theta)} f_\theta(t)dt \le e\big(h_K(\theta) + h_K(-\theta)\big) f_\theta(0).$$

This completes the proof.   □

Lemma 4.2. *Let K be a convex body in $\mathbb{R}^n$ with volume 1 and centre of mass at the origin. For every $\theta \in S^{n-1}$,*

$$\|\langle \cdot, \theta \rangle\|_{\psi_2} \ge \frac{c \max\{h_K(\theta), h_K(-\theta)\}}{\sqrt{n}},$$

where $c > 0$ is an absolute constant.

Proof. Let $\theta \in S^{n-1}$ and define

$$I_p(\theta) := \left(\int_K |\langle x, \theta \rangle|^p dx \right)^{1/p}$$

for every $p \ge 1$. Then, (1.2) shows that

$$\|\langle \cdot, \theta \rangle\|_{\psi_2} \ge \frac{c I_n(\theta)}{\sqrt{n}}.$$

From Lemma 4.1 we easily see that $I_n(\theta) \simeq \max\{h_K(\theta), h_K(-\theta)\}$ and the result follows.   □

Proof of Theorem 4.1. Since K is a ψ_2-body with constant b_2, Lemma 4.2 shows that

$$\frac{c h_K(\theta)}{\sqrt{n}} \le \|\langle \cdot, \theta \rangle\|_{\psi_2} \le b_2 \|\langle \cdot, \theta \rangle\|_1$$

for every $\theta \in S^{n-1}$. Since K is isotropic, we have

$$\|\langle \cdot, \theta \rangle\|_1 \le L_K$$

for every $\theta \in S^{n-1}$. Bourgain's argument in [Bou] (see also [D1]) together with the ψ_2-assumption show that

$$L_K \le c' b_2 \log n.$$

This implies that

$$K \subseteq C b_2^2 \sqrt{n} \log n B_2^n. \qquad □$$

Theorem 4.1 shows that ψ_2-bodies belong to a rather restricted class (their polars have at most logarithmic volume ratio). It would be interesting to decide if zonoids are ψ_2-bodies or not.

References

[A]   Alesker, S. (1995): ψ_2-estimate for the Euclidean norm on a convex body in isotropic position. Oper. Theory Adv. Appl., **77**, 1–4

[AnBP]   Anttila, M., Ball, K.M., Perissinaki, I. The central limit problem for convex bodies. Preprint

[B]   Ball, K.M. (1991): Volume ratios and a reverse isoperimetric inequality. J. London Math. Soc., (2) **44**, 351–359

[BoN]   Bobkov, S.G., Nazarov, F.L. On convex bodies and log-concave probability measures with unconditional basis. GAFA Seminar Volume, to appear

[Bou]   Bourgain, J. (1991): On the distribution of polynomials on high dimensional convex sets. Geometric Aspects of Functional Analysis (1989–1990), Lecture Notes in Mathematics, **1469**, Springer, Berlin, 127–137

[BouL]   Bourgain, J., Lindenstrauss, J. (1988): Projection bodies. Geometric Aspects of Functional Analysis (1986–1987), Lecture Notes in Mathematics, **1317**, Springer, Berlin, 250–270

[D1]   Dar, S. (1995): Remarks on Bourgain's problem on slicing of convex bodies. Geometric Aspects of Functional Analysis, Operator Theory: Advances and Applications, **77**, 61–66

[D2]   Dar, S. (1997): On the isotropic constant of non-symmetric convex bodies. Israel J. Math., **97**, 151–156

[GM1]   Giannopoulos, A., Milman, V.D. (2000): Extremal problems and isotropic positions of convex bodies. Israel J. Math., **117**, 29–60

[GM2]   Giannopoulos, A., Milman, V.D. (2000): Concentration property on probability spaces. Advances in Math., **156**, 77–106

[GMR]   Giannopoulos, A., Milman, V.D., Rudelson, M. (2000): Convex bodies with minimal mean width. Geometric Aspects of Functional Analysis, Lecture Notes in Mathematics, **1745**, Springer, Berlin, 81–93

[GP]   Giannopoulos, A., Papadimitrakis, M. (1999): Isotropic surface area measures. Mathematika, **46**, 1–13

[GoMR]   Gordon, Y., Meyer, M., Reisner, S. (1988): Zonoids with minimal volume product - a new proof. Proc. Amer. Math. Soc., **104**, 273–276

[L]   Lewis, D.R. (1978): Finite dimensional subspaces of L_p. Studia Math., **63**, 207–212

[MM]   Makai, Jr, E., Martini, H. (1996): The cross-section body, plane sections of convex bodies and approximation of convex bodies, I. Geom. Dedicata, **63**, 267–296

[MiP]   Milman, V.D., Pajor, A. (1989): Isotropic position and inertia ellipsoids and zonoids of the unit ball of a normed n-dimensional space. Geometric Aspects of Functional Analysis (1987-1988), Lecture Notes in Mathematics, **1376**, Springer, Berlin, 64–104

[MiS]   Milman, V.D., Schechtman, G. (1986): Asymptotic theory of finite dimensional normed spaces. Lecture Notes in Mathematics, **1200**, Springer, Berlin

[P]   Paouris, G. (2000): On the isotropic constant of non-symmetric convex bodies. Geometric Aspects of Functional Analysis, Lecture Notes in Mathematics, **1745**, Springer, Berlin, 238–243

[Pe]   Petty, C.M. (1961): Surface area of a convex body under affine transformations. Proc. Amer. Math. Soc., **12**, 824–828

[Pi]   Pisier, G. (1989): The Volume of Convex Bodies and Banach Space Geometry. Cambridge Tracts in Mathematics, **94**, Cambridge University Press, Cambridge

[R]   Reisner, S. (1986): Zonoids with minimal volume product. Math. Z., **192**, 339–346

[S]   Schneider, R. (1993): Convex Bodies: The Brunn-Minkowski Theory. Cambridge University Press, Cambridge

Maximal ℓ_p^n-Structures in Spaces with Extremal Parameters

G. Schechtman[1], N. Tomczak-Jaegermann[2] and R. Vershynin[3]

[1] Department of Mathematics, The Weizmann Institute of Science, Rehovot 76100, Israel *gideon@wisdom.weizmann.ac.il*
[2] Department of Mathematical Sciences, University of Alberta, Edmonton, Alberta, Canada, T6G 2G1 *nicole@ellpspace.math.ualberta.ca*
[3] Department of Mathematics, The Weizmann Institute of Science, Rehovot 76100, Israel
Current address: Department of Mathematical Sciences, University of Alberta, Edmonton, Alberta, Canada, T6G 2G1 *vershynin@yahoo.com*

Summary. We prove that every n-dimensional normed space with a type $p < 2$, cotype 2, and (asymptotically) extremal Euclidean distance has a quotient of a subspace, which is well isomorphic to ℓ_p^k and with the dimension k almost proportional to n. A structural result of a similar nature is also proved for a sequence of vectors with extremal Rademacher average inside a space of type p. The proofs are based on new results on restricted invertibility of operators from ℓ_r^n into a normed space X with either type r or cotype r.

1 Introduction

The initial motivation of this paper was the following problem from [J-S]: Let $1 \leq p \leq 2$ and let X be an n-dimensional subspace of L_p whose distance from Euclidean space satisfies the inequality $d(X, \ell_2^n) \geq \alpha n^{1/p-1/2}$. Does X contain a subspace of proportional dimension, which is well isomorphic to ℓ_p^k? For $p = 1$, the answer is positive [J-S], while for $1 < p \leq 2$ the question is still open. The paper [B-T] contains related results and solutions to other problems from [J-S], but left this particular problem open as well. Although the present paper leaves this problem open as well, we do show here that X has an almost proportional *quotient of a subspace* which is almost well isomorphic to ℓ_p^k. The term "almost" above refers to factors of order a power of $\log n$. We actually get the same conclusion for a wider class of spaces. This clearly is implied by Theorem 13.

Not surprisingly our approach involves restricted invertibility methods. We have two kinds of such results. The first is for operators from ℓ_q^n into spaces with cotype q. This is the content of Corollary 6. Section 2 in which it is contained is heavily based on a method developed by Gowers in [G1] and [G2]. The second restricted invertibility result is for operators from either ℓ_2^n or ℓ_p^n into spaces with type p. This is contained in Section 3. Section 4 contains the proof of the structural Theorem 13. Finally, Section 5 contains a related result: Under the same conditions as in Theorem 13 one can get a subspace, rather than quotient of a subspace, almost well isomorphic to ℓ_p^k. However, its dimension k is a certain power of n rather than being close to a proportion of n.

Most of the undefined notions here can be found in [TJ]. We only recall here the definition of the Lorentz spaces $L_{p,q}$.

Let (Ω, Σ, μ) be a measure space, $1 \leq p \leq \infty$, and $1 \leq q \leq \infty$. The Lorentz space $L_{p,q}(\mu)$ consists of all equivalent classes of μ-measurable functions f such that

$$\|f\|_{p,q} = \left(\int_0^\infty (t^{1/p} f^*(t))^q \, dt/t \right)^{1/q} < \infty \qquad \text{if } 1 \leq q < \infty,$$

$$\|f\|_{p,\infty} = \sup_t t^{1/p} f^*(t) < \infty,$$

where f^* is the decreasing rearrangement of $|f|$, i.e. $f^*(t) = \inf \left(a : \mu\{|f| > a\} \leq t \right)$, $0 < t < \infty$.

If $p = q$ then $L_{p,p}(\mu)$ is $L_p(\mu)$. In general $\|f\|_{p,q}$ is a quasi-norm, which for $p > 1$ is equivalent to a norm, the equivalence constant depending on p and q only. So we consider $L_{p,q}(\mu)$ under this norm.

For a positive integer n, one defines the finite dimensional spaces $\ell_{p,q}^n$ to be $L_{p,q}(\mu)$, where μ is the uniform measure on the interval $I = \{1, \ldots, n\}$, $\mu(\{i\}) = 1$.

It can be easily checked for $1 \leq p < \infty$, $1 \leq q \leq \infty$ that $\|x\|_p \leq (\log n)^{1/p} \|x\|_{p,\infty}$ for all $x \in \ell_{p,q}^n$, and that $\|\sum_{i=1}^n e_i\|_{p,q} \sim n^{1/p}$, where e_i are the coordinate vectors in $\ell_{p,q}^n$.

Our estimates often involve "constants" that depend on various parameters. So we write, for example, $c = c(p, M)$ to denote a constant depending on p and M only.

Acknowledgement. The first named author was supported in part by the ISF, and the second named author holds the Canada Research Chair in Mathematics.

2 Restricted Invertibility: Spaces with Cotype

Let us start with a general theorem about finite symmetric block bases which is of independent interest. This theorem (and its proof) is a variant of Gowers' results on the subject and in a sense lies in-between [G1] and [G2].

Theorem 1. *Let $1 \leq q < \infty$, and let $B \geq 1$. Let X be a Banach space, let $n \geq 1$ and $(x_i)_{i \leq n}$ be a sequence of n vectors in X satisfying*

$$\left\| \sum a_i x_i \right\| \leq B \|a\|_q \quad \text{and} \quad \mathbb{E} \left\| \sum \varepsilon_i a_i x_i \right\| \geq \|a\|_q$$

for all $a = (a_i) \in \mathbb{R}^n$. Then for any $\varepsilon > 0$ there exists a block basis $(y_i)_{i \leq m}$ of permutation of (x_i), which is $(1 + \varepsilon)$-symmetric and has cardinality

$$m \geq (c\varepsilon/B)^{2q+2}n/\log n,$$

where $c > 0$ is an absolute constant.

First recall the definition of a symmetric basis in its natural "localized" form used in the proof. Let $m \geq 1$ and consider the group

$$\Psi = \{-1, 1\}^m \times S_m$$

acting on $\mathbb{R}^m$ as follows: for $a \in \mathbb{R}^m$ and $(\eta, \sigma) \in \Psi$, we define $a_{\eta,\sigma} = \sum_{i=1}^m \eta_i e_{\sigma(i)}$.

Definition 2. Let $C \geq 1$. A set of vectors $(y_i)_{i \leq m}$ in X is said to be C-symmetric at $a \in \mathbb{R}^m$ if for every $(\eta, \sigma) \in \Psi$ we have

$$\left\| \sum_i (a_{\eta,\sigma})_i y_i \right\|_X \leq C \left\| \sum_i a_i y_i \right\|_X.$$

A set $(y_i)_{i \leq m}$ is C-symmetric if for every $a \in \mathbb{R}^m$, $(y_i)_{i \leq m}$ is C-symmetric at a.

Proof of Theorem 1. Fix an integer m of the form $m = 2 + (c'\varepsilon/B)^{2q+2}n/\log n$ where $c' > 0$ is an absolute constant to be defined later. As in [G1], we divide the interval of natural numbers $[1, n]$ into m blocks of length h (where $h \sim \log n$), and relabel the indices in $[1, n]$ as follows: the pair (i, j) will be the j-th element in the i-th block, $i = 1, \ldots, m$, $j = 1, \ldots, h$. This identifies $[1, n]$ with the product $[1, m] \times [1, h]$. Consider the group

$$\Omega = \{-1, 1\}^n \times S_n.$$

Here we think of S_n as the group of permutations of the product $[1, m] \times [1, h]$. We write $\pi_{ij} = \pi((i, j))$ for $\pi \in S_n$ and $\theta_{ij} = \theta_{(i,j)}$ for $\theta \in \{-1, 1\}^n$. Define the random operator $\phi_{\theta,\pi} : \mathbb{R}^m \to X$ by setting

$$\phi_{\theta,\pi}(e_i) = \sum_{j=1}^h \theta_{ij} x_{\pi_{ij}}, \qquad i = 1, \ldots, m.$$

We shall show that with high probability the vectors $y_i = \phi_{\theta,\pi}(e_i)$ for $i = 1, \ldots, m$ are $(1 + \varepsilon)$-symmetric.

The first ingredient in the proof is a lemma from [G2], which says that in any normed space the symmetry of a sequence can be verified on a set of a polynomial, not exponential, cardinality.

Lemma 3. [G2]. *Let $\varepsilon > 0$, let $(\mathbb{R}^m, \|\cdot\|)$ be a normed space and set $N = m^D$, where $D = \varepsilon^{-1}\log(3\varepsilon^{-1})$. There exists a set $\mathcal{N}$ of cardinality N in $\mathbb{R}^m$ such that if the standard basis of $\mathbb{R}^m$ is $(1 + \varepsilon)$-symmetric at each element from $\mathcal{N}$, then it is $(1 + \varepsilon)(1 - 6\varepsilon)^{-1}$-symmetric.*

The next lemma is central. For a real valued random variable Z, by $M(Z)$ we denote its median, that is, the number satisfying $\mathbb{P}\{Z \leq M(Z)\} \geq 1/2$ and $\mathbb{P}\{Z \geq M(Z)\} \geq 1/2$.

Lemma 4. *Let $1 < q < \infty$ and $B \geq 1$. Let $(x_j)_{j \leq n}$ be a sequence of vectors satisfying $\left\| \sum a_j x_j \right\| \leq B\|a\|_q$ for all $a \in \mathbb{R}^n$. Fix $a \in \mathbb{R}^m$ and $0 < \beta < 1/2$. Then with the notation above we have*

$$\mathbb{P}_\Omega \left(\max_{\eta,\sigma \in \Psi} \left| \|\phi_{\theta,\pi}(a_{\eta,\sigma})\| - \overline{M}(\|\phi_{\theta,\pi}(a)\|) \right| > \beta\|a\|_q h^{1/q} \right) \leq m^{-(c/\beta)\log(c/\beta)},$$

where $\overline{M}$ denotes the expectation if $q = 1$, or the median if $q > 1$, and provided that

$$m \leq (c\beta/B)^{2q+2} n / \log n,$$

where $c > 0$ is an absolute constant.

This deviation inequality was proved in [G1] (page 195, (iii)) and the form of $\overline{M}$ follows from the proof. Moreover, the inequality is stated in [G1] for a particular value of m although it is clear from the proof that it is valid for all smaller values of m as well.

To successfully apply this lemma we require the estimate

$$\overline{M}(\|\phi_{\theta,\pi}(a)\|) \geq (1/6)\|a\|_q h^{1/q}. \tag{1}$$

For $\overline{M}$ being the expectation, an estimate follows readily from our lower bound assumption in Theorem 1, even with the constant 1 replacing 1/6. This settles the case $q = 1$. For $q > 1$, we will use the following lemma, a version of which will also be needed in Section 5.

Lemma 5. *Let (x_i) be a finite sequence of vectors in a Banach space, and (a_i) be scalars. Then*

$$\mathbb{P}_\Omega \left\{ \left\| \sum \theta_i a_i x_{\pi(i)} \right\| \geq (1/2)\mathbb{E} \left\| \sum \theta_i a_i x_{\pi(i)} \right\| \right\} \geq \delta,$$

where $\delta > 0$ is an absolute constant.

Proof. Define the random variable $Z = \left\| \sum \theta_i a_i x_{\pi(i)} \right\|$, and let $\|Z\|_p = (\mathbb{E}|Z|^p)^{1/p}$. By Kahane's inequality for any $0 < p, r < \infty$ we have $\|Z\|_r \leq A\|Z\|_p$, where $A = A(p, r)$ (see [M-S] 9.2). Then

$$\mathbb{P}_\Omega\{Z \geq 2^{-1/p}\|Z\|_p\} \geq (2A^p)^{r/(p-r)} \tag{2}$$

This estimate follows from the standard argument (see e.g., [Le-Ta], Lemma 4.2) based on Hölder's inequality. For $t > 0$ we have

$$\mathbb{E}Z^p \leq t^p + \int_{Z>t} Z^p d\mathbb{P}_\Omega \leq t^p + \|Z\|_r^p \mathbb{P}_\Omega\{Z > t\}^{1-p/r}.$$

Setting $t = 2^{-1/p}\|Z\|_p$ we get (2). Now the conclusion of the lemma follows from (2) with $p = 1$, $r = 1/2$.

We return to the proof of Theorem 1. First, to complete the proof of (1), let $Z = \|\phi_{\theta,\pi}(a)\|$. It is easy to check that our lower bound assumption implies that $\mathbb{E}Z \geq \|a\|_q h^{1/q}$. Let δ be as in Lemma 5 and let $c > 0$ be a constant from Lemma 4. Fix $0 < \beta_1 < 1/3$ such that $2^{-(c/\beta_1)\log(c/\beta_1)} < \delta$. We shall ensure later that m satisfies the upper bound assumption of Lemma 4. Using Lemma 4 together with the above lower bound for $\mathbb{E}Z$ we get, since $m \geq 2$,

$$\mathbb{P}_\Omega\{|Z - M(Z)| > (1/3)\mathbb{E}Z\}$$
$$\leq \mathbb{P}_\Omega\{|Z - M(Z)| > \beta\mathbb{E}Z\} \leq 2^{-(c/\beta)\log(c/\beta)} < \delta.$$

On the other hand, by Lemma 5 we have $\mathbb{P}_\Omega\{Z \geq (1/2)\mathbb{E}Z\} \geq \delta$. An easy calculation shows $M(Z) \geq (1/6)\mathbb{E}Z \geq (1/6)\|a\|_q h^{1/q}$, which is (1).

Finally, we can now finish the proof of Theorem 1. Fix any $0 < \varepsilon < 1/6$, let D be as in the Lemma 3, given by $D = \varepsilon^{-1}\log(3\varepsilon^{-1})$ and let $\mathcal{N}$ be the set in the conclusion of this lemma. Let $c > 0$ be the constant from Lemma 4. Set $\beta_2 = c\varepsilon/3$. Then $(c/\beta_2)\log(c/\beta_2) > D$. We may additionally assume that $\beta_2 < \varepsilon/2$. By a suitable choice of the constant c' fixed at the beginning of the proof we may ensure that m satisfies the upper bound assumption in Lemma 4 for $\beta = \min(\beta_1, \beta_2)$. By Lemma 4 together with (1) we observe that the vectors $(y_i)_{i\leq m} = (\phi_{\theta,\pi}(e_i))_{i\leq m}$ are $(1+\varepsilon)$-symmetric at any fixed $a \in \mathcal{N}$ with probability at least $1 - m^{-D}$. It follows that there is a choice of $(y_i)_{i\leq m}$ which is $(1+\varepsilon)$-symmetric at each $a \in \mathcal{N}$. Then Lemma 3 yields that $(y_i)_{i\leq m}$ is $(1+\varepsilon)(1-6\varepsilon)^{-1}$-symmetric. This completes the proof of Theorem 1.

As an immediate corollary we get a restricted invertibility result for operators $\ell_q^n \to X$ where X is a Banach space of cotype q.

Corollary 6. *Let $q \geq 2$ and $K, M \geq 1$. Let X be a Banach space with cotype q constant $C_q(X) \leq K$. Let $u : \ell_q^n \to X$ be an operator with $\|u\| \leq M$ and satisfying the non-degeneracy condition $\|ue_i\| \geq 1$ for $i = 1, \dots, n$. Then there exists a subspace E in $\mathbb{R}^n$ spanned by disjointly supported vectors such that*

$$\|ux\| \geq (1/2K)\|x\| \quad \text{for} \quad x \in E,$$

and

$$\dim E \geq (c/MK)^{2q+2}n/\log n,$$

where $c > 0$ is an absolute constant.

3 Restricted Invertibility: Spaces with Type

In this section we prove some restricted invertibility results for operators with values in spaces of type p. The conclusion is slightly weaker than the known results for the more special case of operators between ℓ_p^n spaces ([B-T],

Theorem 5.7). In that case the conclusion holds with the ℓ_p- rather than $\ell_{p,\infty}$- norm. As we will see later such a stronger conclusion does not hold in general under our assumptions (see Remark 2 after Corollary 12).

Theorem 7. *Let $1 < p \le 2$ and $K, M \ge 1$. Let X be a Banach space with type p constant $T_p(X) \le K$. Let $u : \ell_2^n \to X$ be an operator with $\|u\| \le M$ and satisfying the non-degeneracy condition $\ell(u) \ge \sqrt{n}$. Then there exists a subset $\sigma \subset \{1, \ldots, n\}$ of cardinality $|\sigma| \ge cn$ such that*

$$\|ux\|_X \ge (c/K)n^{1/2-1/p}\|x\|_{p,\infty} \quad \text{for} \quad x \in \mathbb{R}^\sigma,$$

where $c = c(p, M) > 0$.

Remark. Let $p = 2$, let X be a space with dual of cotype 2, $C_2(X^*) \le K$ and let u satisfy all the assumptions of Theorem 7. Then the resulting estimate can be improved to the lower ℓ_2 estimate $\|ux\|_X \ge c\|x\|_2$ for all $x \in \mathbb{R}^\sigma$, where $c = c(K, M) > 0$.

The proof of the theorem is based on the following two lemmas. The first one is a reformulation of the generalization of Elton's theorem in [B-T], Theorem 5.2.

Lemma 8. *Let $1 < r < \infty$ and $M \ge 1$. Let $(x_i)_1^n$ be a set of vectors in a Banach space satisfying*
   (1) $\|\sum_\eta x_i\| \le M|\eta|^{1/r}$ *for any subset $\eta \subset \{1, \ldots, n\}$;*
   (2) $\mathbb{E}\|\sum \varepsilon_i x_i\| \ge n^{1/r}$.
Then there exists a subset $\sigma \subset \{1, \ldots, n\}$ of cardinality $|\sigma| \ge cn$ such that

$$\left\|\sum_\sigma a_i x_i\right\| \ge cn^{-1/r'}\|a\|_1 \quad \text{for} \quad a \in \mathbb{R}^\sigma,$$

where $c = c(r, M) > 0$.

The second lemma is a factorization result of Pisier [P] for $(q, 1)$-summing operators. We do not need here the definition of such operators and their norms $\pi_{q,1}$, and the interested reader can find them e.g., in [TJ]. Let us only recall that it is easy to see (e.g., [TJ], the proof of Theorem 21.4) that if Y is a Banach space of cotype $q \ge 2$ and $\mathcal{K}$ is a compact Hausdorff space then every bounded operator $T : C(\mathcal{K}) \to Y$ is $(q, 1)$-summing and $\pi_{q,1}(T) \le C_q(Y)\|T\|$. We shall combine this fact with Pisier's factorization theorem which states [P] (see also [TJ] Theorem 21.2 and (21.6))

Lemma 9. *Let $1 \le q < \infty$, let Y be a Banach space and let $T : C(\mathcal{K}) \to Y$ be a $(q, 1)$-summing operator. There exists a probability measure λ on $\mathcal{K}$ such that T factors as $T = \tilde{T}j$,*

$$T : \; C(\mathcal{K}) \overset{j}{\to} L_{q,1}(\lambda) \overset{\tilde{T}}{\to} Y,$$

where j is the natural inclusion map and $\|\tilde{T}\| \le c\pi_{q,1}(T)$, where c is an absolute constant.

Corollary 10. *Let $q > 2$ and $K \geq 1$. Let Y be a Banach space with $C_q(Y) \leq K$. Let $T : \ell_\infty^n \to Y$. Then there exists a subset $\sigma \subset \{1, \ldots, n\}$ of cardinality $|\sigma| \geq n/2$ such that*

$$\|TR_\sigma : \ell_{q,1}^n \to Y\| \leq cKn^{-1/q}\|T\|,$$

where R_σ denotes the coordinate projection in $\mathbb{R}^n$ onto $\mathbb{R}^\sigma$ and c is an absolute constant.

Proof. Observe that $\pi_{q,1}(T) \leq K\|T\|$. Consider Pisier's factorization

$$T : \ell_\infty^n \xrightarrow{j} L_{q,1}(\lambda) \xrightarrow{\tilde{T}} Y,$$

where λ is a probability measure on $\{1, \ldots, n\}$ and $\|\tilde{T}\| \leq c\pi_{q,1}(T)$. Then the set $\sigma = \{j : \lambda(j) \leq 2/n\}$ has cardinality at least $n/2$. Moreover

$$\|jR_\sigma : \ell_{q,1}^\sigma \to L_{q,1}(\lambda)\| \leq (2/n)^{1/q}.$$

This immediately completes the proof.

In the dual setting, this gives

Corollary 11. *Let $1 < p < 2$ and let $T_p(X) \leq K$. Consider vectors $(y_j)_1^n$ in X such that $\|\sum a_i y_i\| \geq \|a\|_1$ for all $a \in \mathbb{R}^n$. Then there exists a subset $\sigma \subset \{1, \ldots, n\}$ of cardinality $|\sigma| \geq n/2$ such that*

$$\left\|\sum_\sigma a_i y_i\right\| \geq (c/K)n^{1/p'}\|a\|_{p,\infty} \quad \text{for} \quad a \in \mathbb{R}^\sigma,$$

where $c > 0$ is an absolute constant.

Proof. Let X_0 be the span of $(y_j)_1^n$, and define $T : X_0 \to \ell_1^n$ by $Ty_j = e_j$ for $j = 1, \ldots, n$. Then $\|T\| \leq 1$, so $\|T^* : \ell_\infty^n \to X_0^*\| \leq 1$. Apply Corollary 10 with $Y = X_0^*$ and $q = p'$. We get a subset σ of cardinality at least $n/2$ such that

$$\|T^*R_\sigma : \ell_{p',1}^n \to X_0^*\| \leq cKn^{-1/p'}.$$

Thus

$$\|R_\sigma T : X_0 \to \ell_{p,\infty}^n\| \leq cKn^{-1/p'}.$$

Note that $R_\sigma Ty_j = e_j$ for $j \in \sigma$. From this the desired estimate follows.

Now, Theorem 7 is a combination of Lemma 8 (for $r = 2$) and Corollary 11. One needs only to recall that X has cotype q, where $q < \infty$ and $C_q(X)$ both depend only on p and $T_p(X)$ (see [K-T] for quantitative estimates), and that $\ell(u) \leq C\mathbb{E}\|\sum \varepsilon_i ue_i\|$ where C depends on q and $C_q(X)$ only. If $p = 2$, the remark following the theorem is proved by a similar argument, with use of Pisier's factorization in Lemma 9 replaced by Maurey's strengthening of Grothendieck's theorem ([TJ], Theorem 10.4) and Pietsch's factorization for 2-summing operators ([TJ], Theorem 9.2).

As a corollary we have a further invertibility result.

Corollary 12. *Let $1 < p \leq 2$, $K, M \geq 1$ and $\alpha > 0$. Let X be a Banach space with type p constant $T_p(X) \leq K$. Let $u : \ell_p^n \to X$ be an operator with $\|u\| \leq M$ and satisfying the non-degeneracy condition $\ell(u : \ell_2^n \to X) \geq n^{1/p}$. Then there exists a subset $\sigma \subset \{1,\ldots,n\}$ of cardinality $|\sigma| \geq cn$ such that*

$$\|ux\|_X \geq (c/K)\|x\|_{p,\infty} \quad \text{for} \quad x \in \mathbb{R}^\sigma,$$

where $c = c(p, M) > 0$.

The proof is an easy application of Theorem 7 for the operator $w = n^{1/2-1/p}u : \ell_2^n \to X$.

Remarks. 1. The proof above shows that Theorem 7 remains valid with the same estimates if the norm $\|u : \ell_2^n \to X\|$ is replaced by $M = \|u : \ell_{2,1}^n \to X\|$. An analogous fact is true also for Corollary 12. If $p = 1$, both Theorem 7 and Corollary 12 are true (and follow directly from Lemma 8) if the space X is assumed to have cotype q, for some $q < \infty$.

2. The space $\ell_{p,q}$ (with $1 < p < 2$ and $1 < q < \infty$) has type p. This known fact follows for example from the easy fact that $\ell_{p,q}$ has an upper p-estimate for disjoint vectors, together with Theorems 1.e.16 and 1.f.10 in [L-T]. It follows that one cannot improve the conclusions of Theorem 7 and Corollary 12 by replacing $\|\cdot\|_{p,\infty}$ by $\|\cdot\|_p$.

4 Spaces with Extremal Euclidean Distance

In this section we concentrate on the structure of finite-dimensional normed spaces which, while satisfying geometric type-cotype conditions, have the distance to a Euclidean space of maximal order. The maximality of the distance is expressed in terms of the lower estimate which for some $1 \leq p \leq 2$ (depending on of the properties of X) has the form

$$d_X = d(X, \ell_2^n) \geq \alpha n^{1/p-1/2} \tag{3}$$

for some constant $\alpha > 0$.

The main result of this section is

Theorem 13. *Let $1 < p \leq 2$, $K \geq 1$ and $\alpha > 0$. Let X be an n-dimensional normed space with cotype 2 constant $C_2(X) \leq K$ and type p constant $T_p(X) \leq K$, and whose Euclidean distance satisfies (3). Then there exists Y, a quotient of a subspace of X, of dimension $k \geq cn(\log n)^{-b}$ such that $d(Y, \ell_p^k) \leq C(\log n)^{1/p}$, where $c = c(p, K, \alpha) > 0$, $C = C(p, K, \alpha)$ and $b = b(p) > 0$.*

We do not know whether the log-factor can be removed in either the distance or the dimension estimates. We also do not know whether "quotient of a subspace" can be replaced by "subspace" without an essential change to the estimates.

The proof of this theorem depends on two successive steps: the first is the lower estimate result for spaces satisfying our assumptions, and the second is a lower estimate for dual spaces. The latter step is based on Corollary 6, while the former one is contained in the following lower $\ell_{p,\infty}$-estimate for spaces with maximal Euclidean distance.

Theorem 14. *Let $1 < p < 2$, $K \geq 1$ and $\alpha > 0$. Let X be an n-dimensional normed space with cotype 2 constant $C_2(X) \leq K$ and type p constant $T_p(X) \leq K$, and whose Euclidean distance satisfies $d(X, \ell_2^n) \geq \alpha n^{1/p-1/2}$. Then there exist $k \geq cn$ norm one vectors $y_1, \ldots, y_k$ in X such that*

$$\left\| \sum_i a_i y_i \right\|_X \geq c\|a\|_{p,\infty} \quad \text{for} \quad a \in \mathbb{R}^k,$$

where $c = c(p, K, \alpha) > 0$.

Remark. As often happens in such cases, the proof has the unsatisfactory feature that it yields constants tending to 0 as $p \to 2$. Of course, by Kwapien's theorem (see e.g., [TJ] Theorem 13.15) an even stronger statement holds for $p = 2$.

To prove Theorem 14 we require some preliminaries. First recall the definition which has often been used in a similar context (see [TJ], §27). The *relative Euclidean factorization constant* $e_k(X)$ $(k = 1, 2, \ldots)$ of a Banach space X is the smallest C such that for every subspace E of X of dimension k there exists a projection P in X onto E with the ℓ_2 factorable norm satisfying $\gamma_2(P) \leq C$.

Note that the Euclidean distance satisfies

$$d(X, \ell_2^n) \leq e_n(X).$$

We will work with a relaxation of the parameter $e_k(X)$ which will be shown to be comparable to $e_k(X)$ (up to a logarithm of the dimension).

Definition 15. *For $k = 1, 2, \ldots$, we denote by $e_k'(X)$ the smallest C such that for every subspace E of X of dimension k there exists a projection P in X such that $P(X) \subset E$, $\operatorname{rank} P \geq k/2$, and $\gamma_2(P) \leq C$.*

Lemma 16. *Let X be a Banach space and n be a natural number. Then*

$$e_n'(X) \leq e_n(X) \leq \sum_{k=0}^{\infty} e_{n/2^k}'(X).$$

Proof. Assume for simplicity that n is a power of 2; the general case easily follows. It is well known in the theory of 2-factorable operators (see e.g., [TJ],

Theorem 27.1) that the right hand side inequality will follow once we prove that for every $v : \ell_2^n \to X$ such that $\pi_2(v^*) = 1$ we have

$$\pi_2(v) \leq \sum_{k=0}^{\infty} e'_{n/2^k}(X).$$

To this end fix v as above and without loss of generality assume that v is one-to-one. Let P_0 be a projection on X such that $P_0(X) \subset v(\ell_2^n)$, $\mathrm{rank} P_0 \geq n/2$ and $\gamma_2(P_0) \leq e'_n(X)$. Let $H_0 = v^{-1}(P_0 X)$. By passing to a smaller subspace if necessary we may assume that $\dim H_0 = n/2$.

By induction construct $k_0 = \log_2 n$ mutually orthogonal subspaces $H_k \subset \ell_2^n$ with $\dim H_k = n/2^{k+1}$ and projections P_k from X onto $v(H_k)$ such that $\gamma_2(P_k) \leq e'_{n/2^k}(X)$ for $k = 0, \ldots, k_0 - 1$.

For $k = 0, \ldots, k_0 - 1$, denote by $Q_k : \ell_2^n \to H_k$ the orthogonal projection onto H_k. Then

$$\pi_2(v Q_k) = \pi_2(P_k v Q_k) \leq \pi_2(P_k v) \leq \gamma_2(P_k) \pi_2(v^*) \leq e'_{n/2^k}(X).$$

Since $\ell_2^n = H_0 \oplus \ldots \oplus H_{k_0-1}$, then

$$\pi_2(v) = \pi_2\left(\sum_{k=0}^{k_0-1} v Q_k\right) \leq \sum_{k=0}^{k_0-1} e'_{n/2^k}(X),$$

as required.   ◆

Let us recall a standard set-up for finite-dimensional normed spaces. The Euclidean unit ball on $\mathbb{R}^n$ is denoted by B_2^n (and it corresponds to the Euclidean norm $\|\cdot\|_2$). Let $\|\cdot\|_X$ be a norm on $\mathbb{R}^n$, and X be the corresponding normed space. Let Q be an orthogonal projection in $\mathbb{R}^n$. Then by QX we denote the quotient of X with the canonical norm $\|y\|_{QX} = \inf\{\|x\|_X : Qx = y\}$. This way we view QX as the vector space $Q(\mathbb{R}^n)$ with the norm $\|\cdot\|_{QX}$. In particular, QX carries the Euclidean structure inherited from $\mathbb{R}^n$ with the unit ball $Q(B_2^n) = B_2^n \cap Q(\mathbb{R}^n)$.

Lemma 17. Let X be a normed space, $\dim X = n$, and assume that $\pi_2(id : X \to \ell_2^n) \leq A\sqrt{n}$. Let Q be an orthogonal projection in $\mathbb{R}^n$. Let $Y \subset Q(\mathbb{R}^n)$ be an m-dimensional subspace on which we consider two norms: the Euclidean norm $\|\cdot\|_2$ and the norm $\|\cdot\|_{QX}$. Then

$$\ell\big(id : (Y, \|\cdot\|_2) \to (Y, \|\cdot\|_{QX})\big) \geq (1/AT_2(X^*)^2) m/\sqrt{n}.$$

Proof. To shorten the notation, denote the operator $id : (Y, \|\cdot\|_2) \to (Y, \|\cdot\|_{QX})$ by u. We first estimate $\pi_2(u^{-1})$. Recall that for any operator $w : Z \to Z_1$, the norm $\pi_2(w)$ is equal to the supremum of $(\sum \|wve_j\|^2)^{1/2}$ where the supremum runs over all operators $v : \ell_2^k \to Z$ with $\|v\| \leq 1$ and all k (see [TJ] Proposition 9.7).

Thus fix $v : \ell_2^k \to (Y, \|\cdot\|_{QX})$ with $\|v\| \leq 1$. Consider v as an operator into QX. Using Maurey's extension theorem for the dual operator (see [TJ] Theorem 13.13), there exists a lifting $v' : \ell_2^k \to X$, $Qv' = v$, with $\|v'\| \leq T_2(X^*)$ (note that $(QX)^*$ is a subspace of X^*). Therefore

$$\left(\sum_{i=1}^k \|u^{-1}ve_i\|_2^2 \right)^{1/2} = \left(\sum_{i=1}^k \|Qv'e_i\|_2^2 \right)^{1/2} \leq \left(\sum_{i=1}^k \|v'e_i\|_2^2 \right)^{1/2}$$
$$\leq T_2(X^*)\pi_2(id : X \to \ell_2^n) \leq AT_2(X^*)\sqrt{n}.$$

Thus $\pi_2(u^{-1}) \leq AT_2(X^*)\sqrt{n}$.

It is now sufficient to use two well known and easy facts (see [TJ], Proposition 9.10 and Theorem 12.2 (ii)) that $m \leq \pi_2(u)\pi_2(u^{-1})$ and $\pi_2(u) \leq C_2(Y)\ell(u)$, to get $\ell(u) \geq m/AC_2(Y)T_2(X^*)\sqrt{n}$. Since $C_2(Y) \leq T_2(Y^*) \leq T_2(X^*)$, this completes the proof.

Proof of Theorem 14. It is well known and easy to see from Maurey's extension theorem (see [TJ], Prop. 27.4) that for every $k = 1, 2, \ldots$ we have

$$e_k'(X) \leq cC_2(X)T_p(X)k^{1/p-1/2} \leq cK^2k^{1/p-1/2}, \qquad (4)$$

where c is an absolute constant.

Assume again that n is a power of 2, let $A_p = 1 - 2^{1/p-1/2}$ and let k_0 be the smallest k such that $e_{n/2^k}'(X) \geq (A_p\alpha/2)(n/2^k)^{1/p-1/2}$. If no such k exists let $k_0 = \infty$. By the maximality of distance, Lemma 16 and (4) we get

$$\alpha n^{1/p-1/2} \leq e_n(X) \leq \sum_{k=0}^{\infty} e_{n/2^k}'(X)$$
$$\leq n^{1/p-1/2}\left(\frac{A_p\alpha}{2} \sum_{k=0}^{\infty} 2^{-k(1/p-1/2)} + cK^2 \sum_{k=k_0}^{\infty} 2^{-k(1/p-1/2)} \right)$$
$$\leq n^{1/p-1/2}\left((\alpha/2) + cK^2 2^{-k_0(1/p-1/2)}A_p^{-1} \right).$$

This shows that k_0 is finite and $k_0 \leq C$, where $C = C(p, K, \alpha)$.

Set $m = n/2^{k_0}$ and $d = (A_p\alpha/2)m^{1/p-1/2}$. Then $m \geq \beta n$ and $d \geq \beta d_X$, where $\beta = \beta(p, K, \alpha) > 0$. Moreover,

$$e_m'(X) \geq d. \qquad (5)$$

Let $|\cdot|_2$ be a Euclidean norm on X given by a combination of a distance ellipsoid and the maximal volume ellipsoid (see [TJ], Prop. 17.2). Denote the n-dimensional Hilbert space $(X, |\cdot|_2)$ by H and write $\|\cdot\|_X$ for the norm in X. Then we have

$$(\sqrt{2}d_X)^{-1}|x|_2 \leq \|x\|_X \leq \sqrt{2}|x|_2, \quad \text{for} \quad x \in X,$$
$$\pi_2(id : X \to H) \leq \sqrt{2n}.$$

Using (5), we will prove

Lemma 18. *Under the notation above there exist vectors $x_1, \ldots, x_{\beta'm}$ in X with $\|x_i\|_X \le d^{-1}$ and an orthogonal projection R on H with $\mathrm{rank}R \ge m/2$ and such that*

$$\beta'\|a\|_2 \le \left| \sum a_i R x_i \right|_2 \le B\|a\|_2 \quad \text{for} \quad a \in \mathbb{R}^{\beta'm},$$

where $\beta' = \beta'(p, K, \alpha) > 0$ and $B = B(p, K, \alpha)$.

Proof. Estimate (5) implies that there exists a subspace E in X with $\dim E = m$ such that for every projection P in X with $P(X) \subset E$, $\mathrm{rank}P \ge m/2$ we have $\|P : X \to H\| \ge d/\|id : H \to X\| \ge d/\sqrt{2}$.

Our vectors x_i will be chosen among a sequence of vectors constructed by induction as follows. Assume that $1 \le k < m/2$ and that vectors $x_1, \ldots, x_{k-1}$ have already been constructed. Let P be the orthogonal projection in H onto $[x_1, \ldots, x_{k-1}]^{\perp} \cap E$. Then P satisfies the assumptions above, so there exists an $x_k \in X$ such that $\|x_k\|_X = 1/d$ and $|Px_k|_2 \ge 1/\sqrt{2}$. Let also $f_k = Px_k/|Px_k|_2$.

This procedure gives us vectors $x_1, \ldots, x_{m/2}$ with $\|x_i\|_X = 1/d$ and orthonormal vectors $f_1, \ldots, f_{m/2}$ such that

$$\langle x_i, f_i \rangle \ge 1/\sqrt{2} \quad \text{for} \quad 1 \le i \le m/2.$$

Let $\beta = \beta(p, K, \alpha)$ be the constant appearing before (5), and we may clearly assume that $\beta \le 1$.

Note that

$$|x_i|_2 \le \sqrt{2}d_X/d \le \sqrt{2}/\beta \quad \text{for} \quad i \le m/2. \tag{6}$$

A known and easy argument shows that for every $0 < \delta < 1/2$ there exists an orthogonal projection R in $[x_i]_{i \le m/2}$ with $\mathrm{corank}R \le \delta m$ and such that

$$\left| \sum_{1}^{m/2} a_i R x_i \right|_2 \le 1/(\beta^2 \delta)\|a\|_2 \quad \text{for} \quad a \in \mathbb{R}^{m/2}. \tag{7}$$

Indeed, denote by H_1 the space $([x_i], |\cdot|_2)$ and consider the operator $T : \ell_2^{m/2} \to H_1$ defined by $Te_i = x_i$ for $i = 1, \ldots, m/2$. Let $\lambda_1 \ge \lambda_2 \ge \ldots \ge 0$ be the s-numbers of T so that $Tf_i = \lambda_i f_i'$ for some orthonormal bases $\{f_i\}$ and $\{f_i'\}$ in $\ell_2^{m/2}$ and H_1, respectively. We have, by (6),

$$\sum_{i=1}^{m/2} \lambda_i^2 = \|T\|_{HS}^2 = \sum_{i=1}^{m/2} |x_i|_2^2 \le m/\beta^2.$$

This implies that for $i_0 = \delta m$ we have $\lambda_{i_0} \le 1/(\beta^2 \delta)$, and then the projection R onto $[f_{i_0+1}, \ldots, f_{m/2}]$ satisfies (7).

Set $\delta = \beta^2/32$ and let R satisfy (7). Extend R to all of H by setting $Rx = x$ for $x \in [x_i]^{\perp}$. Then for $R' = id - R$ we have $\mathrm{rank}R' \le \delta m$. Therefore

$$\sum_{i=1}^{m/2} |Rx_i|_2 \geq \sum_{i=1}^{m/2} \langle Rx_i, f_i \rangle = \sum_{i=1}^{m/2} \langle x_i, f_i \rangle - \sum_{i=1}^{m/2} \langle R'x_i, f_i \rangle$$

$$\geq 2^{-3/2}m - \sum_{i=1}^{m/2} \langle x_i, R'f_i \rangle \geq 2^{-3/2}m - (\sqrt{2}/\beta) \sum_{i=1}^{m/2} |R'f_i|_2$$

$$\geq 2^{-3/2}m - (\sqrt{2}/\beta)\sqrt{m/2}\,\|R'\|_{\mathrm{HS}}$$

$$\geq 2^{-3/2}m - (\sqrt{\delta}/\beta)m = 2^{-5/2}m.$$

From this inequality and (6) it easily follows that the set $\sigma = \{i : |Rx_i|_2 \geq 1/4\sqrt{2}\}$ has cardinality $|\sigma| \geq \beta m/16$. Applying Theorem 1.2 from [B-T] for the operator $T : \ell_2^\sigma \to H$ defined by $Te_i = Rx_i$ for $i \in \sigma$, we get, by (7) and the definition of σ that there exists a subset $\sigma' \subset \sigma$ such that

$$\left| \sum_{i \in \sigma'} a_i Rx_i \right|_2 \geq \beta'\|a\|_2 \quad \text{for} \quad a \in \mathbb{R}^{\sigma'}.$$

Moreover, $|\sigma'| \geq \beta'm$, where $\beta' = \beta'(p, K, \alpha) > 0$. This together with (7) completes the proof by relabeling the vectors from σ'.

Returning to the proof of Theorem 14, identify X with $\mathbb{R}^n$ in such a way that $|\cdot|_2$ coincides with the usual ℓ_2^n-norm $\|\cdot\|_2$. Let $x_1, \ldots, x_{\beta'n}$ be vectors constructed in Lemma 18. If $(RX, \|\cdot\|_{RX})$ denotes the quotient of X given by R then first note that

$$\left\| \sum a_i Rx_i \right\|_{RX} \leq \sqrt{2} \left| \sum a_i Rx_i \right|_2 \leq \sqrt{2}B\|a\|_2 \quad \text{for} \quad a \in \mathbb{R}^{\beta'n}.$$

Consider the subspace $Y = [Rx_i]_{i=1}^{\beta'm}$ of RX, (i.e., with the norm $\|\cdot\|_{RX}$ inherited from RX), and consider also the norm $\|\cdot\|_2$ on Y inherited from ℓ_2^n. To apply Lemma 17 note that since X has control of the cotype 2 constant and the K-convexity constant (having non-trivial type) then $T_2(X^*)$ is bounded above by a function of K. Thus by the lemma, the ℓ-norm of the identity operator satisfies $\ell(id : (Y, \|\cdot\|_2) \to Y) \geq c\sqrt{n}$, where $c = c(p, K, \alpha) > 0$. On the other hand, since the set of vectors (Rx_i) admits a lower ℓ_2-estimate, then by the ideal property of the ℓ-norm $\ell(id)$ can be estimated using that set, namely,

$$\ell\big(id : (Y, \|\cdot\|_2) \to Y\big) \leq (1/\beta')\mathbb{E}\left\| \sum_{i=1}^{\beta'm} g_i Rx_i \right\|_{RX}.$$

Thus

$$\mathbb{E}\left\| \sum_{i=1}^{\beta'n} g_i Rx_i \right\|_{RX} \geq c_1\sqrt{n},$$

where $c_1 = c_1(p, K, \alpha) > 0$. Then by Theorem 7 there exists a subset η of $\{1, \ldots, \beta'm\}$ of cardinality $|\eta| > c_2 n$ and such that

$$\left\| \sum_\eta a_i Rx_i \right\|_{RX} \geq c_2 n^{1/2-1/p} \|a\|_{p,\infty}, \quad \text{for} \quad a \in \mathbb{R}^\eta,$$

where $c_2 = c_2(p, K, \alpha) > 0$. Recall that $\|x_i\|_X = d^{-1}$. Then for $i \in \eta$ let $y_i = dx_i$. Clearly, y_i's are unit vectors in X and for $a \in \mathbb{R}^\eta$ we have

$$\left\| \sum a_i y_i \right\|_X \geq d \left\| \sum a_i Rx_i \right\|_{RX} \geq c_2(dn^{1/2-1/p}) \|a\|_{p,\infty} \geq c_3 \|a\|_{p,\infty},$$

where $c_3 = c_3(p, K, \alpha) > 0$. This completes the proof of Theorem 14.

Now we are ready to prove Theorem 13, as a combination of Theorem 14 and Corollary 6.

Proof of Theorem 13. We can clearly assume that $1 < p < 2$, because for $p = 2$ the whole space X is K^2-isomorphic to ℓ_2^n by Kwapien's Theorem (see [TJ] Theorem 13.15).

We apply Theorem 14, and let $(y_i)_{i \leq k}$ be the vectors from its conclusion, $k \geq cn$ with $c = c(p, K, \alpha)$. Consider the space $X_1 = [y_i]_{i \leq k}$ as a subspace of X. Since the vectors y_i are necessarily linearly independent, we may define the operator $v : X_1 \to \ell_p^k$ by

$$v y_i = e_i, \quad \text{for } i \leq k.$$

Then by the conclusion of Theorem 14

$$\|v\| \leq \|v : X_1 \to \ell_{p,\infty}^k\| \| id : \ell_{p,\infty}^k \to \ell_p \| \leq C_1 (\log n)^{1/p},$$

where $C_1 = C_1(p, K, \alpha)$. Consider the adjoint operator $v^* : \ell_q^k \to X_1^*$, where $1/q + 1/p = 1$. Then

$$\|v^*\| \leq C_1 (\log n)^{1/p} \tag{8}$$

and for all $i \leq k$,

$$\|v^* e_i\| \geq \langle v^* e_i, y_i \rangle = \langle e_i, e_i \rangle = 1.$$

Applying Corollary 6 we get norm one vectors $(h_i)_{i \leq m}$ in ℓ_q^k with disjoint supports satisfying for all $a \in \mathbb{R}^k$,

$$\left\| v^* \left(\sum_{i=1}^m a_i h_i \right) \right\|_{X_1^*} \geq \frac{1}{2K} \left\| \sum_{i=1}^m a_i h_i \right\|_q = \frac{1}{2K} \|a\|_q.$$

Moreover, $m \geq c_1 (\log n)^{-2(q+1)/p} n / \log n$, where $c_1 = c_1(p, K, \alpha) > 0$.
Also from (8),

$$\left\| v^* \left(\sum_{i=1}^m a_i h_i \right) \right\|_{X_1^*} \leq C_1 (\log n)^{1/p} \|a\|_q.$$

Thus the sequence of vectors $z_i = v^* h_i$, $i \leq m$, spans in X_1^* a subspace Z, which is $C(\log n)^{1/p}$-isomorphic to ℓ_q^m, with $C = C(p, K, \alpha)$. Since Z is a subspace of a quotient X_1^* of X^*, the space Z^* is a quotient of a subspace of X and is $C(\log n)^{1/p}$-isomorphic to ℓ_p^m. This completes the proof of Theorem 13.

5 ℓ_p^n Subspaces in Spaces with Extremal Type p

We show another interesting application of methods discussed here to the structure of subspaces of spaces which attain their best type. More precisely, if a Banach space of type p contains a sequence of vectors with extremal Rademacher average, then it contains a relatively large subspace close to ℓ_p^k.

Proposition 19. *Let $1 < p \leq 2$, $K \geq 1$ and $\alpha > 0$. Let X be a Banach space with type p constant $T_p(X) \leq K$. Assume that there exist norm one vectors $x_1, \ldots, x_n$ in X such that $\mathbb{E}\| \sum_1^n \varepsilon_i x_i \| \geq \alpha n^{1/p}$. Then there is a block basis of permutation of $(x_i)_{i \leq n}$ of cardinality m which is $C(\log n)^{1/p}$-equivalent to the unit vector basis in ℓ_p^m, and $m \geq c(\log n)^{-1} n^{2/p-1}$, where $C = C(p, K, \alpha)$ and $c = c(p, K, \alpha) > 0$.*

The proof combines the main result of [G2] and Corollary 12.

Proof. Fix an $\varepsilon > 0$. By [G2], our assumption on the Rademacher average of (x_i) implies that there exists a block basis $(y_i)_1^m$ of permutation of $(x_i)_1^n$, with blocks of random ± 1 coefficients and equal lengths, which is 2-symmetric with probability larger than $1 - \varepsilon$. Moreover $m \geq c(\log n)^{-1} n^{2/p-1}$, where $c = c(p, \alpha, \varepsilon)$ and we may assume that m is an even number. The precise definition of the random vectors (y_i) is given in the proof of Theorem 1, the underlying probability space being denoted by $\mathbb{P}_\Omega$.

Then with probability larger than $1 - \varepsilon$ the following holds for all subsets σ of $\{1, \ldots, m\}$ of cardinality $|\sigma| = m/2$:

$$2\left\| \sum_{i \in \sigma} y_i \right\| \geq \left\| \sum_{i \in \sigma} y_i \right\| + \frac{1}{2}\left\| \sum_{i \in \sigma^c} y_i \right\| \geq \frac{1}{2}\left\| \sum_{i=1}^m y_i \right\|.$$

On the other hand, by Lemma 5 we have, with probability larger than $\delta > 0$,

$$\left\| \sum_{i=1}^m y_i \right\| = \left\| \sum_{i=1}^n \varepsilon_i x_{\pi(i)} \right\| \geq \frac{1}{2} \mathbb{E}_\Omega \left\| \sum_{i=1}^n \varepsilon_i x_{\pi(i)} \right\|$$

$$= \frac{1}{2} \mathbb{E} \left\| \sum_{i=1}^n \varepsilon_i x_i \right\| \geq \frac{1}{2} \alpha n^{1/p}$$

(where π denotes a random permutation of $\{1, \ldots, n\}$, and $\mathbb{E}$ is the expectation over random signs ε_i). Therefore with probability larger than $\delta - \varepsilon$

$$\left\|\sum_{i\in\sigma} y_i\right\| \geq \frac{1}{8}\alpha n^{1/p} \quad \text{holds for all subsets } \sigma, \quad |\sigma| = m/2. \tag{9}$$

Let $n = mh$ and we can assume that h is integer. By the type assumption and the definition of y_i, for every $i \leq m$ we have $\mathbb{E}_\Omega \|y_i\| \leq Kh^{1/p}$. Then $\mathbb{E}_\Omega\left(\sum_{i=1}^m \|y_i\|\right) \leq Kmh^{1/p}$. So, with probability larger than $1 - \varepsilon$ we have $\left(\sum_{i=1}^m \|y_i\|\right) \leq (1/\varepsilon)Kmh^{1/p}$. This clearly implies that

$$\exists \text{ a subset } \sigma, |\sigma| = m/2, \text{ such that } \|y_i\| \leq \frac{2}{\varepsilon}Kh^{1/p} \quad \text{for } i \in \sigma. \tag{10}$$

With probability at least $\delta - 2\varepsilon$, events (9) and (10) hold simultaneously. For $\varepsilon = \delta/3$ this probability is positive, so we can consider a realization of (y_i) for which both events occur. Let $z_i = h^{-1/p}y_i$, $i \in \sigma$. Then $\|z_i\| \leq (6/\delta)K$, so by the type p and symmetry

$$\left\|\sum_{i\in\sigma} a_i z_i\right\| \leq (12/\delta)K^2\|a\|_p \quad \text{for all } (a_i)_{i\in\sigma}.$$

Next, by (9) and symmetry

$$\left\|\sum_{i\in\sigma} \varepsilon_i z_i\right\| \geq \frac{1}{2}h^{-1/p}\left\|\sum_{i\in\sigma} y_i\right\| \geq \frac{1}{16}\alpha m^{1/p}.$$

Corollary 12 yields then that there exists a subset $\sigma_1 \subset \sigma$ with cardinality $|\sigma_1| \geq cm$ and such that

$$\left\|\sum_{i\in\sigma_1} a_i z_i\right\| \geq c\|a\|_{p,\infty} \quad \text{for all } (a_i)_{i\in\sigma},$$

where $c = c(p, K, \alpha) > 0$. This completes the proof.

Remark. It is not clear whether the exponent $2/p - 1$ in Proposition 19 is optimal. However, for $p = 2$ the optimal exponent must be 0, because the identical vectors $x_i = 1$ in $X = \mathbb{R}^1$ satisfy the assumptions of Proposition 19.

As a corollary we get a variant of Theorem 13 where the conclusion is improved by getting a subspace rather than quotient of a subspace, at the price of a worse estimate on the dimension.

Proposition 20. *Under the assumptions of Theorem 13, there exists a subspace Y of X of dimension $k \geq c(\log n)^{-1}n^{2/p-1}$, with $d(Y, \ell_p^k) \leq C(\log n)^{1/p}$, where $C = C(p, K, \alpha)$ and $c = c(p, K, \alpha) > 0$.*

Proof. By (3) and Kwapien's theorem we get $T_2(X) \geq c_1 K^{-1}\alpha n^{1/p-1/2}$, where $c_1 > 0$ is an absolute constant. By Tomczak-Jaegermann's result (cf. [TJ], Theorems 25.6 and 25.1), the type 2 constant can be essentially computed on n vectors, i.e. there exist vectors $(x_i)_{i\leq n}$ in X such that

$$\mathbb{E}\left\|\sum_{i=1}^{n}\varepsilon_i x_i\right\| \geq c_2 K^{-1}\alpha n^{1/p-1/2}\left(\sum_{i=1}^{n}\|x_i\|^2\right)^{1/2} \tag{11}$$

for some absolute $c_2 > 0$.

Now we employ a known argument to show that the vectors x_i can be essentially chosen of norm one. We can assume that $\sum_{i=1}^{n}\|x_i\|^2 = n$, so that the right side in (11) is $c_2 K^{-1}\alpha n^{1/p}$. Fix a positive number M and let $\sigma = \{i \in [1,n] : \|x_i\| \leq M\}$. Then $|\sigma^c| \leq (\sum_{i=1}^{n}\|x_i\|^2)/M^2 = M^{-2}n$. Therefore, using the type p of X we see that

$$\mathbb{E}\left\|\sum_{i\in\sigma^c}\varepsilon_i x_i\right\| \leq K\left(\sum_{i\in\sigma^c}\|x_i\|^p\right)^{1/p} \leq K|\sigma^c|^{1/p-1/2}\left(\sum_{i\in\sigma^c}\|x_i\|^2\right)^{1/2}$$

$$\leq KM^{1-2/p}n^{1/p}.$$

Define the vectors $y_i = x_i/\|x_i\|$, $i \in \sigma$. By the standard comparison principle it follows that

$$\mathbb{E}\left\|\sum_{i\in\sigma}\varepsilon_i y_i\right\| \geq M^{-1}\mathbb{E}\left\|\sum_{i\in\sigma}\varepsilon_i x_i\right\| \geq M^{-1}\left(\mathbb{E}\left\|\sum_{i=1}^{n}\varepsilon_i x_i\right\| - \mathbb{E}\left\|\sum_{i\in\sigma^c}\varepsilon_i x_i\right\|\right)$$

$$\geq M^{-1}(c_2 K^{-1}\alpha n^{1/p} - KM^{1-2/p}n^{1/p}). \tag{12}$$

Choosing M so that $KM^{1-2/p} = (c_2/2)K^{-1}\alpha$, we make the right hand side in (12) bounded below by $((c_2\alpha/2)K^{2+p})^{2-p}$. This clearly implies that there exist norm one vectors $(z_i)_{i\leq n}$ in X for which

$$\mathbb{E}\left\|\sum_{i=1}^{n}\varepsilon_i z_i\right\| \geq c(K,\alpha)n^{1/p}.$$

An application of Proposition 19 for the vectors (z_i) finishes the proof.

References

[B-T] Bourgain, J., Tzafriri, L. (1987): Invertibility of "large" submatrices with applications to the geometry of Banach spaces and harmonic analysis. Israel J. Math., **57**, 137–224

[G1] Gowers, W.T. (1989): Symmetric block bases in finite dimensional normed spaces. Israel J. Math., **68**, 193–219

[G2] Gowers, W.T. (1990): Symmetric block bases of sequences with large average growth. Israel J. Math., **69**, 129–151

[J-S] Johnson, W.B., Schechtman, G. (1982): On subspaces of L_1 with maximal distances to Euclidean space. Proceedings of Research Workshop on Banach Space Theory (Iowa City, Iowa, 1981), 83–96, Univ. Iowa, Iowa City, Iowa

[K-T]   König, H., Tzafriri, L. (1981): Some estimates for type and cotype constants. Math. Ann., **256**, 85–94

[Le-Ta] Ledoux, M., Talagrand, M. (1991): Probability in Banach Spaces. Springer-Verlag, Berlin

[L-T]   Lindenstrauss, J., Tzafriri, L. (1979): Classical Banach Spaces, Vol II. Springer-Verlag, Berlin

[M-S]   Milman, V.D., Schechtman, G. (1986): Asymptotic theory of finite-dimensional normed spaces. Lecture Notes in Mathematics, **1200**, Springer-Verlag, Berlin

[P]   Pisier, G. (1986): Factorization of operators through $L_{p\infty}$ or L_{p1} and noncommutative generalizations. Math. Ann., **276**, 105–136

[TJ]   Tomczak-Jaegermann, N. (1989): Banach-Mazur distances and finite-dimensional operator ideals. Pitman Monographs and Surveys in Pure and Applied Mathematics, **38**, New York

Polytopes with Vertices Chosen Randomly from the Boundary of a Convex Body

Carsten Schütt[1] * and Elisabeth Werner[2,3] **

[1] Christian Albrechts Universität, Mathematisches Seminar, 24098 Kiel, Germany, *schuett@math.uni-kiel.de*
[2] Department of Mathematics, Case Western Reserve University, Cleveland, Ohio 44106, USA, *emw2@po.cwru.edu*
[3] Université de Lille 1, Ufr de Mathématique, 59655 Villeneuve d'Ascq, France.

Summary. Let K be a convex body in $\mathbb{R}^n$ and let $f : \partial K \to \mathbb{R}_+$ be a continuous, positive function with $\int_{\partial K} f(x) \mathrm{d}\mu_{\partial K}(x) = 1$ where $\mu_{\partial K}$ is the surface measure on ∂K. Let $\mathbb{P}_f$ be the probability measure on ∂K given by $\mathrm{d}\mathbb{P}_f(x) = f(x)\mathrm{d}\mu_{\partial K}(x)$. Let κ be the (generalized) Gauß-Kronecker curvature and $\mathbb{E}(f, N)$ the expected volume of the convex hull of N points chosen randomly on ∂K with respect to $\mathbb{P}_f$. Then, under some regularity conditions on the boundary of K

$$\lim_{N \to \infty} \frac{\mathrm{vol}_n(K) - \mathbb{E}(f, N)}{\left(\frac{1}{N}\right)^{\frac{2}{n-1}}} = c_n \int_{\partial K} \frac{\kappa(x)^{\frac{1}{n-1}}}{f(x)^{\frac{2}{n-1}}} \mathrm{d}\mu_{\partial K}(x),$$

where c_n is a constant depending on the dimension n only.
The minimum at the right-hand side is attained for the normalized affine surface area measure with density

$$f_{as}(x) = \frac{\kappa(x)^{\frac{1}{n+1}}}{\int_{\partial K} \kappa(x)^{\frac{1}{n+1}} \mathrm{d}\mu_{\partial K}(x)}.$$

Contents

* Partially supported by the Schrödinger Institute,Vienna, and by the Edmund Landau Center for Research in Mathematical Analysis and Related Areas, Jerusalem, sponsored by the Minerva Foundation (Germany).
** Partially supported by the Schrödinger Institute,Vienna, by the Edmund Landau Center for Research in Mathematical Analysis and Related Areas, Jerusalem, sponsored by the Minerva Foundation (Germany) and by a grant from the National Science Foundation.

1 Introduction

1.1 Notation and Background. The Main Theorem

How well can a convex body be approximated by a polytope?

This is a central question in the theory of convex bodies, not only because it is a natural question and interesting in itself but also because it is relevant in many applications, for instance in computervision ([SaT1], [SaT2]), tomography [Ga], geometric algorithms [E].
We recall that a convex body K in $\mathbb{R}^n$ is a compact, convex subset of $\mathbb{R}^n$ with non-empty interior and a polytope P in $\mathbb{R}^n$ is the convex hull of finitely many points in $\mathbb{R}^n$.

As formulated above, the question is vague and we need to make it more precise.
Firstly, we need to clarify what we mean by "approximated". There are many metrics which can and have been considered. For a detailed account concerning these metrics we refer to the articles by Gruber [Gr1],[Gr3]. We will concentrate here on the symmetric difference metric d_s which measures the distance between two convex bodies C and K through the volume of the difference set

$$d_s(C, K) = \text{vol}_n(C \triangle K) = \text{vol}_n((C \setminus K) \cup (K \setminus C)).$$

Secondly, various assumptions can be made and have been made on the approximating polytopes P. For instance, one considers only polytopes contained in K or only polytopes containing K, polytopes with a fixed number of verices, polytopes with a fixed number of facets, etc. Again we refer to the articles [Gr1],[Gr3] for details.

We will concentrate here on the question of approximating a convex body K in $\mathbb{R}^n$ by inscribed polytopes P_N with a fixed number of vertices N in the d_s metric. As we deal with inscribed polytopes the d_s metric reduces to the volume difference

$$\mathrm{vol}_n(K) - \mathrm{vol}_n(P_N)$$

and we ask how the (optimal) dependence is in this metric on the various paramenters involved like the dimension n, the number of vertices N and so on.

As a first result in this direction we want to mention a result by Bronshteyn and Ivanov [BrI].
There is a numerical constant $c > 0$ such that for every convex body K in $\mathbb{R}^n$ which is contained in the Euclidean unit ball and for every $N \in \mathbb{N}$ there exists a polytope $P_N \subseteq K$ with N vertices such that

$$\mathrm{vol}_n(K) - \mathrm{vol}_n(P_N) \le c \, \frac{n \, \mathrm{vol}_n(K)}{N^{\frac{2}{n-1}}}.$$

The dependence on N and n in this result is optimal. This can be seen from the next two results. The first is due to Macbeath and says that the Euclidean unit ball B_2^n is worst approximated in the d_s metric by polytopes or more precisely [Ma]:
For every convex body K in $\mathbb{R}^n$ with $\mathrm{vol}_n(K) = \mathrm{vol}_n(B_2^n)$ we have

$$\inf \{d_s(K, P_N) : P_N \subseteq K \text{ and } P_N \text{ has at most N vertices}\} \le$$

$$\inf \{d_s(B_2^n, P_N) : P_N \subseteq B_2^n \text{ and } P_N \text{ has at most N vertices}\}.$$

Notice that $\inf \{d_s(K, P_N) : P_N \subseteq K$ and P_N has at most N vertices$\}$ is the d_s-distance of the best approximating inscribed polytope with N vertices to K. By a compactness argument such a best approximating polytope exists always.

Hence to get an estimate from below for the Bronshteyn Ivanov result, it is enough to check the Euclidean unit ball which was done by Gordon, Reisner and Schütt [GRS1], [GRS2].
There are two positive constants a and b such that for all $n \ge 2$, every $N \ge (bn)^{\frac{2}{n+1}}$, every polytope $P_N \subseteq B_2^n$ with at most N vertices one has

$$\mathrm{vol}_n(B_2^n) - \mathrm{vol}_n(P_N) \ge a \, \frac{n \, \mathrm{vol}_n(B_2^n)}{N^{\frac{2}{n-1}}}.$$

Thus the optimal dependence on the dimension is n and on N it is $N^{\frac{2}{n-1}}$. The next result is about best approximation for large N.
Let K be a convex body in $\mathbb{R}^n$ with C^2-boundary ∂K and everywhere strictly positive curvature κ. Then

$$\lim_{N \to \infty} \frac{\inf\{d_s(K, P_N) | P_N \subseteq K \text{ and } P_N \text{ has at most N vertices}\}}{\left(\frac{1}{N}\right)^{\frac{2}{n-1}}}$$

$$= \tfrac{1}{2} \mathrm{del}_{n-1} \left(\int_{\partial K} \kappa(x)^{\frac{1}{n-1}} d\mu_{\partial K}(x) \right)^{\frac{n+1}{n-1}}.$$

This theorem was proved by McClure and Vitale [McV] in dimension 2 and by Gruber [Gr2] for general n. On the right hand side of the above equation we find the expression $\int_{\partial K} \kappa(x)^{\frac{1}{n-1}} d\mu_{\partial K}(x)$ which is an affine invariant, the so called affine surface area of K which "measures" the boundary behaviour of K. It is natural that such a term should appear in questions of approximation of convex bodies by polytopes. Intuitively we expect that more vertices of the approximating polytope should be put where the boundary of K is very curved and fewer points should be put where the boundary of K is flat to get a good approximation in the d_s-metric. In Section 1.3 we will discuss the affine surface in more detail.

del_{n-1}, which also appears on the right hand side of the above formula, is a constant that depends on n only. The value of this constant is known for for $n = 2, 3$. Putting for K the Euclidean unit ball in the last mentioned theorem, it follows from the result above by Gordon, Reisner and Schütt [GRS1], [GRS2] that del_{n-1} is of the order n. del_{n-1} was determined more precisely by Mankiewicz and Schütt [MaS1], [MaS2]. We refer to Section 1.4. for the exact statements.

Now we want to come to approximation of convex bodies by random polytopes.

A random polytope is the convex hull of finitely many points that are chosen from K with respect to a probability measure $\mathbb{P}$ on K. The expected volume of a random polytope of N points is

$$\mathbb{E}(\mathbb{P}, N) = \int_K \cdots \int_K \mathrm{vol}_n([x_1, \ldots, x_N]) d\mathbb{P}(x_1) \ldots d\mathbb{P}(x_N)$$

where $[x_1, \ldots, x_N]$ is the convex hull of the points $x_1, \ldots, x_N$. Thus the expression $\mathrm{vol}_n(K) - \mathbb{E}(\mathbb{P}, N)$ measures how close a random polytope and the convex body are in the symmetric difference metric. Rényi and Sulanke [ReS1], [ReS2] have investigated this expression for large numbers N of chosen points. They restricted themselves to dimension 2 and the case that the probability measure is the normalized Lebesgue measure on K.

Their results were extended to higher dimensions in case that the probability measure is the normalized Lebesgue measure. Wieacker [Wie] settled the case of the Euclidean ball in dimension n. Bárány proved the result for convex bodies with C^3-boundary and everywhere positive curvature [Ba1]. This result was generalized to arbitrary convex bodies in [Sch1] (see also Section 1.4):

Let K be a convex body in $\mathbb{R}^n$. Then

$$\lim_{N \to \infty} \frac{\mathrm{vol}_n(K) - \mathbb{E}(\mathbb{P}_m, N)}{\left(\frac{\mathrm{vol}_n(K)}{N}\right)^{\frac{2}{n+1}}} = c_1(n) \int_{\partial K} \kappa(x)^{\frac{1}{n-1}} \, \mathrm{d}\mu_{\partial K}(x),$$

where $c_1(n)$ is a constant that depends on n.

We can use this result to obtain an approximation of a convex body by a polytope with at most N vertices. Notice that this does not give the optimal dependence on N. One of the reasons is that not all the points chosen at random from K appear as vertices of the approximating random polytope. We will get back to this point in Section 1.4.

One avoids this problem that not all points chosen appear as vertices of the random polytope by choosing the points at random directly on the boundary of the convex body K.

This is what we do in this paper. We consider convex bodies in dimension n and probability measures that are concentrated on the boundary of the convex body. It is with respect to such probability measures that we choose the points at random on the boundary of K and all those points will then be vertices of the random polytope. This had been done before only in the case of the Euclidean ball by Müller [Mü] who proved the asymptotic formula for the Euclidean ball with the normalized surface measure as probability measure.

Here we treat much more general measures $\mathbb{P}_f$ defined on the boundary of K where we only assume that the measure has a continuous density f with respect to the surface measure $\mu_{\partial K}$ on ∂K. Under some additional technical assumptions we prove an asymptotic formula. This is the content of Theorem 1.1.

In the remainder of Section 1.1 we will introduce further notation used throughout the paper. We conclude Section 1.1 by stating the Theorem 1.1. The whole paper is devoted to prove this main theorem. In doing that, we develop tools that should be helpful in further investigations.

In Section 1.2 we compute which is the optimal f to give the least value in the volume difference

$$\mathrm{vol}_n(K) - \mathbb{E}(\mathbb{P}_f, N).$$

It will turn out that the affine surface area density gives the optimal measure: Choosing points according to this measure gives random polytopes of greatest possible volume. Again, this is intuitively clear: An optimal measure should put more weight on points with higher curvature. Moreover, and this is a crucial observation, if the optimal measure is unique then it must be affine invariant. There are not too many such measures and the affine surface measure is the first that comes to ones mind. This measure satisfies two necessary properties: It is affine invariant and it puts more weight on points with greater curvature.

In Section 1.5 we compare random approximation with best approximation and observe a remarkable fact. Namely, it turns out that -up to a numerical constant- random approximation with the points chosen $\mathbb{P}_f$-randomly from the boundary of K with the optimal f is as good as best approximation.

In Section 1.3 we propose an extension of the p-affine surface area which was introduced by Lutwak [Lu] and Hug [Hu]. We also give a geometric interpretation of the p-affine surface area in terms of random polytopes.

It was a crucial step in the proof of Theorem 1.1 to relate the random polytope to a geometric object. The appropriate geometric object turned out to be the surface body which we introduce in Chapter 2.

In Chapter 3 we review J. Müller's proof for the case of the Euclidean ball. We use his result in our proof.

Chapter 4 is devoted to prove probabilistic inequalities needed for the proof of Theorem 1.1 and finally Chapter 5 gives the proof of Theorem 1.1.

Now we introduce further notations used throughout the paper.

$B_2^n(x, r)$ is the Euclidean ball in $\mathbb{R}^n$ centered at x with radius r. We denote $B_2^n = B_2^n(0, 1)$. S^{n-1} is the boundary ∂B_2^n of the Euclidean unit ball. The norm $\| \cdot \|$ is the Euclidean norm.

The distance $d(A, B)$ of two sets in $\mathbb{R}^n$ is

$$d(A, B) = \inf\{\|x - y\| \,|\, x \in A, y \in B\}.$$

For a convex body K the metric projection $p : \mathbb{R}^n \to K$ maps x onto the unique point $p(x) \in K$ with

$$\|x - p(x)\| = \inf_{y \in K} \|x - y\|.$$

The uniqueness of the point $p(x)$ follows from the convexity of K. If $x \in K$ then $p(x) = x$.

For x, ξ in $\mathbb{R}^n$, $\xi \neq 0$, $H(x, \xi)$ denotes the hyperplane through x and orthogonal to ξ. The two closed halfspaces determined by this hyperplane are denoted by $H^-(x, \xi)$ and $H^+(x, \xi)$. $H^-(x, \xi)$ is usually the halfspace that contains $x + \xi$. Sometimes we write H, H^+ and H^-, if it is clear which are the vectors x and ξ involved.

For points $x_1, \ldots x_N \in \mathbb{R}^n$ we denote by

$$[x_1, \ldots x_N] = \left\{ \lambda_1 x_1 + \cdots + \lambda_N x_N \,\middle|\, 0 \leq \lambda_i \leq 1, 1 \leq i \leq N, \sum_{i=1}^{N} \lambda_i = 1 \right\}$$

the convex hull of these points. In particular, the closed line segment between two points x and y is

$$[x, y] = \{\lambda x + (1 - \lambda)y \,|\, 0 \leq \lambda \leq 1\}.$$

The open line segment is denoted by

$$(x, y) = \{\lambda x + (1 - \lambda)y \,|\, 0 < \lambda < 1\}.$$

$\mu_{\partial K}$ is the surface area measure on ∂K. It equals the restriction of the $n-1$-dimensional Hausdorff measure to ∂K. We write in short μ if it is clear which is the body K involved. Let $f : \partial K \to \mathbb{R}$ be a integrable, nonnegative function with

$$\int_{\partial K} f(x)\mathrm{d}\mu = 1.$$

Then we denote by $\mathbb{P}_f$ the probability measure with $\mathrm{d}\mathbb{P}_f = f\mathrm{d}\mu_{\partial K}$ and $\mathbb{E}(f, N) = \mathbb{E}(\mathbb{P}_f, N)$. If f is the constant function $(\mathrm{vol}_{n-1}(\partial K))^{-1}$ then we write $\mathbb{E}(\partial K, N) = \mathbb{E}(\mathbb{P}_f, N)$. For a measurable subset A of ∂K we write $\mathrm{vol}_{n-1}(A)$ for $\mu_{\partial K}(A)$.

Let K be a convex body in $\mathbb{R}^n$ with boundary ∂K. For $x \in \partial K$ we denote the outer unit normal by $N_{\partial K}(x)$. We write in short $N(x)$ if it is clear which is the body K involved. The normal $N(x)$ may not be unique. $\kappa_{\partial K}(x)$ is the (generalized) Gauß curvature at x (see also Section 1.5 for the precise definition). By a result of Aleksandrov [Al] it exists almost everywhere. Again, we write in short $\kappa(x)$ if it is clear which is the body K involved. The centroid or center of mass cen of K is

$$cen = \frac{\int_K x\mathrm{d}x}{\mathrm{vol}_n(K)}.$$

We conclude Section 1.1 with the main theorem.

Theorem 1.1. *Let K be a convex body in $\mathbb{R}^n$ such that there are r and R in $\mathbb{R}$ with $0 < r \leq R < \infty$ so that we have for all $x \in \partial K$*

$$B_2^n(x - rN_{\partial K}(x), r) \subseteq K \subseteq B_2^n(x - RN_{\partial K}(x), R)$$

and let $f : \partial K \to \mathbb{R}_+$ be a continuous, positive function with $\int_{\partial K} f(x)\mathrm{d}\mu_{\partial K}(x) = 1$. Let $\mathbb{P}_f$ be the probability measure on ∂K given by $\mathrm{d}\mathbb{P}_f(x) = f(x)\mathrm{d}\mu_{\partial K}(x)$. Then we have

$$\lim_{N \to \infty} \frac{\mathrm{vol}_n(K) - \mathbb{E}(f, N)}{\left(\frac{1}{N}\right)^{\frac{2}{n-1}}} = c_n \int_{\partial K} \frac{\kappa(x)^{\frac{1}{n-1}}}{f(x)^{\frac{2}{n-1}}}\mathrm{d}\mu_{\partial K}(x)$$

where κ is the (generalized) Gauß-Kronecker curvature and

$$c_n = \frac{(n-1)^{\frac{n+1}{n-1}} \Gamma\left(n + 1 + \frac{2}{n-1}\right)}{2(n+1)!(\mathrm{vol}_{n-2}(\partial B_2^{n-1}))^{\frac{2}{n-1}}}.$$

The minimum at the right-hand side is attained for the normalized affine surface area measure with density

$$f_{as}(x) = \frac{\kappa(x)^{\frac{1}{n+1}}}{\int_{\partial K} \kappa(x)^{\frac{1}{n+1}}\mathrm{d}\mu_{\partial K}(x)}.$$

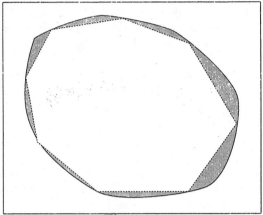

Fig. 1.1.1

The condition: There are r and R in $\mathbb{R}$ with $0 < r \leq R < \infty$ so that we have for all $x \in \partial K$

$$B_2^n(x - rN_{\partial K}(x), r) \subseteq K \subseteq B_2^n(x - RN_{\partial K}(x), R)$$

is satisfied if K has a C^2-boundary with everywhere positive curvature. This follows from Blaschke's rolling theorem ([Bla2] , p.118) and a generalization of it ([Lei], Remark 2.3). Indeed, we can choose

$$r = \min_{x \in \partial K} \min_{1 \leq i \leq n-1} r_i(x) \qquad\qquad R = \max_{x \in \partial K} \max_{1 \leq i \leq n-1} r_i(x)$$

where $r_i(x)$ denotes the i-th principal curvature radius.

By a result of Aleksandrov [Al] the generalized curvature κ exists a.e. on every convex body. It was shown in [SW1] that $\kappa^{\frac{1}{n+1}}$ is an integrable function. Therefore the density

$$f_{as}(x) = \frac{\kappa(x)^{\frac{1}{n+1}}}{\int_{\partial K} \kappa(x)^{\frac{1}{n+1}} \, d\mu_{\partial K}(x)}.$$

exists provided that $\int_{\partial K} \kappa(x)^{\frac{1}{n+1}} \, d\mu_{\partial K}(x) > 0$. This is certainly assured by the assumption on the boundary of K.

1.2 Discussion of Some Measures $\mathbb{P}_f$ and the Optimality of the Affine Surface Area Measure

We want to discuss some measures that are of interest.

1. The most interesting measure is the normalized affine surface area measure as given in the theorem. This measure is affine invariant, i.e. for an affine, volume preserving map T and all measurable subsets A of ∂K

$$\int_A \kappa_{\partial K}^{\frac{1}{n+1}}(x)\mathrm{d}\mu_{\partial K}(x) = \int_{T(A)} \kappa_{\partial T(K)}^{\frac{1}{n+1}}(x)\mathrm{d}\mu_{\partial T(K)}(x).$$

Please note that if the optimal measure is unique it should be affine invariant since the image measure induced by T must also be optimal.

We show that the measure is affine invariant. To do so we introduce the convex floating body. For $t \in \mathbb{R}$, $t > 0$ sufficiently small, the convex floating body $C_{[t]}$ of a convex body C [SW1] is the intersection of all halfspaces whose defining hyperplanes cut off a set of n-dimensional volume t from C. By [SW1] we have for all convex bodies C

$$\lim_{t \to 0} \frac{\mathrm{vol}_n(C) - \mathrm{vol}_n(C_{[t]})}{t^{\frac{2}{n+1}}} = d_n \int_{\partial C} \kappa_{\partial C}(x)^{\frac{1}{n+1}}\mathrm{d}\mu_{\partial C}(x),$$

where $d_n = \frac{1}{2}\left(\frac{n+1}{\mathrm{vol}_{n-1}(B_2^{n-1})}\right)^{2/(n+1)}$. For an affine, volume preserving map T we have

$$\mathrm{vol}_n(C) = \mathrm{vol}_n(T(C)) \quad \text{and} \quad \mathrm{vol}_n(C_{[t]}) = \mathrm{vol}_n(T(C_{[t]})). \tag{1}$$

Thus the expression

$$\int_{\partial C} \kappa_{\partial C}(x)^{\frac{1}{n+1}}\mathrm{d}\mu_{\partial C}(x)$$

is affine invariant. For a closed subset A of ∂K where K is a convex body, we define the convex body C as the convex hull of A. For a point $x \in \partial C$ with $x \notin A$ we have that the curvature must be 0 if it exists. Thus we get by the affine invariance (1) for all closed sets A

$$\int_A \kappa_{\partial C}(x)^{\frac{1}{n+1}}\mathrm{d}\mu_{\partial C}(x) = \int_{\partial T(A)} \kappa_{\partial T(C)}(y)^{\frac{1}{n+1}}\mathrm{d}\mu_{\partial T(C)}(y).$$

This formula extends to all measurable sets. For the affine surface measure we get

$$\lim_{N \to \infty} \frac{\mathrm{vol}_n(K) - \mathbb{E}(f, N)}{\left(\frac{1}{N}\right)^{\frac{2}{n-1}}} = c_n \left(\int_{\partial K} \kappa(x)^{\frac{1}{n+1}}\mathrm{d}\mu_{\partial K}(x)\right)^{\frac{n+1}{n-1}}. \tag{2}$$

We show now that the expression for any other measure given by a density f is greater than or equal to (2). Since $\int_{\partial K} f(x)\mathrm{d}\mu_{\partial K}(x) = 1$, we have

$$\left(\frac{1}{\mathrm{vol}_{n-1}(\partial K)}\int_{\partial K}\left|\frac{\kappa(x)}{f(x)^2}\right|^{\frac{1}{n-1}}\mathrm{d}\mu_{\partial K}(x)\right)^{\frac{1}{n+1}}$$

$$= \left(\frac{1}{\mathrm{vol}_{n-1}(\partial K)}\int_{\partial K}\left|\left(\frac{\kappa(x)}{f(x)^2}\right)^{\frac{1}{n^2-1}}\right|^{n+1}\mathrm{d}\mu_{\partial K}(x)\right)^{\frac{1}{n+1}} \times$$

$$\left(\frac{1}{\mathrm{vol}_{n-1}(\partial K)}\int_{\partial K}\left|f(x)^{\frac{2}{n^2-1}}\right|^{\frac{n^2-1}{2}}\mathrm{d}\mu_{\partial K}(x)\right)^{\frac{-2}{n^2-1}} (\mathrm{vol}_{n-1}(\partial K))^{\frac{2}{n^2-1}}.$$

We have $\frac{1}{n+1} + \frac{2}{n^2-1} = \frac{1}{n-1}$ and we apply Hölder inequality to get

$$\left(\frac{1}{\text{vol}_{n-1}(\partial K)} \int_{\partial K} \left| \frac{\kappa(x)}{f(x)^2} \right|^{\frac{1}{n-1}} d\mu_{\partial K}(x) \right)^{\frac{1}{n+1}}$$

$$\geq \left(\frac{1}{\text{vol}_{n-1}(\partial K)} \int_{\partial K} \kappa(x)^{\frac{1}{n+1}} d\mu_{\partial K}(x) \right)^{\frac{1}{n-1}} (\text{vol}_{n-1}(\partial K))^{\frac{2}{n^2-1}},$$

which gives us

$$\int_{\partial K} \left| \frac{\kappa(x)}{f(x)^2} \right|^{\frac{1}{n-1}} d\mu_{\partial K}(x) \geq \left(\int_{\partial K} \kappa(x)^{\frac{1}{n+1}} d\mu_{\partial K}(x) \right)^{\frac{n+1}{n-1}}.$$

2. The second measure of interest is the surface measure given by the constant density

$$f(x) = \frac{1}{\text{vol}_{n-1}(\partial K)}.$$

This measure is not affine invariant and we get

$$\lim_{N \to \infty} \frac{\text{vol}_n(K) - \mathbb{E}(f, N)}{\left(\frac{\text{vol}_{n-1}(\partial K)}{N} \right)^{\frac{2}{n-1}}} = c_n \int_{\partial K} \kappa(x)^{\frac{1}{n-1}} d\mu_{\partial K}(x).$$

3. The third measure is obtained in the following way. Let K be a convex body, cen its centroid and A a subset of ∂K. Let

$$\mathbb{P}(A) = \frac{\text{vol}_n([cen, A])}{\text{vol}_n(K)}.$$

If the centroid is the origin, then the density is given by

$$f(x) = \frac{<x, N_{\partial K}(x)>}{\int_{\partial K} <x, N_{\partial K}(x)> d\mu_{\partial K}(x)}$$

and the measure is invariant under linear, volume preserving maps. We have $\frac{1}{n} \int_{\partial K} <x, N(x)> d\mu_{\partial K}(x) = \text{vol}_n(K)$ and thus

$$f(x) = \frac{<x, N_{\partial K}(x)>}{n \, \text{vol}_n(K)}.$$

We get

$$\lim_{N \to \infty} \frac{\text{vol}_n(K) - \mathbb{E}(f, N)}{\left(\frac{n \, \text{vol}_n(K)}{N} \right)^{\frac{2}{n-1}}} = c_n \int_{\partial K} \frac{\kappa(x)^{\frac{1}{n-1}}}{<x, N_{\partial K}(x)>^{\frac{2}{n-1}}} d\mu_{\partial K}(x).$$

We recall that for $p > 0$ the p-affine surface area $O_p(K)$ [Lu], [Hu] of a convex body K is defined as (see 1.3 below for more details)

$$O_p(K) = \int_{\partial K} \frac{\kappa(x)^{\frac{p}{n+p}}}{< x, N_{\partial K}(x) >^{\frac{n(p-1)}{n+p}}} d\mu_{\partial K}(x).$$

Note that then for $n > 2$ the right hand expression above is a p-affine surface area with $p = n/(n-2)$.

4. More generally, let K be a convex body in $\mathbb{R}^n$ with centroid at the origin and satisfying the assumptions of Theorem 1.1. Let α and β be real numbers. Let the density be given by

$$f_{\alpha,\beta}(x) = \frac{< x, N_{\partial K}(x) >^\alpha \kappa(x)^\beta}{\int_{\partial K} < x, N_{\partial K}(x) >^\alpha \kappa(x)^\beta d\mu_{\partial K}(x)}.$$

Then by Theorem 1.1

$$\lim_{N \to \infty} \frac{\text{vol}_n(K) - \mathbb{E}(f_{\alpha,\beta}, N)}{\left(\frac{1}{N}\right)^{\frac{2}{n-1}}} =$$

$$c_n \left(\int_{\partial K} \frac{\kappa(x)^{\frac{1-2\beta}{n-1}}}{< x, N_{\partial K}(x) >^{\frac{2\alpha}{n-1}}} d\mu_{\partial K}(x) \right) \left(\int_{\partial K} < x, N_{\partial K}(x) >^\alpha \kappa(x)^\beta d\mu_{\partial K}(x) \right)^{\frac{2}{n-1}}.$$

The second expression on the right hand side of this equation is a p-affine surface area iff

$$\alpha = -\frac{n(p-1)}{n+p} \quad \text{and} \quad \beta = \frac{p}{n+p}.$$

Then

$$\lim_{N \to \infty} \frac{\text{vol}_n(K) - \mathbb{E}(f, N)}{\left(\frac{O_p(K)}{N}\right)^{\frac{2}{n-1}}}$$

$$= c_n \int_{\partial K} \kappa(x)^{\frac{n-p}{(n-1)(n+p)}} < x, N_{\partial K}(x) >^{\frac{2n(p-1)}{(n-1)(n+p)}} d\mu_{\partial K}(x).$$

Note that the right hand side of this equality is a q-affine surface area with $q = \frac{n-p}{n+p-2}$.

5. Another measure of interest is the measure induced by the Gauß map. The Gauß map $N_{\partial K} : \partial K \to \partial B_2^n$ maps a point x to its normal $N_{\partial K}(x)$. As a measure we define

$$\mathbb{P}(A) = \sigma\{N_{\partial K}(x) | x \in A\}$$

where σ is the normalized surface measure on ∂B_2^n. This can also be written as

$$\mathbb{P}(A) = \frac{\int_A \kappa(x) d\mu_{\partial K}(x)}{\text{vol}_{n-1}(\partial B_2^n)}.$$

This measure is not invariant under linear transformations with determinant 1. This can easily be seen by considering the circle with radius 1 in $\mathbb{R}^2$. An affine transformation changes the circle into an ellipse. We consider a small neighborhood of an apex with small curvature. This is the affine image of a small set whose image under the Gauß map is larger. We get

$$\lim_{N \to \infty} \frac{\mathrm{vol}_n(K) - \mathbb{E}(f, N)}{\left(\frac{\mathrm{vol}_{n-1}(\partial B_2^n)}{N} \right)^{\frac{2}{n-1}}} = c_n \int_{\partial K} \kappa(x)^{-\frac{1}{n-1}} \mathrm{d}\mu_{\partial K}(x).$$

1.3   Extensions of the p-Affine Surface Area

The p-affine surface area $O_p(K)$ was introduced by Lutwak [Lu], see also Hug [Hu]. For $p = 1$ we get the affine surface area which is related to curve evolution and computer vision [SaT1, SaT2]. Meyer and Werner [MW1, MW2] gave a geometric interpretation of the p-affine surface area in terms of the Santaló bodies. They also observed that -provided the integrals exist- the definition of Lutwak for the p-affine surface area makes sense for $-n < p \leq 0$ and their geometric interpretation in terms of the Santaló bodies also holds for this range of p. They also gave a definition of the p-affine surface area for $p = -n$ together with its geometric interpretation.

In view of 1.2.4 we propose here to extend the p-range even further, namely to $-\infty \leq p \leq \infty$. Theorem 1.1 then provides a geometric interpretation of the p-affine surface area for this whole p-range. See also [SW2] for another geometric interpretation.

Let K be a convex body in $\mathbb{R}^n$ with the origin in its interior. For p with $p \neq -n$ and $-\infty \leq p \leq \infty$ we put

$$O_{\pm\infty}(K) = \int_{\partial K} \frac{\kappa(x)}{< x, N_{\partial K}(x) >^n} \mathrm{d}\mu_{\partial K}(x)$$

and

$$O_p(K) = \int_{\partial K} \frac{\kappa(x)^{\frac{p}{n+p}}}{< x, N_{\partial K}(x) >^{\frac{n(p-1)}{n+p}}} \mathrm{d}\mu_{\partial K}(x),$$

provided the integrals exist.

If 0 is an interior point of K then there are strictly positive constants a and b such that

$$a \leq < x, N_{\partial K}(x) > \leq b.$$

Assume now that K is such that the assumptions of Theorem 1.1 hold. Then the above integrals are finite. We consider the densities

$$f_{\pm\infty}(x) = \frac{1}{O_{\pm\infty}(K)} \frac{\kappa(x)}{< x, N_{\partial K}(x) >^n}$$

and for $-\infty < p < \infty$, $p \neq -n$

$$f_p(x) = \frac{1}{O_p(K)} \frac{\kappa(x)^{\frac{p}{n+p}}}{<x, N_{\partial K}(x)>^{\frac{n(p-1)}{n+p}}}.$$

As a corollary to Theorem 1.1 we get the following geometric interpretation of the p-affine surface area.

$$\lim_{N \to \infty} \frac{\text{vol}_n(K) - \mathbb{E}(f_{\pm\infty}, N)}{\left(\frac{O_{\pm\infty}(K)}{N}\right)^{\frac{2}{n-1}}} =$$

$$c_n \int_{\partial K} \kappa(x)^{-\frac{1}{n-1}} <x, N_{\partial K}(x)>^{\frac{2n}{n-1}} d\mu_{\partial K}(x) = O_{-1}(K)$$

and

$$\lim_{N \to \infty} \frac{\text{vol}_n(K) - \mathbb{E}(f_p, N)}{\left(\frac{O_p(K)}{N}\right)^{\frac{2}{n-1}}} =$$

$$c_n \int_{\partial K} \kappa(x)^{\frac{n-p}{(n-1)(n+p)}} <x, N_{\partial K}(x)>^{\frac{2n(p-1)}{(n-1)(n+p)}} d\mu_{\partial K}(x) = O_q(K)$$

where $q = \frac{n-p}{n+p-2}$.

Thus each density f_p gives us a q-affine surface area O_q with $q = \frac{n-p}{n+p-2}$ as the expected difference volume. Note that for the density f_{-n+2} we get $O_{\pm\infty}(K)$. Conversely, for each q-affine surface area O_q, $-\infty \le q \le +\infty$, $q \ne -n$, there is a density f_p with $p = \frac{n-nq+2q}{q+1}$ such that

$$\lim_{N \to \infty} \frac{\text{vol}_n(K) - \mathbb{E}(f_p, N)}{\left(\frac{O_p(K)}{N}\right)^{\frac{2}{n-1}}} = c_n O_q(K).$$

1.4 Random Polytopes of Points Chosen from the Convex Body

Whereas random polytopes of points chosen from the boundary of a convex body have up to now only been considered in the case of the Euclidean ball [Mü], random polytopes of points chosen from the convex body and not only from the boundary have been investigated in great detail. This has been done by Rényi and Sulanke [ReS1, ReS2] in dimension 2. Wieacker [Wie] computed the expected difference volume for the Euclidean ball in $\mathbb{R}^n$. Bárány [Ba1] showed for convex bodies K in $\mathbb{R}^n$ with C^3-boundary and everywhere positive curvature that

$$\lim_{N \to \infty} \frac{\text{vol}_n(K) - \mathbb{E}(\mathbb{P}, N)}{\left(\frac{\text{vol}_n(K)}{N}\right)^{\frac{2}{n+1}}} = c_1(n) \int_{\partial K} \kappa(x)^{\frac{1}{n+1}} d\mu_{\partial K}(x)$$

where $\mathbb{P}$ is the normalized Lebesgue measure on K, $\kappa(x)$ is the Gauß-Kronecker curvature, and

$$c_1(n) = \frac{(n+1)^{\frac{2}{n+1}}(n^2+n+2)(n^2+1)\Gamma(\frac{n^2+1}{n+1})}{2(n+3)(n+1)!\mathrm{vol}_{n-1}(B_2^{n-1})^{\frac{2}{n+1}}}.$$

Schütt [Sch1] verified that this formula holds for all convex bodies, where $\kappa(x)$ is the generalized Gauß-Kronecker curvature.

The order of best approximation of convex bodies by polytopes with a given number of vertices N is $N^{-\frac{2}{n-1}}$ (see above). The above formula for random polytopes chosen from the body gives $N^{-\frac{2}{n+1}}$. Thus random approximation by choosing the points from K does not give the optimal order. But one has to take into account that not all points chosen from the convex body turn out to be vertices of a random polytope. Substituting N by the number of expected vertices we get the optimal order [Ba2] for the exponent of N in the case of a convex body with C^3-boundary and everywhere positive curvature. Indeed, for all convex bodies with a C^3-boundary and everywhere positive curvature the expected number of i-dimensional faces is of the order $N^{\frac{n-1}{n+1}}$ [Ba2].

1.5 Comparison between Best and Random Approximation

Now we want to compare random approximation with best approximation in more detail. We will not only consider the exponent of N but also the other factors. It turns out that random approximation and best approximation with the optimal density are very close.

McClure and Vitale [McV] obtained an asymptotic formula for best approximation in the case $n = 2$. Gruber [Gr2] generalized this to higher dimensions. The metric used in these results is the symmetric difference metric d_S. Then these asymptotic best approximation results are (see above for the precise formulation):

If a convex body K in $\mathbb{R}^n$ has a C^2-boundary with everywhere positive curvature, then

$$\inf\{d_S(K, P_N) | P_N \subset K \text{ and } P_N \text{ is a polytope with at most N vertices}\}$$

is asymptotically the same as

$$\tfrac{1}{2}\mathrm{del}_{n-1}\left(\int_{\partial K} \kappa(x)^{\frac{1}{n+1}}\,\mathrm{d}\mu_{\partial K}(x)\right)^{\frac{n+1}{n-1}}\left(\frac{1}{N}\right)^{\frac{2}{n-1}}.$$

where del_{n-1} is a constant that is related to the Delone triangulations and depends only on the dimension n. Equivalently, the result states that if we divide one expression by the other and take the limit for N to ∞ we obtain 1. It was shown by Gordon, Reisner and Schütt in [GRS1, GRS2] that the constant del_{n-1} is of the order of n, which means that there are numerical constants a and b such that we have for all $n \in \mathbb{N}$

$$an \leq \mathrm{del}_{n-1} \leq bn.$$

It is clear from Theorem 1.1 that we get the best random approximation if we choose the affine surface area measure. Then the order of magnitude for random approximation is

$$\frac{(n-1)^{\frac{n+1}{n-1}} \Gamma\left(n+1+\frac{2}{n-1}\right)}{2(n+1)!(\mathrm{vol}_{n-2}(\partial B_2^{n-1}))^{\frac{2}{n-1}}} \left(\int_{\partial K} \kappa(x)^{\frac{1}{n+1}} d\mu_{\partial K}(x)\right)^{\frac{n+1}{n-1}} \left(\frac{1}{N}\right)^{\frac{2}{n-1}}.$$

Since

$$(\mathrm{vol}_{n-2}(\partial B_2^{n-1}))^{\frac{2}{n-1}} \sim \frac{1}{n} \quad \text{and} \quad \Gamma\left(n+1+\frac{2}{n-1}\right) \sim \Gamma(n+1)(n+1)^{\frac{2}{n-1}}$$

random approximation (with randomly choosing the points from the boundary of K) is of the same order as

$$n \left(\int_{\partial K} \kappa(x)^{\frac{1}{n+1}} d\mu_{\partial K}(x)\right)^{\frac{n+1}{n-1}} \left(\frac{1}{N}\right)^{\frac{2}{n-1}},$$

which is the same order as best approximation.

In two papers by Mankiewicz and Schütt the constant del_{n-1} has been better estimated [MaS1, MaS2]. It was shown there

$$\frac{n-1}{n+1}\mathrm{vol}_{n-1}(B_2^{n-1})^{-\frac{2}{n-1}} \leq \mathrm{del}_{n-1} \leq (1+\frac{c\ln n}{n})\frac{n-1}{n+1}\mathrm{vol}_{n-1}(B_2^{n-1})^{-\frac{2}{n-1}},$$

where c is a numerical constant. In particular, $\lim_{n\to\infty}\frac{\mathrm{del}_{n-1}}{n} = \frac{1}{2\pi e} = 0.0585498....$ Thus

$$\left(1-c\frac{\ln n}{n}\right) \lim_{N\to\infty} \frac{\mathrm{vol}_n(K) - \mathbb{E}(f_{as}, N)}{\left(\frac{1}{N}\right)^{\frac{2}{n-1}}}$$

$$\leq \lim_{N\to\infty} N^{\frac{2}{n-1}} \inf\{d_S(K, P_N) | P_N \subset K \text{ and } P_N$$

$$\text{is a polytope with at most N vertices}\}.$$

In order to verify this we have to estimate the quotient

$$\frac{(n-1)^{\frac{n+1}{n-1}} \Gamma\left(n+1+\frac{2}{n-1}\right)}{2(n+1)!(\mathrm{vol}_{n-2}(\partial B_2^{n-1}))^{\frac{2}{n-1}}}(\tfrac{1}{2}\mathrm{del}_{n-1})^{-1}.$$

Since $\frac{n-1}{n+1}\mathrm{vol}_{n-1}(B_2^{n-1})^{-\frac{2}{n-1}} \leq \mathrm{del}_{n-1}$ the quotient is less than $\frac{1}{n!}\Gamma(n+1+\frac{2}{n-1})$. Now we use Stirlings formula to get

$$\frac{\Gamma(n+1+\frac{2}{n-1})}{n!} \leq 1+c\frac{\ln n}{n}.$$

1.6 Subdifferentials and Indicatrix of Dupin

Let $\mathcal{U}$ be a convex, open subset of $\mathbb{R}^n$ and let $f : \mathcal{U} \to \mathbb{R}$ be a convex function. $df(x) \in \mathbb{R}^n$ is called subdifferential at the point $x_0 \in \mathcal{U}$, if we have for all $x \in \mathcal{U}$

$$f(x_0)+ < df(x_0), x - x_0 >\le f(x).$$

A convex function has a subdifferential at every point and it is differentiable at a point if and only if the subdifferential is unique. Let $\mathcal{U}$ be an open, convex subset in $\mathbb{R}^n$ and $f : \mathcal{U} \to \mathbb{R}$ a convex function. f is said to be twice differentiable in a generalized sense in $x_0 \in \mathcal{U}$, if there is a linear map $d^2 f(x_0)$ and a neighborhood $\mathcal{U}(x_0) \subseteq \mathcal{U}$ such that we have for all $x \in \mathcal{U}(x_0)$ and for all subdifferentials $df(x)$

$$\|df(x) - df(x_0) - d^2 f(x_0)(x - x_0)\| \le \Theta(\|x - x_0\|)\|x - x_0\|,$$

where Θ is a monotone function with $\lim_{t\to 0} \Theta(t) = 0$. $d^2 f(x_0)$ is called generalized Hesse-matrix. If $f(0) = 0$ and $df(0) = 0$ then we call the set

$$\{x \in \mathbb{R}^n | x^t d^2 f(0)x = 1\}$$

the indicatrix of Dupin at 0. Since f is convex this set is an ellipsoid or a cylinder with a base that is an ellipsoid of lower dimension. The eigenvalues of $d^2 f(0)$ are called principal curvatures and their product is called the Gauß-Kronecker curvature κ. Geometrically the eigenvalues of $d^2 f(0)$ that are different from 0 are the lengths of the principal axes of the indicatrix raised to the power -2.

The following lemma can be found in e.g. [SW1].

Lemma 1.1. *Let $\mathcal{U}$ be an open, convex subset of $\mathbb{R}^n$ and $0 \in \mathcal{U}$. Suppose that $f : \mathcal{U} \to \mathbb{R}$ is twice differentiable in the generalized sense at 0 and that $f(0) = 0$ and $df(0) = 0$.*
(i) Suppose that the indicatrix of Dupin at 0 is an ellipsoid. Then there is a monotone, increasing function $\psi : [0, 1] \to [1, \infty)$ with $\lim_{s\to 0} \psi(s) = 1$ such that

$$\left\{ (x, s) \left| x^t d^2 f(0)x \le \frac{2s}{\psi(s)} \right. \right\}$$
$$\subseteq \{(x, s) | f(x) \le s\} \subseteq \{(x, s) | x^t d^2 f(0)x \le 2s\psi(s)\}.$$

(ii) Suppose that the indicatrix of Dupin is an elliptic cylinder. Then for every $\epsilon > 0$ there is $s_0 > 0$ such that we have for all s with $s < s_0$

$$\left\{ (x, s) \left| x^t d^2 f(0)x + \epsilon\|x\|^2 \le 2s \right. \right\} \subseteq \{(x, s) | f(x) \le s\}.$$

Lemma 1.2. *Let K be a convex body in $\mathbb{R}^n$ with $0 \in \partial K$ and $N(0) = -e_n$. Suppose that the indicatrix of Dupin at 0 is an ellipsoid. Suppose that the principal axes $b_i e_i$ of the indicatrix are multiples of the unit vectors e_i, $i = 1, \ldots, n-1$. Let $\mathcal{E}$ be the n-dimensional ellipsoid*

$$\mathcal{E} = \left\{ x \in \mathbb{R}^n \,\middle|\, \sum_{i=1}^{n-1} \frac{x_i^2}{b_i^2} + \frac{\left(x_n - \left(\prod_{i=1}^{n-1} b_i \right)^{\frac{2}{n-1}} \right)^2}{\left(\prod_{i=1}^{n-1} b_i \right)^{\frac{2}{n-1}}} \leq \left(\prod_{i=1}^{n-1} b_i \right)^{\frac{2}{n-1}} \right\}.$$

Then there is an increasing, continuous function $\phi : [0, \infty) \to [1, \infty)$ with $\phi(0) = 1$ such that we have for all t

$$\left\{ \left(\frac{x_1}{\phi(t)}, \ldots, \frac{x_{n-1}}{\phi(t)}, t \right) \,\middle|\, x \in \mathcal{E}, x_n = t \right\}$$
$$\subseteq K \cap H((0, \ldots, 0, t), N(0))$$
$$\subseteq \left\{ (\phi(t) x_1, \ldots, \phi(t) x_{n-1}, t) \,\middle|\, x \in \mathcal{E}, x_n = t \right\}.$$

We call $\mathcal{E}$ the standard approximating ellipsoid .

Proof. Lemma 1.2 follows from Lemma 1.1. Let f be a function whose graph is locally the boundary of the convex body. Consider (x, s) with

$$x^t \mathrm{d}^2 f(0) x = 2s$$

which is the same as

$$\sum_{i=1}^{n-1} \frac{x_i^2}{b_i^2} = 2s.$$

Then

$$\sum_{i=1}^{n-1} \frac{x_i^2}{b_i^2} + \frac{\left(x_n - \left(\prod_{i=1}^{n-1} b_i \right)^{\frac{2}{n-1}} \right)^2}{\left(\prod_{i=1}^{n-1} b_i \right)^{\frac{2}{n-1}}}$$
$$= 2s + \frac{\left(s - \left(\prod_{i=1}^{n-1} b_i \right)^{\frac{2}{n-1}} \right)^2}{\left(\prod_{i=1}^{n-1} b_i \right)^{\frac{2}{n-1}}} = -\frac{s^2}{\left(\prod_{i=1}^{n-1} b_i \right)^{\frac{2}{n-1}}} + \left(\prod_{i=1}^{n-1} b_i \right)^{\frac{2}{n-1}}.$$

$\square$

Let us denote the lengths of the principal axes of the indicatrix of Dupin by b_i, $i = 1, \ldots, n-1$. Then the lengths a_i, $i = 1, \ldots, n$ of the principal axes of the standard approximating ellipsoid $\mathcal{E}$ are

$$a_i = b_i \left(\prod_{i=1}^{n-1} b_i \right)^{\frac{1}{n-1}} \qquad i = 1, \ldots, n-1 \qquad \text{and} \qquad a_n = \left(\prod_{i=1}^{n-1} b_i \right)^{\frac{2}{n-1}}.$$

$$(3)$$

This follows immediately from Lemma 1.2. For the Gauß-Kronecker curvature we get

$$\prod_{i=1}^{n-1} \frac{a_n}{a_i^2}.$$

$$(4)$$

This follows as the Gauß-Kronecker curvature equals the product of the eigenvalues of the Hesse matrix. The eigenvalues are b_i^{-2}, $i = 1, \ldots, n-1$. Thus

$$\prod_{i=1}^{n-1} b_i^{-2} = \left(\prod_{i=1}^{n-1} b_i \right)^2 \prod_{i=1}^{n-1} \left(b_i \left(\prod_{k=1}^{n-1} b_k \right)^{\frac{1}{n-1}} \right)^{-2} = \prod_{i=1}^{n-1} \frac{a_n}{a_i^2}.$$

In particular, if the indicatrix of Dupin is a sphere of radius $\sqrt{\rho}$ then the standard approximating ellipsoid is a Euclidean ball of radius ρ.

We consider the transform $T : \mathbb{R}^n \to \mathbb{R}^n$

$$T(x) = \left(\frac{x_1}{a_1} \left(\prod_{i=1}^{n-1} b_i \right)^{\frac{2}{n-1}}, \ldots, \frac{x_{n-1}}{a_{n-1}} \left(\prod_{i=1}^{n-1} b_i \right)^{\frac{2}{n-1}}, x_n \right).$$

$$(5)$$

This transforms the standard approximating ellipsoid $\mathcal{E}$ into a Euclidean ball $T(\mathcal{E})$ with radius $r = (\prod_{i=1}^{n-1} b_i)^{2/(n-1)}$. This is obvious since the principal axes of the standard approximating ellipsoid are given by (3). The map T is volume preserving.

Lemma 1.3. *Let*

$$\mathcal{E} = \left\{ x \in \mathbb{R}^n \ \middle| \ \sum_{i=1}^{n} \left| \frac{x_i}{a_i} \right|^2 \leq 1 \right\}$$

and let $H = H((a_n - \Delta)e_n, e_n)$. Then for all Δ with $\Delta \leq \frac{1}{2} a_n$ the intersection $\mathcal{E} \cap H$ is an ellipsoid whose principal axes have lengths

$$\frac{a_i}{a_n} \left(2 a_n \Delta - \Delta^2 \right)^{\frac{1}{2}} \qquad i = 1, \ldots, n-1.$$

Moreover,

$$\mathrm{vol}_{n-1}(\mathcal{E} \cap H) \leq \mathrm{vol}_{n-1}(\partial \mathcal{E} \cap H^-)$$

$$\leq \sqrt{1 + \frac{2 \Delta a_n^3}{(a_n - \Delta)^2 \min_{1 \leq i \leq n-1} a_i^2}} \, \mathrm{vol}_{n-1}(\mathcal{E} \cap H)$$

and

$$\text{vol}_{n-1}(\mathcal{E} \cap H) = \text{vol}_{n-1}(B_2^{n-1}) \left(\prod_{i=1}^{n-1} a_i \right) \left(\frac{2\Delta}{a_n} - \left| \frac{\Delta}{a_n} \right|^2 \right)^{\frac{n-1}{2}}$$

$$= \frac{\text{vol}_{n-1}(B_2^{n-1})}{\sqrt{\kappa(a_n e_n)}} \left(2\Delta - \frac{\Delta^2}{a_n} \right)^{\frac{n-1}{2}},$$

where κ is the Gauß-Kronecker curvature.

Proof. The left hand inequality is trivial. We show the right hand inequality. Let p_{e_n} be the orthogonal projection onto the subspace orthogonal to e_n. We have

$$\text{vol}_{n-1}(\partial\mathcal{E} \cap H^-) = \int_{\mathcal{E} \cap H} \frac{1}{< e_n, N_{\partial\mathcal{E}}(\bar{y}) >} dy \qquad (6)$$

where $\bar{y}_i = y_i$, $i = 1, \ldots, n-1$, and

$$\bar{y}_n = a_n \sqrt{1 - \sum_{i=1}^{n-1} \left| \frac{y_i}{a_i} \right|^2}.$$

Therefore we get

$$\text{vol}_{n-1}(\partial\mathcal{E} \cap H^-) \leq \frac{\text{vol}_{n-1}(\mathcal{E} \cap H)}{\min_{x \in \partial\mathcal{E} \cap H^-} < e_n, N_{\partial\mathcal{E}}(x) >}.$$

We have

$$N_{\partial\mathcal{E}}(x) = \frac{\left(\frac{x_i}{a_i^2} \right)_{i=1}^n}{\sqrt{\sum_{i=1}^n \left| \frac{x_i}{a_i^2} \right|^2}}.$$

Therefore we get

$$< e_n, N_{\partial\mathcal{E}}(x) > = \frac{\frac{x_n}{a_n^2}}{\sqrt{\sum_{i=1}^n \left| \frac{x_i}{a_i^2} \right|^2}} = \left(1 + \frac{a_n^4}{x_n^2} \sum_{i=1}^{n-1} \frac{x_i^2}{a_i^4} \right)^{-\frac{1}{2}}$$

$$\geq \left(1 + \frac{a_n^4}{x_n^2 \min_{1 \leq i \leq n-1} a_i^2} \sum_{i=1}^{n-1} \frac{x_i^2}{a_i^2} \right)^{-\frac{1}{2}}$$

$$= \left(1 + \frac{a_n^4}{x_n^2 \min_{1 \leq i \leq n-1} a_i^2} \left(1 - \left| \frac{x_n}{a_n} \right|^2 \right) \right)^{-\frac{1}{2}}$$

$$= \left(1 + \frac{a_n^2}{\min_{1 \leq i \leq n-1} a_i^2} \left(\frac{a_n^2}{x_n^2} - 1 \right) \right)^{-\frac{1}{2}}.$$

The last expression is smallest for $x_n = a_n - \Delta$. We get

$$< e_n, N_{\partial \mathcal{E}}(x) > \geq \left(1 + \frac{a_n^2(2\Delta a_n - \Delta^2)}{(a_n - \Delta)^2 \min_{1 \leq i \leq n-1} a_i^2}\right)^{-\frac{1}{2}}$$

$$\geq \left(1 + \frac{2\Delta a_n^3}{(a_n - \Delta)^2 \min_{1 \leq i \leq n-1} a_i^2}\right)^{-\frac{1}{2}}.$$

The equalities are proved using

$$\kappa(a_n e_n) = \prod_{i=1}^{n-1} \frac{a_n}{a_i^2}.$$

$\square$

Lemma 1.4. *Let K be a convex body in $\mathbb{R}^n$ and $x_0 \in \partial K$. Suppose that the indicatrix of Dupin at x_0 exists and is an ellipsoid. Let $\mathcal{E}$ be the standard approximating ellipsoid at x_0. Then for all $\epsilon > 0$ there is Δ_0 such that for all $\Delta < \Delta_0$*

$$\text{vol}_{n-1}(K \cap H(x_0 - \Delta N_{\partial K}(x_0), N_{\partial K}(x_0))) \leq$$

$$\text{vol}_{n-1}(\partial K \cap H^-(x_0 - \Delta N_{\partial K}(x_0), N_{\partial K}(x_0))) \leq$$

$$(1+\epsilon)\sqrt{1 + \frac{2\Delta a_n^3}{(a_n - \Delta)^2 \min_{1 \leq i \leq n-1} a_i^2}} \text{vol}_{n-1}(K \cap H(x_0 - \Delta N_{\partial K}(x_0), N_{\partial K}(x_0))).$$

where $a_1, \ldots, a_n$ are the lengths of the principal axes of $\mathcal{E}$.

Proof. We can assume that K is in such a position that $N_{\partial K}(x_0)$ coincides with the n-th unit vector e_n and that the equation of the approximating ellipsoid is

$$\mathcal{E} = \left\{x \in \mathbb{R}^n \,\middle|\, \sum_{i=1}^{n} \left|\frac{x_i}{a_i}\right|^2 \leq 1\right\}.$$

Then the proof follows from Lemma 1.2 and Lemma 1.3. $\square$

Lemma 1.5. *Let H be a hyperplane with distance p from the origin and s the area of the cap cut off by H from B_2^n. r denotes the radius of the $n-1$-dimensional Euclidean ball $H \cap B_2^n$. We have*

$$\frac{dp}{ds} = -\left(r^{n-3}\text{vol}_{n-2}(\partial B_2^{n-1})\right)^{-1} = -\left((1 - p^2)^{\frac{n-3}{2}}\text{vol}_{n-2}(\partial B_2^{n-1})\right)^{-1}.$$

Proof. Using (6) and polar coordinates, we get for the surface area s of a cap of the Euclidean ball of radius 1

$$s = \text{vol}_{n-2}(\partial B_2^{n-1}) \int_0^r \frac{t^{n-2}}{(1-t^2)^{\frac{1}{2}}} dt = \text{vol}_{n-2}(\partial B_2^{n-1}) \int_0^{\sqrt{1-p^2}} \frac{t^{n-2}}{(1-t^2)^{\frac{1}{2}}} dt.$$

This gives

$$\frac{ds}{dp} = -\frac{\text{vol}_{n-2}(\partial B_2^{n-1})(1-p^2)^{\frac{n-2}{2}}}{p} \frac{p}{\sqrt{1-p^2}} = -r^{n-3}\text{vol}_{n-2}(\partial B_2^{n-1}).$$

□

Lemma 1.6. *(Aleksandrov [Al]) Let K be a convex body in $\mathbb{R}^n$. Then its boundary is almost everywhere twice differentiable in the generalized sense.*

For a proof of this result see [Ban], [EvG], [BCP].

At each point where ∂K is twice differentiable in the generalized sense the indicatrix of Dupin exists. Therefore the indicatrix of Dupin exists almost everywhere.

Lemma 1.7. *(John [J]) Let K be a convex body in $\mathbb{R}^n$ that is centrally symmetric with respect to the origin. Then there exists an ellipsoid $\mathcal{E}$ with center 0 such that*

$$\mathcal{E} \subseteq K \subseteq \sqrt{n}\,\mathcal{E}.$$

Lemma 1.8. *Let K and C be convex bodies in $\mathbb{R}^n$ such that $C \subseteq K$ and 0 is an interior point of C. Then we have for all integrable functions f*

$$\int_{\partial C} f(x) d\mu_{\partial C}(x) = \int_{\partial K} f(x(y)) \frac{\|x(y)\|^n < y, N(y) >}{\|y\|^n < x(y), N(x(y)) >} d\mu_{\partial K}(y)$$

where $\{x(y)\} = [0, y] \cap \partial C$.

2 The Surface Body

2.1 Definitions and Properties of the Surface Body

Let $0 < s$ and let $f : \partial K \to \mathbb{R}$ be a nonnegative, integrable function with $\int_{\partial K} f d\mu = 1$.

The surface body $K_{f,s}$ is the intersection of all the closed half-spaces H^+ whose defining hyperplanes H cut off a set of $\mathbb{P}_f$-measure less than or equal to s from ∂K. More precisely,

$$K_{f,s} = \bigcap_{\mathbb{P}_f(\partial K \cap H^-) \leq s} H^+.$$

We write usually K_s for $K_{f,s}$ if it is clear which function f we are considering. It follows from the Hahn-Banach theorem that $K_0 \subseteq K$. If in addition f is almost everywhere nonzero, then $K_0 = K$. This is shown in Lemma 2.1.(iv).

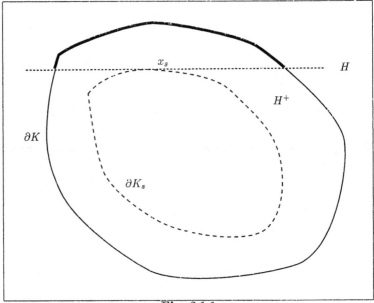

Fig. 2.1.1

We say that a sequence of hyperplanes H_i, $i \in \mathbb{N}$, in $\mathbb{R}^n$ converges to a hyperplane H if we have for all $x \in H$ that

$$\lim_{i \to \infty} d(x, H_i) = 0,$$

where $d(x, H) = \inf\{\|x - y\| : y \in H\}$. This is equivalent to: The sequence of the normals of H_i converges to the normal of H and there is a point $x \in H$ such that

$$\lim_{i \to \infty} d(x, H_i) = 0.$$

Lemma 2.1. *Let K be a convex body in $\mathbb{R}^n$ and let $f : \partial K \to \mathbb{R}$ be a a.e. positive, integrable function with $\int_{\partial K} f d\mu = 1$. Let $\xi \in S^{n-1}$.*
(i) Let $x_0 \in \partial K$. Then

$$\mathbb{P}_f(\partial K \cap H^-(x_0 - t\xi, \xi))$$

is a continuous function of t on

$$\left[0, \max_{y \in K} < x_0 - y, \xi >\right).$$

(ii) Let $x \in \mathbb{R}^n$. Then the function

$$\mathbb{P}_f(\partial K \cap H^-(x - t\xi, \xi))$$

is strictly increasing on

$$\left[\min_{y \in K} < x - y, \xi >, \max_{y \in K} < x - y, \xi >\right].$$

(iii) Let H_i, $i \in \mathbb{N}$, be a sequence of hyperplanes that converge to the hyperplane H_0. Assume that the hyperplane H_0 intersects the interior of K. Then we have

$$\lim_{i \to \infty} \mathbb{P}_f(\partial K \cap H_i^-) = \mathbb{P}_f(\partial K \cap H_0^-).$$

(iv)

$$\overset{\circ}{K} \subseteq \bigcup_{0 < s} K_s$$

In particular, $K = K_0$.

Proof. (i)

$$\mathrm{vol}_{n-1}(\partial K \cap H^-(x_0 - t\xi, \xi))$$

is a continuous function on

$$\left[0, \max_{y \in K} < x_0 - y, \xi >\right).$$

Then (i) follows as f is an integrable function.

(ii) Since $H^-(x, \xi)$ is the half space containing $x + \xi$ we have for $t_1 < t_2$

$$H^-(x - t_1\xi, \xi) \subsetneq H^-(x - t_2\xi, \xi).$$

If

$$\mathbb{P}_f(\partial K \cap H^-(x - t_1\xi, \xi)) = \mathbb{P}_f(\partial K \cap H^-(x - t_2\xi, \xi))$$

then f is a.e. 0 on $\partial K \cap H^-(x - t_2\xi, \xi) \cap H^+(x - t_1\xi, \xi)$. This is not true.

(iii) Let $H_i = H_i(x_i, \xi_i)$, $i = 0, 1, \ldots$. We have that

$$\lim_{i \to \infty} x_i = x_0 \qquad \lim_{i \to \infty} \xi_i = \xi_0,$$

where x_0 is an interior point of K. Therefore

$$\forall \epsilon > 0 \; \exists \, i_0 \; \forall \, i > i_0 :$$
$$\partial K \cap H^-(x_0 + \epsilon\xi_0, \xi_0) \subseteq \partial K \cap H^-(x_i, \xi_i) \subseteq \partial K \cap H^-(x_0 - \epsilon\xi_0, \xi_0).$$

This implies

$$\mathbb{P}_f\left(\partial K \cap H^-(x_0 + \epsilon\xi_0, \xi_0)\right) \leq \mathbb{P}_f\left(\partial K \cap H^-(x_i, \xi_i)\right)$$
$$\leq \mathbb{P}_f\left(\partial K \cap H^-(x_0 - \epsilon\xi_0, \xi_0)\right).$$

Since x_0 is an interior point of K, for ϵ small enough $x_0 - \epsilon\xi_0$ and $x_0 + \epsilon\xi_0$ are interior points of K. Therefore,

$$H(x_0 - \epsilon\xi_0, \xi_0) \quad \text{and} \quad H(x_0 + \epsilon\xi_0, \xi_0)$$

intersect the interior of K. The claim now follows from (i).

(iv) Suppose the inclusion is not true. Then there is $x \in \overset{\circ}{K}$ with $x \notin \bigcup_{0 < s} K_s$. Therefore, for every $s > 0$ there is a hyperplane H_s with $x \in H_s$ and

$$\mathbb{P}_f(\partial K \cap H_s^-) \leq s.$$

By compactness and by (iii) there is a hyperplane H with $x \in H$ and

$$\mathbb{P}_f(\partial K \cap H^-) = 0.$$

On the other hand, $\text{vol}_{n-1}(\partial K \cap H^-) > 0$ which implies

$$\mathbb{P}_f(\partial K \cap H^-) > 0$$

since f is a.e. positive.

We have $K = K_0$ because K_0 is a closed set that contains $\overset{\circ}{K}$. $\quad\square$

Lemma 2.2. *Let K be a convex body in $\mathbb{R}^n$ and let $f : \partial K \to \mathbb{R}$ be a a.e. positive, integrable function with $\int_{\partial K} f \, d\mu = 1$.*

(i) For all s such that $K_s \neq \emptyset$, and all $x \in \partial K_s \cap \overset{\circ}{K}$ there exists a supporting hyperplane H to ∂K_s through x such that $\mathbb{P}_f(\partial K \cap H^-) = s$.
(ii) Suppose that for all $x \in \partial K$ there is $R(x) < \infty$ so that

$$K \subseteq B_2^n(x - R(x)N_{\partial K}(x), R(x)).$$

Then we have for all $0 < s$ that $K_s \subset \overset{\circ}{K}$.

Proof. (i) There is a sequence of hyperplanes H_i with $K_s \subseteq H_i^+$ and $\mathbb{P}_f(\partial K \cap H_i^-) \leq s$ such that the distance between x and H_i is less than $\frac{1}{i}$. We check this.

Since $x \in \partial K_s$ there is $z \notin K_s$ with $\|x - z\| < \frac{1}{i}$. There is a hyperplane H_i separating z from K_s satisfying

$$\mathbb{P}_f(\partial K \cap H_i^-) \leq s \qquad \text{and} \qquad K_s \subseteq H_i^+.$$

We have

$$d(x, H_i) \leq \|x - z\| < \tfrac{1}{i}.$$

By compactness and by Lemma 2.1.(iii) there is a subsequence that converges to a hyperplane H with $x \in H$ and $\mathbb{P}_f(\partial K \cap H^-) \leq s$.

If $\mathbb{P}_f(\partial K \cap H^-) < s$ then we choose a hyperplane $\tilde{H}$ parallel to H such that $\mathbb{P}_f(\partial K \cap \tilde{H}^-) = s$. By Lemma 2.1.(i) there is such a hyperplane. Consequently, x is not an element of K_s. This is a contradiction.

(ii) Suppose there is $x \in \partial K$ with $x \in K_s$ and $0 < s$. By $K \subseteq B_2^n(x - R(x)N_{\partial K}(x), R(x))$ we get

$$\mathrm{vol}_{n-1}(\partial K \cap H(x, N_{\partial K}(x))) = 0.$$

By Lemma 2.1.(i) we can choose a hyperplane H parallel to $H(x, N_{\partial K}(x))$ that cuts off a set with $\mathbb{P}_f(\partial K \cap \tilde{H}^-) = s$. This means that $x \notin K_s$. $\square$

Lemma 2.3. *Let K be a convex body in $\mathbb{R}^n$ and let $f : \partial K \to \mathbb{R}$ be a a.e positive, integrable function with $\int_{\partial K} f \mathrm{d}\mu = 1$.*
(i) Let s_i, $i \in \mathbb{N}$, be a strictly increasing sequence of positive numbers with $\lim_{i \to \infty} s_i = s_0$. Then we have

$$K_{s_0} = \bigcap_{i=1}^{\infty} K_{s_i}.$$

(ii) There exists T with $0 < T \leq \frac{1}{2}$ such that K_T is nonempty and $\mathrm{vol}_n(K_T) = 0$ and $\mathrm{vol}_n(K_t) > 0$ for all $t < T$.

Proof. (i) Since we have for all $i \in \mathbb{N}$ that $K_{s_0} \subseteq K_{s_i}$, we get

$$K_{s_0} \subseteq \bigcap_{i=1}^{\infty} K_{s_i}.$$

We show now that both sets are in fact equal. Let us consider $x \notin K_{s_0}$. If $x \notin K$, then $x \notin \bigcap_{i=1}^{\infty} K_{s_i}$, as

$$K = K_0 \supseteq \bigcap_{i=1}^{\infty} K_{s_i}.$$

If $x \in K$ and $x \notin K_{s_0}$ then there is a hyperplane H with $x \in \overset{\circ}{H^-}$, $K_{s_0} \subseteq H^+$, and

$$\mathbb{P}_f(K \cap H^-) \leq s_0.$$

There is a hyperplane H_1 that is parallel to H and that contains x. There is another hyperplane H_2 that is parallel to both these hyperplanes and whose distance to H equals its distance to H_1. By Lemma 2.1.(ii) we get

$$0 \leq \mathbb{P}_f(\partial K \cap H_1^-) < \mathbb{P}_f(\partial K \cap H_2^-) < \mathbb{P}_f(\partial K \cap H^-) \leq s_0.$$

Let $s_0' = \mathbb{P}_f(\partial K \cap H_2^-)$. It follows that

$$x \notin \bigcap_{\mathbb{P}_f(H^- \cap \partial K) \leq s_0'} H^+ = K_{s_0'}.$$

Therefore $x \notin K_{s_i}$, for $s_i \geq s_0'$.

(ii) We put

$$T = \sup\{s | \text{vol}_n(K_s) > 0\}.$$

Since the sets K_s are compact, convex, nonempty sets,

$$\bigcap_{\text{vol}_n(K_s) > 0} K_s$$

is a compact, convex, nonempty set. On the other hand, by (i) we have

$$K_T = \bigcap_{s < T} K_s = \bigcap_{\text{vol}_n(K_s) > 0} K_s.$$

Now we show that $\text{vol}_n(K_T) = 0$. Suppose that $\text{vol}_n(K_T) > 0$. Then there is $x_0 \in \overset{\circ}{K}_T$. Let

$$t_0 = \inf\{\mathbb{P}_f(\partial K \cap H^-) | x_0 \in H\}.$$

Since we require that $x_0 \in H$ we have that $\mathbb{P}_f(\partial K \cap H^-)$ is only a function of the normal of H. By Lemma 2.1.(iii), $\mathbb{P}_f(\partial K \cap H^-)$ is a continuous function of the normal of H. By compactness this infimum is attained and there is H_0 with $x_0 \in H_0$ and

$$\mathbb{P}_f(\partial K \cap H_0^-) = t_0.$$

Moreover, $t_0 > T$. If not, then $K_T \subseteq H_0^+$ and $x_0 \in H_0$, which means that $x_0 \in \partial K_T$, contradicting the assumption that $x_0 \in \overset{\circ}{K}_T$.

Now we consider $K_{(1/2)(T+t_0)}$. We claim that x_0 is an interior point of this set and therefore

$$\text{vol}_n(K_{\frac{1}{2}(T+t_0)}) > 0,$$

contradicting the fact that T is the supremum of all t with

$$\text{vol}_n(K_t) > 0.$$

We verify now that x_0 is an interior point of $K_{(1/2)(T+t_0)}$. Suppose x_0 is not an interior point of this set. Then in every neighborhood of x_0 there is $x \notin K_{\frac{1}{2}(T+t_0)}$. Therefore for every $\epsilon > 0$ there is a hyperplane H_ϵ such that

$$\mathbb{P}_f(\partial K \cap H_\epsilon^-) \leq \tfrac{1}{2}(T + t_0), \qquad x \in H_\epsilon \qquad \text{and} \qquad \|x - x_0\| < \epsilon.$$

By Lemma 2.1.(iii) we conclude that there is a hyperplane H with $x_0 \in H$ and

$$\mathbb{P}_f(\partial K \cap H^-) \leq \tfrac{1}{2}(T + t_0).$$

But this contradicts the definition of t_0.   □

In the next lemma we need the Hausdorff distance d_H which for two convex bodies K and L in $\mathbb{R}^n$ is

$$d_H(K, L) = \max \left\{ \max_{x \in L} \min_{y \in K} \|x - y\|, \ \max_{y \in K} \min_{x \in L} \|x - y\| \right\}.$$

Lemma 2.4. *Let K be a convex body in $\mathbb{R}^n$ and let $f : \partial K \to \mathbb{R}$ be a positive, continuous function with $\int_{\partial K} f \mathrm{d}\mu = 1$.*
(i) Suppose that K has a C^1-boundary. Let s be such that $K_s \neq \emptyset$ and let $x \in \partial K_s \cap \overset{\circ}{K}$. Let H be a supporting hyperplane of K_s at x such that $\mathbb{P}_f(\partial K \cap H^-) = s$. Then x is the center of gravity of $\partial K \cap H$ with respect to the measure

$$\frac{f(y)\mu_{\partial K \cap H}(y)}{< N_{\partial K \cap H}(y), N_{\partial K}(y) >}$$

i.e.

$$x = \frac{\int_{\partial K \cap H} \frac{y f(y) \mathrm{d}\mu_{\partial K \cap H}(y)}{<N_{\partial K \cap H}(y), N_{\partial K}(y)>}}{\int_{\partial K \cap H} \frac{f(y) \mathrm{d}\mu_{\partial K \cap H}(y)}{<N_{\partial K \cap H}(y), N_{\partial K}(y)>}},$$

where $N_{\partial K}(y)$ is the unit outer normal to ∂K at y and $N_{\partial K \cap H}(y)$ is the unit outer normal to $\partial K \cap H$ at y in the plane H.
(ii) If K has a C^1-boundary and $K_s \subset \overset{\circ}{K}$, then K_s is strictly convex.
(iii) Suppose that K has a C^1-boundary and $K_T \subset \overset{\circ}{K}$. Then K_T consists of one point $\{x_T\}$ only. This holds in particular, if for every $x \in \partial K$ there are $r(x) > 0$ and $R(x) < \infty$ such that $B_2^n(x - r(x)N_{\partial K}(x), r(x)) \subseteq K \subseteq B_2^n(x - R(x)N_{\partial K}(x), R(x))$.
(iv) For all s with $0 \leq s < T$ and $\epsilon > 0$ there is $\delta > 0$ such that $d_H(K_s, K_{s+\delta}) < \epsilon$.

We call the point x_T of Lemma 2.4.(iii) the surface point. If K_T does not consist of one point only, then we define x_T to be the centroid of K_T.

Proof. (i) By Lemma 2.2.(i) there is a hyperplane H with $s = \mathbb{P}_f(\partial K \cap H^-)$. Let $\tilde{H}$ be another hyperplane passing through x and ϵ the angle between the two hyperplanes. Then we have

$$s = \mathbb{P}_f(\partial K \cap H^-) \leq \mathbb{P}_f(\partial K \cap \tilde{H}^-).$$

Let ξ be one of the two vectors in H with $\|\xi\| = 1$ that are orthogonal to $H \cap \tilde{H}$. Then

$$0 \leq \mathbb{P}_f(\partial K \cap \tilde{H}^-) - \mathbb{P}_f(\partial K \cap H^-)$$
$$= \int_{\partial K \cap H} \frac{<y-x,\xi> f(y)\tan\epsilon}{<N_{\partial K \cap H}(y), N_{\partial K}(y)>} d\mu_{\partial K \cap H}(y) + o(\epsilon).$$

We verify the latter equality. First observe that for $y \in \partial K \cap H$ the "height" is $<y-x,\xi> \tan\epsilon$. This follows from the following two graphics.

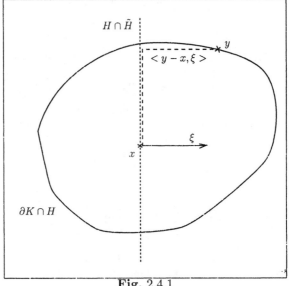

Fig. 2.4.1

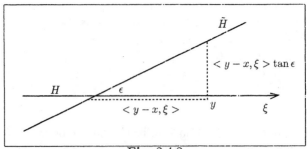

Fig. 2.4.2

A surface element at y equals, up to an error of order $o(\epsilon)$, the product of a volume element at y in $\partial K \cap H$ and the length of the tangential line segment between H and $\tilde{H}$ at y. The length of this tangential line segment is, up to an error of order $o(\epsilon)$,

$$\frac{< y - x, \xi > \tan \epsilon}{< N_{\partial K \cap H}(y), N_{\partial K}(y) >}.$$

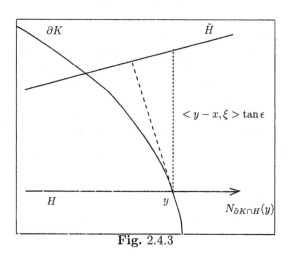

Fig. 2.4.3

Therefore,

$$0 \leq \int_{\partial K \cap H} \frac{< y - x, \xi > f(y) \tan \epsilon}{< N_{\partial K \cap H}(y), N_{\partial K}(y) >} d\mu_{\partial K \cap H}(y) + o(\epsilon).$$

We divide both sides by ϵ and pass to the limit for ϵ to 0. Thus we get for all ξ

$$0 \leq \int_{\partial K \cap H} \frac{< y - x, \xi > f(y)}{< N_{\partial K \cap H}(y), N_{\partial K}(y) >} d\mu_{\partial K \cap H}(y).$$

Since this inequality holds for ξ as well as $-\xi$ we get for all ξ

$$0 = \int_{\partial K \cap H} \frac{< y - x, \xi > f(y)}{< N_{\partial K \cap H}(y), N_{\partial K}(y) >} d\mu_{\partial K \cap H}(y)$$

or

$$0 = \left\langle \int_{\partial K \cap H} \frac{(y - x) f(y)}{< N_{\partial K \cap H}(y), N_{\partial K}(y) >} d\mu_{\partial K \cap H}(y), \xi \right\rangle.$$

Therefore,

$$x = \frac{\int_{\partial K \cap H} \frac{y f(y) d\mu_{\partial K \cap H}(y)}{< N_{\partial K \cap H}(y), N_{\partial K}(y) >}}{\int_{\partial K \cap H} \frac{f(y) d\mu_{\partial K \cap H}(y)}{< N_{\partial K \cap H}(y), N_{\partial K}(y) >}}.$$

(ii) Suppose that K_s is not strictly convex. Then ∂K_s contains a line-segment $[u,v]$. Let $x \in (u,v)$. As $K_s \subseteq \overset{\circ}{K}$ it follows from Lemma 2.2.(i) that there exists a support-hyperplane $H = H(x, N_{K_s}(x))$ of K_s such that $\mathbb{P}_f(\partial K \cap H^-) = s$. Moreover, we have that $u,v \in H$.

By (i)

$$x = u = v = \frac{\int_{\partial K \cap H} \frac{yf(y)d\mu_{\partial K \cap H}(y)}{<N_{\partial K \cap H}(y), N_{\partial K}(y)>}}{\int_{\partial K \cap H} \frac{f(y)d\mu_{\partial K \cap H}(y)}{<N_{\partial K \cap H}(y), N_{\partial K}(y)>}}.$$

(iii) By Lemma 2.3.(ii) there is T such that K_T has volume 0. Suppose that K_T consists of more than one point. All these points are elements of the boundary of K_T since the volume of K_T is 0. Therefore ∂K_T contains a line-segment $[u,v]$ and cannot be strictly convex, contradicting (ii).

The condition: For every $x \in \partial K$ there is $r(x) < \infty$ such that $K \supseteq B_2^n(x - r(x)N_{\partial K}(x), r(x))$, implies that K has everywhere unique normals. This is equivalent to differentiability of ∂K. By Corollary 25.5.1 of [Ro] ∂K is continuously differentiable. The remaining assertion of (iii) now follows from Lemma 2.2.(ii). $\quad\square$

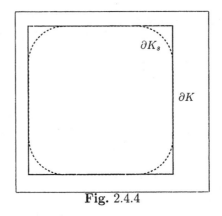

Fig. 2.4.4

We have the following remarks.

(i) The assertion of Lemma 2.2.(i) is not true if $x \in \partial K$. As an example consider the square S with sidelength 1 in $\mathbb{R}^2$ and $f(x) = \frac{1}{4}$ for all $x \in \partial S$. For $s = \frac{1}{16}$ the midpoints of the sides of the square are elements of $S_{1/16}$, but the tangent hyperplanes through these points contain one side and therefore cut off a set of $\mathbb{P}_f$-volume $\frac{1}{4}$ (compare Figure 2.4.4). The construction in higher dimensions for the cube is done in the same way. This example also shows that the surface body is not necessarily strictly convex and it shows that the assertion of Lemma 2.2.(ii) does not hold without additional assumptions.

(ii) If K is a symmetric convex body and f is symmetric (i.e. $f(x) = f(-x)$ if the center of symmetry is 0), then the surface point x_T coincides with the center of symmetry.

If K is not symmetric then $T < \frac{1}{2}$ is possible. An example for this is a regular triangle C in $\mathbb{R}^2$. If the sidelength is 1 and $f = \frac{1}{3}$, then $T = \frac{4}{9}$ and $C_{\frac{4}{9}}$ consists of the barycenter of C.

(iii) In Lemma 2.4 we have shown that under certain assumptions the surface body reduces to a point. In general this is not the case. We give an example. Let K be the Euclidean ball B_2^n and

$$f = \frac{\chi_C + \chi_{-C}}{2\mathrm{vol}_{n-1}(C)}$$

where C is a cap of the Euclidean ball with surface area equal to $\frac{1}{4}\mathrm{vol}_{n-1}(\partial B_2^n)$. Then we get that for all s with $s < \frac{1}{2}$ that K_s contains a Euclidean ball with positive radius. On the other hand $K_{1/2} = \emptyset$.

2.2 Surface Body and the Indicatrix of Dupin

The indicatrix of Dupin was introduced in section 1.5.

Lemma 2.5. *Let K be a convex body in $\mathbb{R}^n$ and let $f : \partial K \to \mathbb{R}$ be a a.e. positive, integrable function with $\int_{\partial K} f d\mu = 1$. Let $x_0 \in \partial K$. Suppose that the indicatrix of Dupin exists at x_0 and is an ellipsoid (and not a cylinder). For all s such that $K_s \neq \emptyset$, let the point x_s be defined by*

$$\{x_s\} = [x_T, x_0] \cap \partial K_s.$$

Then for every $\epsilon > 0$ there is s_ϵ so that for all s with $0 < s \leq s_\epsilon$ the points x_s are interior points of K and for all normals $N_{\partial K_s}(x_s)$ (if not unique)

$$< N_{\partial K}(x_0), N_{\partial K_s}(x_s) > \geq 1 - \epsilon.$$

If x_0 is an interior point of an $(n-1)$-dimensional face, then, as in the example of the cube, there is $s_0 > 0$ such that we have for all s with $0 \leq s \leq s_0$ that $x_0 \in \partial K_s$. Thus $x_s = x_0$.

Proof. Let us first observe that for all s with $0 < s < T$ where T is given by Lemma 2.3.(ii) the point x_s is an interior point of K. First we observe that $x_0 \neq x_T$ since the indicatrix of Dupin at x_0 is an ellipsoid. Again (see Figure 2.5.1), since the indicatrix of Dupin at x_0 is an ellipsoid, (x_T, x_0) is a subset of the convex hull of a cap contained in K and x_T. Thus $(x_T, x_0) \subset \overset{\circ}{K}$. Lemma 2.1. (i) assures that

$$\mathbb{P}_f(\partial K \cap H(x_0 - t N_{\partial K}(x_0), N_{\partial K}(x_0)))$$

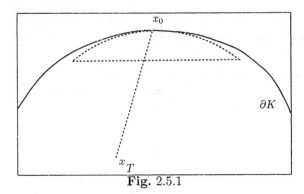

x_0

∂K

x_T

Fig. 2.5.1

is a continuous function on $[0, \max_{y \in K} < x_0 - y, N_{\partial K}(x_0) >)$.
We claim now

$$\forall \delta > 0 \exists s_\delta > 0 \forall s, 0 \le s \le s_\delta :< N_{\partial K}(x_0), N_{\partial K_s}(x_s) > \ge 1 - \delta.$$

Suppose that is not true. Then there is a sequence s_n, $n \in \mathbb{N}$, such that

$$\lim_{n \to \infty} s_n = 0 \qquad \lim_{n \to \infty} N_{\partial K_{s_n}}(x_{s_n}) = \xi$$

where $\xi \ne N_{\partial K}(x_0)$. By Lemma 2.1.(iv) $\lim_{n \to \infty} x_{s_n} = x_0$. Thus we get

$$\lim_{n \to \infty} s_n = 0 \qquad \lim_{n \to \infty} x_{s_n} = x_0 \qquad \lim_{n \to \infty} N_{\partial K_{s_n}}(x_{s_n}) = \xi.$$

Since the normal at x_0 is unique and $\xi \ne N_{\partial K}(x_0)$ the hyperplane $H(x_0, \xi)$
contains an interior point of K. There is $y \in \partial K$ and a supporting hyperplane
$H(y, \xi)$ to K at y that is parallel to $H(x_0, \xi)$. There is $\epsilon > 0$ and n_0 such
that for all n with $n \ge n_0$

$$B_2^n(y, \epsilon) \cap H^+(x_{s_n}, N_{\partial K_{s_n}}(x_{s_n})) = \emptyset.$$

Thus we get

$$B_2^n(y, \epsilon) \cap \bigcup_{n \ge n_0} K_{s_n} = \emptyset.$$

On the other hand, by Lemma 2.1.(iv) we have

$$\bigcup_{s > 0} K_s \supseteq \overset{\circ}{K}.$$

This is a contradiction. □

Lemma 2.6. *Let* $A : \mathbb{R}^n \to \mathbb{R}^n$ *be a diagonal matrix with* $a_i > 0$ *for all*
$i = 1, \ldots, n$. *Then we have for all* $x, y \in \mathbb{R}^n$ *with* $\|x\| = \|y\| = 1$

$$\left\| \frac{Ax}{\|Ax\|} - \frac{Ay}{\|Ay\|} \right\| \leq 2 \left(\frac{\max_{1 \leq i \leq n} a_i}{\min_{1 \leq i \leq n} a_i} \right) \|x - y\|.$$

In particular we have

$$1 - \left\langle \frac{Ax}{\|Ax\|}, \frac{Ay}{\|Ay\|} \right\rangle \leq 2 \left(\frac{\max_{1 \leq i \leq n} a_i}{\min_{1 \leq i \leq n} a_i} \right)^2 \|x - y\|^2.$$

Proof. We have

$$\|Ax - Ay\| \leq (\max_{1 \leq i \leq n} a_i) \|x - y\|$$

and

$$\begin{aligned}
\left\| \frac{Ax}{\|Ax\|} - \frac{Ay}{\|Ay\|} \right\| &\leq \left\| \frac{Ax}{\|Ax\|} - \frac{Ay}{\|Ax\|} \right\| + \left\| \frac{Ay}{\|Ax\|} - \frac{Ay}{\|Ay\|} \right\| \\
&\leq \frac{(\max_{1 \leq i \leq n} a_i)\|x - y\|}{\|Ax\|} + \frac{|\|Ax\| - \|Ay\||}{\|Ax\|\|Ay\|} \|Ay\| \\
&\leq 2 \frac{(\max_{1 \leq i \leq n} a_i)\|x - y\|}{\|Ax\|}.
\end{aligned}$$

Since $\|x\| = 1$ we have $\|Ax\| \geq \min_{1 \leq i \leq n} |a_i| \|x\|$. $\quad\square$

By Lemma 2.5 the normal to ∂K_s at x_s differs little from the normal to K at x_0 if s is small. Lemma 2.7 is a strengthening of this result.

Lemma 2.7. *Let K be a convex body in $\mathbb{R}^n$ and $x_0 \in \partial K$. Let $f : \partial K \to \mathbb{R}$ be an integrable, a.e. positive function with $\int_{\partial K} f \, d\mu = 1$ that is continuous at x_0. Suppose that the indicatrix of Dupin exists at x_0 and is an ellipsoid (and not a cylinder). For all s such that $K_s \neq \emptyset$, let x_s be defined by $\{x_s\} = [x_T, x_0] \cap \partial K_s$.*
(i) Then for every $\epsilon > 0$ there is s_ϵ so that for all s with $0 < s \leq s_\epsilon$ the points x_s are interior points of K and

$$s \leq \mathbb{P}_f(\partial K \cap H^-(x_s, N_{\partial K}(x_0))) \leq (1 + \epsilon)s.$$

(ii) Then for every $\epsilon > 0$ there is s_ϵ so that for all s with $0 < s \leq s_\epsilon$ and all normals $N_{\partial K_s}(x_s)$ at x_s

$$s \leq \mathbb{P}_f(\partial K \cap H^-(x_s, N_{\partial K_s}(x_s))) \leq (1 + \epsilon)s.$$

Proof. We position K so that $x_0 = 0$ and $N_{\partial K}(x_0) = e_n$. Let b_i, $i = 1, \ldots, n - 1$ be the lengths of the principal axes of the indicatrix of Dupin. Then, by Lemma 1.2 and (3) the lengths of the principal axes of the standard approximating ellipsoid $\mathcal{E}$ at x_0 are given by

$$a_i = b_i \left(\prod_{i=1}^{n-1} b_i \right)^{\frac{1}{n-1}} \qquad i = 1, \ldots, n-1 \qquad \text{and} \qquad a_n = \left(\prod_{i=1}^{n-1} b_i \right)^{\frac{2}{n-1}}.$$

We consider the transform $T : \mathbb{R}^n \to \mathbb{R}^n$ (5)

$$T(x) = \left(\frac{x_1}{a_1} \left(\prod_{i=1}^{n-1} b_i \right)^{\frac{2}{n-1}}, \ldots, \frac{x_{n-1}}{a_{n-1}} \left(\prod_{i=1}^{n-1} b_i \right)^{\frac{2}{n-1}}, x_n \right). \qquad (7)$$

This transforms the standard approximating ellipsoid into a Euclidean ball with radius $r = (\prod_{i=1}^{n-1} b_i)^{2/(n-1)}$. T is a diagonal map with diagonal elements $\frac{\sqrt{a_n}}{b_1}, \ldots, \frac{\sqrt{a_n}}{b_{n-1}}, 1$.

Let $\epsilon > 0$ be given. Let $\delta > 0$ be such that

$$\frac{(1+\delta)^{\frac{5}{2}}}{(1-\delta)(1-c^2\delta)^3} \leq 1 + \epsilon,$$

where

$$c = 2 \frac{\max\left\{ \max_{1 \leq i \leq n-1} \frac{b_i}{\sqrt{a_n}}, \ 1 \right\}}{\min\left\{ \min_{1 \leq i \leq n-1} \frac{b_i}{\sqrt{a_n}}, \ 1 \right\}}.$$

As f is continuous at x_0 there exists a neighborhood $B_2^n(x_0, \alpha)$ of x_0 such that for all $x \in B_2^n(x_0, \alpha) \cap \partial K$

$$f(x_0)\,(1-\delta) \leq f(x) \leq f(x_0)\,(1+\delta). \qquad (8)$$

By Lemma 2.5, for all $\rho > 0$ there exists $s(\rho)$ such that for all s with $0 < s \leq s(\rho)$

$$< N_{\partial K}(x_0), N_{\partial K_s}(x_s) >\, \geq 1 - \rho \qquad (9)$$

and the points x_s are interior points of K.

Therefore, for $\delta > 0$ given, it is possible to choose $s(\delta)$ such that for all s with $0 < s \leq s(\delta)$, $N_{\partial K}(x_0)$ and $N_{\partial K_s}(x_s)$ differ so little that both of the following hold

$$\partial K \cap H^-(x_s, N_{\partial K_s}(x_s)) \subseteq B_2^n(x_0, \alpha) \qquad (10)$$

and

$$< N_{\partial K}(x_0), N_{\partial K_s}(x_s) > \, \geq 1 - \delta. \qquad (11)$$

Indeed, in order to obtain (11) we have to choose ρ smaller than δ. In order to satisfy (10) we choose $s(\delta)$ so small that the distance of x_s to x_0 is less than one half of the height of the biggest cap of K with center x_0 that is contained in the set $K \cap B_2^n(x_0, \alpha)$. Now we choose ρ in (9) sufficiently small so that (10) holds.

As the points x_s are interior points of K, by Lemma 2.2.(i), for all s with $0 < s \leq s(\delta)$ there is $N_{\partial K_s}(x_s)$ such that

$$s = \mathbb{P}_f(\partial K \cap H(x_s, N_{\partial K_s}(x_s))). \tag{12}$$

Please note that

$$\frac{T^{-1t}(N_{\partial K_s}(x_s))}{\|T^{-1t}(N_{\partial K_s}(x_s))\|} \tag{13}$$

is the normal of the hyperplane

$$T(H(x_s, N_{\partial K_s}(x_s))).$$

We observe next that (9) implies that for all $\rho > 0$ there exists $s(\rho)$ such that for all $s \leq s(\rho)$

$$\left\langle N_{\partial K}(x_0), \frac{T^{-1t}(N_{\partial K_s}(x_s))}{\|T^{-1t}(N_{\partial K_s}(x_s))\|} \right\rangle \geq 1 - c^2 \rho, \tag{14}$$

where T^{-1t} is the transpose of the inverse of T and c the constant above.

Indeed, since

$$< N_{\partial K}(x_0), N_{\partial K_s}(x_s) > \; \geq 1 - \rho$$

we have

$$\|N_{\partial K}(x_0) - N_{\partial K_s}(x_s)\| \leq \sqrt{2\rho}.$$

Now we apply Lemma 2.6 to the map T^{-1t}. Since $N_{\partial K}(x_0) = e_n = T^{-1t}(e_n) = T^{-1t}(N_{\partial K}(x_0))$ we obtain with

$$c = 2 \frac{\max\{\max_{1 \leq i \leq n-1} \frac{b_i}{\sqrt{a_n}}, \; 1\}}{\min\{\min_{1 \leq i \leq n-1} \frac{b_i}{\sqrt{a_n}}, \; 1\}}$$

that

$$\left\| N_{\partial K}(x_0) - \frac{T^{-1t}(N_{\partial K_s}(x_s))}{\|T^{-1t}(N_{\partial K_s}(x_s))\|} \right\| \leq c\sqrt{2\rho}$$

which is the same as

$$1 - c^2 \rho \leq \left\langle N_{\partial K}(x_0), \frac{T^{-1t}(N_{\partial K_s}(x_s))}{\|T^{-1t}(N_{\partial K_s}(x_s))\|} \right\rangle.$$

By Lemma 1.4, for δ given there exists t_1 such that for all t with $t \leq t_1$

$$\mathrm{vol}_{n-1}(K \cap H(x_0 - t\, N_{\partial K}(x_0), N_{\partial K}(x_0)))$$
$$\leq \mathrm{vol}_{n-1}(\partial K \cap H^-(x_0 - t\, N_{\partial K}(x_0), N_{\partial K}(x_0))) \tag{15}$$
$$\leq (1 + \delta) \sqrt{1 + \frac{2t a_n^3}{(a_n - t)^2 \min_{1 \leq i \leq n-1} a_i^2}}$$
$$\times \mathrm{vol}_{n-1}(K \cap H(x_0 - t\, N(x_0), N(x_0))).$$

Recall that r is the radius of the approximating Euclidean ball for $T(K)$ at $x_0 = 0$. For δ given, we choose $\eta = \eta(\delta)$ such that

$$\eta < \min\left\{r\,\frac{1-(1-c^2\delta)^{\frac{2}{n-1}}}{1+(1-c^2\delta)^{\frac{2}{n-1}}},\ \delta\right\}. \tag{16}$$

Then, for such an η, by Lemma 1.2, there is $t_2 > 0$ so that we have for all t with $0 \leq t \leq t_2$

$$B_2^n(x_0 - (r-\eta)N_{\partial K}(x_0), r-\eta) \cap T(H(x_0 - t\,N_{\partial K}(x_0), N_{\partial K}(x_0)))$$
$$\subseteq T(K) \cap T(H(x_0 - t\,N_{\partial K}(x_0), N_{\partial K}(x_0))) \tag{17}$$
$$\subseteq B_2^n(x_0 - (r+\eta)N_{\partial K}(x_0), r+\eta) \cap T(H(x_0 - t\,N_{\partial K}(x_0), N_{\partial K}(x_0))).$$

Let $t_0 = \min\{t_1, t_2\}$.

By (14) we can choose $s(\eta)$ such that for all $s \leq s(\eta)$, $N_{\partial K}(x_0)$ and the normal to $T(H(x_s, N_{\partial K_s}(x_s)))$ differ so little that both of the following hold

$$\left\langle N_{\partial K}(x_0),\ \frac{T^{-1t}(N_{\partial K_s}(x_s))}{\|T^{-1t}(N_{\partial K_s}(x_s))\|}\right\rangle \geq 1 - c^2\eta \geq 1 - c^2\delta \tag{18}$$

and

$$\min\{y_n | y = (y_1, \ldots, y_n) \in T(H(x_s, N_{\partial K_s}(x_s))) \tag{19}$$
$$\cap B_2^n(x_0 - (r-\eta)N_{\partial K}(x_0), r-\eta)\} \geq -t_0.$$

Then we get by (17) for all s with $0 < s \leq s(\eta)$

$$B_2^n(x_0 - (r-\eta)N_{\partial K}(x_0), r-\eta) \cap T(H(x_s, N_{\partial K_s}(x_s)))$$
$$\subseteq T(K) \cap T(H(x_s, N_{\partial K_s}(x_s))) \tag{20}$$
$$\subseteq B_2^n(x_0 - (r+\eta)N_{\partial K}(x_0), r+\eta) \cap T(H(x_s, N_{\partial K_s}(x_s))).$$

The set on the left hand side of (20) is a $(n-1)$-dimensional Euclidean ball whose radius is greater or equal

$$\sqrt{2(r-\eta)h_s - h_s^2} \tag{21}$$

where h_s is the distance of $T(x_s)$ to the boundary of the Euclidean ball $B_2^n(x_0 - (r-\eta)N_{\partial K}(x_0), r-\eta)$. See Figure 2.7.1. The height of the cap

$$K \cap H^-(x_s, N_{\partial K}(x_0))$$

is denoted by Δ_s. It is also the height of the cap

$$K \cap H^-(T(x_s), N_{\partial K}(x_0))$$

because T does not change the last coordinate. Let θ be the angle between $x_0 - T(x_T)$ and $N_{\partial K}(x_0)$. Then we have by the Pythagorean theorem

$$((r - \eta) - h_s)^2 = ((r - \eta) - \Delta_s)^2 + (\Delta_s \tan \theta)^2$$

and consequently

$$h_s = (r - \eta) \left[1 - \sqrt{\left(1 - \frac{\Delta_s}{r - \eta}\right)^2 + \left(\frac{\Delta_s \tan \theta}{r - \eta}\right)^2} \right].$$

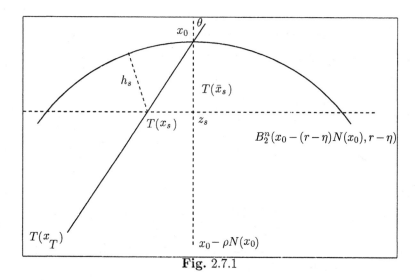

Fig. 2.7.1

x_0 and $T'(x_s)$ are in the plane that can be seen in Figure 2.7.1. We use now $\sqrt{1-t} \le 1 - \frac{1}{2}t$ to get that

$$h_s \ge \Delta_s - \frac{1}{2} \frac{\Delta_s^2}{r - \eta} \left(1 + \tan^2 \theta\right). \tag{22}$$

Now we prove (i). The inequality

$$s \le \mathbb{P}_f(\partial K \cap H^-(x_s, N_{\partial K}(x_0)))$$

holds because H passes through x_s. We show the right hand inequality. Let ϵ, δ and η be as above. We choose s_δ such that

1. $s_\delta \le \min \{s(\delta), s(\eta)\}$

2. $\Delta_{s_\delta} \le \min \left\{ t_0, \frac{a_n}{2}, (r - \eta), \frac{a_n^2 \delta}{8 \min_{1 \le i \le (n-1)} b_i}, \frac{4c^2 \delta (r - \eta)}{(n-1)(1 + \tan^2 \theta)}, \right.$

$$\left. 2 \left(r - \eta \frac{1 + (1 - c^2\delta)^{\frac{2}{n-1}}}{1 - (1 - c^2\delta)^{\frac{2}{n-1}}} \right) \right\}.$$

We have for all $s \leq s_\delta$

$$\text{vol}_{n-1}(\partial K \cap H^-(x_s, N_{\partial K_s}(x_s))) \geq \text{vol}_{n-1}(K \cap H(x_s, N_{\partial K_s}(x_s))).$$

Now note that

$$\text{vol}_{n-1}(K \cap H(x_s, N_{\partial K_s}(x_s))) = \frac{\text{vol}_{n-1}(p_{e_n}(K \cap H(x_s, N_{\partial K_s}(x_s))))}{< N_{\partial K}(x_0), N_{\partial K_s}(x_s) >} \quad (23)$$

$$\geq \text{vol}_{n-1}(p_{e_n}(K \cap H(x_s, N_{\partial K_s}(x_s)))) \quad (24)$$

where p_{e_n} is the orthogonal projection onto the first $n-1$ coordinates.

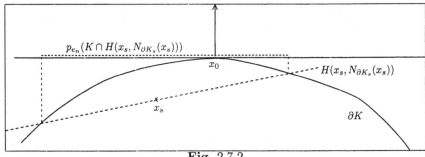

Fig. 2.7.2

Since $T \circ p_{e_n} = p_{e_n} \circ T$ and since T is volume preserving in hyperplanes that are orthogonal to e_n we get

$$\text{vol}_{n-1}(\partial K \cap H^-(x_s, N_{\partial K_s}(x_s)))$$
$$\geq \text{vol}_{n-1}(p_{e_n}(T(K) \cap T(H(x_s, N_{\partial K_s}(x_s)))))$$
$$= \left\langle N_{\partial K}(x_0), \frac{T^{-1t}(N_{\partial K_s}(x_s))}{\|T^{-1t}(N_{\partial K_s}(x_s))\|} \right\rangle \text{vol}_{n-1}(T(K) \cap T(H(x_s, N_{\partial K_s}(x_s)))).$$

The last equality follows from (13) and (23). By (18) we then get that the latter is greater than or equal to

$$(1 - c^2\delta)\, \text{vol}_{n-1}(T(K) \cap T(H(x_s, N_{\partial K_s}(x_s)))),$$

which, in turn, by (20) and (21) is greater than or equal to

$$(1 - c^2\delta)\text{vol}_{n-1}(B_2^{n-1})\, \left(2(r - \eta)h_s - h_s^2\right)^{\frac{n-1}{2}}.$$

By (22) and as the function $\left(2(r - \eta)\Delta - \Delta^2\right)^{\frac{n-1}{2}}$ is increasing in Δ for $\Delta \leq r - \eta$, the latter is greater or equal

$$(1 - c^2\delta)\text{vol}_{n-1}(B_2^{n-1}) \left(1 - \frac{(1 + \tan^2\theta)\Delta_s}{2(r - \eta)}\right)^{\frac{n-1}{2}} \left(2(r - \eta)\Delta_s - \Delta_s^2\right)^{\frac{n-1}{2}}. \quad (25)$$

In the last inequality we have also used that $(1 - \frac{(1+\tan^2\theta)\Delta_s}{2(r-\eta)})^{\frac{n-1}{2}} \le 1$. $\Delta_s \le \frac{4c^2\delta(r-\eta)}{(n-1)(1+\tan^2\theta)}$ implies that

$$\left(1 - \frac{(1+\tan^2\theta)\Delta_s}{2(r-\eta)}\right)^{\frac{n-1}{2}} \ge \left(1 - \frac{2c^2}{n-1}\delta\right)^{\frac{n-1}{2}} \ge 1 - c^2\delta.$$

$\Delta_s \le 2\left(r - \eta\frac{1+(1-c^2\delta)^{\frac{2}{n-1}}}{1-(1-c^2\delta)^{\frac{2}{n-1}}}\right)$ implies that

$$2(r-\eta) - 2(r+\eta)(1-c^2\delta)^{\frac{2}{n-1}} \ge \Delta_s(1 - (1-c^2\delta)^{\frac{2}{n-1}})$$

which is equivalent to

$$(2(r-\eta) - \Delta_s) \ge (1-c^2\delta)^{\frac{2}{n-1}}(2(r+\eta) - \Delta_s)$$

and

$$\left(2(r-\eta)\Delta_s - \Delta_s^2\right)^{\frac{n-1}{2}} \ge (1-c^2\delta)\left(2(r+\eta)\Delta_s - \Delta_s^2\right)^{\frac{n-1}{2}}.$$

Hence we get for all $s \le s_\delta$ that (25) is greater than

$$(1-c^2\delta)^3\mathrm{vol}_{n-1}(B_2^{n-1})\left(2(r+\eta)\Delta_s - \Delta_s^2\right)^{\frac{n-1}{2}}$$
$$= (1-c^2\delta)^3\mathrm{vol}_{n-1}(B_2^n(x_0 - (r+\eta)N_{\partial K}(x_0), r+\eta)$$
$$\cap H(x_0 - \Delta_s N_{\partial K}(x_0), N_{\partial K}(x_0)))$$
$$= (1-c^2\delta)^3\mathrm{vol}_{n-1}(B_2^n(x_0 - (r+\eta)N_{\partial K}(x_0), r+\eta)$$
$$\cap T(H(x_0 - \Delta_s N_{\partial K}(x_0), N_{\partial K}(x_0)))),$$

as T does not change the last coordinate. By (17) the latter is greater than

$$(1-c^2\delta)^3\mathrm{vol}_{n-1}(T(K) \cap T(H(x_0 - \Delta_s N_{\partial K}(x_0), N_{\partial K}(x_0)))$$
$$= (1-c^2\delta)^3\mathrm{vol}_{n-1}(K \cap H(x_0 - \Delta_s N_{\partial K}(x_0), N_{\partial K}(x_0)))$$
$$\ge \frac{(1-c^2\delta)^3}{1+\delta}\frac{\mathrm{vol}_{n-1}(\partial K \cap H^-(x_0 - \Delta_s N_{\partial K}(x_0), N_{\partial K}(x_0)))}{\left(1 + \frac{2\Delta_s a_n^3}{(a_n-\Delta_s)^2 \min_{1\le i\le(n-1)} a_i^2}\right)^{\frac{1}{2}}}$$
$$\ge \frac{(1-c^2\delta)^3}{(1+\delta)^{\frac{3}{2}}}\mathrm{vol}_{n-1}(\partial K \cap H^-(x_0 - \Delta_s N_{\partial K}(x_0), N_{\partial K}(x_0))).$$

The second last inequality follows with (15) and the last inequality follows as $\Delta_s \le \frac{a_n^2\delta}{8 \min_{1\le i\le(n-1)} b_i}$.
Therefore we get altogether that

$$\mathrm{vol}_{n-1}(\partial K \cap H^-(x_s, N_{\partial K_s}(x_s))) \tag{26}$$
$$\ge \frac{(1-c^2\delta)^3}{(1+\delta)^{\frac{3}{2}}}\mathrm{vol}_{n-1}(\partial K \cap H^-(x_0 - \Delta_s N_{\partial K}(x_0), N_{\partial K}(x_0))).$$

Hence, by (12)

$$s = \mathbb{P}_f(\partial K \cap H^-(x_s, N_{\partial K_s}(x_s))) = \int_{\partial K \cap H^-(x_s, N_{\partial K_s}(x_s))} f(x)d\mu.$$

By (8)

$$s \geq (1 - \delta)f(x_0)\mathrm{vol}_{n-1}(\partial K \cap H^-(x_s, N_{\partial K_s}(x_s))).$$

By (26)

$$s \geq \frac{(1 - \delta)(1 - c^2\delta)^3}{(1+\delta)^{\frac{3}{2}}} f(x_0)\mathrm{vol}_{n-1}(\partial K \cap H^-(x_0 - \Delta_s N_{\partial K}(x_0), N_{\partial K}(x_0)))).$$

By (8) and (10)

$$s \geq \frac{(1 - \delta)(1 - c^2\delta)^3}{(1+\delta)^{\frac{5}{2}}} \int_{\partial K \cap H^-(x_0 - \Delta_s N_{\partial K}(x_0), N_{\partial K}(x_0)))} f(x)d\mu$$

$$= \frac{(1 - \delta)(1 - c^2\delta)^3}{(1+\delta)^{\frac{5}{2}}} \mathbb{P}_f(\partial K \cap H^-(x_0 - \Delta_s N_{\partial K}(x_0), N_{\partial K}(x_0)))).$$

For ϵ given, we choose now $s_\epsilon = s_\delta$. By our choice of δ, this finishes (i).

(ii) We assume that the assertion is not true. Then

$$\exists \epsilon > 0 \forall s_\epsilon > 0 \exists s, 0 < s < s_\epsilon \exists N_{\partial K_s}(x_s) : \mathbb{P}_f(\partial K \cap H(x_s, N_{\partial K_s}(x_s))) \geq (1+\epsilon)s.$$

We consider $y_s \in H(x_s, N_{\partial K_s}(x_s))$ such that $T(y_s)$ is the center of the $n-1$-dimensional Euclidean ball

$$B_2^n(x_0 - (r - \eta)N(x_0), r - \eta) \cap T(H(x_s, N_{\partial K_s}(x_s))).$$

Since $y_s \in H(x_s, N_{\partial K_s}(x_s))$ we have $y_s \notin \overset{\circ}{K}_s$. Consequently, by the definition of K_s there is a hyperplane H such that $y_s \in H$ and $\mathbb{P}_f(\partial K \cap H^-) \leq s$.

On the other hand, we shall show that for all hyperplanes H with $y_s \in H$ we have $\mathbb{P}_f(\partial K \cap H^-) > s$ which gives a contradiction.

We choose δ as in the proof of (i) and moreover so small that $\epsilon > 10\delta$ and s_δ small enough so that the two following estimates hold.

$$(1 + \epsilon)s \leq \mathbb{P}_f(\partial K \cap H^-(x_s, N_{\partial K_s}(x_s)))$$
$$\leq (1 + \delta)f(x_0)\mathrm{vol}_{n-1}(\partial K \cap H^-(x_s, N_{\partial K_s}(x_s)))$$

We verify this. As f is continuous at x_0, for all $\delta > 0$ there exists α such that for all $x \in B_2^n(x_0, \alpha) \cap \partial K$

$$(1 - \delta)f(x_0) \leq f(x) \leq (1 + \delta)f(x_0).$$

By Lemma 2.5, for all $\rho > 0$ there is s_ρ such that for all s with $0 < s \leq s_\rho$

$$< N_{\partial K}(x_0), N_{\partial K_s}(x_s) > \geq 1 - \rho.$$

Moreover, the indicatrix at x_0 exists and is an ellipsoid. Therefore we can choose s_ρ sufficiently small so that for all s with $0 < s \leq s_\rho$

$$\partial K \cap H^-(x_s, N_{\partial K_s}(x_s)) \subseteq B_2^n(x_0, \alpha).$$

Thus there is s_δ such that for all s with $0 < s \leq s_\delta$

$$\mathbb{P}_f(\partial K \cap H^-(x_s, N_{\partial K_s}(x_s))) = \int_{\partial K \cap H^-(x_s, N_{\partial K_s}(x_s))} f(x) d\mu(x)$$
$$\leq (1 + \delta) f(x_0) \mathrm{vol}_{n-1}(\partial K \cap H^-(x_s, N_{\partial K_s}(x_s))).$$

Thus

$$(1 + \epsilon)s \leq (1 + \delta) f(x_0) \mathrm{vol}_{n-1}(\partial K \cap H^-(x_s, N_{\partial K_s}(x_s))).$$

Since the indicatrix at x_0 exists and is an ellipsoid for all ρ there is s_ρ such that for all $x \in \partial K \cap H^-(x_s, N_{\partial K_s}(x_s))$

$$< N_{\partial K}(x), N_{\partial K_s}(x_s) > \geq 1 - \rho.$$

Therefore

$$(1 + \epsilon)s \leq (1 + 2\delta) f(x_0) \mathrm{vol}_{n-1}(K \cap H(x_s, N_{\partial K_s}(x_s)))$$

which by (23) equals

$$(1 + 2\delta) f(x_0) \frac{\mathrm{vol}_{n-1}(p_{e_n}(K \cap H(x_s, N_{\partial K_s}(x_s))))}{< N_{\partial K}(x_0), N_{\partial K_s}(x_s) >}.$$

By Lemma 2.5 for all s with $0 < s \leq s_\delta$

$$(1 + \epsilon)s \leq (1 + 3\delta) f(x_0) \mathrm{vol}_{n-1}(p_{e_n}(K \cap H(x_s, N_{\partial K_s}(x_s)))).$$

Since $T \circ p_{e_n} = p_{e_n} \circ T$ and since T is volume preserving in hyperplanes that are orthogonal to e_n we get

$$(1 + \epsilon)s \leq (1 + 3\delta) f(x_0) \mathrm{vol}_{n-1}(p_{e_n}(T(K) \cap T(H(x_s, N_{\partial K_s}(x_s))))).$$

Since

$$T(K) \cap T(H(x_s, N_{\partial K_s}(x_s))))$$
$$\subseteq B_2^n(x_0 - (r + \eta)N_{\partial K}(x_0), r + \eta) \cap T(H(x_s, N_{\partial K_s}(x_s))))$$

we get

$$(1 + \epsilon)s$$
$$\leq (1 + 3\delta) f(x_0) \mathrm{vol}_{n-1}(p_{e_n}(B_2^n(x_0 - (r + \eta)N_{\partial K}(x_0), r + \eta))$$
$$\cap T(H(x_s, N_{\partial K_s}(x_s))))$$

and thus

$$(1+\epsilon)s$$
$$\leq (1+4\delta)f(x_0)\mathrm{vol}_{n-1}(p_{e_n}(B_2^n(x_0-(r-\eta)N_{\partial K}(x_0),r-\eta)$$
$$\cap T(H(x_s,N_{\partial K_s}(x_s)))))).$$

Since $T(y_s)$ is the center of

$$B_2^n(x_0-(r-\eta)N_{\partial K}(x_0),r-\eta)\cap T(H(x_s,N_{\partial K_s}(x_s))))$$

we have for all hyperplanes H with $y_s\in H$

$$(1+\epsilon)s$$
$$\leq (1+4\delta)f(x_0)\mathrm{vol}_{n-1}(p_{e_n}(B_2^n(x_0-(r-\eta)N_{\partial K}(x_0),r-\eta)\cap T(H)).$$

Thus we get for all hyperplanes H with $y_s\in H$ and

$$B_2^n(x_0-(r-\eta)N_{\partial K}(x_0),r-\eta)\cap T(H)\subseteq T(K)\cap T(H)$$

that

$$(1+\epsilon)s\leq (1+5\delta)\mathbb{P}_f(\partial K\cap H^-).$$

Please note that $\epsilon>10\delta$. We can choose s_δ so small that we have for all s with $0<s\leq s_\delta$ and all hyperplanes H with $y_s\in H$ and

$$B_2^n(x_0-(r-\eta)N_{\partial K}(x_0),r-\eta)\cap T(H)\not\subseteq T(K)\cap T(H)$$

that

$$s<\mathbb{P}_f(\partial K\cap H^-).$$

Thus we have $s<\mathbb{P}_f(\partial K\cap H^-)$ for all H which is a contradiction.   □

Lemma 2.8. *Let K be a convex body in $\mathbb{R}^n$ and $x_0\in\partial K$. Suppose that the indicatrix of Dupin at x_0 exists and is an ellipsoid. Let $f:\partial K\to\mathbb{R}$ be a a.e. positive, integrable function with $\int f d\mu=1$ that is continuous at x_0. Let $\mathcal{E}$ be the standard approximating ellipsoid at x_0. For $0\leq s\leq T$ let x_s be given by*

$$\{x_s\}=[x_T,x_0]\cap\partial K_s$$

and $\bar{x}_s$ by

$$\{\bar{x}_s\}=H(x_s,N_{\partial K_s}(x_s))\cap\{x_0+tN_{\partial K}(x_0)|t\in\mathbb{R}\}.$$

The map $\Phi:\partial K\cap H(x_s,N_{\partial K_s}(x_s))\to\partial\mathcal{E}\cap H(x_s,N_{\partial K_s}(x_s))$ is defined by

$$\{\Phi(y)\}=\partial\mathcal{E}\cap\{\bar{x}_s+t(y-x_s)|t\geq 0\}.$$

Then, for every $\epsilon > 0$ there is s_ϵ such that we have for all s with $0 < s < s_\epsilon$ and all $z \in \partial\mathcal{E} \cap H(x_s, N_{\partial K_s}(x_s))$

$$\left| \frac{1}{\sqrt{1 - <N_{\partial\mathcal{E}}(z), N_{\partial K_s}(x_s)>^2}} - \frac{1}{\sqrt{1 - <N_{\partial K}(\Phi^{-1}(z)), N_{\partial K_s}(x_s)>^2}} \right|$$
$$\leq \frac{\epsilon}{\sqrt{1 - <N_{\partial\mathcal{E}}(z), N_{\partial K_s}(x_s)>^2}}.$$

Proof. During this proof several times we choose the number s_ϵ sufficiently small in order to assure certain properties. Overall, we take the minimum of all these numbers.

Note that $\bar{x}_s \in K$ and by Lemma 2.7.(i) x_s is an interior point of K for s with $0 < s \leq s_\epsilon$. Therefore the angles between any of the normals are strictly larger than 0 and the expressions are well-defined.

Let z_s be given by

$$\{z_s\} = \{x_0 + tN_{\partial K}(x_0)|t \in \mathbb{R}\} \cap H(x_s, N_{\partial K}(x_0)).$$

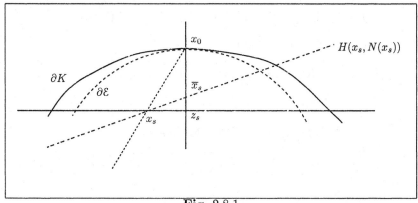

Fig. 2.8.1

In Figure 2.8.1 we see the plane through x_0 spanned by $N_{\partial K}(x_0)$ and $N_{\partial K_s}(x_s)$. The point x_s is not necessarily in this plane, but z_s is. The point x_s is contained in the intersection of the planes $H(x_s, N_{\partial K_s}(x_s))$ and $H(x_s, N_{\partial K}(x_0))$.

As in the proof of Lemma 2.7 let b_i, $i = 1, \ldots, n-1$ be the lenghts of the principal axes of the indicatrix of Dupin. Then, by Lemma 1.2 and by (3) in the standard approximating ellipsoid $\mathcal{E}$ at x_0 the lengths of the principal axes are given by

$$a_i = b_i \left(\prod_{i=1}^{n-1} b_i \right)^{\frac{1}{n-1}} \quad i = 1, \ldots, n-1 \quad \text{and} \quad a_n = \left(\prod_{i=1}^{n-1} b_i \right)^{\frac{2}{n-1}}.$$

We can assume that $x_0 = 0$ and $N_{\partial K}(x_0) = e_n$. The standard approximating ellipsoid $\mathcal{E}$ is centered at $x_0 - a_n N_{\partial K}(x_0)$ and given by

$$\sum_{i=1}^{n-1} \left| \frac{x_i}{a_i} \right|^2 + \left| \frac{x_n}{a_n} + 1 \right|^2 \leq 1.$$

We consider the transform $T : \mathbb{R}^n \to \mathbb{R}^n$

$$T(x) = \left(\frac{x_1}{a_1} \left(\prod_{i=1}^{n-1} b_i \right)^{\frac{2}{n-1}} , \ldots, \frac{x_{n-1}}{a_{n-1}} \left(\prod_{i=1}^{n-1} b_i \right)^{\frac{2}{n-1}} , x_n \right).$$

See (5) and (7). This transforms the ellipsoid into a Euclidean sphere with radius $\rho = \left(\prod_{i=1}^{n-1} b_i \right)^{\frac{2}{n-1}}$, i.e.

$$T(\mathcal{E}) = B_2^n \left((0, \ldots, 0, -\rho), \rho \right).$$

Let $\delta > 0$ be given. Then there exists s_δ such that for all s with $0 < s \leq s_\delta$ and all normals $N_{\partial K_s}(x_s)$ at x_s (the normal may not be unique)

$$f(x_0) \operatorname{vol}_{n-1}(T(\mathcal{E}) \cap T(H(x_s, N_{\partial K_s}(x_s)))) \leq (1+\delta)s. \qquad (27)$$

Indeed, by Lemma 2.7.(ii) we have

$$\mathbb{P}_f(\partial K \cap H^-(x_s, N_{\partial K_s}(x_s))) \leq (1+\delta)s.$$

Now

$$(1+\delta)s \geq \mathbb{P}_f(\partial K \cap H^-(x_s, N_{\partial K_s}(x_s)))$$
$$= \int_{\partial K \cap H^-(x_s, N_{\partial K_s}(x_s))} f(x) \mathrm{d}\mu_{\partial K}(x).$$

By continuity of f at x_0

$$(1+\delta)^2 s \geq f(x_0)\operatorname{vol}_{n-1}(\partial K \cap H^-(x_s, N_{\partial K_s}(x_s)))$$
$$\geq f(x_0)\operatorname{vol}_{n-1}(K \cap H(x_s, N_{\partial K_s}(x_s))).$$

We have $N_{\partial K}(x_0) = e_n$. By (23) we see that the latter equals

$$f(x_0) \frac{\operatorname{vol}_{n-1}(p_{e_n}(K \cap H(x_s, N_{\partial K_s}(x_s))))}{< N_{\partial K}(x_0), N_{\partial K_s}(x_s) >}.$$

Since $< N_{\partial K}(x_0), N_{\partial K_s}(x_s) > \leq 1$

$$(1+\delta)^2 s \geq f(x_0)\operatorname{vol}_{n-1}(p_{e_n}(K \cap H(x_s, N_{\partial K_s}(x_s)))).$$

Since T is volume preserving in all hyperplanes orthogonal to $N_{\partial K}(x_0)$

$$(1 + \delta)^2 s \geq f(x_0)\text{vol}_{n-1}(T(p_{e_n}(K \cap H(x_s, N_{\partial K_s}(x_s))))).$$

Since $T \circ p_{e_n} = p_{e_n} \circ T$

$$(1 + \delta)^2 s \geq f(x_0)\text{vol}_{n-1}(p_{e_n}(T(K) \cap T(H(x_s, N_{\partial K_s}(x_s)))))$$

$$= f(x_0)\left\langle N_{\partial K}(x_0), \frac{T^{-1t}(N_{\partial K_s}(x_s))}{\|T^{-1t}(N_{\partial K_s}(x_s))\|} \right\rangle$$

$$\times \text{vol}_{n-1}(T(K) \cap T(H(x_s, N_{\partial K_s}(x_s)))).$$

The latter equality follows since $e_n = N_{\partial K}(x_0)$. As in the proof of Lemma 2.7. (i) we get

$$(1 + \delta)^3 s \geq f(x_0)\text{vol}_{n-1}(T(K) \cap T(H(x_s, N_{\partial K_s}(x_s)))).$$

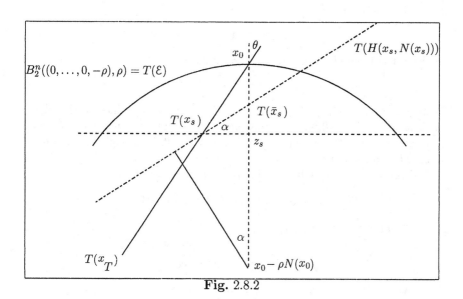

Fig. 2.8.2

$T(\mathcal{E})$ approximates $T(K)$ well as $\mathcal{E}$ approximates K well. By Lemma 2.5 we have $< N_{\partial K}(x_0), N_{\partial K_s}(x_s) >\,\geq 1 - \delta$. This and Lemma 1.2 give

$$(1 + \delta)^4 s \geq f(x_0)\text{vol}_{n-1}(T(\mathcal{E}) \cap T(H(x_s, N_{\partial K_s}(x_s)))).$$

Now we pass to a new δ and establish (27).

$\bar{x}_s$ is the point where the plane $H(x_s, N_{\partial K_s}(x_s))$ and the line through x_0 with direction $N_{\partial K}(x_0)$ intersect.

$$\{\bar{x}_s\} = H(x_s, N_{\partial K_s}(x_s)) \cap \{x_0 + tN_{\partial K}(x_0)|t \in \mathbb{R}\}$$

In Figure 2.8.2 we see the plane through x_0 spanned by the vectors $N_{\partial K}(x_0)$ and $T^{-1t}(N_{\partial K_s}(x_s))$. The point z_s is also contained in this plane. The line through x_0, $T(x_s)$, and $T(x_{T'})$ is not necessarily in this plane. We see only its projection onto this plane. Also the angle θ is not necessarily measured in this plane. θ is measured in the plane spanned by $N_{\partial K}(x_0)$ and $x_0 - T(x_{T'})$.

α is the angle between the hyperplanes

$$T(H(x_s, N_{\partial K}(x_s))) \quad \text{and} \quad H(z_s, N_{\partial K}(x_0)).$$

Please observe that $\bar{x}_s = T(\bar{x}_s)$, $z_s = T(z_s)$ and that the plane

$$T(H(x_s, N_{\partial K_s}(x_s)))$$

is orthogonal to $T^{-1t}(N_{\partial K_s}(x_s))$.

We observe that for small enough s_δ we have for s with $0 < s \le s_\delta$

$$\|x_0 - \bar{x}_s\| \ge (1 - \delta)\|x_0 - z_s\| \tag{28}$$

which is the same as

$$\|x_0 - T(\bar{x}_s)\| \ge (1 - \delta)\|x_0 - z_s\|.$$

We check the inequality. Figure 2.8.2 gives us that

$$\|\bar{x}_s - z_s\| \le \tan\theta \tan\alpha \|x_0 - z_s\|.$$

We would have equality here if the angle θ would be contained in the plane that is seen in Figure 2.8.2. The angle θ is fixed, but we can make sure that the angle α is arbitrarily small. By Lemma 2.5 it is enough to choose s_δ sufficiently small. Thus (28) is established.

By Figure 2.8.2 the radius of the $n - 1$-dimensional ball

$$B_2^n(x_0 - \rho N_{\partial K}(x_0), \rho) \cap T(H(x_s, N_{\partial K_s}(x_s)))$$

with $\rho = \left(\prod_{i=1}^{n-1} b_i\right)^{\frac{2}{n-1}}$ equals

$$\sqrt{\rho^2 - (\rho - \|x_0 - \bar{x}_s\|)^2 \cos^2\alpha}$$

which by (28) is greater than or equal to

$$\sqrt{\rho^2 - (\rho - (1 - \delta)\|x_0 - z_s\|)^2 \cos^2\alpha}$$

$$= \sqrt{\rho^2 - (\rho - (1 - \delta)\|x_0 - z_s\|)^2 \left\langle N_{\partial K}(x_0), \frac{T^{-1t}(N_{\partial K_s}(x_s))}{\|T^{-1t}(N_{\partial K_s}(x_s))\|} \right\rangle^2}.$$

By (27) we get with a new δ

$$\left[\rho^2 - (\rho - (1 - \delta)\|x_0 - z_s\|)^2 \left\langle N_{\partial K}(x_0), \frac{T^{-1t}(N_{\partial K_s}(x_s))}{\|T^{-1t}(N_{\partial K_s}(x_s))\|} \right\rangle^2\right]^{\frac{n-1}{2}}$$

$$\times \mathrm{vol}_{n-1}(B_2^{n-1})$$

$$\le \mathrm{vol}_{n-1}(T(\mathcal{E}) \cap H(T(x_s), T^{-1t}(N_{\partial K_s}(x_s)))) \le \frac{(1 + \delta)s}{f(x_0)}. \tag{29}$$

On the other hand,

$$s \le \mathbb{P}_f(\partial K \cap H^-(x_s, N_{\partial K}(x_0))) = \mathbb{P}_f(\partial K \cap H^-(z_s, N_{\partial K}(x_0)))$$
$$= \int_{\partial K \cap H^-(z_s, N_{\partial K}(x_0))} f(x) \mathrm{d}\mu(x).$$

Now we use the continuity of f at x_0 and Lemma 1.4 to estimate the latter.

$$s \le (1+\delta) f(x_0) \mathrm{vol}_{n-1}(K \cap H(z_s, N_{\partial K}(x_0)))$$

As above we use that T is volume-preserving in hyperplanes orthogonal to $N_{\partial K}(x_0)$. Note that $T(H(z_s, N_{\partial K}(x_0))) = H(z_s, N_{\partial K}(x_0))$.

$$s \le (1+\delta) f(x_0) \mathrm{vol}_{n-1}(T(K) \cap H(z_s, N_{\partial K}(x_0)))$$

Since $T(\mathcal{E})$ approximates $T(K)$ well (Lemma 1.2)

$$s \le (1+\delta)^2 f(x_0) \mathrm{vol}_{n-1}(T(\mathcal{E}) \cap H(z_s, N_{\partial K}(x_0))).$$

Therefore (29) is less than

$$(1+\delta)^3 \mathrm{vol}_{n-1}(T(\mathcal{E}) \cap H(z_s, N_{\partial K}(x_0)))$$
$$= (1+\delta)^3 (\rho^2 - (\rho - \|x_0 - z_s\|)^2)^{\frac{n-1}{2}} \mathrm{vol}_{n-1}(B_2^{n-1})$$
$$= (1+\delta)^3 (2\rho\|x_0 - z_s\| - \|x_0 - z_s\|^2)^{\frac{n-1}{2}} \mathrm{vol}_{n-1}(B_2^{n-1}).$$

From this we get

$$\rho^2 - (\rho - (1-\delta)\|x_0 - z_s\|)^2 \left\langle N_{\partial K}(x_0), \frac{T^{-1t}(N_{\partial K_s}(x_s))}{\|T^{-1t}(N_{\partial K_s}(x_s))\|} \right\rangle^2$$
$$\le (1+\delta)^{\frac{6}{n-1}} (2\rho\|x_0 - z_s\| - \|x_0 - z_s\|^2)$$

which gives us

$$(\rho - (1-\delta)\|x_0 - z_s\|)^2 \left(1 - \left\langle N_{\partial K}(x_0), \frac{T^{-1t}(N_{\partial K_s}(x_s))}{\|T^{-1t}(N_{\partial K_s}(x_s))\|} \right\rangle^2 \right)$$
$$\le (1+\delta)^{\frac{6}{n-1}} (2\rho\|x_0 - z_s\| - \|x_0 - z_s\|^2)$$
$$-2(1-\delta)\rho\|x_0 - z_s\| + (1-\delta)^2\|x_0 - z_s\|^2.$$

This is less than $c\delta\rho\|x_0 - z_s\|$ where c is a numerical constant. Thus we have

$$1 - \left\langle N_{\partial K}(x_0), \frac{T^{-1t}(N_{\partial K_s}(x_s))}{\|T^{-1t}(N_{\partial K_s}(x_s))\|} \right\rangle^2 \le c\delta \frac{\rho\|x_0 - z_s\|}{(\rho - \|x_0 - z_s\|)^2}.$$

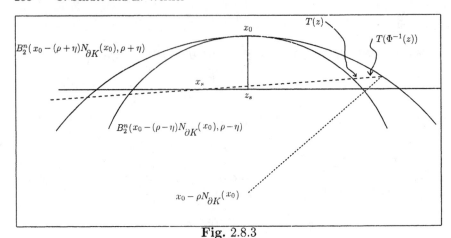

Fig. 2.8.3

If we choose s_δ sufficiently small we get for all s with $0 < s \leq s_\delta$

$$1 - \left\langle N_{\partial K}(x_0), \frac{T^{-1t}(N_{\partial K_s}(x_s))}{\|T^{-1t}(N_{\partial K_s}(x_s))\|} \right\rangle^2 \leq \delta\|x_0 - z_s\|. \qquad (30)$$

This is equivalent to

$$1 - \left\langle N_{\partial K}(x_0), \frac{T^{-1t}(N_{\partial K_s}(x_s))}{\|T^{-1t}(N_{\partial K_s}(x_s))\|} \right\rangle \leq \delta\|x_0 - z_s\| \qquad (31)$$

which is the same as

$$\left\| N_{\partial K}(x_0) - \frac{T^{-1t}(N_{\partial K_s}(x_s))}{\|T^{-1t}(N_{\partial K_s}(x_s))\|} \right\| \leq \sqrt{2\delta\|x_0 - z_s\|}. \qquad (32)$$

Now we show that for every $\epsilon > 0$ there is s_ϵ such that we have for all s with $0 < s \leq s_\epsilon$

$$\|N_{\partial K}(\Phi^{-1}(z)) - N_{\partial \mathcal{E}}(z)\| \leq \epsilon\sqrt{\|x_0 - z_s\|}. \qquad (33)$$

By Lemma 2.6 it is enough to show

$$\left\| \frac{T^{-1t}(N_{\partial K}(\Phi^{-1}(z)))}{\|T^{-1t}(N_{\partial K}(\Phi^{-1}(z)))\|} - \frac{T^{-1t}(N_{\partial \mathcal{E}}(z))}{\|T^{-1t}(N_{\partial \mathcal{E}}(z))\|} \right\| \leq \epsilon\sqrt{\|x_0 - z_s\|}.$$

T transforms the approximating ellipsoid $\mathcal{E}$ into the Euclidean ball $T(\mathcal{E}) = B_2^n(x_0 - \rho N_{\partial K}(x_0), \rho)$. We have

$$N_{\partial TK}(T(\Phi^{-1}(z))) = \frac{T^{-1t}(N_{\partial K}(\Phi^{-1}(z)))}{\|T^{-1t}(N_{\partial K}(\Phi^{-1}(z)))\|}$$

and

$$N_{\partial T\mathcal{E}}(T(z)) = \frac{T^{-1t}(N_{\partial\mathcal{E}}(z))}{\|T^{-1t}(N_{\partial\mathcal{E}}(z))\|}.$$

Therefore, the above inequality is equivalent to

$$\|N_{\partial TK}(T(\Phi^{-1}(z))) - N_{\partial T\mathcal{E}}(T(z))\| \le \epsilon\sqrt{\|x_0 - z_s\|}.$$

$T(z)$ and $T(\Phi^{-1}(z)))$ are elements of the hyperplane $T(H(x_s, N_{\partial K_s}(x_s)))$ that is orthogonal to $T^{-1t}(N_{\partial K_s}(x_s))$. We want to verify now this inequality. It follows from Lemma 1.2 that for every η there is a δ so that

$$B_2^n(x_0 - (\rho - \eta)N_{\partial K}(x_0), \rho - \eta) \cap H^-(x_0 - \delta N_{\partial K}(x_0), N_{\partial K}(x_0))$$
$$\subseteq T(K) \cap H^-(x_0 - \delta N_{\partial K}(x_0), N_{\partial K}(x_0)) \tag{34}$$
$$\subseteq B_2^n(x_0 - (\rho + \eta)N_{\partial K}(x_0), \rho + \eta) \cap H^-(x_0 - \delta N_{\partial K}(x_0), N_{\partial K}(x_0)).$$

For s_η sufficiently small we get for all s with $0 < s \le s_\eta$

$$T(H^-(x_s, N_{\partial K_s}(x_s))) \cap B_2^n(x_0 - (\rho + \eta)N_{\partial K}(x_0), \rho + \eta)$$
$$\subseteq H^-(x_0 - 2\|x_0 - z_s\|N_{\partial K}(x_0), N_{\partial K}(x_0)) \tag{35}$$
$$\cap B_2^n(x_0 - (\rho + \eta)N_{\partial K}(x_0), \rho + \eta).$$

We verify this. By (30) the angle β between the vectors

$$N_{\partial K}(x_0) \qquad \text{and} \qquad \frac{T^{-1t}(N_{\partial K_s}(x_s))}{\|T^{-1t}(N_{\partial K_s}(x_s))\|}$$

satisfies $\sin^2\beta \le \delta\|x_0 - z_s\|$. In case (35) does not hold we have

$$\tan\beta \ge \frac{1}{4}\sqrt{\frac{\|x_0 - z_s\|}{\rho + \eta}}.$$

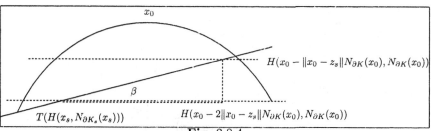

Fig. 2.8.4

This is true since $T(H(x_s, N_{\partial K_s}(x_s)))$ intersects the two hyperplanes $H(x_0 - \|x_0 - z_s\|N_{\partial K}(x_0), N_{\partial K}(x_0))$ and $H(x_0 - 2\|x_0 - z_s\|N_{\partial K}(x_0), N_{\partial K}(x_0))$. Compare Figure 2.8.4. This is impossible if we choose δ sufficiently small.

Let s_η be such that (35) holds. The distance of $T(\Phi^{-1}(z)))$ to the boundary of $B_2^n(x_0 - (\rho - \eta)N_{\partial K}(x_0), \rho - \eta)$ is less than $\frac{4\eta}{\rho - \eta}\|x_0 - z_s\|$. We check this. $T(\Phi^{-1}(z)))$ is contained in $B_2^n(x_0 - (\rho + \eta)N_{\partial K}(x_0), \rho + \eta)$ but not in $B_2^n(x_0 - (\rho - \eta)N_{\partial K}(x_0), \rho - \eta)$. See Figure 2.8.5.

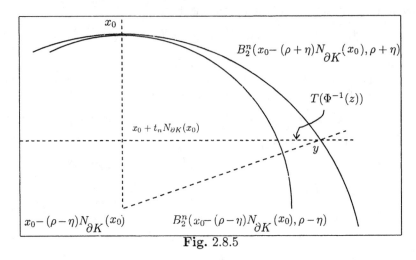

Fig. 2.8.5

Let t_n denote the n-th coordinate of $T(\Phi^{-1}(z)))$. By Figure 2.8.5 we get

$$\|(x_0 - (\rho - \eta)N_{\partial K}(x_0)) - y\|^2$$
$$= (\rho - \eta - |t_n|)^2 + (2|t_n|(\rho + \eta) - t_n^2)$$
$$= (\rho - \eta)^2 + 4\eta|t_n|.$$

Thus the distance of $T(\Phi^{-1}(z)))$ to the boundary of

$$B_2^n(x_0 - (\rho - \eta)N_{\partial K}(x_0), \rho - \eta)$$

is less than

$$\|(x_0 - (\rho - \eta)N_{\partial K}(x_0)) - y\| - (\rho - \eta)$$
$$= \sqrt{(\rho - \eta)^2 + 4\eta|t_n|} - (\rho - \eta)$$
$$= (\rho - \eta)\left\{ \sqrt{1 + \frac{4\eta|t_n|}{(\rho - \eta)^2}} - 1 \right\}$$
$$\le (\rho - \eta)\frac{2\eta|t_n|}{(\rho - \eta)^2} = \frac{2\eta|t_n|}{\rho - \eta}.$$

By (35) we have $|t_n| \le 2\|x_0 - z_s\|$. Thus we get

$$\|(x_0 - (\rho - \eta)N_{\partial K}(x_0)) - y\| - (\rho - \eta) \le \frac{4\eta\|x_0 - z_s\|}{\rho - \eta}.$$

Thus the distance of $T(\Phi^{-1}(z)))$ to the boundary of

$$B_2^n(x_0 - (\rho - \eta)N_{\partial K}(x_0), \rho - \eta)$$

is less than

$$\frac{4\eta}{\rho - \eta}\|x_0 - z_s\|. \tag{36}$$

By (34)

$$B_2^n(x_0 - (\rho - \eta)N_{\partial K}(x_0), \rho - \eta) \cap H^-(x_0 - \delta N_{\partial K}(x_0), N_{\partial K}(x_0))$$
$$\subseteq T(K) \cap H^-(x_0 - \delta N_{\partial K}(x_0), N_{\partial K}(x_0)).$$

Therefore a supporting hyperplane of $\partial T(K)$ at $T(\Phi^{-1}(z)))$ cannot intersect

$$B_2^n(x_0 - (\rho - \eta)N_{\partial K}(x_0), \rho - \eta) \cap H^-(x_0 - \delta N_{\partial K}(x_0), N_{\partial K}(x_0)).$$

Therefore, if we choose s_ϵ small enough a supporting hyperplane of $\partial T(K)$ at $T(\Phi^{-1}(z)))$ cannot intersect

$$B_2^n(x_0 - (\rho - \eta)N_{\partial K}(x_0), \rho - \eta).$$

We consider now a supporting hyperplane of $B_2^n(x_0 - (\rho - \eta)N_{\partial K}(x_0), \rho - \eta)$ that is parallel to $T(H(\Phi^{-1}(z), N_{\partial K}(\Phi^{-1}(z))))$. Let w be the contact point of this supporting hyperplane and $B_2^n(x_0 - (\rho - \eta)N_{\partial K}(x_0), \rho - \eta)$.

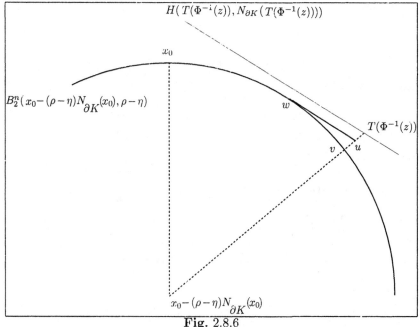

Fig. 2.8.6

Thus the hyperplane is $H(w, N_{\partial K}(\Phi^{-1}(z))))$ and

$$N_{\partial B_2^n(x_0-(\rho-\eta)N_{\partial K}(x_0),\rho-\eta)}(w) = N_{\partial TK}(T(\Phi^{-1}(z))). \qquad (37)$$

We introduce two points $v \in \partial B_2^n(x_0 - (\rho - \eta)N_{\partial K}(x_0), \rho - \eta)$ and u.

$$v = x_0 - (\rho-\eta)N_{\partial K}(x_0) + (\rho-\eta)\frac{T(\Phi^{-1}(z))) - (x_0 - (\rho-\eta)N_{\partial K}(x_0))}{\|T(\Phi^{-1}(z))) - (x_0 - (\rho-\eta)N_{\partial K}(x_0))\|}$$

$$\{u\} = [x_0 - (\rho-\eta)N_{\partial K}(x_0), T(\Phi^{-1}(z))] \cap H(w, T^{-1t}(N_{\partial K}(\Phi^{-1}(z))))$$

We claim that

$$\|w - u\| \le \epsilon\sqrt{\|x_0 - z_s\|}.$$

We check this inequality. By the Pythagorean theorem (see Figure 2.8.6)

$$\|w - u\| = \sqrt{\|u - (x_0 - (\rho-\eta)N_{\partial K}(x_0))\|^2 - (\rho-\eta)^2}.$$

By (36) the distance $\|T(\Phi^{-1}(z))) - v\|$ of $T(\Phi^{-1}(z)))$ to the boundary of $B_2^n(x_0 - (\rho - \eta)N_{\partial K}(x_0), \rho - \eta)$ is less than $\frac{4\eta}{\rho-\eta}\|x - z_s\|$. Since $\|v - u\| \le \|v - T(\Phi^{-1}(z)))\|$ we get with $\epsilon = \frac{4\eta}{\rho-\eta}$

$$\|w - u\| \le \sqrt{(\rho-\eta+\epsilon\|x_0 - z_s\|)^2 - (\rho-\eta)^2}$$
$$\doteq \sqrt{2\epsilon\rho\|x_0 - z_s\|) + (\epsilon\|x_0 - z_s\|)^2}.$$

This implies

$$\|w - u\| \le \epsilon\sqrt{\|x_0 - z_s\|}$$

and also

$$\|w - v\| \le \epsilon\sqrt{\|x_0 - z_s\|}.$$

Since

$$N(w) = N_{\partial B_2^n(x_0-(\rho-\eta)N_{\partial K}(x_0),\rho-\eta)}(w)$$
$$N(v) = N_{\partial B_2^n(x_0-(\rho-\eta)N_{\partial K}(x_0),\rho-\eta)}(v)$$

we get

$$\|N(w) - N(v)\| = \frac{\|w - v\|}{\rho-\eta} \le \epsilon\cdot\frac{\sqrt{\|x_0 - z_s\|}}{\rho-\eta}.$$

Since $N(w) = N_{\partial K}(T(\Phi^{-1}(z))))$ we get

$$\|N_{\partial T(K)}(T(\Phi^{-1}(z)))) - N(v)\| \le \epsilon\frac{\sqrt{\|x_0 - z_s\|}}{\rho-\eta}.$$

We observe that

$$\|v - T(z)\| \le \frac{\epsilon}{\rho}\sqrt{\|x_0 - z_s\|}.$$

This is done as above. Both points are located between the two Euclidean balls $B_2^n(x_0 - (\rho - \eta)N_{\partial K}(x_0), \rho - \eta)$ and $B_2^n(x_0 - (\rho + \eta)N_{\partial K}(x_0), \rho + \eta)$.

The line passing through both points also intersects both balls and thus the distance between both points must be smaller than $\frac{\epsilon}{\rho}\sqrt{\|x_0 - z_s\|}$.

From this we conclude in the same way as we have done for $N(v)$ and $N_{\partial K}(T(\Phi^{-1}(z))))$ that we have with a new ϵ

$$\|N(v) - N_{\partial T\mathcal{E}}(T(z))\| \leq \frac{\epsilon}{\rho}\sqrt{\|x_0 - z_s\|}.$$

Therefore we get by triangle inequality

$$\|N_{\partial TK}(T(\Phi^{-1}(z)))) - N_{\partial T\mathcal{E}}(T(z))\| \leq \frac{\epsilon}{\rho}\sqrt{\|x_0 - z_s\|}$$

and thus finally the claimed inequality (33) with a new ϵ

$$\|N_{\partial K}(\Phi^{-1}(z)) - N_{\partial \mathcal{E}}(z)\| \leq \epsilon\sqrt{\|x_0 - z_s\|}.$$

Now we show

$$1- <N_{\partial K}(\Phi^{-1}(z)), N_{\partial K_s}(x_s)>^2 \geq c\|x_0 - z_s\|. \tag{38}$$

For all s with $0 < s \leq s_\epsilon$ the distance of $T(x_s)$ to the boundary of $T\mathcal{E} = B_2^n(x_0 - \rho N_{\partial K}(x_0), \rho)$ is larger than $c\|x_0 - z_s\|$. Thus the height of the cap

$$T\mathcal{E} \cap H^-(x_s, N_{\partial K_s}(x_s))$$

is larger than $c\|x_0 - z_s\|$. The radius of the cap is greater than $\sqrt{2c\rho\|x_0 - z_s\|}$. By Figure 2.8.2 there is a c such that we have for all s with $0 < s \leq s_\eta$

$$\|T(x_s) - x_0\| \leq c\|x_0 - z_s\|.$$

By triangle inequality we get with a new c

$$\|x_0 - T(z)\| \geq c\sqrt{\rho\|x_0 - z_s\|}.$$

We have

$$N_{\partial T\mathcal{E}}(T(z)) = \tfrac{1}{\rho}(T(z)) - (x_0 - \rho N_{\partial K}(x_0))).$$

We get

$$
\begin{aligned}
c\sqrt{\rho\|x_0 - z_s\|} &\leq \|x_0 - T(z)\| \\
&= \|\rho N_{\partial K}(x_0) - (T(z) - (x_0 - \rho N_{\partial K}(x_0))))\| \\
&= \rho\|N_{\partial K}(x_0) - N_{\partial T\mathcal{E}}(T(z))\|.
\end{aligned}
$$

Since $T(N_{\partial K}(x_0))) = N_{\partial K}(x_0)$ we get by Lemma 2.6 with a new c

$$c\sqrt{\|x_0 - z_s\|} \leq \|N_{\partial K}(x_0) - N_{\partial \mathcal{E}}(z))\|.$$

We have by (32) and Lemma 2.6

$$\|N_{\partial K}(x_0) - N_{\partial K_s}(x_s)\| \leq \delta\sqrt{\|x_0 - z_s\|}. \tag{39}$$

Now we get by triangle inequality

$$c\sqrt{\|x_0 - z_s\|} \leq \|N_{\partial K_s}(x_s) - N_{\partial \mathcal{E}}(z))\|.$$

By (33) and triangle inequality we get

$$c\sqrt{\|x_0 - z_s\|} \leq \|N_{\partial K_s}(x_s) - N_{\partial \mathcal{E}}(\Phi^{-1}(z))\|.$$

Therefore we get with a new constant c

$$c\|x_0 - z_s\| \leq 1 - < N_{\partial K_s}(x_s), N_{\partial K}(\Phi^{-1}(z)) >$$
$$\leq 1 - < N_{\partial K_s}(x_s), N_{\partial K}(\Phi^{-1}(z)) >^2.$$

We have

$$| < N_{\partial K}(\Phi^{-1}(z)), N_{\partial K_s}(x_s) >^2 - < N_{\partial \mathcal{E}}(z), N_{\partial K_s}(x_s) >^2 |$$
$$= | < N_{\partial K}(\Phi^{-1}(z)) + N_{\partial \mathcal{E}}(z), N_{\partial K_s}(x_s) > \times$$
$$< N_{\partial K}(\Phi^{-1}(z)) - N_{\partial \mathcal{E}}(z), N_{\partial K_s}(x_s) > |$$
$$\leq 2| < N_{\partial K}(\Phi^{-1}(z)) - N_{\partial \mathcal{E}}(z), N_{\partial K_s}(x_s) > |$$
$$\leq 2| < N_{\partial K}(\Phi^{-1}(z)) - N_{\partial \mathcal{E}}(z), N_{\partial K_s}(x_s) - N_{\partial \mathcal{E}}(z) > |$$
$$+ 2| < N_{\partial K}(\Phi^{-1}(z)) - N_{\partial \mathcal{E}}(z), N_{\partial \mathcal{E}}(z) > |$$
$$\leq 2\|N_{\partial K}(\Phi^{-1}(z)) - N_{\partial \mathcal{E}}(z)\| \, \|N_{\partial K_s}(x_s) - N_{\partial \mathcal{E}}(z)\|$$
$$+ 2|1 - < N_{\partial K}(\Phi^{-1}(z)), N_{\partial \mathcal{E}}(z) > |.$$

By (33)

$$\|N_{\partial K}(\Phi^{-1}(z)) - N_{\partial \mathcal{E}}(z)\| \leq \epsilon\sqrt{\|x_0 - z_s\|}$$

which is the same as

$$1 - < N_{\partial K}(\Phi^{-1}(z)), N_{\partial \mathcal{E}}(z) > \leq \tfrac{1}{2}\epsilon^2\|x - z_s\|.$$

We get

$$| < N_{\partial K}(\Phi^{-1}(z)), N_{\partial K_s}(x_s) >^2 - < N_{\partial \mathcal{E}}(z), N_{\partial K_s}(x_s) >^2 | \tag{40}$$
$$\leq 2\epsilon\sqrt{\|x_0 - z_s\|} \, \|N_{\partial K_s}(x_s) - N_{\partial \mathcal{E}}(z)\| + \epsilon^2\|x_0 - z_s\|.$$

We show

$$\|N_{\partial K_s}(x_s) - N_{\partial \mathcal{E}}(z)\| \leq c\sqrt{\|x_0 - z_s\|}. \tag{41}$$

By (35) we have

$$\|N_{\partial T K_s}(T x_s) - N_{\partial T \mathcal{E}}(T z)\| \leq c\sqrt{\|x_0 - z_s\|}.$$

(41) follows now from this and Lemma 2.6. (40) and (41) give now

$$| < N_{\partial K}(\Phi^{-1}(z)), N_{\partial K_s}(x_s) >^2 \; - \; < N_{\partial \mathcal{E}}(z), N_{\partial K_s}(x_s) >^2 |$$
$$\leq 2\epsilon\sqrt{\|x_0 - z_s\|}\sqrt{\|x_0 - z_s\|} + \epsilon^2\|x_0 - z_s\| \leq 3\epsilon\|x_0 - z_s\|.$$

With this we get

$$\left| \frac{1}{\sqrt{1- < N_{\partial \mathcal{E}}(z), N_{\partial K_s}(x_s) >^2}} - \frac{1}{\sqrt{1- < N_{\partial K}(\Phi^{-1}(z)), N_{\partial K_s}(x_s) >^2}} \right|$$

$$= \frac{\left| \sqrt{1- < N_{\partial K}(\Phi^{-1}(z)), N_{\partial K_s}(x_s) >^2} - \sqrt{1- < N_{\partial \mathcal{E}}(z), N_{\partial K_s}(x_s) >^2} \right|}{\sqrt{1- < N_{\partial \mathcal{E}}(z), N_{\partial K_s}(x_s) >^2}\sqrt{1- < N_{\partial K}(\Phi^{-1}(z)), N_{\partial K_s}(x_s) >^2}}$$

$$\leq \frac{\left| < N_{\partial K}(\Phi^{-1}(z)), N_{\partial K_s}(x_s) >^2 - < N_{\partial \mathcal{E}}(z), N_{\partial K_s}(x_s) >^2 \right|}{\sqrt{1- < N_{\partial \mathcal{E}}(z), N_{\partial K_s}(x_s) >^2}\,(1- < N_{\partial K}(\Phi^{-1}(z)), N_{\partial K_s}(x_s) >^2)}$$

$$\leq \frac{1}{\sqrt{1- < N_{\partial \mathcal{E}}(z), N_{\partial K_s}(x_s) >^2}} \frac{3\epsilon\|x_0 - z_s\|}{(1- < N_{\partial K}(\Phi^{-1}(z)), N_{\partial K_s}(x_s) >^2)}.$$

By (38) we have that $1- < N_{\partial K}(\Phi^{-1}(z)), N_{\partial K_s}(x_s) >^2 \geq c\|x_0 - z_s\|$. Therefore we get

$$\left| \frac{1}{\sqrt{1- < N_{\partial \mathcal{E}}(z), N_{\partial K_s}(x_s) >^2}} - \frac{1}{\sqrt{1- < N_{\partial K}(\Phi^{-1}(z)), N_{\partial K_s}(x_s) >^2}} \right|$$

$$\leq \frac{3\epsilon}{c\sqrt{1- < N_{\partial \mathcal{E}}(z), N_{\partial K_s}(x_s) >^2}}.$$

$\square$

Lemma 2.9. *Let K be a convex body in $\mathbb{R}^n$ and $x_0 \in \partial K$. Suppose that the indicatrix of Dupin at x_0 exists and is an ellipsoid. Let $f : \partial K \to \mathbb{R}$ be a integrable, a.e. positive function with $\int f d\mu = 1$ that is continuous at x_0. Let $\bar{x}_s$ and Φ be as given in Lemma 2.8 and z_s as given in the proof of Lemma 2.8.*
(i) For every ϵ there is s_ϵ so that we have for all s with $0 < s \leq s_\epsilon$

$$(1 - \epsilon) \sup_{y \in \partial K \cap H(x_s, N_{\partial K_s}(x_s))} | < N_{\partial K}(x_0), y - x_0 > |$$
$$\leq \|x_0 - z_s\|$$
$$\leq (1 + \epsilon) \inf_{y \in \partial K \cap H(x_s, N_{\partial K_s}(x_s))} | < N_{\partial K}(x_0), y - x_0 > |.$$

(ii) For every ϵ there is s_ϵ so that we have for all s with $0 < s \leq s_\epsilon$ and all $z \in \partial \mathcal{E} \cap H(x_s, N_{\partial K_s}(x_0))$

$$(1 - \epsilon) < N_{\partial K \cap H}(\Phi^{-1}(z)), z - x_s >$$
$$\leq < N_{\partial \mathcal{E} \cap H}(z), z - x_s >$$
$$\leq (1 + \epsilon) < N_{\partial K \cap H}(\Phi^{-1}(z))), z - x_s >$$

where $H = H(x_s, N_{\partial K_s}(x_s))$ and the normals are taken in the plane H.
(iii) Let $\phi : \partial K \cap H \to \mathbb{R}$ be the real valued, positive function such that

$$\Phi(y) = \bar{x}_s + \phi(y)(y - \bar{x}_s).$$

For every ϵ there is s_ϵ such that we have for all s with $0 < s \leq s_\epsilon$ and all $y \in \partial K \cap H(x_s, N_{\partial K_s}(x_s))$

$$1 - \epsilon \leq \phi(y) \leq 1 + \epsilon.$$

Proof. We may suppose that $x_0 = 0$ and $N_{\partial K}(x_0) = e_n$.
(i) We put

$$m_s = \inf_{y \in \partial K \cap H(x_s, N_{\partial K_s}(x_s))} | < N_{\partial K}(x_0), y - x_0 > |.$$

We show now the right hand inequality. Let ρ be strictly greater than all the lengths of the principal axes of the standard approximating ellipsoid $\mathcal{E}$. Then there is $\eta > 0$

$$\mathcal{E} \cap H(x_0 - \eta N_{\partial K}(x_0), N_{\partial K}(x_0))$$
$$\subseteq B_2^n(x_0 - \rho N_{\partial K}(x_0), \rho) \cap H(x_0 - \eta N_{\partial K}(x_0), N_{\partial K}(x_0)).$$

Let α_s denote the angle between $N_{\partial K}(x_0)$ and $N_{\partial K_s}(x_s)$. Recall that in the proof of Lemma 2.8 we put

$$\{z_s\} = \{x_0 + t N_{\partial K}(x_0) | t \in \mathbb{R}\} \cap H(x_s, N_{\partial K}(x_0)).$$

Then we have

$$\tan \alpha_s \geq \frac{\|x_0 - z_s\| - m_s}{c\|x_0 - z_s\| + \sqrt{\rho^2 - (\rho - \|x_0 - z_s\|)^2}}$$
$$\geq \frac{\|x_0 - z_s\| - m_s}{c\|x_0 - z_s\| + \sqrt{2\rho\|x_0 - z_s\| - \|x_0 - z_s\|^2}}$$
$$\geq \frac{\|x_0 - z_s\| - m_s}{c\|x_0 - z_s\| + \sqrt{2\rho\|x_0 - z_s\|}}.$$

To see this consult Figure 2.9.1. In Figure 2.9.1 we see the plane through x_0 that is spanned by $N_{\partial K}(x_0)$ and $N_{\partial K_s}(x_s)$. The point x_s is not necessarily in this plane.

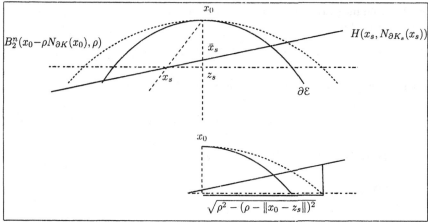

Fig. 2.9.1

On the other hand, by (39)

$$\sin^2 \alpha_s = 1 - <N_{\partial K}(x_0), N_{\partial K_s}(x_s)>^2 \leq \epsilon \|x_0 - z_s\|$$

which implies for sufficiently small ϵ

$$\tan \alpha_s \leq \sqrt{2\epsilon \|x_0 - z_s\|}.$$

Altogether we get

$$\sqrt{2\epsilon \|x_0 - z_s\|} \geq \frac{\|x_0 - z_s\| - m_s}{c\|x_0 - z_s\| + \sqrt{2\rho\|x_0 - z_s\|}}$$

and thus

$$(c\sqrt{2\epsilon} + 4\sqrt{\epsilon\rho})\|x_0 - z_s\| \geq \|x_0 - z_s\| - m_s.$$

Finally we get with a new constant c

$$(1 - 2c\sqrt{\epsilon})\|x_0 - z_s\| \leq m_s.$$

The left hand inequality is proved similarly.

(ii) By (i) we have for all s with $0 < s \leq s_\epsilon$

$$\partial K \cap H^-(x_s + \epsilon\|x_0 - z_s\|N_{\partial K}(x_0), N_{\partial K}(x_0))$$
$$\subseteq \partial K \cap H^-(x_s, N_{\partial K_s}(x_s))$$
$$\subseteq \partial K \cap H^-(x_s - \epsilon\|x_0 - z_s\|N_{\partial K}(x_0), N_{\partial K}(x_0)).$$

$p_{N_{\partial K}(x_0)}$ is the orthogonal projection onto the subspace orthogonal to $N_{\partial K}(x_0)$. From this we get

$$p_{N_{\partial K}(x_0)}(K \cap H(x_s + \epsilon\|x_0 - z_s\|N_{\partial K}(x_0), N_{\partial K}(x_0)))$$
$$\subseteq p_{N_{\partial K}(x_0)}(K \cap H(x_s, N_{\partial K_s}(x_s)))$$
$$\subseteq p_{N_{\partial K}(x_0)}(K \cap H(x_s - \epsilon\|x_0 - z_s\|N_{\partial K}(x_0), N_{\partial K}(x_0))).$$

Let $\mathcal{D}$ be the indicatrix of Dupin at x_0. By Lemma 1.1 for every ϵ there is t_ϵ so that for all t with $0 < t \leq t_\epsilon$

$$(1 - \epsilon)\mathcal{D} \subseteq \frac{1}{\sqrt{2t}} p_{N_{\partial K}(x_0)}(K \cap H(x_0 - t N_{\partial K}(x_0), N_{\partial K}(x_0))) \subseteq (1 + \epsilon)\mathcal{D}.$$

By choosing a proper s_ϵ we get for all s with $0 < s \leq s_\epsilon$

$$(1 - \epsilon)\mathcal{D} \subseteq \frac{1}{\sqrt{2\|x_0 - z_s\|}} p_{N_{\partial K}(x_0)}(K \cap H(x_s, N_{\partial K_s}(x_s))) \subseteq (1 + \epsilon)\mathcal{D}. \quad (42)$$

We get the same inclusions for $\mathcal{E}$ instead of K.

$$(1 - \epsilon)\mathcal{D} \subseteq \frac{1}{\sqrt{2\|x_0 - z_s\|}} p_{N_{\partial \mathcal{E}}(x_0)}(\mathcal{E} \cap H(x_s, N_{\partial K_s}(x_s))) \subseteq (1 + \epsilon)\mathcal{D} \quad (43)$$

Consider now $y \in \partial K \cap H(x_s, N_{\partial K_s}(x_s))$ and $\Phi(y)$. Since

$$p_{N_{\partial K}(x_0)}(\bar{x}_s) = x_0 = 0$$

there is $\lambda > 0$ so that

$$p_{N_{\partial K}(x_0)}(y) = \lambda p_{N_{\partial K}(x_0)}(\Phi(y)).$$

By (42) and (43) we get with a new s_ϵ

$$\|N_{p_{N_{\partial K}(x_0)}(\partial K \cap H)}(p_{N_{\partial K}(x_0)}(y)) - N_{p_{N_{\partial K}(x_0)}(\partial \mathcal{E} \cap H)}(p_{N_{\partial K}(x_0)}(\Phi(y)))\| < \epsilon$$

where $H = H(x_s, N_{\partial K_s}(x_s))$ and the normals are taken in the subspace of the first $n - 1$ coordinates. The projection $p_{N_{\partial K}(x_0)}$ is an isomorphism between $\mathbb{R}^{n-1}$ and $H(x_s, N_{\partial K_s}(x_s))$. The norm of this isomorphism equals 1 and the norm of its inverse is less than $1 + \epsilon$ if we choose s_ϵ sufficiently small. Therefore, if we choose a new s_ϵ we get for all s with $0 < s \leq s_\epsilon$

$$\|N_{\partial K \cap H}(y) - N_{\partial \mathcal{E} \cap H}(\Phi(y))\| < \epsilon.$$

(iii) follows from (42) and (43) and from the fact that the projection $p_{N_{\partial K}(x_0)}$ is an isomorphism between $\mathbb{R}^{n-1}$ and $H(x_s, N_{\partial K_s}(x_s))$ whose norm equals 1 and the norm of its inverse is less than $1 + \epsilon$. Indeed, the norm of the inverse depends only on the angle between $\mathbb{R}^n$ and $H(x_s, N_{\partial K_s}(x_s))$. The angle between these two planes will be as small as we wish if we choose s_ϵ small enough. $\square$

Lemma 2.10. *(i) Let K be a convex body in $\mathbb{R}^n$ and $x_0 \in \partial K$. Suppose that the indicatrix of Dupin at x_0 exists and is an ellipsoid. Let $f : \partial K \to \mathbb{R}$ be a integrable, a.e. positive function with $\int f d\mu = 1$. Suppose that f is continuous at x_0 and $f(x_0) > 0$. Let x_s and Φ as given by Lemma 2.8 and let z_s be given as in the proof of Lemma 2.8 by*

$$\{z_s\} = \{x_0 + tN_{\partial K}(x_0)|t \in \mathbb{R}\} \cap H(x_s, N_{\partial K}(x_0)).$$

For every $x_0 \in \partial K$ and every $\epsilon > 0$ there is s_ϵ so that we have for all s with $0 < s < s_\epsilon$

$$\left| \int_{\partial K \cap H(x_s, N_{\partial K_s}(x_s))} \frac{f(y)}{\sqrt{1 - < N_{\partial K}(y), N_{\partial K_s}(x_s) >^2}} d\mu_{\partial K \cap H(x_s, N(x_s))}(y) \right.$$

$$\left. - \int_{\partial \mathcal{E} \cap H(z_s, N_{\partial K}(x_0))} \frac{f(\Phi^{-1}(z))}{\sqrt{1 - < N_{\partial \mathcal{E}}(z), N_{\partial K}(x_0) >^2}} d\mu_{\partial \mathcal{E} \cap H(x_s, N(x_0))}(z) \right|$$

$$\leq \epsilon \int_{\partial \mathcal{E} \cap H(z_s, N_{\partial K}(x_0))} \frac{f(\Phi^{-1}(z))}{\sqrt{1 - < N_{\partial \mathcal{E}}(z), N_{\partial K}(x_0) >^2}} d\mu_{\partial \mathcal{E} \cap H(x_s, N_{\partial K}(x_0))}(z).$$

(ii) Let B_2^n denote the Euclidean ball and $(B_2^n)_s$ its surface body with respect to the constant density $(vol_{n-1}(\partial B_2^n))^{-1}$. Let $\{x_s\} = \partial(B_2^n)_s \cap [0, e_n]$ and H_s the tangent hyperplane to $(B_2^n)_s$ at x_s. For every $\epsilon > 0$ there is s_ϵ so that we have for all s with $0 < s < s_\epsilon$

$$(1 - \epsilon) \left(s \frac{vol_{n-1}(\partial B_2^n)}{vol_{n-1}(B_2^{n-1})} \right)^{\frac{n-3}{n-1}} vol_{n-2}(\partial B_2^{n-1})$$

$$\leq \int_{\partial B_2^n \cap H_s} \frac{1}{\sqrt{1 - < N_{\partial(B_2^n)_s}(x_s), N_{\partial B_2^n}(y) >^2}} d\mu_{\partial B_2^n \cap H_s}(y)$$

$$\leq \left(s \frac{vol_{n-1}(\partial B_2^n)}{vol_{n-1}(B_2^{n-1})} \right)^{\frac{n-3}{n-1}} vol_{n-2}(\partial B_2^{n-1}).$$

(iii) Let $a_1, \ldots, a_n > 0$ and

$$\mathcal{E} = \left\{ x \left| \sum_{i=1}^{n} \left| \frac{x(i)}{a_i} \right|^2 \leq 1 \right. \right\}.$$

Let $\mathcal{E}_s$, $0 < s \leq \frac{1}{2}$, be the surface bodies with respect to the constant density $(vol_{n-1}(\partial \mathcal{E}))^{-1}$. Moreover, let $\lambda_\mathcal{E} : \mathbb{R}^+ \to [0, a_n]$ be such that $\lambda_\mathcal{E}(s)e_n \in \partial \mathcal{E}_s$ and H_s the tangent hyperplane to $\mathcal{E}_s$ at $\lambda_\mathcal{E}(s)e_n$. Then, for all $\epsilon > 0$ there is s_ϵ such that for all s and t with $0 \leq s, t \leq \frac{1}{2}$

$$\int_{\partial \mathcal{E} \cap H_s} \frac{1}{\sqrt{1 - < N_{\partial \mathcal{E}_s}(x_s), N_{\partial \mathcal{E}}(y) >^2}} d\mu_{\partial \mathcal{E} \cap H_s}(y)$$

$$\leq (1 + \epsilon)(\tfrac{s}{t})^{\frac{n-3}{n-1}} \int_{\partial \mathcal{E} \cap H_t} \frac{1}{\sqrt{1 - < N_{\partial \mathcal{E}_t}(x_t), N_{\partial \mathcal{E}}(y) >^2}} d\mu_{\partial \mathcal{E} \cap H_t}(y).$$

Please note that $N_{\partial \mathcal{E}_s}(\lambda_\mathcal{E}(s)e_n) = N_{\partial \mathcal{E}}(a_n e_n) = e_n$.

Proof. (i) In the first part of the proof H denotes $H(x_s, N_{\partial K_s}(x_s))$. We prove first that for every ϵ there is s_ϵ so that we have for all s with $0 < s \le s_\epsilon$

$$\left| \int_{\partial K \cap H} \frac{f(y)}{\sqrt{1- < N_{\partial K}(y), N_{\partial K_s}(x_s) >^2}} \, d\mu_{\partial K \cap H}(y) \right.$$
$$\left. - \int_{\partial \mathcal{E} \cap H} \frac{f(\Phi^{-1}(z))}{\sqrt{1- < N_{\partial \mathcal{E}}(z), N_{\partial K_s}(x_s) >^2}} d\mu_{\partial \mathcal{E} \cap H}(z) \right| \qquad (44)$$
$$\le \epsilon \int_{\partial \mathcal{E} \cap H} \frac{f(\Phi^{-1}(z))}{\sqrt{1- < N_{\partial \mathcal{E}}(z), N_{\partial K_s}(x_s) >^2}} d\mu_{\partial \mathcal{E} \cap H}(z).$$

$\bar{x}_s$ and Φ are as given in Lemma 2.8. There is a real valued, positive function $\phi : \partial K \cap H \to \mathbb{R}$ such that

$$\Phi(y) = \bar{x}_s + \phi(y)(y - \bar{x}_s).$$

By Lemma 1.8 we have with $y = \Phi^{-1}(z)$

$$\int_{\partial K \cap H} \frac{f(y)}{\sqrt{1- < N_{\partial K}(y), N_{\partial K_s}(x_s) >^2}} d\mu_{\partial K \cap H}(y)$$
$$= \int_{\partial \mathcal{E} \cap H} \frac{f(\Phi^{-1}(z))\phi^{-n+2}(\Phi^{-1}(z))}{\sqrt{1- < N_{\partial K}(\Phi^{-1}(z)), N_{\partial K_s}(x_s) >^2}}$$
$$\times \frac{< N_{\partial \mathcal{E} \cap H}(z), \frac{z}{\|z\|} >}{< N_{\partial K \cap H}(\Phi^{-1}(z)), \frac{z}{\|z\|} >} d\mu_{\partial \mathcal{E} \cap H}(z)$$
$$= \int_{\partial \mathcal{E} \cap H} \frac{f(\Phi^{-1}(z))\phi^{-n+2}(\Phi^{-1}(z))}{\sqrt{1- < N_{\partial K}(\Phi^{-1}(z)), N_{\partial K_s}(x_s) >^2}}$$
$$\times \frac{< N_{\partial \mathcal{E} \cap H}(z), z >}{< N_{\partial K \cap H}(\Phi^{-1}(z)), z >} d\mu_{\partial \mathcal{E} \cap H}(z).$$

With this we get

$$\left| \int_{\partial K \cap H} \frac{f(y)}{\sqrt{1- < N_{\partial K}(y), N_{\partial K_s}(x_s) >^2}} \, d\mu_{\partial K \cap H}(y) \right.$$
$$\left. - \int_{\partial \mathcal{E} \cap H} \frac{f(\Phi^{-1}(z))}{\sqrt{1- < N_{\partial \mathcal{E}}(z), N_{\partial K_s}(x_s) >^2}} d\mu_{\partial \mathcal{E} \cap H}(z) \right|$$
$$\le \left| \int_{\partial \mathcal{E} \cap H} \frac{f(\Phi^{-1}(z))}{\sqrt{1- < N_{\partial \mathcal{E}}(z), N_{\partial K_s}(x_s) >^2}} \right.$$
$$\left. - \frac{f(\Phi^{-1}(z))}{\sqrt{1- < N_{\partial K}(\Phi^{-1}(z)), N_{\partial K_s}(x_s) >^2}} d\mu_{\partial \mathcal{E} \cap H}(z) \right|$$
$$+ \left| \int_{\partial \mathcal{E} \cap H} \frac{f(\Phi^{-1}(z)) \left(1 - \phi^{-n+2}(\Phi^{-1}(z)) \frac{< N_{\partial \mathcal{E} \cap H}(z), z >}{< N_{\partial K \cap H}(\Phi^{-1}(z)), z >}\right)}{\sqrt{1- < N_{\partial K}(\Phi^{-1}(z)), N_{\partial K_s}(x_s) >^2}} d\mu_{\partial \mathcal{E} \cap H}(z) \right|.$$

By Lemma 2.8 we have

$$\left| \frac{1}{\sqrt{1- <N_{\partial\mathcal{E}}(z), N_{\partial K_s}(x_s)>^2}} - \frac{1}{\sqrt{1- <N_{\partial K}(\Phi^{-1}(z)), N_{\partial K_s}(x_s)>^2}} \right|$$

$$\leq \frac{\epsilon}{\sqrt{1- <N_{\partial\mathcal{E}}(z), N_{\partial K_s}(x_s)>^2}}$$

which gives the right estimate of the first summand.

We apply Lemma 2.9.(ii) and (iii) to the second summand. The second summand is less than

$$\epsilon \int_{\partial\mathcal{E}\cap H} \frac{f(\Phi^{-1}(z))}{\sqrt{1- <N_{\partial K}(\Phi^{-1}(z)), N_{\partial K_s}(x_s)>^2}} d\mu_{\partial\mathcal{E}\cap H}(z).$$

Now we apply Lemma 2.8 and get that this is less than or equal to

$$3\epsilon \int_{\partial\mathcal{E}\cap H} \frac{f(\Phi^{-1}(z))}{\sqrt{1- <N_{\partial\mathcal{E}}(z), N_{\partial K_s}(x_s)>^2}} d\mu_{\partial\mathcal{E}\cap H}(z).$$

This establishes (44). Now we show

$$\left| \int_{\partial\mathcal{E}\cap H(x_s, N_{\partial K_s}(x_s))} \frac{f(\Phi^{-1}(y))}{\sqrt{1- <N_{\partial\mathcal{E}}(y), N_{\partial K_s}(x_s)>^2}} d\mu_{\partial\mathcal{E}\cap H(x_s, N_{\partial K_s}(x_s))}(y) \right.$$

$$\left. - \int_{\partial\mathcal{E}\cap H(z_s, N_{\partial K}(x_0))} \frac{f(\Phi^{-1}(z))}{\sqrt{1- <N_{\partial\mathcal{E}}(z), N_{\partial K_s}(x_s)>^2}} d\mu_{\partial\mathcal{E}\cap H(z_s, N_{\partial K}(x_0))}(z) \right|$$

$$\leq \epsilon \int_{\partial\mathcal{E}\cap H(z_s, N_{\partial K}(x_0))} \frac{f(\Phi^{-1}(z))}{\sqrt{1- <N_{\partial\mathcal{E}}(z), N_{\partial K_s}(x_s)>^2}} d\mu_{\partial\mathcal{E}\cap H(z_s, N_{\partial K}(x_0))}(z).$$

$$(45)$$

Since f is continuous at x_0 and $f(x_0) > 0$ it is equivalent to show

$$\left| \int_{\partial\mathcal{E}\cap H(x_s, N_{\partial K_s}(x_s))} \frac{f(x_0)}{\sqrt{1- <N_{\partial\mathcal{E}}(y), N_{\partial K_s}(x_s)>^2}} d\mu_{\partial\mathcal{E}\cap H(x_s, N_{\partial K_s}(x_s))}(y) \right.$$

$$\left. - \int_{\partial\mathcal{E}\cap H(z_s, N_{\partial K}(x_0))} \frac{f(x_0)}{\sqrt{1- <N_{\partial\mathcal{E}}(z), N_{\partial K_s}(x_s)>^2}} d\mu_{\partial\mathcal{E}\cap H(z_s, N_{\partial K}(x_0))}(z) \right|$$

$$\leq \epsilon \int_{\partial\mathcal{E}\cap H(z_s, N_{\partial K}(x_0))} \frac{f(x_0)}{\sqrt{1- <N_{\partial\mathcal{E}}(z), N_{\partial K_s}(x_s)>^2}} d\mu_{\partial\mathcal{E}\cap H(z_s, N_{\partial K}(x_0))}(z)$$

which is of course the same as

$$\left| \int_{\partial\mathcal{E} \cap H(x_s, N_{\partial K_s}(x_s))} \frac{1}{\sqrt{1 - <N_{\partial\mathcal{E}}(y), N_{\partial K_s}(x_s)>^2}} \, d\mu_{\partial\mathcal{E} \cap H(x_s, N_{\partial K_s}(x_s))}(y) \right.$$

$$\left. - \int_{\partial\mathcal{E} \cap H(z_s, N_{\partial K}(x_0))} \frac{1}{\sqrt{1 - <N_{\partial\mathcal{E}}(z), N_{\partial K_s}(x_s)>^2}} \, d\mu_{\partial\mathcal{E} \cap H(z_s, N_{\partial K}(x_0))}(z) \right|$$

$$\leq \epsilon \int_{\partial\mathcal{E} \cap H(z_s, N_{\partial K}(x_0))} \frac{1}{\sqrt{1 - <N_{\partial\mathcal{E}}(z), N_{\partial K_s}(x_s)>^2}} \, d\mu_{\partial\mathcal{E} \cap H(z_s, N_{\partial K}(x_0))}(z).$$

$$(46)$$

We put $\mathcal{E}$ in such a position that $N_{\partial K}(x_0) = e_n$, $x_0 = r_n e_n$, and such that $\mathcal{E}$ is given by the equation

$$\sum_{i=1}^{n} \left| \frac{y_i}{r_i} \right|^2 = 1.$$

Let $\xi \in \partial B_2^n$ and $y = (r(\xi, y_n)\xi, y_n) \in \partial\mathcal{E}$. Then

$$N_{\partial\mathcal{E}}(y) = \frac{\left(\frac{y_1}{r_1^2}, \ldots, \frac{y_n}{r_n^2} \right)}{\sqrt{\sum_{i=1}^{n} \frac{y_i^2}{r_i^4}}} = \frac{\left(\frac{r(\xi, y_n)\xi_1}{r_1^2}, \ldots, \frac{r(\xi, y_n)\xi_{n-1}}{r_{n-1}^2}, \frac{y_n}{r_n^2} \right)}{\sqrt{\frac{y_n^2}{r_n^4} + r(\xi, y_n)^2 \sum_{i=1}^{n-1} \frac{\xi_i^2}{r_i^4}}}$$

with

$$r(\xi, y_n) = \frac{\sqrt{r_n^2 - y_n^2}}{r_n \sqrt{\sum_{i=1}^{n-1} \frac{\xi_i^2}{r_i^2}}}. \qquad (47)$$

As $N_{\partial K}(x_0) = e_n$ we get

$$<N_{\partial\mathcal{E}}(y), N_{\partial K}(x_0)> = -\frac{y_n}{r_n^2 \sqrt{\sum_{i=1}^{n} \frac{y_i^2}{r_i^4}}}.$$

Therefore

$$\frac{1}{1 - <N_{\partial\mathcal{E}}(y), N_{\partial K}(x_0)>^2} = \frac{\sum_{i=1}^{n} \frac{y_i^2}{r_i^4}}{\sum_{i=1}^{n-1} \frac{y_i^2}{r_i^4}}.$$

For $y, z \in \partial\mathcal{E}$ we get

$$\frac{1 - <N_{\partial\mathcal{E}}(z), N_{\partial K}(x_0)>^2}{1 - <N_{\partial\mathcal{E}}(y), N_{\partial K}(x_0)>^2} = \frac{\sum_{i=1}^{n} \frac{y_i^2}{r_i^4} \sum_{i=1}^{n-1} \frac{z_i^2}{r_i^4}}{\sum_{i=1}^{n} \frac{z_i^2}{r_i^4} \sum_{i=1}^{n-1} \frac{y_i^2}{r_i^4}}.$$

For $y, z \in \partial\mathcal{E}$ with the same direction ξ we get by (47)

$$\frac{1 - <N_{\partial\mathcal{E}}(z), N_{\partial K}(x_0)>^2}{1 - <N_{\partial\mathcal{E}}(y), N_{\partial K}(x_0)>^2} = \frac{\sum_{i=1}^{n} \frac{y_i^2}{r_i^4}}{\sum_{i=1}^{n} \frac{z_i^2}{r_i^4}} \left(\frac{r_n^2 - z_n^2}{r_n^2 - y_n^2} \right).$$

We can choose s_ϵ sufficiently small so that we have for all s with $0 < s \leq s_\epsilon$, and all $y \in \partial\mathcal{E} \cap H(x_s, N_{\partial K_s}(x_s))$, $z \in \partial\mathcal{E} \cap H(x_s, N_{\partial K}(x_0))$

$$|y_n - r_n| < \epsilon \qquad\qquad |z_n - r_n| < \epsilon$$

and by Lemma 2.9.(i)

$$1 - \epsilon \leq \frac{r_n - z_n}{r_n - y_n} \leq 1 + \epsilon.$$

We pass to a new ϵ and obtain: We can choose s_ϵ sufficiently small so that we have for all s with $0 < s \leq s_\epsilon$, and all $y \in \partial\mathcal{E} \cap H(x_s, N_{\partial K_s}(x_s))$, $z \in \partial\mathcal{E} \cap H(x_s, N_{\partial K}(x_0))$ such that $p_{e_n}(y)$ and $p_{e_n}(z)$ are colinear

$$1 - \epsilon \leq \frac{1 - <N_{\partial\mathcal{E}}(z), N_{\partial K}(x_0)>^2}{1 - <N_{\partial\mathcal{E}}(y), N_{\partial K}(x_0)>^2} \leq 1 + \epsilon. \tag{48}$$

By Lemma 2.5 we have

$$<N_{\partial K}(x_0), N_{\partial K_s}(x_s)> \geq 1 - \epsilon.$$

Therefore, the orthogonal projection p_{e_n} restricted to the hyperplane

$$H(x_s, N_{\partial K_s}(x_s))$$

is a linear isomorphism between this hyperplane and $\mathbb{R}^{n-1}$ and moreover, $\|p_{e_n}\| = 1$ and $\|p_{e_n}^{-1}\| \leq \frac{1}{1-\epsilon}$. By this, there is s_ϵ such that for all s with $0 < s \leq s_\epsilon$

$$(1-\epsilon) \int_{\partial\mathcal{E} \cap H(x_s, N_{\partial K_s}(x_s))} \frac{d\mu_{\partial\mathcal{E} \cap H(x_s, N_{\partial K_s}(x_s))}(y)}{\sqrt{1 - <N_{\partial\mathcal{E}}(y), N_{\partial K_s}(x_s)>^2}}$$

$$\leq \int_{p_{e_n}(\partial\mathcal{E} \cap H(x_s, N_{\partial K_s}(x_s)))} \frac{d\mu_{p_{e_n}(\partial\mathcal{E} \cap H(x_s, N_{\partial K_s}(x_s)))}(z)}{\sqrt{1 - <N_{\partial\mathcal{E}}(p_{e_n}^{-1}(z)), N_{\partial K_s}(x_s)>^2}}$$

$$\leq \int_{\partial\mathcal{E} \cap H(x_s, N_{\partial K_s}(x_s))} \frac{d\mu_{\partial\mathcal{E} \cap H(x_s, N_{\partial K_s}(x_s))}(y)}{\sqrt{1 - <N_{\partial\mathcal{E}}(y), N_{\partial K_s}(x_s)>^2}}$$

where $z = p_{e_n}(y)$. Let q_{e_n} denote the orthogonal projection from

$$H(x_s, N_{\partial K}(x_0))$$

to $\mathbb{R}^{n-1}$. q_{e_n} is an isometry. Therefore

$$\int_{\partial\mathcal{E} \cap H(x_s, N_{\partial K_s}(x_0))} \frac{d\mu_{\partial\mathcal{E} \cap H(x_s, N_{\partial K_s}(x_0))}(y)}{\sqrt{1 - <N_{\partial\mathcal{E}}(y), N_{\partial K_s}(x_s)>^2}}$$

$$= \int_{q_{e_n}(\partial\mathcal{E} \cap H(x_s, N_{\partial K}(x_0)))} \frac{d\mu_{q_{e_n}(\partial\mathcal{E} \cap H(x_s, N_{\partial K}(x_0)))}(y)}{\sqrt{1 - <N_{\partial\mathcal{E}}(q_{e_n}^{-1}(y)), N_{\partial K_s}(x_s)>^2}}.$$

Thus, in order to show (46) it suffices to show

$$\left| \int_{p_{e_n}(\partial\mathcal{E}\cap H(x_s, N_{\partial K_s}(x_s)))} \frac{d\mu_{p_{e_n}(\partial\mathcal{E}\cap H(x_s, N_{\partial K_s}(x_s)))}(y)}{\sqrt{1- < N_{\partial\mathcal{E}}(p_{e_n}^{-1}(y)), N_{\partial K_s}(x_s) >^2}} \right.$$

$$\left. - \int_{q_{e_n}(\partial\mathcal{E}\cap H(x_s, N_{\partial K}(x_0)))} \frac{d\mu_{q_{e_n}(\partial\mathcal{E}\cap H(x_s, N_{\partial K}(x_0)))}(y)}{\sqrt{1- < N_{\partial\mathcal{E}}(q_{e_n}^{-1}(y)), N_{\partial K_s}(x_s) >^2}} \right|$$

$$\leq \epsilon \int_{q_{e_n}(\partial\mathcal{E}\cap H(x_s, N_{\partial K}(x_0)))} \frac{d\mu_{q_{e_n}(\partial\mathcal{E}\cap H(x_s, N_{\partial K}(x_0)))}(y)}{\sqrt{1- < N_{\partial\mathcal{E}}(q_{e_n}^{-1}(y)), N_{\partial K_s}(x_s) >^2}}.$$

Let $\rho: q_{e_n}(\partial\mathcal{E}\cap H(x_s, N_{\partial K}(x_0))) \rightarrow p_{e_n}(\partial\mathcal{E}\cap H(x_s, N_{\partial K_s}(x_s)))$ be the radial map defined by

$$\{\rho(y)\} = \{ty | t \geq 0\} \cap p_{e_n}(\partial\mathcal{E}\cap H(x_s, N_{\partial K}(x_0))).$$

We have

$$(1-\epsilon) \int_{p_{e_n}(\partial\mathcal{E}\cap H(x_s, N_{\partial K_s}(x_s)))} \frac{d\mu_{p_{e_n}(\partial\mathcal{E}\cap H(x_s, N_{\partial K_s}(x_s)))}(y)}{\sqrt{1- < N_{\partial\mathcal{E}}(p_{e_n}^{-1}(y)), N_{\partial K_s}(x_s) >^2}}$$

$$\leq \int_{q_{e_n}(\partial\mathcal{E}\cap H(x_s, N_{\partial K}(x_0)))} \frac{d\mu_{q_{e_n}(\partial\mathcal{E}\cap H(x_s, N_{\partial K}(x_0)))}(y)}{\sqrt{1- < N_{\partial\mathcal{E}}(p_{e_n}^{-1}(\rho(y))), N_{\partial K_s}(x_s) >^2}}$$

$$\leq (1+\epsilon) \int_{p_{e_n}(\partial\mathcal{E}\cap H(x_s, N_{\partial K_s}(x_s)))} \frac{d\mu_{p_{e_n}(\partial\mathcal{E}\cap H(x_s, N_{\partial K_s}(x_s)))}(y)}{\sqrt{1- < N_{\partial\mathcal{E}}(p_{e_n}^{-1}(y)), N_{\partial K_s}(x_s) >^2}}.$$

To see this, consider the indicatrix of Dupin $\mathcal{D}$ of K at x_0. We have by (43)

$$(1-\epsilon)\mathcal{D} \subseteq \frac{1}{\sqrt{2}\|x_0 - z_s\|} q_{e_n}(\mathcal{E}\cap H(x_s, N_{\partial K}(x_0))) \subseteq (1+\epsilon)\mathcal{D}$$

$$(1-\epsilon)\mathcal{D} \subseteq \frac{1}{\sqrt{2}\|x_0 - z_s\|} p_{e_n}(\mathcal{E}\cap H(x_s, N_{\partial K}(x_s))) \subseteq (1+\epsilon)\mathcal{D}.$$

They imply that with a new s_ϵ the surface element changes at most by a factor $(1+\epsilon)$. Thus, in order to verify (46), it is enough to show

$$\left| \int_{q_{e_n}(\partial\mathcal{E}\cap H(x_s, N_{\partial K}(x_0)))} \frac{d\mu_{q_{e_n}(\partial\mathcal{E}\cap H(x_s, N_{\partial K}(x_0)))}(y)}{\sqrt{1- < N_{\partial\mathcal{E}}(p_{e_n}^{-1}(\rho(y))), N_{\partial K_s}(x_s) >^2}} \right.$$

$$\left. - \int_{q_{e_n}(\partial\mathcal{E}\cap H(x_s, N_{\partial K}(x_0)))} \frac{d\mu_{q_{e_n}(\partial\mathcal{E}\cap H(x_s, N_{\partial K}(x_0)))}(y)}{\sqrt{1- < N_{\partial\mathcal{E}}(q_{e_n}^{-1}(y)), N_{\partial K_s}(x_s) >^2}} \right| \quad (49)$$

$$\leq \epsilon \int_{q_{e_n}(\partial\mathcal{E}\cap H(x_s, N_{\partial K}(x_0)))} \frac{d\mu_{q_{e_n}(\partial\mathcal{E}\cap H(x_s, N_{\partial K}(x_0)))}(y)}{\sqrt{1- < N_{\partial\mathcal{E}}(q_{e_n}^{-1}(y)), N_{\partial K_s}(x_s) >^2}}.$$

We verify this. By (48) there is s_ϵ so that we have for all s with $0 < s \leq s_\epsilon$, and all $y \in \partial\mathcal{E} \cap H(x_s, N_{\partial K_s}(x_s))$, $z \in \partial\mathcal{E} \cap H(x_s, N_{\partial K}(x_0))$ such that $p_{e_n}(y)$ and $p_{e_n}(z)$ are colinear

$$1 - \epsilon \leq \frac{\|N_{\partial\mathcal{E}}(z) - N_{\partial K}(x_0)\|}{\|N_{\partial\mathcal{E}}(y) - N_{\partial K}(x_0)\|} \leq 1 + \epsilon.$$

By (39) for every ϵ there is s_ϵ such that for all s with $0 < s \leq s_\epsilon$

$$\|N_{\partial K}(x_0) - N_{\partial K_s}(x_s)\| \leq \epsilon\sqrt{\|x_0 - z_s\|}$$

and by the formula following (2.8.13) for all $y \in \partial\mathcal{E} \cap H(x_s, N_{\partial K_s}(x_s))$ and $z \in \partial\mathcal{E} \cap H(x_s, N_{\partial K}(x_0))$

$$\|N_{\partial\mathcal{E}}(y) - N_{\partial K_s}(x_s)\| \geq c\sqrt{\|x_0 - z_s\|}$$
$$\|N_{\partial\mathcal{E}}(z) - N_{\partial K_s}(x_s)\| \geq c\sqrt{\|x_0 - z_s\|}.$$

Therefore,

$$\|N_{\partial K}(x_0) - N_{\partial K_s}(x_s)\| \leq \epsilon\sqrt{\|x_0 - z_s\|} \leq \tfrac{\epsilon}{c}\|N_{\partial\mathcal{E}}(z) - N_{\partial K_s}(x_s)\|.$$

By triangle inequality

$$\|N_{\partial\mathcal{E}}(z) - N_{\partial K_s}(x_s)\| \leq (1 + \tfrac{\epsilon}{c})\|N_{\partial\mathcal{E}}(z) - N_{\partial K}(x_0)\| \tag{50}$$

and the same inequality for y. In the same way we get the estimates from below. Thus there is s_ϵ so that we have for all s with $0 < s \leq s_\epsilon$, and all $y \in \partial\mathcal{E} \cap H(x_s, N_{\partial K_s}(x_s))$, $z \in \partial\mathcal{E} \cap H(x_s, N_{\partial K}(x_0))$ such that $p_{e_n}(y)$ and $p_{e_n}(z)$ are colinear

$$1 - \epsilon \leq \frac{\|N_{\partial\mathcal{E}}(z) - N_{\partial K_s}(x_s)\|}{\|N_{\partial\mathcal{E}}(y) - N_{\partial K_s}(x_s)\|} \leq 1 + \epsilon$$

which is the same as

$$1 - \epsilon \leq \frac{1- < N_{\partial\mathcal{E}}(z), N_{\partial K_s}(x_s) >^2}{1- < N_{\partial\mathcal{E}}(y), N_{\partial K_s}(x_s) >^2} \leq 1 + \epsilon.$$

This establishes (49) and consequently (45). Combining the formulas (44) and (45) gives

$$\left| \int_{\partial K\cap H(x_s, N_{\partial K_s}(x_s))} \frac{f(y)d\mu_{\partial K\cap H(x_s, N(x_s))}(y)}{\sqrt{1- < N_{\partial K}(y), N_{\partial K_s}(x_s) >^2}} \right.$$
$$\left. - \int_{\partial\mathcal{E}\cap H(z_s, N_{\partial K}(x_0))} \frac{f(\Phi^{-1}(z))d\mu_{\partial\mathcal{E}\cap H(x_s, N(x_0))}(z)}{\sqrt{1- < N_{\partial\mathcal{E}}(z), N_{\partial K_s}(x_s) >^2}} \right|$$
$$\leq \epsilon \int_{\partial\mathcal{E}\cap H(z_s, N_{\partial K}(x_0))} \frac{f(\Phi^{-1}(z))d\mu_{\partial\mathcal{E}\cap H(x_s, N_{\partial K}(x_0))}(z)}{\sqrt{1- < N_{\partial\mathcal{E}}(z), N_{\partial K_s}(x_s) >^2}}.$$

It is left to replace $N_{\partial K_s}(x_s)$ by $N_{\partial K}(x_0)$. This is done by using the formula (50) relating the two normals.

(ii) For every $\epsilon > 0$ there is s_ϵ such that for all s with $0 < s \leq s_\epsilon$

$$(1 - \epsilon)s \leq \frac{\mathrm{vol}_{n-1}(B_2^n \cap H_s)}{\mathrm{vol}_{n-1}(\partial B_2^n)} \leq \frac{\mathrm{vol}_{n-1}(\partial B_2^n \cap H_s^-)}{\mathrm{vol}_{n-1}(\partial B_2^n)} = s.$$

$B_2^n \cap H_s$ is the boundary of a $n - 1$-dimensional Euclidean ball with radius

$$r = \left(\frac{\mathrm{vol}_{n-1}(B_2^n \cap H_s)}{\mathrm{vol}_{n-1}(B_2^{n-1})} \right)^{\frac{1}{n-1}}.$$

Therefore

$$\left((1 - \epsilon)s \frac{\mathrm{vol}_{n-1}(\partial B_2^n)}{\mathrm{vol}_{n-1}(B_2^{n-1})} \right)^{\frac{1}{n-1}} \leq r \leq \left(s \frac{\mathrm{vol}_{n-1}(\partial B_2^n)}{\mathrm{vol}_{n-1}(B_2^{n-1})} \right)^{\frac{1}{n-1}}.$$

We have $N(x_s) = e_n$ and $\sqrt{1- < e_n, N_{\partial B_2^n}(y) >^2}$ is the sine of the angle between e_n and $N_{\partial B_2^n}(y)$. This equals the radius r of $B_2^n \cap H_s$. Altogether we get

$$\int_{\partial B_2^n \cap H_s} \frac{\mathrm{d}\mu_{\partial B_2^n \cap H_s}(y)}{\sqrt{1- < N(x_s), N_{\partial B_2^n}(y) >^2}}$$

$$= r^{n-3} \mathrm{vol}_{n-2}(\partial B_2^{n-1}) \leq \left(s \frac{\mathrm{vol}_{n-1}(\partial B_2^n)}{\mathrm{vol}_{n-1}(B_2^{n-1})} \right)^{\frac{n-3}{n-1}} \mathrm{vol}_{n-2}(\partial B_2^{n-1}).$$

(iii) $\mathcal{E} \cap H_s$ and $\mathcal{E} \cap H_t$ are homothetic, $n - 1$-dimensional ellipsoids. The factor ϕ_0 by which we have to multiply $\mathcal{E} \cap H_s$ in order to recover $\mathcal{E} \cap H_t$ is

$$\phi_0 = \left(\frac{\mathrm{vol}_{n-1}(\mathcal{E} \cap H_t)}{\mathrm{vol}_{n-1}(\mathcal{E} \cap H_s)} \right)^{\frac{1}{n-1}}.$$

On the other hand, for all $\epsilon > 0$ there is s_ϵ such that for all s with $0 < s \leq s_\epsilon$

$$(1 - \epsilon)s \leq \frac{\mathrm{vol}_{n-1}(\mathcal{E} \cap H_s)}{\mathrm{vol}_{n-1}(\partial \mathcal{E})} \leq \frac{\mathrm{vol}_{n-1}(\partial \mathcal{E} \cap H_s^-)}{\mathrm{vol}_{n-1}(\partial \mathcal{E})} = s.$$

Therefore

$$\left(\frac{(1 - \epsilon)t}{s} \right)^{\frac{1}{n-1}} \leq \phi_0 \leq \left(\frac{t}{(1 - \epsilon)s} \right)^{\frac{1}{n-1}}.$$

The volume of a volume element of $\partial \mathcal{E} \cap H_s$ that is mapped by the homothety onto one in $\partial \mathcal{E} \cap H_t$ increases by ϕ_0^{n-2}.

Now we estimate how much the angle between $N_{\partial \mathcal{E}}(y)$ and $N_{\partial \mathcal{E}_s}(x_s) = e_n$ changes. The normal to $\mathcal{E}$ at y is

$$\left(\frac{y_i}{a_i^2 \sqrt{\sum_{k=1}^n \frac{y_k^2}{a_k^4}}} \right)_{i=1}^n .$$

Thus

$$< N_{\partial \mathcal{E}}(y), e_n > = \frac{y_n}{a_n^2 \sqrt{\sum_{k=1}^n \frac{y_k^2}{a_k^4}}}$$

and

$$1 - < N_{\partial \mathcal{E}}(y), e_n >^2 = \frac{\sum_{k=1}^{n-1} \frac{y_k^2}{a_k^4}}{\sum_{k=1}^n \frac{y_k^2}{a_k^4}} .$$

Let $y(s) \in \mathcal{E} \cap H_s$ and $y(t) \in \mathcal{E} \cap H_t$ be vectors such that $(y_1(s), \ldots, y_{n-1}(s))$ and $(y_1(t), \ldots, y_{n-1}(t))$ are colinear. Then

$$(y_1(t), \ldots, y_{n-1}(t)) = \phi_0 (y_1(s), \ldots, y_{n-1}(s))$$

Thus

$$\frac{1 - < N_{\partial \mathcal{E}}(y(t)), e_n >^2}{1 - < N_{\partial \mathcal{E}}(y(s)), e_n >^2} = \frac{\sum_{k=1}^{n-1} \frac{y_k^2(t)}{a_k^4} \sum_{k=1}^n \frac{y_k^2(s)}{a_k^4}}{\sum_{k=1}^n \frac{y_k^2(t)}{a_k^4} \sum_{k=1}^{n-1} \frac{y_k^2(s)}{a_k^4}} = \phi_0^2 \frac{\sum_{k=1}^n \frac{y_k^2(s)}{a_k^4}}{\sum_{k=1}^n \frac{y_k^2(t)}{a_k^4}} .$$

For every $\epsilon > 0$ there is s_ϵ such that for all s with $0 < s \le s_\epsilon$ we have $a_n - \epsilon \le y_n(s) \le a_n$. Therefore there is an appropriate s_ϵ such that for all s with $0 < s \le s_\epsilon$

$$1 - \epsilon \le \frac{\sum_{k=1}^n \frac{y_k^2(t)}{a_k^4}}{\sum_{k=1}^n \frac{y_k^2(s)}{a_k^4}} \le 1 + \epsilon.$$

Thus

$$(1 - \epsilon)\phi_0 \le \frac{\sqrt{1 - < N_{\partial \mathcal{E}}(y(t)), e_n >^2}}{\sqrt{1 - < N_{\partial \mathcal{E}}(y(s)), e_n >^2}} \le (1 + \epsilon)\phi_0.$$

Consequently, with a new s_ϵ

$$\int_{\partial \mathcal{E} \cap H_s} \frac{d\mu_{\partial \mathcal{E} \cap H_s}(y)}{\sqrt{1 - < N_{\partial \mathcal{E}_s}(x_s), N_{\partial \mathcal{E}}(y) >^2}}$$

$$\le (1 + \epsilon)\phi_0^{-(n-3)} \int_{\partial \mathcal{E} \cap H_t} \frac{d\mu_{\partial \mathcal{E} \cap H_t}(y)}{\sqrt{1 - < N_{\partial \mathcal{E}_t}(x_t), N_{\partial \mathcal{E}}(y) >^2}}$$

$$\le (1 + \epsilon) \left(\frac{s}{t} \right)^{\frac{n-3}{n-1}} \int_{\partial \mathcal{E} \cap H_t} \frac{d\mu_{\partial \mathcal{E} \cap H_t}(y)}{\sqrt{1 - < N_{\partial \mathcal{E}_t}(x_t), N_{\partial \mathcal{E}}(y) >^2}} .$$

$\square$

Lemma 2.11. *Let K be a convex body in $\mathbb{R}^n$ such that for all $t > 0$ the inclusion $K_t \subseteq \overset{\circ}{K}$ holds and that K has everywhere a unique normal. Let $f : \partial K \to \mathbb{R}$ a continuous, positive function with $\int_{\partial K} f(x)\mathrm{d}\mu_{\partial K}(x) = 1$.*
(i) Let $t < T$ and $\epsilon > 0$ such that $t + \epsilon < T$. Let $x \in \partial K_t$ and let $H(x, N_{\partial K_t}(x))$ be a hyperplane such that

$$\mathbb{P}_f(\partial K \cap H^-(x, N_{\partial K_t}(x))) = t.$$

Let $h(x, \epsilon)$ be defined by

$$\mathbb{P}_f(\partial K \cap H^-(x - h(x,\epsilon)N_{\partial K_t}(x), N_{\partial K_t}(x))) = t + \epsilon.$$

Then we have for sufficiently small ϵ

$$\epsilon - o(\epsilon) = \int_{\partial K \cap H(x, N_{\partial K_t}(x))} \frac{f(y)h(x,\epsilon)\mathrm{d}\mu_{\partial K \cap H(x, N_{\partial K_t}(x))}(y)}{\sqrt{1 - <N_{\partial K_t}(x), N_{\partial K}(y)>^2}}.$$

(ii) Let $t + \epsilon < T$, $x \in \partial K_{t+\epsilon}$, and $H(x, N_{\partial K_{t+\epsilon}}(x))$ a hyperplane such that

$$\mathbb{P}_f(\partial K \cap H^-(x, N_{\partial K_{t+\epsilon}}(x))) = t + \epsilon.$$

Let $k(x, \epsilon)$ be defined

$$\mathbb{P}_f(\partial K \cap H(x + k(x,\epsilon)N_{\partial K_{t+\epsilon}}(x), N_{\partial K_{t+\epsilon}}(x))) = t.$$

Then we have

$$\epsilon + o(\epsilon) = \int_{\partial K \cap H(x, N_{\partial K_{t+\epsilon}}(x))} \frac{f(y)k(x,\epsilon)\mathrm{d}\mu_{\partial K \cap H(x, N_{\partial K_{t+\epsilon}}(x))}(y)}{\sqrt{1 - <N_{\partial K_{t+\epsilon}}(x), N_{\partial K}(y)>^2}}.$$

(iii) Let $\mathcal{E}$ be an ellipsoid

$$\mathcal{E} = \left\{ x \left| \sum_{i=1}^{n} \left| \frac{x_i}{a_i} \right|^2 \le 1 \right. \right\}$$

and $\mathcal{E}_s$, $0 < s \le \frac{1}{2}$ surface bodies with respect to the constant density. $\{x_s\} = [0, a_n e_n] \cap \partial \mathcal{E}_s$. Let $\Delta : (0, T) \to [0, \infty)$ be such that $\Delta(s)$ is the height of the cap $\mathcal{E} \cap H^-(x_s, N_{\partial \mathcal{E}_s}(x_s))$. Then Δ is a differentiable, increasing function and

$$\frac{\mathrm{d}\Delta}{\mathrm{d}s}(s) = \left(\int_{\partial \mathcal{E} \cap H_s} \frac{(\mathrm{vol}_{n-1}(\partial \mathcal{E}))^{-1}}{\sqrt{1 - <N_{\partial \mathcal{E}_s}(x_s), N_{\partial \mathcal{E}}(y)>^2}} \mathrm{d}\mu(y) \right)^{-1}$$

where $H_s = H(x_s, N_{\partial \mathcal{E}_s}(x_s))$.

Proof. (i) As $K_t \subset \overset{\circ}{K}$ we can apply Lemma 2.2 and assure that for all $0 < t < T$ and all $x \in \partial K_t$ there is a normal $N_{\partial K_t}(x)$ with

$$t = \int_{\partial K \cap H^-(x, N_{\partial K_t}(x))} f(z) \mathrm{d}\mu_{\partial K}(z).$$

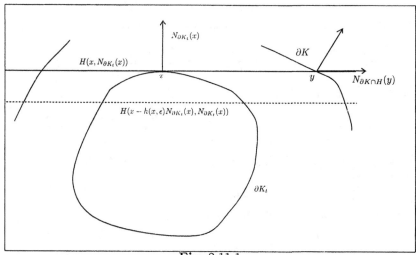

Fig. 2.11.1

We have

$$\epsilon = \int_{\partial K \cap H^-(x - h(x,\epsilon) N_{\partial K_t}(x)), N_{\partial K_t}(x))} f(z) \mathrm{d}\mu_{\partial K}(z)$$

$$- \int_{\partial K \cap H^-(x, N_{\partial K_t}(x))} f(z) \mathrm{d}\mu_{\partial K}(z)$$

$$= \int_{\partial K \cap H^-(x - h(x,\epsilon) N_{\partial K_t}(x)), N_{\partial K_t}(x)) \cap H^+(x, N_{\partial K_t}(x))} f(z) \mathrm{d}\mu_{\partial K}(z).$$

Consider now small ϵ. Since K has everywhere a unique normal a surface element of

$$\partial K \cap H^-(x - h(x,\epsilon) N_{\partial K_t}(x)), N_{\partial K_t}(x)) \cap H^+(x, N_{\partial K_t}(x))$$

at y has approximately the area

$$h(x,\epsilon) \mathrm{d}\mu_{\partial K \cap H(x, N_{\partial K_t}(x))}(y)$$

divided by the cosine of the angle between $N_{\partial K}(y)$ and $N_{\partial K \cap H(x, N_{\partial K_t}(x))}(y)$. The latter normal is taken in the plane $H(x, N_{\partial K_t}(x))$. The vector $N_{\partial K}(y)$ is contained in the plane spanned by $N_{\partial K \cap H(x,N_{\partial K_t}(x))}(y)$ and $N_{\partial K_t}(x)$. Thus we have

$$N_{\partial K}(y) = \; < N_{\partial K}(y), N_{\partial K \cap H(x,N_{\partial K_t}(x))}(y) > N_{\partial K \cap H(x,N_{\partial K_t}(x))}(y)$$
$$+ < N_{\partial K}(y), N_{\partial K_t}(x) > N_{\partial K_t}(x)$$

which implies

$$1 = < N_{\partial K}(y), N_{\partial K \cap H(x,N_{\partial K_t}(x))}(y) >^2 + < N_{\partial K}(y), N_{\partial K_t}(x) >^2 .$$

We get for the approximate area of the surface element

$$\frac{h(x,\epsilon)d\mu_{\partial K \cap H(x,N_{\partial K_t}(x))}(y)}{< N_{\partial K}(y), N_{\partial K \cap H(x,N_{\partial K_t}(x))}(y) >} = \frac{h(x,\epsilon)d\mu_{\partial K \cap H(x,N_{\partial K_t}(x))}(y)}{\sqrt{1 - < N_{\partial K}(y), N_{\partial K_t}(x) >^2}}.$$

Since f is a continuous function

$$\epsilon + o(\epsilon) = \int_{\partial(K \cap H(x,N_{\partial K_t}(x)))} \frac{f(y)h(x,\epsilon)d\mu_{\partial K \cap H(x,N_{\partial K_t}(x))}(y)}{\sqrt{1 - < N_{\partial K_t}(x), N_{\partial K}(y) >^2}}.$$

(iii) By the symmetries of the ellipsoids e_n is a normal to the surface body $\mathcal{E}_s$. In fact we have

$$\mathbb{P}\{\partial \mathcal{E} \cap H^-(x_s, e_n)\} = s.$$

This follows from Lemma 2.4. Moreover,

$$h(x_s, \epsilon) \leq \Delta(s + \epsilon) - \Delta(s) \leq k(x_s, \epsilon).$$

□

Lemma 2.12. *Let K be a convex body in $\mathbb{R}^n$ that has everywhere a unique normal and let $f : \partial K \to \mathbb{R}$ be a continuous, positive function with $\int_{\partial K} f(x)d\mu_{\partial K}(x) = 1$. K_s, $0 \leq s \leq T$, are the surface bodies of K with respect to the density f. Suppose that for all t with $0 < t \leq T$ we have $K_t \subseteq \overset{\circ}{K}$. Let $G : K \to \mathbb{R}$ be a continuous function. Then*

$$\int_{K_s} G(x)dx$$

is a continuous, decreasing function of s on the interval $[0,T]$ and a differentiable function on $(0,T)$. Its derivative is

$$\frac{d}{ds}\int_{K_s} G(x)dx = -\int_{\partial K_s} \frac{G(x_s)d\mu_{\partial K_s}(x_s)}{\int_{\partial K \cap H_s} \frac{f(y)}{\sqrt{1 - <N_{\partial K_s}(x_s), N_{\partial K}(y)>^2}}d\mu_{\partial K \cap H_s}(y)}.$$

where $H_s = H(x_s, N_{\partial K_s}(x_s))$. The derivative is bounded on all intervals $[a,T]$ with $[a,T] \subset (0,T)$ and

$$\int_K G(x)dx = \int_0^T \int_{\partial K_s} \frac{G(x_s)d\mu_{\partial K_s}(x_s)ds}{\int_{\partial K \cap H_s} \frac{f(y)}{\sqrt{1 - <N_{\partial K_s}(x_s), N_{\partial K}(y)>^2}}d\mu_{\partial K \cap H_s}(y)}.$$

Proof. We have

$$\frac{\mathrm{d}}{\mathrm{d}s}\int_{K_s}G(x)\mathrm{d}x = \lim_{\epsilon\to 0}\frac{1}{\epsilon}\left(\int_{K_{s+\epsilon}}G(x)\mathrm{d}x - \int_{K_s}G(x)\mathrm{d}x\right)$$

$$= -\lim_{\epsilon\to 0}\frac{1}{\epsilon}\int_{K_s\setminus K_{s+\epsilon}}G(x)\mathrm{d}x$$

provided that the right hand side limit exists.

Let $\Delta(x_s,\epsilon)$ be the distance of x_s to $\partial K_{s+\epsilon}$. By Lemma 2.4.(iv), for all s and $\delta > 0$ there is $\epsilon > 0$ such that $d_H(K_s,K_{s+\epsilon}) < \delta$. By this and the continuity of G we get

$$\frac{\mathrm{d}}{\mathrm{d}s}\int_{K_s}G(x)\mathrm{d}x = -\lim_{\epsilon\to 0}\frac{1}{\epsilon}\int_{\partial K_s}G(x_s)\Delta(x_s,\epsilon)\mathrm{d}\mu_{\partial K_s}(x_s).$$

We have to show that the right hand side limit exists. By Lemma 2.11.(i) we have

$$\epsilon - o(\epsilon) = \int_{\partial(K\cap H(x,N_{\partial K_t}(x)))}\frac{f(y)h(x,\epsilon)\mathrm{d}\mu_{\partial K\cap H(x,N_{\partial K_t}(x))}(y)}{\sqrt{1-<N_{\partial K_t}(x),N_{\partial K}(y)>^2}}.$$

Since $h(x_s,\epsilon)\leq \Delta(x_s,\epsilon)$ we get

$$\liminf_{\epsilon\to 0}\frac{1}{\epsilon}\int_{\partial K_s}G(x_s)\Delta(x_s,\epsilon)\mathrm{d}\mu_{\partial K_s}(x_s)$$

$$\geq \int_{\partial K_s}\frac{G(x_s)}{\int_{\partial K\cap H_s}\frac{f(y)}{\sqrt{1-<N_{\partial K_s}(x_s),N_{\partial K}(y)>^2}}\mathrm{d}\mu_{\partial K\cap H_s}(y)}\mathrm{d}\mu_{\partial K_s}(x_s)$$

where $H_s = H(x_s,N_{\partial K_s}(x_s))$. We show the inverse inequality for the Limes Superior. This is done by using Lemma 2.11.(ii).

We show now that the function satisfies the fundamental theorem of calculus.

$$\int_{\partial K\cap H_s}\frac{f(y)\mathrm{d}\mu_{\partial K\cap H_s}(y)}{\sqrt{1-<N_{\partial K_s}(x_s),N_{\partial K}(y)>^2}}$$

$$\geq \int_{\partial K\cap H_s}f(y)\mathrm{d}\mu_{\partial K\cap H_s}(y) \geq \min_{y\in\partial K}f(y)\mathrm{vol}_{n-2}(\partial K\cap H_s).$$

By the isoperimetric inequality there is a constant $c > 0$ such that

$$\int_{\partial K\cap H_s}\frac{f(y)\mathrm{d}\mu_{\partial K\cap H_s}(y)}{\sqrt{1-<N_{\partial K_s}(x_s),N_{\partial K}(y)>^2}} \geq c\min_{y\in\partial K}f(y)\mathrm{vol}_{n-1}(K\cap H_s).$$

By our assumption $K_s\subseteq\overset{\circ}{K}$ the distance between ∂K and ∂K_s is strictly larger than 0. From this we conclude that there is a constant $c > 0$ such that for all $x_s\in\partial K_s$

$$\mathrm{vol}_{n-1}(K \cap H_s) \geq c.$$

This implies that for all s with $0 < s < T$ there is a constant $c_s > 0$

$$\left| \frac{\mathrm{d}}{\mathrm{d}s} \int_{K_s} G(x)\mathrm{d}x \right| \leq c_s.$$

Thus, on all intervals $[a, T) \subset (0, T)$ the derivative is bounded and therefore the function is absolutely continuous. We get for all t_0, t with $0 < t_0 \leq t < T$

$$\int_{t_0}^{t} \frac{\mathrm{d}}{\mathrm{d}s} \int_{K_s} G(x)\mathrm{d}x = \int_{K_t} G(x)\mathrm{d}x - \int_{K_{t_0}} G(x)\mathrm{d}x.$$

We take the limit of $t_0 \to 0$. By Lemma 2.3.(iii) we have $\bigcup_{t>0} K_t \supseteq \overset{\circ}{K}$. The monotone convergence theorem implies

$$\int_{0}^{t} \frac{\mathrm{d}}{\mathrm{d}s} \int_{K_s} G(x)\mathrm{d}x = \int_{K_t} G(x)\mathrm{d}x - \int_{K} G(x)\mathrm{d}x.$$

Now we take the limit $t \to T$. By Lemma 2.3 we have $K_T = \bigcap_{t<T} K_t$. The monotone convergence theorem implies

$$\int_{0}^{T} \frac{\mathrm{d}}{\mathrm{d}s} \int_{K_s} G(x)\mathrm{d}x = \int_{K_T} G(x)\mathrm{d}x - \int_{K} G(x)\mathrm{d}x.$$

Since the volume of K_T equals 0 we get

$$\int_{0}^{T} \frac{\mathrm{d}}{\mathrm{d}s} \int_{K_s} G(x)\mathrm{d}x = - \int_{K} G(x)\mathrm{d}x.$$

□

3 The Case of the Euclidean Ball

We present here a proof of the main theorem in case that the convex body is the Euclidean ball. This result was proven by J. Müller [Mü]. We include the results of chapter 3 for the sake of completeness. Most of them are known.

Proposition 3.1. (Müller [Mü]) *We have*

$$\lim_{N \to \infty} \frac{\mathrm{vol}_n(B_2^n) - \mathbb{E}(\partial B_2^n, N)}{N^{-\frac{2}{n-1}}}$$

$$= \frac{\mathrm{vol}_{n-2}(\partial B_2^{n-1})}{2(n+1)!} \left(\frac{(n-1)\mathrm{vol}_{n-1}(\partial B_2^n)}{\mathrm{vol}_{n-2}(\partial B_2^{n-1})} \right)^{\frac{n+1}{n-1}} \Gamma\left(n+1+\frac{2}{n-1}\right)$$

$$= \frac{(n-1)^{\frac{n+1}{n-1}} (\mathrm{vol}_{n-1}(\partial B_2^n))^{\frac{n+1}{n-1}}}{(\mathrm{vol}_{n-2}(\partial B_2^{n-1}))^{\frac{2}{n-1}}} \frac{\Gamma\left(n+1+\frac{2}{n-1}\right)}{2(n+1)!}.$$

We want to show first that almost all random polytopes are simplicial.

Lemma 3.1. *The n^2-dimensional Hausdorff measure of the real $n \times n$-matrices with determinant 0 equals 0.*

Proof. We use induction. For $n = 1$ the only matrix with determinant 0 is the zeromatrix. Let A_{11} be the submatrix of the matrix A that is obtained by deleting the first row and column. We have

$$\{A|\det(A) = 0\} \subseteq \{A|\det(A_{11}) = 0\} \cup \{A|\det(A) = 0 \text{ and } \det(A_{11}) \neq 0\}.$$

Since

$$\{A|\det(A_{11}) = 0\} = \mathbb{R}^{n^2-(n-1)^2} \times \{B \in M_{n-1}|\det(B) = 0\}$$

we get by the induction assumption that $\{A|\det(A_{11}) = 0\}$ is a nullset. We have

$$\{A|\det(A) = 0 \text{ and } \det(A_{11}) \neq 0\}$$
$$= \left\{ A \,\middle|\, a_{11} = \frac{1}{\det(A_{11})} \sum_{i=1}^{n} a_{1i}(-1)^{1+i} \det(A_{1i})) \right\}.$$

Since this is the graph of a function it is a nullset. $\square$

Lemma 3.2. *The $n(n-1)$-dimensional Hausdorff measure of the real $n \times n$-matrices whose determinant equal 0 and whose columns have Euclidean norm equal to 1 is 0.*

Proof. Let $A_{i,j}$ be the submatrix of the matrix A that is obtained by deleting the i-th row and j-th column. We have

$$\{A|\det(A) = 0\} \subseteq \{A|\det(A_{11}) = 0\} \cup \{A|\det(A) = 0 \text{ and } \det(A_{11}) \neq 0\}.$$

By Lemma 3.1 the set of all $(n-1) \times (n-1)$ matrices with determinant equal to 0 has $(n-1)^2$-dimensional Hausdorff measure 0. Therefore, the set

$$\{(a_1, \dots, a_{n-1})|\det(\bar{a}_1, \dots, \bar{a}_{n-1}) = 0\}$$

has $(n-1)^2$-dimensional Hausdorff measure 0 where $\bar{a}_i$ is the vector a_i with the first coordinate deleted. From this we conclude that $\{A|\det(A_{11}) = 0\}$ has $n(n-1)$-dimensional Hausdorff measure 0.

As in Lemma 3.1 we have

$$\{A|\det(A) = 0 \text{ and } \det(A_{11}) \neq 0\}$$

$$= \left\{ A \,\middle|\, a_{11} = \frac{1}{\det(A_{11})} \sum_{i=1}^{n} a_{1i}(-1)^{1+i} \det(A_{1i})) \right\}.$$

By this and since the columns of the matrix have Euclidean length 1 the above set is the graph of a differentiable function of $n(n-1)-1$ variables. Thus the $n(n-1)$-dimensional Hausdorff measure is 0.   □

The next lemma says that almost all random polytopes of points chosen from a convex body are simplicial. Intuitively this is obvious. Suppose that we have chosen $x_1, \ldots, x_n$ and we want to choose x_{n+1} so that it is an element of the hyperplane spanned by $x_1, \ldots, x_n$, then we are choosing it from a nullset.

Lemma 3.3. *Let K be a convex body in $\mathbb{R}^n$ and $\mathbb{P}$ the normalized Lebesgue measure on K. Let $\mathbb{P}_K^N$ the N-fold probability measure of $\mathbb{P}$. Then*
(i)

$$\mathbb{P}_K^N\{(x_1, \ldots, x_N)|\exists i_1, \ldots, i_{n+1} \exists H : x_{i_1}, \ldots, x_{i_{n+1}} \in H\} = 0$$

where H denotes a hyperplane in $\mathbb{R}^n$.
(ii)

$$\mathbb{P}_K^N\{(x_1, \ldots, x_N)| \; \exists i_1, \ldots, i_n : x_{i_1}, \ldots, x_{i_n} \text{ are linearly dependent}\} = 0$$

Proof. (i) It suffices to show that

$$\mathbb{P}_K^N\{(x_1, \ldots, x_N)|\exists H : x_1, \ldots, x_{n+1} \in H\} = 0.$$

Let $X = (x_1, \ldots, x_n)$. We have that

$$\{(x_1, \ldots, x_N)|\exists H : x_1, \ldots, x_{n+1} \in H\} = \{(x_1, \ldots, x_N)| \det(X) = 0\}$$

$$\cup \left\{ (x_1, \ldots, x_N)| \det(X) \neq 0 \text{ and } \exists t_1, \ldots, t_{n-1} : \right.$$

$$\left. x_{n+1} = x_n + \sum_{i=1}^{n-1} t_i(x_i - x_n) \right\}.$$

The set with $\det(X) = 0$ has measure 0 by Lemma 3.1. Now we consider the second set. $\det(X) \neq 0$ and $x_{n+1} = x_n + \sum_{i=1}^{n-1} t_i(x_i - x_n)$ imply that

$$X^{-1}(x_{n+1}) = X^{-1}\left(x_n + \sum_{i=1}^{n-1} t_i(x_i - x_n) \right) = e_n + \sum_{i=1}^{n-1} t_i(e_i - e_n).$$

We get

$$t_i = < X^{-1}(x_{n+1}), e_i > \qquad i = 1, \ldots, n-1.$$

Therefore we get

$$\left\{ (x_1, \ldots, x_N) \,\middle|\, \det(X) \neq 0 \text{ and } \exists t_1, \ldots, t_{n-1} : x_{n+1} = x_n + \sum_{i=1}^{n-1} t_i(x_i - x_n) \right\}$$

$$\subseteq \left\{ (x_1, \ldots, x_n, z, x_{n+2}, \ldots, x_N) \,\middle|\, \det(X) \neq 0 \text{ and} \right.$$

$$\left. z = x_n + \sum_{i=1}^{n-1} < X^{-1}(x_{n+1}), e_i > (x_i - x_n) \right\}.$$

We have that

$$\frac{\partial z}{\partial x_{n+1}(j)} = \sum_{i=1}^{n-1} < X^{-1}(e_j), e_i > (x_i - x_n).$$

Since all the vectors $\frac{\partial z}{\partial x_{n+1}(j)}$, $j = 1, \ldots, n$ are linear combinations of the vectors $x_i - x_n$, $i = 1, \ldots, n-1$, the rank of the matrix

$$\left(\frac{\partial z}{\partial x_{n+1}(j)} \right)_{j=1}^{n}$$

is at most $n-1$. Therefore, the determinant of the Jacobian of the function mapping $(x_1, \ldots, x_N)$ onto $(x_1, \ldots, x_n, z, x_{n+2}, \ldots, x_N)$ is 0. Thus the set

$$\left\{ (x_1, \ldots, x_N) \,\middle|\, \det(X) \neq 0 \text{ and } \exists t_1, \ldots, t_{n-1} : x_{n+1} = x_n + \sum_{i=1}^{n-1} t_i(x_i - x_n) \right\}$$

has measure 0. □

Lemma 3.4. *Let $\mathbb{P}_{\partial B_2^n}$ be the normalized surface measure on ∂B_2^n. Let $\mathbb{P}_{\partial B_2^n}^N$ the N-fold probability measure of $\mathbb{P}_{\partial B_2^n}$. Then we have*
(i)

$$\mathbb{P}_{\partial B_2^n}^N \{(x_1, \ldots, x_N) | \exists i_1, \ldots, i_{n+1} \exists H : x_{i_1}, \ldots, x_{i_{n+1}} \in H\} = 0$$

where H denotes a hyperplane in $\mathbb{R}^n$.
(ii)

$$\mathbb{P}_{\partial B_2^n}^N \{(x_1, \ldots, x_N) | \exists i_1, \ldots, i_n : x_{i_1}, \ldots, x_{i_n} \text{ are linearly dependent}\} = 0$$

Proof. Lemma 3.4 is shown in the same way as Lemma 3.3. We use in addition the Cauchy-Binet formula ([EvG], p. 89). □

Lemma 3.5. *Almost all random polytopes of points chosen from the boundary of the Euclidean ball with respect to the normalized surface measure are simplicial.*

Lemma 3.5 follows from Lemma 3.4.(i).

Let F be a $n-1$-dimensional face of a polytope. Then dist(F) is the distance of the hyperplane containing F to the origin 0. We define

$$\Phi_{j_1,\ldots,j_k}(x) = \frac{1}{n}\text{vol}_{n-1}([x_{j_1},\ldots,x_{j_k}])\text{dist}(x_{j_1},\ldots,x_{j_k})$$

if $[x_{j_1},\ldots,x_{j_k}]$ is a $n-1$-dimensional face of the polytope $[x_1,\ldots,x_N]$ and if $0 \in H^+$ where H denotes the hyperplane containing the face $[x_{j_1},\ldots,x_{j_k}]$ and H^+ the halfspace containing $[x_1,\ldots,x_N]$. We define

$$\Phi_{j_1,\ldots,j_k}(x) = -\frac{1}{n}\text{vol}_{n-1}([x_{j_1},\ldots,x_{j_k}])\text{dist}(x_{j_1},\ldots,x_{j_k})$$

if $[x_{j_1},\ldots,x_{j_k}]$ is a $n-1$-dimensional face of the polytope $[x_1,\ldots,x_N]$ and if $0 \in H^-$. We put

$$\Phi_{j_1,\ldots,j_k}(x) = 0$$

if $[x_{j_1},\ldots,x_{j_k}]$ is not a $n-1$-dimensional face of the polytope $[x_1,\ldots,x_N]$.

Lemma 3.6. *Let $x_1,\ldots,x_N \in \mathbb{R}^n$ such that $[x_1,\ldots,x_N]$ is a simplicial polytope. Then we have*

$$\text{vol}_n([x_1,\ldots,x_N]) = \sum_{\{j_1,\ldots,j_n\}\subseteq\{1,\ldots,N\}} \Phi_{j_1,\ldots,j_n}(x).$$

Note that the above formula holds if $0 \in [x_1,\ldots,x_N]$ and if $0 \notin [x_1,\ldots,x_N]$.

dL_k^n is the measure on all k-dimensional affine subspaces of $\mathbb{R}^n$ and $dL_k^n(0)$ is the measure on all k-dimensional subspaces of $\mathbb{R}^n$ [San].

Lemma 3.7. [Bla1, San]

$$\bigwedge_{i=0}^{k} dx_i^n = (k!vol_k([x_0,\ldots,x_k]))^{n-k}\bigwedge_{i=0}^{k} dx_i^k dL_k^n$$

where dx_i^n is the volume element in $\mathbb{R}^n$ and dx_i^k is the volume element in L_k^n.

The above formula can be found as formula (12.22) on page 201 in [San]. We need this formula here only in the case $k = n-1$. It can be found as formula (12.24) on page 201 in [San]. The general formula can also be found in [Mil]. See also [Ki] and [Pe].

Lemma 3.8.

$$dL^n_{n-1} = dp d\mu_{\partial B^n_2}(\xi)$$

where p is the distance of the hyperplane from the origin and ξ is the normal of the hyperplane.

This lemma is formula (12.40) in [San].

Let X be a metric space. Then a sequence of probability measures $\mathbb{P}_n$ converges weakly to a probability measure $\mathbb{P}$ if we have for all $\phi \in C(X)$ that

$$\lim_{n \to \infty} \int_X \phi d\mathbb{P}_n = \int_X \phi d\mathbb{P}_n.$$

See ([Bil], p.7). In fact, we have that two probability measures $\mathbb{P}_1$ and $\mathbb{P}_2$ coincide on the underlying Borel σ-algebra if we have for all continuous functions ϕ that

$$\int_X \phi d\mathbb{P}_1 = \int_X \phi d\mathbb{P}_2.$$

Lemma 3.9. *We put*

$$A_\epsilon = B^n_2(0, r + \epsilon) \setminus B^n_2(0, r)$$

and as probability measure $\mathbb{P}_\epsilon$ on $A_\epsilon \times A_\epsilon \times \cdots \times A_\epsilon$

$$\mathbb{P}_\epsilon = \frac{\chi_{A_\epsilon} \times \cdots \times \chi_{A_\epsilon}(x_1) dx_1 \ldots dx_k}{((r + \epsilon)^n - r^n)^k (\text{vol}_n(B^n_2))^k}.$$

Then $\mathbb{P}_\epsilon$ converges weakly for ϵ to 0 to the k-fold product of the normalized surface measure on $\partial B^n_2(0, r)$

$$\frac{\mu_{\partial B^n_2(0,r)}(x_1) \ldots \mu_{\partial B^n_2(0,r)}(x_k)}{r^{k(n-1)}(\text{vol}_{n-1}(\partial B^n_2))^k}.$$

Proof. All the measures are being viewed as measures on $\mathbb{R}^n$, otherwise it would not make sense to talk about convergence. For the proof we consider a continuous function ϕ on $\mathbb{R}^n$ and Riemann sums for the Euclidean sphere. □

Lemma 3.10. [Mil]

$$d\mu_{\partial B^n_2}(x_1) \cdots d\mu_{\partial B^n_2}(x_n)$$

$$= (n - 1)! \frac{\text{vol}_{n-1}([x_1, \ldots, x_n])}{(1 - p^2)^{\frac{n}{2}}} d\mu_{\partial B^n_2 \cap H}(x_1) \cdots d\mu_{\partial B^n_2 \cap H}(x_n) dp d\mu_{\partial B^n_2}(\xi)$$

where ξ is the normal to the plane H through $x_1, \ldots, x_n$ and p is the distance of the plane H to the origin.

Proof. We put

$$A_\epsilon = B_2^n(0, 1 + \epsilon) \setminus B_2^n(0, 1)$$

and as probability measure $\mathbb{P}_\epsilon$ on $A_\epsilon \times A_\epsilon \times \cdots \times A_\epsilon$

$$\mathbb{P}_\epsilon = \frac{\chi_{A_\epsilon} \times \cdots \times \chi_{A_\epsilon}(x_1) dx_1 \ldots dx_n}{((1 + \epsilon)^n - 1)^n (\mathrm{vol}_n(B_2^n))^n}.$$

Then, by Lemma 3.9, $\mathbb{P}_\epsilon$ converges for ϵ to 0 to the n-fold product of the normalized surface measure on ∂B_2^n

$$\frac{\mu_{\partial B_2^n}(x_1) \ldots \mu_{\partial B_2^n}(x_n)}{(\mathrm{vol}_{n-1}(\partial B_2^n))^n}.$$

By Lemma 3.7 we have

$$\bigwedge_{i=1}^n dx_i^n = (n-1)! \mathrm{vol}_{n-1}([x_1, \ldots, x_n]) dL_{n-1}^n \bigwedge_{i=1}^n dx_i^{n-1}$$

and by Lemma 3.8

$$dL_{n-1}^n = dp d\mu_{\partial B_2^n}(\xi).$$

We get

$$\bigwedge_{i=1}^n dx_i^n = (n-1)! \mathrm{vol}_{n-1}([x_1, \ldots, x_n]) \bigwedge_{i=1}^n dx_i^{n-1} dp d\mu_{\partial B_2^n}(\xi).$$

Thus we get

$$\mathbb{P}_\epsilon = \chi_{A_\epsilon} \times \cdots \times \chi_{A_\epsilon}(n-1)! \mathrm{vol}_{n-1}([x_1, \ldots, x_n])$$
$$\times \frac{dx_1^{n-1} \ldots dx_n^{n-1} dp d\mu_{\partial B_2^n}(\xi)}{((1 + \epsilon)^n - 1)^n (\mathrm{vol}_n(B_2^n))^n}.$$

This can also be written as

$$\mathbb{P}_\epsilon = (n-1)! \mathrm{vol}_{n-1}([x_1, \ldots, x_n])$$
$$\times \frac{\chi_{A_\epsilon \cap H} \times \cdots \times \chi_{A_\epsilon \cap H} dx_1^{n-1} \ldots dx_n^{n-1} dp d\mu_{\partial B_2^n}(\xi)}{((1 + \epsilon)^n - 1)^n (\mathrm{vol}_n(B_2^n))^n}$$

where H is the hyperplane with normal ξ that contains the points $x_1, \ldots, x_n$. p is the distance of H to 0. $A_\epsilon \cap H$ is the set-theoretic difference of a Euclidean ball of dimension $n - 1$ with radius $(1 - p^2 + 2\epsilon + \epsilon^2)^{\frac{1}{2}}$ and a ball with radius $(1 - p^2)^{\frac{1}{2}}$. By Lemma 3.9 we have that

$$\frac{\chi_{A_\epsilon \cap H} \times \cdots \times \chi_{A_\epsilon \cap H} dx_1^{n-1} \ldots dx_n^{n-1}}{((1 - p^2 + 2\epsilon + \epsilon^2)^{\frac{n-1}{2}} - (1 - p^2)^{\frac{n-1}{2}})^n (\mathrm{vol}_{n-1}(B_2^{n-1}))^n}$$

converges weakly to the n-fold product of the normalized surface measure on $\partial B_2^n \cap H$

$$\frac{\mathrm{d}\mu_{\partial B_2^n \cap H} \ldots \mathrm{d}\mu_{\partial B_2^n \cap H}}{(1-p^2)^{n\frac{n-2}{2}}(\mathrm{vol}_{n-2}(\partial B_2^{n-1}))^n}.$$

Therefore we get that

$$\frac{\chi_{A_\epsilon \cap H} \times \cdots \times \chi_{A_\epsilon \cap H} \mathrm{d}x_1^{n-1} \ldots \mathrm{d}x_n^{n-1}}{((1+\epsilon)^n - 1)^n (\mathrm{vol}_n(B_2^n))^n}$$

converges to

$$\left(\frac{(n-1)\mathrm{vol}_{n-1}(B_2^{n-1})}{n \, \mathrm{vol}_n(B_2^n)}\right)^n (1-p^2)^{n\frac{n-1}{2}-n} \frac{\mathrm{d}\mu_{\partial B_2^n \cap H} \ldots \mathrm{d}\mu_{\partial B_2^n \cap H}}{(1-p^2)^{n\frac{n-2}{2}}(\mathrm{vol}_{n-2}(\partial B_2^{n-1}))^n}$$

$$= \frac{\mathrm{d}\mu_{\partial B_2^n \cap H} \ldots \mathrm{d}\mu_{\partial B_2^n \cap H}}{(1-p^2)^{\frac{n}{2}}(\mathrm{vol}_{n-1}(\partial B_2^n))^n}.$$

□

Lemma 3.11. [Mil]

$$\int_{\partial B_2^n(0,r)} \cdots \int_{\partial B_2^n(0,r)} (\mathrm{vol}_n([x_1,\ldots,x_{n+1}]))^2$$

$$\times \mathrm{d}\mu_{\partial B_2^n(0,r)}(x_1)\cdots \mathrm{d}\mu_{\partial B_2^n(0,r)}(x_{n+1})$$

$$= \frac{(n+1)r^{2n}}{n!n^n}(\mathrm{vol}_{n-1}(\partial B_2^n(r)))^{n+1} = \frac{(n+1)r^{n^2+2n-1}}{n!n^n}(\mathrm{vol}_{n-1}(\partial B_2^n))^{n+1}$$

We just want to refer to [Mil] for the proof. But we want to indicate an alternative proof here. One can use

$$\lim_{N \to \infty} \mathbb{E}(\partial B_2^n, N) = \mathrm{vol}_n(B_2^n)$$

and the computation in the proof of Proposition 3.1.

Lemma 3.12. *Let C be a cap of a Euclidean ball with radius 1. Let s be the surface area of this cap and r its radius. Then we have*

$$\left(\frac{s}{\mathrm{vol}_{n-1}(B_2^{n-1})}\right)^{\frac{1}{n-1}} - \frac{1}{2(n+1)}\left(\frac{s}{\mathrm{vol}_{n-1}(B_2^{n-1})}\right)^{\frac{3}{n-1}}$$

$$-c\left(\frac{s}{\mathrm{vol}_{n-1}(B_2^{n-1})}\right)^{\frac{5}{n-1}} \le r(s) \le \left(\frac{s}{\mathrm{vol}_{n-1}(B_2^{n-1})}\right)^{\frac{1}{n-1}}$$

$$- \frac{1}{2(n+1)}\left(\frac{s}{\mathrm{vol}_{n-1}(B_2^{n-1})}\right)^{\frac{3}{n-1}} + c\left(\frac{s}{\mathrm{vol}_{n-1}(B_2^{n-1})}\right)^{\frac{5}{n-1}}$$

where c is a numerical constant.

Proof. The surface area s of a cap of the Euclidean ball of radius 1 is

$$s = \text{vol}_{n-2}(\partial B_2^{n-1}) \int_0^\alpha \sin^{n-2} t\, dt$$

where α is the angle of the cap. Then $\alpha = \arcsin r$ where r is the radius of the cap. For all t with $t \geq 0$

$$t - \tfrac{1}{3!}t^3 \leq \sin t \leq t - \tfrac{1}{3!}t^3 + \tfrac{1}{5!}t^5.$$

Therefore we get for all t with $t \geq 0$

$$\sin^{n-2} t \geq (t - \tfrac{1}{3!}t^3)^{n-2} = t^{n-2}(1 - \tfrac{1}{3!}t^2)^{n-2} \geq t^{n-2}(1 - \tfrac{n-2}{3!}t^2) = t^{n-2} - \tfrac{n-2}{3!}t^n.$$

Now we use $(1-u)^k \leq 1 - ku + \tfrac{1}{2}k(k-1)u^2$ and get for all $t \geq 0$

$$\sin^{n-2} t \leq t^{n-2} - \tfrac{n-2}{3!}t^n + ct^{n+2}.$$

Thus

$$
\begin{aligned}
s &\geq \text{vol}_{n-2}(\partial B_2^{n-1}) \int_0^\alpha t^{n-2} - \tfrac{n-2}{3!}t^n\, dt \\
&= \text{vol}_{n-2}(\partial B_2^{n-1}) \left(\tfrac{1}{n-1}\alpha^{n-1} - \tfrac{n-2}{6(n+1)}\alpha^{n+1} \right) \\
&= \text{vol}_{n-2}(\partial B_2^{n-1}) \left(\tfrac{1}{n-1}(\arcsin r)^{n-1} - \tfrac{n-2}{6(n+1)}(\arcsin r)^{n+1} \right)
\end{aligned}
$$

and

$$
s \leq \text{vol}_{n-2}(\partial B_2^{n-1}) \times \\
\left(\tfrac{1}{n-1}(\arcsin r)^{n-1} - \tfrac{n-2}{6(n+1)}(\arcsin r)^{n+1} + \tfrac{c}{n+3}(\arcsin r)^{n+3} \right).
$$

We have

$$\arcsin r = r + \frac{1}{2}\frac{r^3}{3} + \frac{1\cdot 3}{2\cdot 4}\frac{r^5}{5} + \frac{1\cdot 3\cdot 5}{2\cdot 4\cdot 6}\frac{r^7}{7} + \cdots$$

Thus we have for all sufficiently small r that

$$r + \tfrac{1}{3!}r^3 \leq \arcsin r \leq r + \tfrac{1}{3!}r^3 + r^5.$$

We get with a new constant c

$$
\begin{aligned}
s &\geq \text{vol}_{n-2}(\partial B_2^{n-1}) \left(\tfrac{1}{n-1}(r + \tfrac{1}{3!}r^3)^{n-1} - \tfrac{n-2}{6(n+1)}(r + \tfrac{1}{3!}r^3 + r^5)^{n+1} \right) \\
&\geq \text{vol}_{n-2}(\partial B_2^{n-1}) \left(\tfrac{1}{n-1}r^{n-1} + \tfrac{1}{3!}r^{n+1} - \tfrac{n-2}{6(n+1)}r^{n+1} - cr^{n+3} \right) \\
&= \text{vol}_{n-2}(\partial B_2^{n-1}) \left(\tfrac{1}{n-1}r^{n-1} + \tfrac{1}{2(n+1)}r^{n+1} - cr^{n+3} \right) \\
&= \text{vol}_{n-1}(B_2^{n-1}) \left(r^{n-1} + \tfrac{n-1}{2(n+1)}r^{n+1} - c(n-1)r^{n+3} \right).
\end{aligned}
$$

We get the inverse inequality

$$s \leq \mathrm{vol}_{n-1}(B_2^{n-1}) \left(r^{n-1} + \tfrac{n-1}{2(n+1)} r^{n+1} + c(n-1)r^{n+3} \right)$$

in the same way. We put now

$$u = \frac{s}{\mathrm{vol}_{n-1}(B_2^{n-1})}$$

and get

$$u^{\frac{1}{n-1}} - \frac{1}{2(n+1)} u^{\frac{3}{n-1}} - c u^{\frac{5}{n-1}}$$

$$\leq \left(r^{n-1} + \tfrac{n-1}{2(n+1)} r^{n+1} + c(n-1)r^{n+3} \right)^{\frac{1}{n-1}}$$

$$- \frac{1}{2(n+1)} \left(r^{n-1} + \tfrac{n-1}{2(n+1)} r^{n+1} - c(n-1)r^{n+3} \right)^{\frac{3}{n-1}}$$

$$- a \left(r^{n-1} + \tfrac{n-1}{2(n+1)} r^{n+1} - c(n-1)r^{n+3} \right)^{\frac{5}{n-1}}.$$

If we choose a big enough then this can be estimated with a new constant c by

$$r - cr^5 \leq r$$

provided r is small enough. The opposite inequality is shown in the same way. Altogether we have with an appropriate constant c

$$\left(\frac{s}{\mathrm{vol}_{n-1}(B_2^{n-1})} \right)^{\frac{1}{n-1}} - \frac{1}{2(n+1)} \left(\frac{s}{\mathrm{vol}_{n-1}(B_2^{n-1})} \right)^{\frac{3}{n-1}}$$

$$- c \left(\frac{s}{\mathrm{vol}_{n-1}(B_2^{n-1})} \right)^{\frac{5}{n-1}} \leq r(s) \leq \left(\frac{s}{\mathrm{vol}_{n-1}(B_2^{n-1})} \right)^{\frac{1}{n-1}}$$

$$- \frac{1}{2(n+1)} \left(\frac{s}{\mathrm{vol}_{n-1}(B_2^{n-1})} \right)^{\frac{3}{n-1}} + c \left(\frac{s}{\mathrm{vol}_{n-1}(B_2^{n-1})} \right)^{\frac{5}{n-1}}.$$

□

Proof. (Proof of Proposition 3.1) We have

$$\mathbb{P} = \frac{\mu_{\partial B_2^n}}{\mathrm{vol}_{n-1}(B_2^n)}$$

and

$$\mathbb{E}(\partial B_2^n, N) = \int_{\partial B_2^n} \cdots \int_{\partial B_2^n} \mathrm{vol}_n([x_1, \ldots, x_N]) \mathrm{d}\mathbb{P}(x_1) \cdots \mathrm{d}\mathbb{P}(x_N).$$

By Lemma 3.5 almost all random polytopes are simplicial. Therefore we get with Lemma 3.6

$$
\mathbb{E}(\partial B_2^n, N)
$$
$$
= \int_{\partial B_2^n} \cdots \int_{\partial B_2^n} \sum_{\{j_1,\ldots,j_n\}\subseteq\{1,\ldots,N\}} \Phi_{j_1,\ldots,j_n}(x_1,\ldots,x_N)d\mathbb{P}(x_1)\cdots d\mathbb{P}(x_N)
$$
$$
= \binom{N}{n} \int_{\partial B_2^n} \cdots \int_{\partial B_2^n} \Phi_{1,\ldots,n}(x_1,\ldots,x_N)d\mathbb{P}(x_1)\cdots d\mathbb{P}(x_N).
$$

H is the hyperplane containing the points $x_1,\ldots,x_n$. The set of points where H is not well defined has measure 0. H^+ is the halfspace containing the polytope $[x_1,\ldots,x_N]$. We have

$$
\mathbb{P}^{N-n}\{(x_{n+1},\ldots,x_N)|\Phi_{1,\ldots,n}(x_1,\ldots,x_N)
$$
$$
= \tfrac{1}{n}\mathrm{vol}_{n-1}([x_1,\ldots,x_n])\mathrm{dist}(x_1,\ldots,x_n)\}
$$
$$
= \left(\frac{\mathrm{vol}_{n-1}(\partial B_2^n \cap H^+)}{\mathrm{vol}_{n-1}(\partial B_2^n)}\right)^{N-n}
$$

and

$$
\mathbb{P}^{N-n}\{(x_{n+1},\ldots,x_N)|\Phi_{1,\ldots,n}(x_1,\ldots,x_N)
$$
$$
= -\tfrac{1}{n}\mathrm{vol}_{n-1}([x_1,\ldots,x_n])\mathrm{dist}(x_1,\ldots,x_n)\}
$$
$$
= \left(\frac{\mathrm{vol}_{n-1}(\partial B_2^n \cap H^-)}{\mathrm{vol}_{n-1}(\partial B_2^n)}\right)^{N-n}.
$$

Therefore

$$
\mathbb{E}(\partial B_2^n, N) = \binom{N}{n}\frac{1}{n}\int_{\partial B_2^n}\cdots\int_{\partial B_2^n}\mathrm{vol}_{n-1}([x_1,\ldots,x_n])\mathrm{dist}(x_1,\ldots,x_n)
$$
$$
\times\left\{\left(\frac{\mathrm{vol}_{n-1}(\partial B_2^n \cap H^+)}{\mathrm{vol}_{n-1}(\partial B_2^n)}\right)^{N-n} - \left(\frac{\mathrm{vol}_{n-1}(\partial B_2^n \cap H^-)}{\mathrm{vol}_{n-1}(\partial B_2^n)}\right)^{N-n}\right\}
$$
$$
\times d\mathbb{P}(x_1)\cdots d\mathbb{P}(x_n).
$$

By Lemma 3.10 we get

$$
\mathbb{E}(\partial B_2^n, N) = \frac{1}{n}\binom{N}{n}\frac{(n-1)!}{(\mathrm{vol}_{n-1}(\partial B_2^n))^n}\int_{\partial B_2^n}\int_0^1 p(1-p^2)^{-\frac{n}{2}}
$$
$$
\times\left\{\left(\frac{\mathrm{vol}_{n-1}(\partial B_2^n \cap H^+)}{\mathrm{vol}_{n-1}(\partial B_2^n)}\right)^{N-n} - \left(\frac{\mathrm{vol}_{n-1}(\partial B_2^n \cap H^-)}{\mathrm{vol}_{n-1}(\partial B_2^n)}\right)^{N-n}\right\}
$$
$$
\times\int_{\partial B_2^n\cap H}\cdots\int_{\partial B_2^n\cap H}(\mathrm{vol}_{n-1}([x_1,\ldots,x_n]))^2
$$
$$
\times d\mu_{\partial B_2^n\cap H}(x_1)\cdots d\mu_{\partial B_2^n\cap H}(x_n)dp d\mu_{\partial B_2^n}(\xi).
$$

We apply Lemma 3.11 for the dimension $n - 1$

$$\mathbb{E}(\partial B_2^n, N) = \frac{1}{n}\binom{N}{n}\frac{(n-1)!}{(\mathrm{vol}_{n-1}(\partial B_2^n))^n}\int_{\partial B_2^n}\int_0^1 p(1-p^2)^{-\frac{n}{2}}$$

$$\times\left\{\left(\frac{\mathrm{vol}_{n-1}(\partial B_2^n \cap H^+)}{\mathrm{vol}_{n-1}(\partial B_2^n)}\right)^{N-n} - \left(\frac{\mathrm{vol}_{n-1}(\partial B_2^n \cap H^-)}{\mathrm{vol}_{n-1}(\partial B_2^n)}\right)^{N-n}\right\}$$

$$\times\frac{nr^{n^2-2}}{(n-1)!(n-1)^{n-1}}(\mathrm{vol}_{n-2}(\partial B_2^{n-1}))^n dp d\mu_{\partial B_2^n}(\xi).$$

Since $r(p) = \sqrt{1-p^2}$ we get

$$\mathbb{E}(\partial B_2^n, N) = \binom{N}{n}\frac{(\mathrm{vol}_{n-2}(\partial B_2^{n-1}))^n}{(\mathrm{vol}_{n-1}(\partial B_2^n))^{n-1}}\frac{1}{(n-1)^{n-1}}\int_0^1 r^{n^2-n-2}\sqrt{1-r^2}$$

$$\left\{\left(\frac{\mathrm{vol}_{n-1}(\partial B_2^n \cap H^+)}{\mathrm{vol}_{n-1}(\partial B_2^n)}\right)^{N-n} - \left(\frac{\mathrm{vol}_{n-1}(\partial B_2^n \cap H^-)}{\mathrm{vol}_{n-1}(\partial B_2^n)}\right)^{N-n}\right\} dp.$$

Now we introduce the surface area s of a cap with height $1 - p$ as a new variable. By Lemma 1.5 we have

$$\frac{dp}{ds} = -\left(r^{n-3}\mathrm{vol}_{n-2}(\partial B_2^{n-1})\right)^{-1}.$$

Thus we get

$$\mathbb{E}(\partial B_2^n, N) = \binom{N}{n}\frac{(\mathrm{vol}_{n-2}(\partial B_2^{n-1}))^{n-1}}{(\mathrm{vol}_{n-1}(\partial B_2^n))^{n-1}}\frac{1}{(n-1)^{n-1}}$$

$$\times\int_0^{\frac{1}{2}\mathrm{vol}_{n-1}(\partial B_2^n)} r^{(n-1)^2}\sqrt{1-r^2}$$

$$\times\left\{\left(1-\frac{s}{\mathrm{vol}_{n-1}(\partial B_2^n)}\right)^{N-n} - \left(\frac{s}{\mathrm{vol}_{n-1}(\partial B_2^n)}\right)^{N-n}\right\} ds.$$

Now we introduce the variable

$$u = \frac{s}{\mathrm{vol}_{n-1}(\partial B_2^n)}$$

and obtain

$$\mathbb{E}(\partial B_2^n, N) = \binom{N}{n}\frac{(\mathrm{vol}_{n-2}(\partial B_2^{n-1}))^{n-1}}{(\mathrm{vol}_{n-1}(\partial B_2^n))^{n-2}}\frac{1}{(n-1)^{n-1}}$$

$$\times\int_0^{\frac{1}{2}} r^{(n-1)^2}\sqrt{1-r^2}\left\{(1-u)^{N-n} - u^{N-n}\right\} du.$$

By Lemma 3.12 we get

$$\mathbb{E}(\partial B_2^n, N)$$

$$\leq \binom{N}{n} \frac{(\mathrm{vol}_{n-2}(\partial B_2^{n-1}))^{n-1}}{(\mathrm{vol}_{n-1}(\partial B_2^n))^{n-2}} \frac{1}{(n-1)^{n-1}} \int_0^{\frac{1}{2}} \left\{ (1-u)^{N-n} - u^{N-n} \right\}$$

$$\times \left\{ \left(\frac{u\, \mathrm{vol}_{n-1}(\partial B_2^n)}{\mathrm{vol}_{n-1}(B_2^{n-1})} \right)^{\frac{1}{n-1}} - \frac{1}{2(n+1)} \left(\frac{u\, \mathrm{vol}_{n-1}(\partial B_2^n)}{\mathrm{vol}_{n-1}(B_2^{n-1})} \right)^{\frac{3}{n-1}} \right.$$

$$\left. + c \left(\frac{u\, \mathrm{vol}_{n-1}(\partial B_2^n)}{\mathrm{vol}_{n-1}(B_2^{n-1})} \right)^{\frac{5}{n-1}} \right\}^{(n-1)^2}$$

$$\times \left\{ 1 - \left[\left(\frac{u\, \mathrm{vol}_{n-1}(\partial B_2^n)}{\mathrm{vol}_{n-1}(B_2^{n-1})} \right)^{\frac{1}{n-1}} - \frac{1}{2(n+1)} \left(\frac{u\, \mathrm{vol}_{n-1}(\partial B_2^n)}{\mathrm{vol}_{n-1}(B_2^{n-1})} \right)^{\frac{3}{n-1}} \right. \right.$$

$$\left. \left. - c \left(\frac{u\, \mathrm{vol}_{n-1}(\partial B_2^n)}{\mathrm{vol}_{n-1}(B_2^{n-1})} \right)^{\frac{5}{n-1}} \right]^2 \right\}^{\frac{1}{2}} du.$$

From this we get

$$\mathbb{E}(\partial B_2^n, N)$$

$$\leq \binom{N}{n} \mathrm{vol}_{n-1}(\partial B_2^n) \int_0^{\frac{1}{2}} \left\{ (1-u)^{N-n} - u^{N-n} \right\} u^{n-1} \times$$

$$\left\{ 1 - \frac{1}{2(n+1)} \left(\frac{u\, \mathrm{vol}_{n-1}(\partial B_2^n)}{\mathrm{vol}_{n-1}(B_2^{n-1})} \right)^{\frac{2}{n-1}} + c \left(\frac{u\, \mathrm{vol}_{n-1}(\partial B_2^n)}{\mathrm{vol}_{n-1}(B_2^{n-1})} \right)^{\frac{4}{n-1}} \right\}^{(n-1)^2}$$

$$\times \left\{ 1 - \left[\left(\frac{u\, \mathrm{vol}_{n-1}(\partial B_2^n)}{\mathrm{vol}_{n-1}(B_2^{n-1})} \right)^{\frac{1}{n-1}} - \frac{1}{2(n+1)} \left(\frac{u\, \mathrm{vol}_{n-1}(\partial B_2^n)}{\mathrm{vol}_{n-1}(B_2^{n-1})} \right)^{\frac{3}{n-1}} \right. \right.$$

$$\left. \left. - c \left(\frac{u\, \mathrm{vol}_{n-1}(\partial B_2^n)}{\mathrm{vol}_{n-1}(B_2^{n-1})} \right)^{\frac{5}{n-1}} \right]^2 \right\}^{\frac{1}{2}} du.$$

This implies that we get for a new constant c

$$\mathbb{E}(\partial B_2^n, N)$$

$$\leq \binom{N}{n} \mathrm{vol}_{n-1}(\partial B_2^n) \int_0^{\frac{1}{2}} \left\{ (1-u)^{N-n} - u^{N-n} \right\} u^{n-1}$$

$$\times \left\{ 1 - \frac{(n-1)^2}{2(n+1)} \left(\frac{u\, \mathrm{vol}_{n-1}(\partial B_2^n)}{\mathrm{vol}_{n-1}(B_2^{n-1})} \right)^{\frac{2}{n-1}} + c \left(\frac{u\, \mathrm{vol}_{n-1}(\partial B_2^n)}{\mathrm{vol}_{n-1}(B_2^{n-1})} \right)^{\frac{4}{n-1}} \right\}$$

$$\times \left(1 - \left\{ \frac{1}{2} \left(\frac{u\, \mathrm{vol}_{n-1}(\partial B_2^n)}{\mathrm{vol}_{n-1}(B_2^{n-1})} \right)^{\frac{2}{n-1}} - c \left(\frac{u\, \mathrm{vol}_{n-1}(\partial B_2^n)}{\mathrm{vol}_{n-1}(B_2^{n-1})} \right)^{\frac{4}{n-1}} \right\} \right) du.$$

This gives, again with a new constant c

$$\mathbb{E}(\partial B_2^n, N) \le \binom{N}{n} \mathrm{vol}_{n-1}(\partial B_2^n) \int_0^{\frac{1}{2}} \left\{ (1-u)^{N-n} - u^{N-n} \right\} u^{n-1} du$$

$$- \binom{N}{n} \frac{n^2 - n + 2}{2(n+1)} \frac{\mathrm{vol}_{n-1}(\partial B_2^n)^{\frac{n+1}{n-1}}}{\mathrm{vol}_{n-1}(B_2^{n-1})^{\frac{2}{n-1}}}$$

$$\times \int_0^{\frac{1}{2}} \left\{ (1-u)^{N-n} - u^{N-n} \right\} u^{n-1+\frac{2}{n-1}} du$$

$$+ c \binom{N}{n} \int_0^{\frac{1}{2}} \left\{ (1-u)^{N-n} - u^{N-n} \right\} u^{n-1+\frac{4}{n-1}} du.$$

From this we get

$$\mathbb{E}(\partial B_2^n, N) \le \binom{N}{n} \mathrm{vol}_{n-1}(\partial B_2^n) B(N-n+1, n)$$

$$- \binom{N}{n} \frac{n^2 - n + 2}{2(n+1)} \frac{\mathrm{vol}_{n-1}(\partial B_2^n)^{\frac{n+1}{n-1}}}{\mathrm{vol}_{n-1}(B_2^{n-1})^{\frac{2}{n-1}}} B(N-n+1, n+\tfrac{2}{n-1})$$

$$+ c \binom{N}{n} B(N-n+1, n+\tfrac{4}{n-1}) + c \left(\frac{1}{2}\right)^{-N+\frac{2}{n-1}}.$$

This implies

$$\mathbb{E}(\partial B_2^n, N) \le \mathrm{vol}_n(B_2^n)$$

$$- \binom{N}{n} \frac{n^2 - n + 2}{2(n+1)} \frac{\mathrm{vol}_{n-1}(\partial B_2^n)^{\frac{n+1}{n-1}}}{\mathrm{vol}_{n-1}(B_2^{n-1})^{\frac{2}{n-1}}} \frac{\Gamma(N-n+1)\Gamma(n+\frac{2}{n-1})}{\Gamma(N+1+\frac{2}{n-1})}$$

$$+ c \binom{N}{n} \frac{\Gamma(N-n+1)\Gamma(n+\frac{4}{n-1})}{\Gamma(N+1+\frac{4}{n-1})} + c \left(\frac{1}{2}\right)^{-N+\frac{2}{n-1}}.$$

We have the asymptotic formula

$$\lim_{k \to \infty} \frac{\Gamma(k+\beta)}{\Gamma(k)k^\beta} = 1.$$

Therefore we get that $\mathbb{E}(\partial B_2^n, N)$ is asymptotically less than

$$\mathrm{vol}_n(B_2^n) - \frac{n^2 - n + 2}{2(n+1)} \frac{\mathrm{vol}_{n-1}(\partial B_2^n)^{\frac{n+1}{n-1}}}{\mathrm{vol}_{n-1}(B_2^{n-1})^{\frac{2}{n-1}}} \frac{\Gamma(n+\frac{2}{n-1})}{n! N^{\frac{2}{n-1}}}$$

$$+ c \frac{\Gamma(n+\frac{4}{n-1})}{n! N^{\frac{4}{n-1}}} + c \left(\frac{1}{2}\right)^{-N+\frac{2}{n-1}}.$$

We apply now $x\Gamma(x) = \Gamma(x+1)$ to $x = n + \frac{2}{n-1}$.

$$\mathbb{E}(\partial B_2^n, N) \le \mathrm{vol}_n(B_2^n) - \frac{n-1}{2(n+1)!} \frac{\mathrm{vol}_{n-1}(\partial B_2^n)^{\frac{n+1}{n-1}}}{\mathrm{vol}_{n-1}(B_2^{n-1})^{\frac{2}{n-1}}} \frac{\Gamma(n+1+\frac{2}{n-1})}{N^{\frac{2}{n-1}}}$$

$$+ c \frac{\Gamma(n+\frac{4}{n-1})}{n! N^{\frac{4}{n-1}}} + c \left(\frac{1}{2}\right)^{-N+\frac{2}{n-1}}.$$

The other inequality is proved similarly.   □

4 Probabilistic Estimates

4.1 Probabilistic Estimates for General Convex Bodies

Lemma 4.1. *Let K be a convex body in $\mathbb{R}^n$ with 0 as an interior point. The $n(n-1)$-dimensional Hausdorff measure of the real $n \times n$-matrices whose determinant equal 0 and whose columns are elements of ∂K is 0.*

Proof. We deduce this lemma from Lemma 3.2. We consider the map $rp : \partial B_2^n \to \partial K$

$$rp^{-1}(x) = \frac{x}{\|x\|}$$

and $Rp : \partial B_2^n \times \cdots \times \partial B_2^n \to \partial K \times \cdots \times \partial K$ with

$$Rp(x_1, \ldots, x_n) = (rp(x_1), \ldots, rp(x_n)).$$

Rp is a Lipschitz-map and the image of a nullset is a nullset.   □

Lemma 4.2. *Let K be a convex body in $\mathbb{R}^n$ and let $f : \partial K \to \mathbb{R}$ be a continuous, positive function with $\int f d\mu = 1$. Then we have for all $x \in \overset{\circ}{K}$*

$$\mathbb{P}_f^N \{(x_1, \ldots, x_N) | x \in \partial[x_1, \ldots, x_N]\} = 0.$$

Let $\epsilon = (\epsilon(i))_{1 \leq i \leq n}$ be a sequence of signs, that is $\epsilon(i) = \pm 1$, $1 \leq i \leq n$. We denote, for a given sequence ϵ of signs, by K^ϵ the following subset of K

$$K^\epsilon = \{x = (x(1), x(2), \ldots, x(n)) \in K | \ \forall i = 1, \ldots, n : \text{sgn}(x(i)) = \epsilon(i)\}.$$

Lemma 4.3. *(i) Let K be a convex body in $\mathbb{R}^n$, a, b positive constants and $\mathcal{E}$ an ellipsoid with center 0 such that $a\mathcal{E} \subseteq K \subseteq b\mathcal{E}$. Then we have*

$$\mathbb{P}_{\partial K}^N \{(x_1, \ldots, x_N) | 0 \notin [x_1, \ldots, x_N]\} \leq 2^n \left(1 - \frac{1}{2^n} \left(\frac{a}{b}\right)^{n-1}\right)^N.$$

(ii) Let K be a convex body in $\mathbb{R}^n$, 0 an interior point of K, and let $f : \partial K \to \mathbb{R}$ be a continuous, nonnegative function with $\int_{\partial K} f(x) d\mu = 1$. Then we have

$$\mathbb{P}_f^N \{(x_1, \ldots, x_N) | 0 \notin [x_1, \ldots, x_N]\} \leq 2^n \left(1 - \min_\epsilon \int_{\partial K^\epsilon} f(x) d\mu\right)^N.$$

(Here we do not assume that the function f is strictly positive.)

Proof. (i) A rotation puts K into such a position that

$$\mathcal{E} = \left\{ x \left| \sum_{i=1}^{n} \left| \frac{x(i)}{a_i} \right|^2 \leq 1 \right. \right\}.$$

We have for all ϵ

$$\frac{a^{n-1}}{2^n} \mathrm{vol}_{n-1}(\partial\mathcal{E}) \leq \mathrm{vol}_{n-1}(\partial K^{\epsilon}).$$

We show this. Let $p_{K,a\mathcal{E}}$ be the metric projection from ∂K onto $\partial a\mathcal{E}$. We have $p_{K,a\mathcal{E}}(\partial K^{\epsilon}) = \partial a\mathcal{E}^{\epsilon}$. Thus we get

$$\frac{a^{n-1}}{2^n} \mathrm{vol}_{n-1}(\partial\mathcal{E}) = a^{n-1}\mathrm{vol}_{n-1}(\partial\mathcal{E}^{\epsilon}) \leq \mathrm{vol}_{n-1}(\partial K^{\epsilon}).$$

We have

$$\{(x_1,\ldots,x_N)| \ \forall\epsilon \ \exists i : x_i \in \partial K^{\epsilon}\} \subseteq \{(x_1,\ldots,x_N)|0 \in [x_1,\ldots,x_N]\}$$

and therefore

$$\{(x_1,\ldots,x_N)| \ \exists\epsilon \ \forall i : x_i \notin \partial K^{\epsilon}\} \supseteq \{(x_1,\ldots,x_N)|0 \notin [x_1,\ldots,x_N]\}.$$

Consequently

$$\bigcup_{\epsilon}\{(x_1,\ldots,x_N)| \ \forall i : x_i \notin \partial K^{\epsilon}\} \supseteq \{(x_1,\ldots,x_N)|0 \notin [x_1,\ldots,x_N]\}.$$

Therefore we get

$$\mathbb{P}_f^N\{(x_1,\ldots,x_N)|0 \notin [x_1,\ldots,x_N]\} \leq \sum_{\epsilon} \left(1 - \frac{\mathrm{vol}_{n-1}(\partial K^{\epsilon})}{\mathrm{vol}_{n-1}(\partial K)} \right)^N$$

$$\leq 2^n \left(1 - \frac{\min_{\epsilon} \mathrm{vol}_{n-1}(\partial K^{\epsilon})}{\mathrm{vol}_{n-1}(\partial K)} \right)^N$$

$$\leq 2^n \left(1 - \frac{a^{n-1}}{2^n} \frac{\mathrm{vol}_{n-1}(\partial\mathcal{E})}{\mathrm{vol}_{n-1}(\partial K)} \right)^N$$

$$\leq 2^n \left(1 - \frac{1}{2^n} \left(\frac{a}{b} \right)^{n-1} \right)^N.$$

(ii) As in (i)

$$\mathbb{P}_f^N\{(x_1,\ldots,x_N)|0 \notin [x_1,\ldots,x_N]\} \leq \mathbb{P}_f^N\{(x_1,\ldots,x_N)|\exists\epsilon\forall i : x_i \notin \partial K^{\epsilon}\}$$

$$\leq 2^n \left(1 - \min_{\epsilon} \int_{\partial K^{\epsilon}} f(x)\mathrm{d}\mu(x) \right)^N.$$

$\square$

Lemma 4.4. *Let K be a convex body in $\mathbb{R}^n$ and $x_0 \in \partial K$. Let $f : \partial K \to \mathbb{R}$ be a strictly positive, continuous function with $\int_{\partial K} f d\mu = 1$. Suppose that for all $0 < t \le T$ we have $K_t \subseteq \overset{\circ}{K}$ and that there are $r, R > 0$ with*

$$B_2^n(x_0 - rN_{\partial K}(x_0), r) \subseteq K \subseteq B_2^n(x_0 - RN_{\partial K}(x_0), R)$$

and let $N_{\partial K_s}(x_s)$ be a normal such that $s = \mathbb{P}_f(\partial K \cap H^-(x_s, N_{\partial K_s}(x_s)))$. Then there is s_0 that depends only on r, R, and f such that we have for all s with $0 < s \le s_0$ and for all sequences of signs ϵ, δ

$$\text{vol}_{n-1}((K \cap H(x_s, N_{\partial K_s}(x_s)))^\delta)$$
$$\le C(r, R, f, \theta, n)\text{vol}_{n-1}((K \cap H(x_s, N_{\partial K_s}(x_s)))^\epsilon)$$

where the signed sets are taken in the plane $H(x_s, N_{\partial K_s}(x_s))$ with x_s as the origin and any orthogonal coordinate system. θ is the angle between $N_{\partial K}(x_0)$ and $x_0 - x_T$.

The important point in Lemma 4.4 is that s_0 and the constant in the inequality depend only on r, R, and f.

Another approach is to use that x_s is the center of gravity of $K \cap H(x_s, N_{\partial K_s}(x_s))$ with respect to the weight

$$\frac{f(y)}{< N_{\partial K \cap H}(y), N_{\partial K}(y) >}$$

where $H = H(x_s, N_{\partial K_s}(x_s))$. See Lemma 2.4.

Proof. We choose s_0 so small that $x_0 - rN_{\partial K}(x_0) \in K_{s_0}$. We show first that there is s_0 that depends only on r and R such that we have for all s with $0 \le s \le s_0$

$$\sqrt{1 - \frac{2R\Delta}{r^2}\left(\frac{\max_{x \in \partial K} f(x)}{\min_{x \in \partial K} f(x)}\right)^{\frac{2}{n-1}}} \le \langle N_{\partial K}(x_0), N_{\partial K_s}(x_s)\rangle \qquad (51)$$

where Δ is the distance of x_0 to the hyperplane $H(x_s, N_{\partial K}(x_0))$

$$\Delta = < N_{\partial K}(x_0), x_0 - x_s > .$$

Let α denote the angle between $N_{\partial K}(x_0)$ and $N_{\partial K_s}(x_s)$. From Figure 4.4.1 and 4.4.2 we deduce that the height of the cap

$$B_2^n(x_0 - rN_{\partial K}(x_0), r) \cap H^-(x_s, N_{\partial K_s}(x_s))$$

is greater than

$$r(1 - \cos\alpha) = r(1 - \langle N_{\partial K}(x_0), N_{\partial K_s}(x_s)\rangle).$$

Here we use that $x_T \in K_{s_0}$ and $x_0 - rN_{\partial K}(x_0) \in K_{s_0}$. We have

$$\mathbb{P}_f(\partial K \cap H^-(x_s, N_{\partial K_s}(x_s))) = \int_{\partial K \cap H^-(x_s, N_{\partial K_s}(x_s))} f(x) d\mu_{\partial K}(x)$$
$$\geq \min_{x \in \partial K} f(x) \mathrm{vol}_{n-1}(\partial K \cap H^-(x_s, N_{\partial K_s}(x_s))).$$

Since $B_2^n(x_0 - rN_{\partial K}(x_0), r) \subseteq K$ we get

$$\mathbb{P}_f(\partial K \cap H^-(x_s, N_{\partial K_s}(x_s)))$$
$$\geq \min_{x \in \partial K} f(x) \mathrm{vol}_{n-1}(\partial B_2^n(x_0 - rN_{\partial K}(x_0), r) \cap H^-(x_s, N_{\partial K_s}(x_s)))$$
$$\geq \min_{x \in \partial K} f(x) \mathrm{vol}_{n-1}(B_2^n(x_0 - rN_{\partial K}(x_0), r) \cap H(x_s, N_{\partial K_s}(x_s))).$$

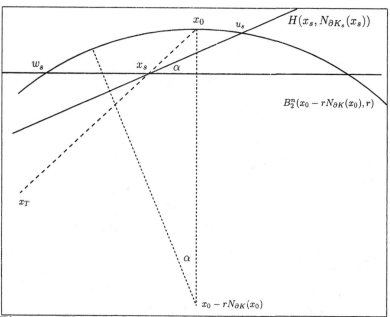

Fig. 4.4.1: We see the plane through x_0 that is spanned by $N_{\partial K}(x_0)$ and $N_{\partial K_s}(x_s)$. The points x_s and x_T are not necessarily in this plane. Since the height of the cap is greater than $r(1 - \cos \alpha)$ we get

$$\mathbb{P}_f(\partial K \cap H^-(x_s, N_{\partial K_s}(x_s)))$$
$$\geq \min_{x \in \partial K} f(x) \mathrm{vol}_{n-1}(B_2^{n-1}) \left(2r^2(1 - \cos \alpha) - r^2(1 - \cos \alpha)^2\right)^{\frac{n-1}{2}}$$
$$= \min_{x \in \partial K} f(x) \mathrm{vol}_{n-1}(B_2^{n-1}) \left(r^2(1 - \cos^2 \alpha)\right)^{\frac{n-1}{2}}. \tag{52}$$

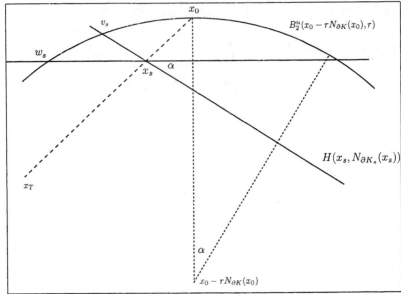

Fig. 4.4.2

On the other hand

$$s = \mathbb{P}_f(\partial K \cap H^-(x_s, N_{\partial K_s}(x_s)))$$

$$= \int_{\partial K \cap H^-(x_s, N_{\partial K_r}(x_s))} f(x)\mathrm{d}\mu_{\partial K}(x)$$

$$= \int_{\partial K \cap H^-(x_s, N_{\partial K}(x_0))} f(x)\mathrm{d}\mu_{\partial K}(x)$$

$$\leq \max_{x \in \partial K} f(x)\mathrm{vol}_{n-1}(\partial K \cap H^-(x_s, N_{\partial K}(x_0))).$$

Since $B_2^n(x_0 - rN_{\partial K}(x_0), r) \subseteq K \subseteq B_2^n(x_0 - RN_{\partial K}(x_0), R)$ we get for sufficiently small s_0

$$\mathbb{P}_f(\partial K \cap H^-(x_s, N_{\partial K_s}(x_s)))$$

$$\leq \max_{x \in \partial K} f(x)\mathrm{vol}_{n-1}(\partial B_2^n(x_0 - RN_{\partial K}(x_0), R) \cap H^-(x_s, N_{\partial K}(x_0)))$$

$$\leq \max_{x \in \partial K} f(x)\mathrm{vol}_{n-1}(B_2^{n-1})(2R\Delta)^{\frac{n-1}{2}}. \tag{53}$$

Since

$$s = \mathbb{P}_f(\partial K \cap H^-(x_s, N_{\partial K_s}(x_s))) \leq \mathbb{P}_f(\partial K \cap H^-(x_s, N_{\partial K}(x_0)))$$

we get by (52) and (53)

$$\min_{x \in \partial K} f(x)\mathrm{vol}_{n-1}(B_2^{n-1})\left(r^2(1 - \cos^2 \alpha)\right)^{\frac{n-1}{2}}$$

$$\leq \max_{x \in \partial K} f(x)\mathrm{vol}_{n-1}(B_2^{n-1})(2R\Delta)^{\frac{n-1}{2}}.$$

This implies
$$\cos \alpha \geq \sqrt{1 - \frac{2R\Delta}{r^2}\left(\frac{\max_{x \in \partial K} f(x)}{\min_{x \in \partial K} f(x)}\right)^{\frac{2}{n-1}}}.$$

Thus we have established (51).

The distance of x_s to $\partial K \cap H(x_s, N_{\partial K_s}(x_s))$ is greater than the distance of x_s to $\partial B_2^n(x_0 - rN_{\partial K}(x_0), r) \cap H(x_s, N_{\partial K_s}(x_s))$. We have $\|x_s - (x_0 - \Delta N_{\partial K}(x_0))\| = \Delta \tan\theta$. Let $\bar{x}_s$ be the image of x_s under the orthogonal projection onto the 2-dimensional plane seen in Figures 4.4.1 and 4.4.2. Then $\|\bar{x}_s - x_s\| \leq \Delta \tan\theta$. There is a $n - 1$-dimensional ball with center $\bar{x}_s$ and radius $\min\{\|\bar{x}_s - u_s\|, \|\bar{x}_s - v_s\|\}$ that is contained in $K \cap H(x_s, N_{\partial K_s}(x_s))$.

We can choose s_0 small enough so that for all s with $0 < s \leq s_0$ we have $\cos \alpha \geq \frac{1}{2}$.

$$\tan \alpha = \frac{\sqrt{1 - \cos^2 \alpha}}{\cos \alpha} \leq 2\frac{\sqrt{2R\Delta}}{r}\left(\frac{\max_{x \in \partial K} f(x)}{\min_{x \in \partial K} f(x)}\right)^{\frac{1}{n-1}} \tag{54}$$

We compute the point of intersection of the line through v_s and $\bar{x}_s$ and the line through x_0 and w_s. Formula (54) and the fact that the height of the cap $B_2^n(x_0 - rN_{\partial K}(x_0), N_{\partial K}(x_0)) \cap H^-(x_s, N_{\partial K}(x_0))$ is Δ and its radius $2r\Delta - \Delta^2$ give further
$$c\sqrt{\Delta} \leq \min\{\|\bar{x}_s - u_s\|, \|\bar{x}_s - v_s\|\}$$

where c is a constant depending only on r, R, f, n. Thus $K \cap H(x_s, N_{\partial K_s}(x_s))$ contains a Euclidean ball with center $\bar{x}_s$ and radius greater $c\sqrt{\Delta}$. Therefore, $K \cap H(x_s, N_{\partial K_s}(x_s))$ contains a Euclidean ball with center x_s and radius greater $c\sqrt{\Delta} - \Delta \tan\theta$. On the other hand,

$$K \cap H(x_s, N_{\partial K_s}(x_s)) \subseteq B_2^n(x_0 - RN_{\partial K}(x_0), R) \cap H(x_s, N_{\partial K_s}(x_s)).$$

Following arguments as above we find that $K \cap H(x_s, N_{\partial K_s}(x_s))$ is contained in a Euclidean ball with center x_s and radius $C\sqrt{\Delta}$ where C is a constant that depends only on r, R, f, n. Therefore, with new constants c, C we get for all sequences of signs δ

$$c\Delta^{\frac{n-1}{2}} \leq \mathrm{vol}_{n-1}((K \cap H(x_s, N_{\partial K_s}(x_s)))^\delta) \leq C\Delta^{\frac{n-1}{2}}.$$

$\square$

Lemma 4.5. *Let K be a convex body in $\mathbb{R}^n$ and $x_0 \in \partial K$. Let $f : \partial K \to \mathbb{R}$ be a strictly positive, continuous function with $\int_{\partial K} f d\mu = 1$. For all t with $0 < t \leq T$ we have $K_t \subseteq \overset{\circ}{K}$. Suppose that there are $r, R > 0$ with*

$$B_2^n(x_0 - rN_{\partial K}(x_0), r) \subseteq K \subseteq B_2^n(x_0 - RN_{\partial K}(x_0), R)$$

and let $N_{\partial K_s}(x_s)$ be a normal such that $s = \mathbb{P}_f(\partial K \cap H^-(x_s, N_{\partial K_s}(x_s)))$. Then there is s_0 that depends only on r, R, and f such that we have for all s with $0 < s \leq s_0$

$$\mathrm{vol}_{n-1}(\partial K \cap H^-(x_s, N_{\partial K_s}(x_s))) \leq 3\,\mathrm{vol}_{n-1}(K \cap H(x_s, N_{\partial K_s}(x_s))).$$

Proof. Since

$$B_2^n(x_0 - rN_{\partial K}(x_0), r) \subseteq K \subseteq B_2^n(x_0 - RN_{\partial K}(x_0), R)$$

we can choose Δ sufficiently small so that we have for all $y \in \partial K \cap H^-(x_0 - \Delta N_{\partial K}(x_0), N_{\partial K}(x_0))$

$$< N_{\partial K}(x_0), N_{\partial K}(y) > \geq 1 - \tfrac{1}{8} \tag{55}$$

and Δ depends only on r and R. Since f is strictly positive we find s_0 that depends only on r, R, and f such that we have for all s with $0 < s \leq s_0$

$$K \cap H(x_s, N_{\partial K_s}(x_s)) \subseteq K \cap H^-(x_0 - \Delta N_{\partial K}(x_0), N_{\partial K}(x_0)). \tag{56}$$

By (55) and (56)
$$< N_{\partial K}(x_0), N_{\partial K_s}(x_s) > \geq 1 - \tfrac{1}{8}.$$

Thus

$$\begin{aligned}
< N_{\partial K_s}&(x_s), N_{\partial K}(y) > \\
&= < N_{\partial K}(x_0), N_{\partial K}(y) > + < N_{\partial K_s}(x_s) - N_{\partial K}(x_0), N_{\partial K}(y) > \\
&\geq 1 - \tfrac{1}{8} - \|N_{\partial K_s}(x_s) - N_{\partial K}(x_0)\| \\
&= 1 - \tfrac{1}{8} - \sqrt{2 - 2 < N_{\partial K_s}(x_s), N_{\partial K}(x_0) >} \geq 1 - \tfrac{3}{8}.
\end{aligned}$$

Altogether
$$< N_{\partial K_s}(x_s), N_{\partial K}(y) > \geq 1 - \tfrac{3}{8}.$$

Let $p_{N_{\partial K_s}(x_s)}$ be the metric projection from $\partial K \cap H^-(x_s, N_{\partial K_s}(x_s))$ onto the plane $H(x_s, N_{\partial K_s}(x_s))$. With this we get now

$$\begin{aligned}
\mathrm{vol}_{n-1}&(\partial K \cap H^-(x_s, N_{\partial K_s}(x_s))) \\
&= \int_{K \cap H(x_s, N_{\partial K_s}(x_s))} \frac{1}{< N_{\partial K_s}(x_s), N_{\partial K}(p_{N_{\partial K_s}(x_s)}^{-1}(z)) >} dz \\
&\leq 3\,\mathrm{vol}_{n-1}(K \cap H(x_s, N_{\partial K_s}(x_s))).
\end{aligned}$$

$\square$

Lemma 4.6. *Let K be a convex body in $\mathbb{R}^n$ and $x_0 \in \partial K$. x_s is defined by $\{x_s\} = [x_0, x_T] \cap K_s$. Let $f : \partial K \to \mathbb{R}$ be a strictly positive, continuous function with $\int_{\partial K} f d\mu = 1$. For all t with $0 < t \leq T$ we have $K_t \subseteq \overset{\circ}{K}$. Suppose that there are $r, R > 0$ such that*

$$B_2^n(x_0 - rN_{\partial K}(x_0), r) \subseteq K \subseteq B_2^n(x_0 - RN_{\partial K}(x_0), R)$$

and let $N_{\partial K_s}(x_s)$ be a normal such that $s = \mathbb{P}_f(\partial K \cap H^-(x_s, N_{\partial K_s}(x_s)))$. Then there is s_0 that depends only on r, R, and f such that for all s with $0 < s \leq s_0$ there are hyperplanes $H_1, \ldots, H_{n-1}$ containing x_T and x_s such that the angle between two $n-2$-dimensional hyperplanes $H_i \cap H(x_s, N_{\partial K_s}(x_s))$ is $\frac{\pi}{2}$ and such that for

$$\partial K_{H,\epsilon} = \partial K \cap H^-(x_s, N_{\partial K_s}(x_s)) \cap \bigcap_{i=1}^{n-1} H_i^{\epsilon_i}$$

and all sequences of signs ϵ and δ we have

$$\operatorname{vol}_{n-1}(\partial K_{H,\epsilon}) \leq c \operatorname{vol}_{n-1}(\partial K_{H,\delta})$$

where c depends on n, r, R, f and $d(x_T, \partial K)$ only.

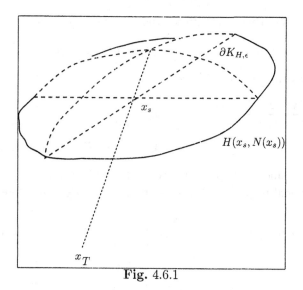

Fig. 4.6.1

Proof. Since x_T is an interior point of K we have $d(x_T, \partial K) > 0$. We choose s_0 so small that

$$B_2^n(x_T, \tfrac{1}{2}d(x_T, \partial K)) \subseteq K_{s_0}. \tag{57}$$

We choose hyperplanes H_i, $i = 1, ..., n-1$, such that they contain x_T and x_s and such that the angles between the hyperplanes $H_i \cap H(x_s, N_{\partial K_s}(x_s))$, $i = 1, ..., n-1$ is $\frac{\pi}{2}$.

By Lemma 4.4 there is s_0 so that we have for all s with $0 < s \leq s_0$ and for all sequences of signs ϵ and δ

$$\text{vol}_{n-1}((K \cap H(x_s, N_{\partial K_s}(x_s)))^\epsilon) \leq c\, \text{vol}_{n-1}((K \cap H(x_s, N_{\partial K_s}(x_s)))^\delta)$$

where c depends only on r, R, and n. Then we have by Lemma 4.5

$$\begin{aligned}
\text{vol}_{n-1}(\partial K_{H,\epsilon}) &\leq \text{vol}_{n-1}(\partial K \cap H^-(x_s, N_{\partial K_s}(x_s))) \\
&\leq c\, \text{vol}_{n-1}(K \cap H(x_s, N_{\partial K_s}(x_s))).
\end{aligned}$$

Therefore we get with a new constant c that depends only on n, f, r and R

$$\text{vol}_{n-1}(\partial K_{H,\epsilon}) \leq c\, \text{vol}_{n-1}((K \cap H(x_s, N_{\partial K_s}(x_s)))^\delta).$$

We consider the affine projections $q : \mathbb{R}^n \to H(x_s, N_{\partial K_s}(x_s))$ and $p : \mathbb{R}^n \to H(x_s, \frac{x_s - x_T}{\|x_s - x_T\|})$ given by $q(t(x_s - x_T) + y) = y$ where $y \in H(x_s, N_{\partial K_s}(x_s))$ and $p(t(x_s - x_T) + y) = y$ where $y \in H(x_s, \frac{x_s - x_T}{\|x_s - x_T\|})$. Please note that p is a metric projection and $q \circ p = q$. Since p is a metric projection we have

$$\text{vol}_{n-1}(p(\partial K_{H,\delta})) \leq \text{vol}_{n-1}(\partial K_{H,\delta}).$$

q is an affine, bijective map between the two hyperplanes and

$$q \circ p(\partial K_{H,\delta}) = q(\partial K_{H,\delta}) \supseteq (K \cap H(x_s, N_{\partial K_s}(x_s)))^\delta.$$

By this (compare the proof of Lemma 2.7)

$$\frac{\text{vol}_{n-1}(\partial K_{H,\delta})}{< N_{\partial K_s}(x_s), \frac{x_s - x_T}{\|x_s - x_T\|} >} \geq \text{vol}_{n-1}(q(\partial K_{H,\delta}))$$

$$\geq \text{vol}_{n-1}((K \cap H(x_s, N_{\partial K_s}(x_s)))^\delta).$$

By (57) the cosine of the angle between the plane $H(x_s, N_{\partial K_s}(x_s))$ and the plane orthogonal to $x_s - x_T$ is greater than $\frac{1}{2} \frac{d(x_T, \partial K)}{\|x_s - x_T\|}$. Therefore we get

$$\text{vol}_{n-1}(\partial K_{H,\delta}) \geq \frac{1}{2} \frac{d(x_T, \partial K)}{\|x_s - x_T\|} \text{vol}_{n-1}((K \cap H(x_s, N_{\partial K_s}(x_s)))^\delta).$$

□

Lemma 4.7. *Let K be a convex body in $\mathbb{R}^n$ and $x_0 \in \partial K$. x_s is defined by $\{x_s\} = [x_0, x_T] \cap K_s$. Let $f : \partial K \to \mathbb{R}$ be a strictly positive, continuous function with $\int_{\partial K} f d\mu = 1$. Suppose that there are $r, R > 0$ such that we have for all $x \in \partial K$*

$$B_2^n(x - rN_{\partial K}(x), r) \subseteq K \subseteq B_2^n(x - RN_{\partial K}(x), R)$$

and let $N_{\partial K_s}(x_s)$ be a normal such that $s = \mathbb{P}_f(\partial K \cap H^-(x_s, N_{\partial K_s}(x_s)))$. Then there are constants s_0, a, and b with $0 \leq a, b < 1$ that depend only on r, R, and f such that we have for all s with $0 < s \leq s_0$ and for all $N \in \mathbb{N}$ and all $k = 1, \ldots, N$

$$\mathbb{P}_f^N\{(x_1, \ldots, x_N) \mid x_s \notin [x_1, \ldots, x_N], \ x_1, \ldots, x_k \in \partial K \cap H^-(x_s, N_{\partial K_s}(x_s))$$
$$\text{and } x_{k+1}, \ldots, x_N \in \partial K \cap H^+(x_s, N_{\partial K_s}(x_s))\}$$
$$\leq (1 - s)^{N-k} s^k 2^n (a^{N-k} + b^k).$$

Proof. Let $H_1, \ldots, H_{n-1}$ be hyperplanes and $\partial K_{H, \epsilon}$ as specified in Lemma 4.6:

$$\partial K_{H, \epsilon} = \partial K \cap H^-(x_s, N_{\partial K_s}(x_s)) \cap \bigcap_{i=1}^{n-1} H_i^{\epsilon_i}.$$

We have by Lemma 4.6 that for all sequences of signs ϵ and δ

$$\text{vol}_{n-1}(\partial K_{H, \epsilon}) \leq c \, \text{vol}_{n-1}(\partial K_{H, \delta})$$

where c depends on n, f, r, R and $d(x_T, \partial K)$. As

$$\{(x_1, \ldots, x_N) \mid x_s \in [x_1, \ldots, x_N]\}$$
$$\supseteq \{(x_1, \ldots, x_N) \mid x_T \in [x_1, \ldots, x_N] \text{ and } [x_s, x_0] \cap [x_1, \ldots, x_N] \neq \emptyset\}$$

we get

$$\{(x_1, \ldots, x_N) \mid x_s \notin [x_1, \ldots, x_N]\}$$
$$\subseteq \{(x_1, \ldots, x_N) \mid x_T \notin [x_1, \ldots, x_N] \text{ or } [x_s, x_0] \cap [x_1, \ldots, x_N] = \emptyset\}.$$

Therefore we get

$$\{(x_1, \ldots, x_N) \mid x_s \notin [x_1, \ldots, x_N], \ x_1, \ldots, x_k \in \partial K \cap H^-(x_s, N_{\partial K_s}(x_s))$$
$$\text{and } x_{k+1}, \ldots, x_N \in \partial K \cap H^+(x_s, N_{\partial K_s}(x_s))\}$$
$$\subseteq \{(x_1, \ldots, x_N) \mid x_T \notin [x_1, \ldots, x_N], \ x_1, \ldots, x_k \in \partial K \cap H^-(x_s, N_{\partial K_s}(x_s))$$
$$\text{and } x_{k+1}, \ldots, x_N \in \partial K \cap H^+(x_s, N_{\partial K_s}(x_s))\}$$
$$\cup \{(x_1, \ldots, x_N) \mid [x_s, x_0] \cap [x_1, \ldots, x_N] = \emptyset, \ x_1, \ldots, x_k \in \partial K \cap$$
$$H^-(x_s, N_{\partial K_s}(x_s)) \text{ and } x_{k+1}, \ldots, x_N \in \partial K \cap H^+(x_s, N_{\partial K_s}(x_s))\}.$$

With $H_s = H(x_s, N_{\partial K_s}(x_s))$

$$\mathbb{P}_f^N\{(x_1, \ldots, x_N) \mid x_s \notin [x_1, \ldots, x_N], \ x_1, \ldots, x_k \in \partial K \cap H^-(x_s, N_{\partial K_s}(x_s))$$
$$\text{and } x_{k+1}, \ldots, x_N \in \partial K \cap H^+(x_s, N_{\partial K_s}(x_s))\}$$
$$\leq (1 - s)^{N-k} s^k \, \mathbb{P}_{f, \partial K \cap H_s^+}^{N-k}\{(x_{k+1}, \ldots, x_N) \mid x_T \notin [x_{k+1}, \ldots, x_N]\}$$
$$+ (1 - s)^{N-k} s^k \mathbb{P}_{f, \partial K \cap H_s^-}^k\{(x_1, \ldots, x_k) \mid [x_s, x_0] \cap [x_1, \ldots, x_k] = \emptyset\}$$

where we obtain $\mathbb{P}_{f,\partial K \cap H_s^+}$ from $\mathbb{P}_f$ by restricting it to the subset $\partial K \cap H_s^+$ and then normalizing it. The same for $\mathbb{P}_{f,\partial K \cap H_s^-}$. We have

$$\mathbb{P}_{f,\partial K \cap H_s^+}^{N-k}\{(x_{k+1},\dots,x_N)|x_T \notin [x_{k+1},\dots,x_N]\} \tag{58}$$
$$= \mathbb{P}_{\tilde{f}}^{N-k}\{(x_{k+1},\dots,x_N)|x_T \notin [x_{k+1},\dots,x_N]\}$$

where $\tilde{f} : \partial(K \cap H^+(x_s, N_{\partial K_s}(x_s))) \to \mathbb{R}$ is given by

$$\tilde{f}(x) = \begin{cases} \dfrac{f(x)}{\mathbb{P}_f(\partial K \cap H_s^+)} & x \in \partial K \cap H^+(x_s, N_{\partial K_s}(x_s)) \\[2mm] 0 & x \in \overset{\circ}{K} \cap H(x_s, N_{\partial K_s}(x_s)). \end{cases}$$

We apply Lemma 4.3.(ii) to $K \cap H^+(x_s, N_{\partial K_s}(x_s))$, $\tilde{f}$, and x_T as the origin. We get

$$\mathbb{P}_{\tilde{f}}^{N-k}\{(x_{k+1},\dots,x_N)|x_T \notin [x_{k+1},\dots,x_N]\} \tag{59}$$
$$\leq 2^n \left(1 - \min_\epsilon \int_{\partial(K \cap H_s^+)^\epsilon} \tilde{f}(x)\mathrm{d}\mu\right)^{N-k}.$$

Since
$$B_2^n(x_0 - rN_{\partial K}(x_0), r) \subseteq K \subseteq B_2^n(x_0 - RN_{\partial K}(x_0), R)$$

we can choose s_0 sufficiently small so that for all s with $0 < s \leq s_0$

$$\min_\epsilon \int_{\partial(K \cap H_s^+)^\epsilon} \tilde{f}(x)\mathrm{d}\mu \geq c > 0$$

where c depends only on s_0 and s_0 can be chosen in such a way that it depends only on r, R, and f. Indeed, we just have to make sure that the surface area of the cap $K \cap H^-(x_s, N_{\partial K_s}(x_s))$ is sufficiently small. We verify the inequality. Since we have for all $x \in \partial K$

$$B_2^n(x - rN_{\partial K}(x), r) \subseteq K \subseteq B_2^n(x - RN_{\partial K}(x), R)$$

the point x_T is an interior point. We consider

$$B_2^n(x_T, \tfrac{1}{2}d(x_T, \partial K)).$$

Then, by considering the metric projection

$$\tfrac{1}{2^n}\mathrm{vol}_{n-1}(\partial B_2^n(x_T, \tfrac{1}{2}d(x_T.\partial K)))$$
$$= \mathrm{vol}_{n-1}(\partial B_2^n(x_T, \tfrac{1}{2}d(x_T, \partial K))^\epsilon) \leq \mathrm{vol}_{n-1}(\partial K^\epsilon).$$

We choose now

$$s_0 = \tfrac{1}{2^{n+1}}\mathrm{vol}_{n-1}(\partial B_2^n(x_T, \tfrac{1}{2}d(x_T, \partial K))) \min_{x \in \partial K} f(x).$$

Then we get

$$\mathbb{P}_f(\partial K \cap H_s^+) \int_{\partial(K \cap H_s^+)^\epsilon} \tilde{f}(x)\mathrm{d}\mu(x)$$

$$= \int_{\partial(K \cap H_s^+)^\epsilon} f(x)\mathrm{d}\mu(x)$$

$$= \int_{\partial K^\epsilon} f(x)\mathrm{d}\mu(x) - \int_{\partial K^\epsilon \cap H_s^-} f(x)\mathrm{d}\mu.$$

Since $\int_{\partial K^\epsilon \cap H_s^-} f(x)\mathrm{d}\mu = s \leq s_0$

$$\mathbb{P}_f(\partial K \cap H_s^+) \int_{\partial(K \cap H_s^+)^\epsilon} \tilde{f}(x)\mathrm{d}\mu(x)$$

$$\geq \int_{\partial K^\epsilon} f(x)\mathrm{d}\mu(x) - s_0$$

$$\geq \mathrm{vol}_{n-1}(\partial K^\epsilon) \min_{x \in \partial K} f(x) - s_0$$

$$\geq \tfrac{1}{2^{n+1}}\mathrm{vol}_{n-1}(\partial B_2^n(x_T, \tfrac{1}{2}d(x_T, \partial K))) \min_{x \in \partial K} f(x).$$

We put

$$a = 1 - \min_\epsilon \int_{\partial(K \cap H_s^+)^\epsilon} \tilde{f}(x)\mathrm{d}\mu.$$

We get by (58) and (59)

$$\mathbb{P}_{f,\partial K \cap H_s^+}^{N-k}\{(x_{k+1}, \ldots, x_N)|x_T \notin [x_{k+1}, \ldots, x_N]\} \leq 2^n a^{N-k}.$$

Moreover, since

$$\{(x_1, \ldots, x_k)| \ [x_s, x_0] \cap [x_1, \ldots, x_k] \neq \emptyset\} \supseteq \{(x_1, \ldots, x_k)| \ \forall \epsilon \ \exists i : x_i \in \partial K_{H,\epsilon}\}$$

we get

$$\{(x_1, \ldots, x_k)| \ [x_s, x_0] \cap [x_1, \ldots, x_k] = \emptyset\} \subseteq \{(x_1, \ldots, x_k)| \ \exists \epsilon \ \forall i : x_i \notin \partial K_{H,\epsilon}\}.$$

By Lemma 4.6 there is b with $0 \leq b < 1$ so that

$$\mathbb{P}_{f,\partial K \cap H_s^-}^k\{(x_1, \ldots, x_k)|[x_s, x_0] \cap [x_1, \ldots, x_k] = \emptyset\} \leq 2^{n-1}b^k.$$

Thus we get

$$\mathbb{P}_{\partial K}^N\{(x_1, \ldots, x_N)| \ x_s \notin [x_1, \ldots, x_N], \ x_1, \ldots, x_k \in \partial K \cap H^-(x_s, N_{\partial K_s}(x_s))$$
$$\text{and } x_{k+1}, \ldots, x_N \in \partial K \cap H^+(x_s, N_{\partial K_s}(x_s))\}$$
$$\leq (1-s)^{N-k} s^k 2^n (a^{N-k} + b^k).$$

$\square$

Lemma 4.8. *Let K be a convex body in $\mathbb{R}^n$ and $x_0 \in \partial K$. x_s is defined by $\{x_s\} = [x_0, x_T] \cap K_s$. Let $f : \partial K \to \mathbb{R}$ be a strictly positive, continuous function with $\int_{\partial K} f d\mu = 1$. Suppose that there are $r, R > 0$ such that we have for all $x \in \partial K$*

$$B_2^n(x - rN_{\partial K}(x), r) \subseteq K \subseteq B_2^n(x - RN_{\partial K}(x), R)$$

and let $N_{\partial K_s}(x_s)$ be a normal such that $s = \mathbb{P}_f(\partial K \cap H^-(x_s, N_{\partial K_s}(x_s)))$. Then there are constants s_0, a and b with $0 \le a, b < 1$ that depend only on r, R, and f such that we have for all s with $0 < s \le s_0$ and for all $N \in \mathbb{N}$ and all $k = 1, \dots, N$

$$\mathbb{P}_f^N\{(x_1, \dots, x_N)|\ x_s \notin [x_1, \dots, x_N]\} \le 2^n \left(a - as + s\right)^N + 2^n (1 - s + bs)^N.$$

s_0, a, and b are as given in Lemma 4.7.

Proof. We have

$$\mathbb{P}_f^N\{(x_1, \dots, x_N)|\ x_s \notin [x_1, \dots, x_N]\}$$

$$= \sum_{k=0}^{N} \binom{N}{k} \mathbb{P}_f^N\{(x_1, \dots, x_N)|\ x_s \notin [x_1, \dots, x_N],\ x_1, \dots, x_k \in \partial K \cap$$

$$H^-(x_s, N_{\partial K_s}(x_s))\ \text{and}\ x_{k+1}, \dots, x_N \in \partial K \cap H^+(x_s, N_{\partial K_s}(x_s))\}.$$

By Lemma 4.7 we get

$$\mathbb{P}_f^N\{(x_1, \dots, x_N)|\ x_s \notin [x_1, \dots, x_N]\}$$

$$\le 2^n \sum_{k=0}^{N} \binom{N}{k} (1-s)^{N-k} s^k (a^{N-k} + b^k)$$

$$= 2^n \left(a - as + s\right)^N + 2^n (1 - s + bs)^N.$$

$\square$

Lemma 4.9. *Let K be a convex body in $\mathbb{R}^n$ and $x_0 \in \partial K$. x_s is defined by $\{x_s\} = [x_0, x_T] \cap K_s$. Let $f : \partial K \to \mathbb{R}$ be a strictly positive, continuous function with $\int_{\partial K} f d\mu = 1$. Suppose that there are $r, R > 0$ such that we have for all $x \in \partial K$*

$$B_2^n(x - rN_{\partial K}(x), r) \subseteq K \subseteq B_2^n(x - RN_{\partial K}(x), R)$$

and let $N_{\partial K_s}(x_s)$ be a normal such that $s = \mathbb{P}_f(\partial K \cap H^-(x_s, N_{\partial K_s}(x_s)))$. Then for all s_0 with $0 < s_0 \le T$

$$\lim_{N \to \infty} N^{\frac{2}{n-1}} \int_{s_0}^{T} \int_{\partial K_s} \frac{\mathbb{P}_f^N\{(x_1, \dots, x_N)|\ x_s \notin [x_1, \dots, x_N]\} d\mu_{\partial K_s}(x_s) ds}{\int_{\partial K \cap H_s} \frac{f(y)}{\sqrt{1 - \langle N_{\partial K_s}(x_s), N_{\partial K}(y) \rangle^2}} d\mu_{\partial K \cap H_s}(y)} = 0$$

where $H_s = H(x_s, N_{\partial K_s}(x_s))$.

Proof. Since $< N_{\partial K_s}(x_s), N_{\partial K}(y) > \leq 1$

$$N^{\frac{2}{n-1}} \int_{s_0}^{T} \int_{\partial K_s} \frac{\mathbb{P}_f^N\{(x_1,\ldots,x_N)| \ x_s \notin [x_1,\ldots,x_N]\}}{\int_{\partial K \cap H_s} \frac{f(y)}{\sqrt{1-<N_{\partial K_s}(x_s),N_{\partial K}(y)>^2}} d\mu_{\partial K \cap H_s}(y)} d\mu_{\partial K_s}(x_s)ds$$

$$\leq \frac{N^{\frac{2}{n-1}}}{\min_{x \in \partial K} f(x)} \int_{s_0}^{T} \int_{\partial K_s} \frac{\mathbb{P}_f^N\{(x_1,\ldots,x_N)| \ x_s \notin [x_1,\ldots,x_N]\}}{\text{vol}_{n-2}(\partial(K \cap H(x_s, N_{\partial K_s}(x_s))))} d\mu_{\partial K_s}(x_s)ds.$$

We observe that there is a constant $c_1 > 0$ such that

$$c_1 = d(\partial K, \partial K_{s_0}) = \inf\{\|x - x_{s_0}\| | x \in \partial K, x_{s_0} \in \partial K_{s_0}\}. \tag{60}$$

If not, there is $x_{s_0} \in \partial K \cap \partial K_{s_0}$. This cannot be because the condition

$$\forall x \in \partial K : B_2^n(x - rN_{\partial K}(x), r) \subseteq K \subseteq B_2^n(x - RN_{\partial K}(x), R)$$

implies that K_{s_0} is contained in the interior of K. It follows that there is a constant $c_2 > 0$ that depends on K and f only such that for all $s \geq s_0$ and all $x_s \in \partial K_s$

$$\text{vol}_{n-2}(\partial(K \cap H(x_s, N_{\partial K_s}(x_s)))) \geq c_2. \tag{61}$$

Therefore

$$N^{\frac{2}{n-1}} \int_{s_0}^{T} \int_{\partial K_s} \frac{\mathbb{P}_f^N\{(x_1,\ldots,x_N)| \ x_s \notin [x_1,\ldots,x_N]\}}{\int_{\partial K \cap H_s} \frac{f(y)}{\sqrt{1-<N(x_s),N(y)>^2}} d\mu_{\partial K \cap H_s}(y)} d\mu_{\partial K_s}(x_s)ds$$

$$\leq \frac{N^{\frac{2}{n-1}}}{c_2 \min_{x \in \partial K} f(x)} \times$$

$$\int_{s_0}^{T} \int_{\partial K_s} \mathbb{P}_f^N\{(x_1,\ldots,x_N)| \ x_s \notin [x_1,\ldots,x_N]\} d\mu_{\partial K_s}(x_s)ds.$$

Now we apply Lemma 4.3.(ii) to K with x_s as the origin. Let

$$\partial K^{\epsilon}(x_s) = \{x \in \partial K | \forall i = 1,\ldots,n : \text{sgn}(x(i) - x_s(i)) = \epsilon_i\}.$$

With the notation of Lemma 4.3 we get that the latter expression is less than

$$\frac{2^n N^{\frac{2}{n-1}}}{c_2 \min_{x \in \partial K} f(x)} \int_{s_0}^{T} \int_{\partial K_s} \left(1 - \min_{\epsilon} \int_{\partial K^{\epsilon}(x_s)} f(x)d\mu\right)^N d\mu_{\partial K_s}(x_s)ds$$

$$\leq \frac{2^n N^{\frac{2}{n-1}}}{c_2 \min_{x \in \partial K} f(x)} \times$$

$$\int_{s_0}^{T} \int_{\partial K_s} \left(1 - \min_{x \in \partial K} f(x) \min_{\epsilon} \text{vol}_{n-1}(\partial K^{\epsilon}(x_s))\right)^N d\mu_{\partial K_s}(x_s)ds$$

$$\leq \frac{2^n N^{\frac{2}{n-1}} \text{vol}_{n-1}(\partial K)(T - s_0)}{c_2 \min_{x \in \partial K} f(x)} \times$$

$$\left(1 - \min_{x \in \partial K} f(x) \inf_{s_0 \leq s \leq T} \min_{\epsilon} \text{vol}_{n-1}(\partial K^{\epsilon}(x_s))\right)^N.$$

By (60) the ball with center x_s and radius c_1 is contained in K

$$c_1^{n-1} 2^{-n} \mathrm{vol}_{n-1}(\partial B_2^n) = c_1^{n-1} \mathrm{vol}_{n-1}(\partial (B_2^n)^\epsilon) \leq \mathrm{vol}_{n-1}(\partial K^\epsilon(x_s)).$$

Thus we obtain

$$N^{\frac{2}{n-1}} \int_{s_0}^T \int_{\partial K_s} \frac{\mathbb{P}_f^N \{(x_1, \ldots, x_N) \mid x_s \notin [x_1, \ldots, x_N]\}}{\int_{\partial K \cap H_s} \frac{f(y)}{\sqrt{1 - <N(x_s), N(y)>^2}} d\mu_{\partial K \cap H_s}(y)} d\mu_{\partial K_s}(x_s) ds \qquad (62)$$

$$\leq \frac{2^n N^{\frac{2}{n-1}} \mathrm{vol}_{n-1}(\partial K)(T - s_0)}{c_2 \min_{x \in \partial K} f(x)} \left(1 - \min_{x \in \partial K} f(x) c_1^{n-1} 2^{-n} \mathrm{vol}_{n-1}(\partial B_2^n) \right)^N.$$

Since f is strictly positive the latter expression tends to 0 for N to infinity. $\square$

Lemma 4.10. *Let K be a convex body in $\mathbb{R}^n$ and $x_0 \in \partial K$. Let $x_s \in \partial K_s$ be given by the equation $\{x_s\} = [x_0, x_T] \cap \partial K_s$. Suppose that there are r, R with $0 < r, R < \infty$ and*

$$B_2^n(x_0 - rN_{\partial K}(x_0), r) \subseteq K \subseteq B_2^n(x_0 - RN_{\partial K}(x_0), R).$$

Let $f : \partial K \to \mathbb{R}$ be a strictly positive, continuous function with $\int_{\partial K} f d\mu = 1$. Suppose that for all t with $0 < t \leq T$ we have $K_t \subseteq \overset{\circ}{K}$. Let the normals $N_{\partial K_s}(x_s)$ be such that

$$s = \mathbb{P}_f(\partial K \cap H^-(x_s, N_{\partial K_s}(x_s))).$$

Let Θ be the angle between $N_{\partial K}(x_0)$ and $x_0 - x_T$ and s_0 the minimum of

$$\frac{1}{2} \left(\frac{r}{8R} \right)^{\frac{n-1}{2}} \frac{(\min_{x \in \partial K} f(x))^2}{\max_{x \in \partial K} f(x)} \mathrm{vol}_{n-1}(B_2^{n-1}) r^{n-1} \left(\tfrac{1}{4} \cos^3 \Theta \right)^{\frac{n-1}{2}}$$

and the constant $C(r, R, f, \Theta, n)$ of Lemma 4.4. Then we have for all s with $0 < s < s_0$ and all $y \in \partial K \cap H^-(x_s, N_{\partial K_s}(x_s))$

$$\sqrt{1 - <N_{\partial K_s}(x_s), N_{\partial K}(y) >^2} \leq \frac{30R}{r^2} \left(\frac{s \max_{x \in \partial K} f(x)}{(\min_{x \in \partial K} f(x))^2 \mathrm{vol}_{n-1}(B_2^{n-1})} \right)^{\frac{1}{n-1}}$$

Proof. Θ is the angle between $N_{\partial K}(x_0)$ and $x_0 - x_T$. Let $\Delta_r(s)$ be the height of the cap

$$B_2^n(x_0 - rN_{\partial K}(x_0), r) \cap H^-(x_s, N_{\partial K_s}(x_s))$$

and $\Delta_R(s)$ the one of

$$B_2^n(x_0 - RN_{\partial K}(x_0), R) \cap H^-(x_s, N_{\partial K_s}(x_s)).$$

By assumption

$$s_0 \leq \frac{1}{2} \left(\frac{r}{8R}\right)^{\frac{n-1}{2}} \frac{(\min_{x \in \partial K} f(x))^2}{\max_{x \in \partial K} f(x)} \mathrm{vol}_{n-1}(B_2^{n-1}) r^{n-1} \left(\tfrac{1}{4} \cos^3 \Theta\right)^{\frac{n-1}{2}}. \quad (63)$$

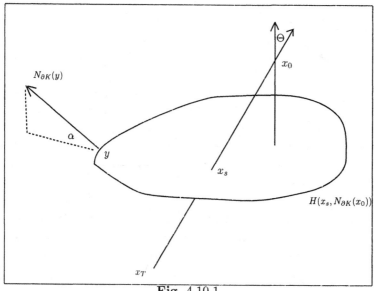

Fig. 4.10.1

First we want to make sure that for s with $0 < s < s_0$ the number $\Delta_r(s)$ is well-defined, i.e. the above cap is not the empty set. For this we have to show that $H(x_s, N_{\partial K_s}(x_s))$ intersects $B_2^n(x_0 - r N_{\partial K}(x_0), r)$. It is enough to show that for all s with $0 < s \leq s_0$ we have $x_s \in B_2^n(x_0 - r N_{\partial K}(x_0), r)$. This follows provided that there is s_0 such that for all s with $0 < s \leq s_0$

$$\|x_0 - x_s\| \leq \tfrac{1}{2} r \cos^2 \Theta. \quad (64)$$

See Figure 4.10.2. We are going to verify this inequality. We consider the point $z \in [x_T, x_0]$ with $\|x_0 - z\| = \tfrac{1}{2} r \cos^2 \Theta$. Let H be any hyperplane with $z \in H$. Then

$$\mathbb{P}_f(\partial K \cap H^-) = \int_{\partial K \cap H^-} f(x) \mathrm{d}\mu_{\partial K}(x) \geq \left(\min_{x \in \partial K} f(x)\right) \mathrm{vol}_{n-1}(\partial K \cap H^-).$$

The set $K \cap H^-$ contains a cap of $B_2^n(x_0 - r N_{\partial K}(x_0), r)$ with height greater than $\tfrac{3}{8} r \cos^2 \Theta$. We verify this. By Figure 4.10.3 we have

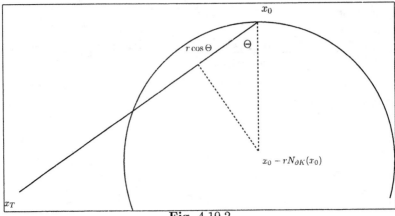

Fig. 4.10.2

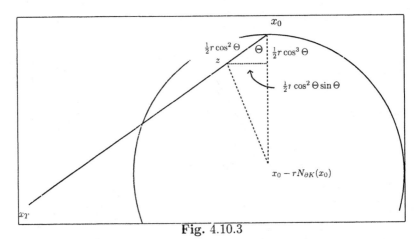

Fig. 4.10.3

$$\|z - (x_0 - rN_{\partial K}(x_0))\| = \sqrt{|r - \tfrac{1}{2}r\cos^3\Theta|^2 + \tfrac{1}{4}r^2\cos^4\Theta\sin^2\Theta}$$

$$= \sqrt{r^2 - r^2\cos^3\Theta + \tfrac{1}{4}r^2\cos^6\Theta + \tfrac{1}{4}r^2\cos^4\Theta\sin^2\Theta}$$

$$= \sqrt{r^2 - r^2\cos^3\Theta + \tfrac{1}{4}r^2\cos^4\Theta}$$

$$\leq r\sqrt{1 - \tfrac{3}{4}\cos^3\Theta}.$$

Therefore the height of a cap is greater than

$$r - \|z - (x_0 - rN_{\partial K}(x_0))\| \geq r\left(1 - \sqrt{1 - \tfrac{3}{4}\cos^3\Theta}\right) \geq \tfrac{3}{8}r\cos^3\Theta.$$

By Lemma 1.3 a cap of a Euclidean ball of radius r with height $h = \tfrac{3}{8}r\cos^3\Theta$ has surface area greater than

$$\mathrm{vol}_{n-1}(B_2^{n-1})r^{\frac{n-1}{2}}\left(2h-\frac{h^2}{r}\right)^{\frac{n-1}{2}}$$

$$= \mathrm{vol}_{n-1}(B_2^{n-1})r^{\frac{n-1}{2}}\left(\tfrac{3}{4}r\cos^3\Theta - \tfrac{9}{64}r\cos^6\Theta\right)^{\frac{n-1}{2}}$$

$$\geq \mathrm{vol}_{n-1}(B_2^{n-1})r^{n-1}\left(\tfrac{1}{4}\cos^3\Theta\right)^{\frac{n-1}{2}}.$$

By our choice of s_0 (63) we get

$$\mathbb{P}_f(\partial K \cap H^-) \geq \left(\min_{x\in\partial K} f(x)\right)\mathrm{vol}_{n-1}(B_2^{n-1})r^{n-1}\left(\tfrac{1}{4}\cos^3\Theta\right)^{\frac{n-1}{2}} > s_0.$$

Therefore we have for all s with $0 < s < s_0$ that $z \in K_{s_0}$. By convexity we get

$$\partial K_s \cap [z, x_0] \neq \emptyset.$$

Thus (64) is shown.

Next we show that for all s with $0 < s < s_0$ we have

$$\sqrt{1 - \frac{8R}{3r^3}\left(s\frac{\max_{x\in\partial K} f(x)}{(\min_{x\in\partial K} f(x))^2 \, \mathrm{vol}_{n-1}(B_2^{n-1})}\right)^{\frac{2}{n-1}}} \leq \langle N_{\partial K}(x_0), N_{\partial K_s}(x_s)\rangle.$$

$$(65)$$

By the same consideration for showing (64) we get for all s with $0 < s < s_0$

$$\Delta_r(s) \leq \tfrac{3}{8}r\cos^3\Theta$$

and by Lemma 1.3

$$s = \mathbb{P}_f(\partial K \cap H^-(x_s, N_{\partial K_s}(x_s)))$$

$$\geq \left(\min_{x\in\partial K} f(x)\right)\mathrm{vol}_{n-1}(B_2^{n-1})r^{\frac{n-1}{2}}\left(2\Delta_r(s) - \frac{(\Delta_r(s))^2}{r}\right)^{\frac{n-1}{2}}.$$

Since $\Delta_r(s) \leq \tfrac{3}{8}r\cos^3\Theta$

$$s \geq \left(\min_{x\in\partial K} f(x)\right)\mathrm{vol}_{n-1}(B_2^{n-1})r^{\frac{n-1}{2}}\left(2\Delta_r(s) - \Delta_r(s)\tfrac{3}{8}\cos^3\Theta\right)^{\frac{n-1}{2}}$$

$$\geq \left(\min_{x\in\partial K} f(x)\right)\mathrm{vol}_{n-1}(B_2^{n-1})(r\Delta_r(s))^{\frac{n-1}{2}}.$$

Thus we have

$$s \geq \left(\min_{x\in\partial K} f(x)\right)\mathrm{vol}_{n-1}(B_2^{n-1})(r\Delta_r(s))^{\frac{n-1}{2}}$$

or equivalently

$$\Delta_r(s) \leq \frac{1}{r} \left(\frac{s}{\min_{x \in \partial K} f(x) \mathrm{vol}_{n-1}(B_2^{n-1})} \right)^{\frac{2}{n-1}}. \tag{66}$$

Next we show

$$\tfrac{3}{4}\Delta(s) \leq \Delta_r(s)$$

where $\Delta(s)$ is the distance of x_0 to the hyperplane $H(x_s, N_{\partial K}(x_0))$

$$\Delta(s) = \, <N_{\partial K}(x_0), x_0 - x_s> .$$

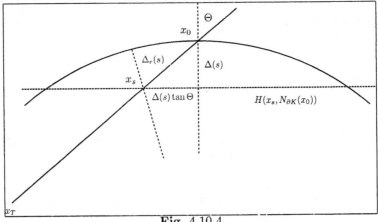

Fig. 4.10.4

By the Pythagorean Theorem, see Figure 4.10.4,

$$(r - \Delta_r(s))^2 = (r - \Delta(s))^2 + (\Delta(s)\tan\Theta)^2.$$

Thus

$$\begin{aligned}
\Delta_r(s) &= r - \sqrt{(r - \Delta(s))^2 + (\Delta(s)\tan\Theta)^2} \\
&= r \left(1 - \sqrt{1 - \frac{1}{r^2}(2r\Delta(s) - \Delta^2(s) - (\Delta(s)\tan\Theta)^2)} \right).
\end{aligned}$$

We use $\sqrt{1-t} \leq 1 - \tfrac{1}{2}t$

$$\begin{aligned}
\Delta_r(s) &\geq \frac{1}{2r}\left(2r\Delta(s) - \Delta^2(s) - (\Delta(s)\tan\Theta)^2 \right) \\
&= \Delta_r(s)\left[1 - \tfrac{1}{2}\frac{\Delta_r(s)}{r}(1 + \tan^2\Theta) \right].
\end{aligned}$$

By (64) we get $\Delta(s) = \|x_0 - x_s\| \cos\Theta \leq \tfrac{1}{2}r\cos^3\Theta$ and thus $\Delta(s) \leq \tfrac{1}{2}r\cos^3\Theta$. With this

$$\Delta_r(s) = \Delta_r(s) \left[1 - \tfrac{1}{2} \frac{\Delta_r(s)}{r} (1 + \tan^2 \Theta) \right]$$

$$= \Delta_r(s) \left[1 - \frac{1}{2r} (1 + \tan^2 \Theta) \tfrac{1}{2} r \cos^3 \Theta \right]$$

$$= \Delta_r(s) \left[1 - \tfrac{1}{4} \cos^4 \Theta \right] \geq \tfrac{3}{4} \Delta_r(s).$$

By formula (51) of the proof of Lemma 4.4 we have

$$\sqrt{1 - \frac{2R\Delta(s)}{r^2} \left(\frac{\max_{x \in \partial K} f(x)}{\min_{x \in \partial K} f(x)} \right)^{\frac{2}{n-1}}} \leq \langle N_{\partial K}(x_0), N_{\partial K_s}(x_s) \rangle.$$

By $\tfrac{3}{4} \Delta(s) \leq \Delta_r(s)$

$$\sqrt{1 - \frac{8R\Delta_r(s)}{3r^2} \left(\frac{\max_{x \in \partial K} f(x)}{\min_{x \in \partial K} f(x)} \right)^{\frac{2}{n-1}}} \leq \langle N_{\partial K}(x_0), N_{\partial K_s}(x_s) \rangle.$$

By (66) we get

$$\sqrt{1 - \frac{8R}{3r^3} \left(s \frac{\max_{x \in \partial K} f(x)}{(\min_{x \in \partial K} f(x))^2 \operatorname{vol}_{n-1}(B_2^{n-1})} \right)^{\frac{2}{n-1}}} \leq \langle N_{\partial K}(x_0), N_{\partial K_s}(x_s) \rangle.$$

Thus we have shown (65).

Next we show that for all $y \in \partial B_2^n(x_0 - RN_{\partial K}(x_0), R) \cap H^-(x_s, N_{\partial K_s}(x_s))$

$$1 - \frac{\Delta_r(s)}{r} \leq \left\langle N_{\partial K_s}(x_s), \frac{y - (x_0 - RN_{\partial K}(x_0))}{\|y - (x_0 - RN_{\partial K}(x_0))\|} \right\rangle. \tag{67}$$

For this we show first that for all s with $0 < s < s_0$

$$\Delta_R(s) \leq \frac{R}{r} \Delta_r(s). \tag{68}$$

By our choice (63) of s_0 and by (65)

$$\langle N_{\partial K_s}(x_s), N_{\partial K}(x_0) \rangle \geq \sqrt{1 - \tfrac{1}{12} \cos^3 \Theta}$$

and by (64) we have $\|x_s - x_0\| < \tfrac{1}{2} r \cos^2 \Theta$. Therefore we have for all s with $0 < s < s_0$ that the hyperplane $H(x_s, N_{\partial K_s}(x_s))$ intersects the line segment

$$[x_0, x_0 - rN_{\partial K}(x_0)].$$

Let r_1 be the distance of x_0 to the point defined by the intersection

$$[x_0, x_0 - rN_{\partial K}(x_0)] \cap H(x_s, N_{\partial K_s}(x_s)).$$

We get by Figure 4.10.5

$$\frac{r - \Delta_r(s)}{R - \Delta_R(s)} = \frac{r - r_1}{R - r_1} \le \frac{r}{R}.$$

The right hand side inequality follows from the monotonicity of the function $(r - t)/(R - t)$.

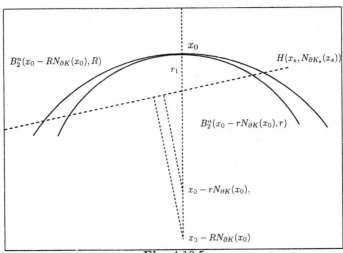

Fig. 4.10.5

Thus

$$r - \Delta_r(s) \le \frac{r}{R}(R - \Delta_R(s)) = r - \frac{r}{R}\Delta_R(s)$$

and therefore

$$\frac{r}{R}\Delta_R(s) \le \Delta_r(s).$$

For all $y \in \partial B_2^n(x_0 - RN_{\partial K}(x_0), R) \cap H^-(x_s, N_{\partial K_s}(x_s))$ the cosine of the angle between $N_{\partial K_s}(x_s)$ and $y - (x_0 - RN_{\partial K}(x_0))$ is greater than $1 - \frac{\Delta_R(s)}{R}$. This holds since y is an element of a cap of a Euclidean ball with radius R and with height $\Delta_R(s)$. Thus we have

$$1 - \frac{\Delta_R(s)}{R} \le \left\langle N_{\partial K_s}(x_s), \frac{y - (x_0 - RN_{\partial K}(x_0))}{\|y - (x_0 - RN_{\partial K}(x_0))\|} \right\rangle.$$

By (68)

$$1 - \frac{\Delta_r(s)}{r} \le \left\langle N_{\partial K_s}(x_s), \frac{y - (x_0 - RN_{\partial K}(x_0))}{\|y - (x_0 - RN_{\partial K}(x_0))\|} \right\rangle$$

and we have verified (67).

We show now that this inequality implies that for all s with $0 < s < s_0$ and all $y \in \partial B_2^n(x_0 - RN_{\partial K}(x_0), R) \cap H^-(x_s, N_{\partial K_s}(x_s))$

$$1 - \Delta_r(s)\frac{R^2}{r^3} \le \left\langle N_{\partial K_s}(x_s), \frac{y - (x_0 - rN_{\partial K}(x_0))}{\|y - (x_0 - rN_{\partial K}(x_0))\|} \right\rangle. \tag{69}$$

Let α be the angle between $N_{\partial K_s}(x_s)$ and $y - (x_0 - RN_{\partial K}(x_0))$ and let β be the angle between $N_{\partial K_s}(x_s)$ and $y - (x_0 - rN_{\partial K}(x_0))$.

$$\cos\alpha = \left\langle N_{\partial K_s}(x_s), \frac{y - (x_0 - RN_{\partial K}(x_0))}{\|y - (x_0 - RN_{\partial K}(x_0))\|} \right\rangle$$

$$\cos\beta = \left\langle N_{\partial K_s}(x_s), \frac{y - (x_0 - rN_{\partial K}(x_0))}{\|y - (x_0 - rN_{\partial K}(x_0))\|} \right\rangle$$

We put

$$a = \|y - (x_0 - rN_{\partial K}(x_0))\| \qquad b = \|y - x_0\|.$$

See Figure 4.10.6.

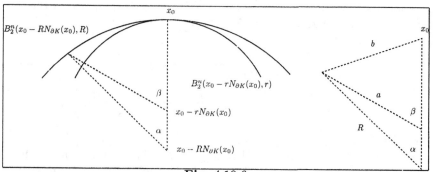

Fig. 4.10.6

By elementary trigonometric formulas we get

$$b^2 = 2R^2(1 - \cos\alpha) \qquad\qquad b^2 = a^2 + r^2 - 2ar\cos\beta$$

and

$$a^2 = R^2 + (R - r)^2 - 2R(R - r)\cos\alpha = r^2 + 2R(R - r)(1 - \cos\alpha).$$

From these equations we get

$$\begin{aligned}
\cos\beta &= \frac{a^2 + r^2 - b^2}{2ar} = \frac{a^2 + r^2 - 2R^2(1 - \cos\alpha)}{2ar} \\
&= \frac{2r^2 - 2Rr(1 - \cos\alpha)}{2r\sqrt{r^2 + 2R(R - r)(1 - \cos\alpha)}} = \frac{r - R(1 - \cos\alpha)}{\sqrt{r^2 + 2R(R - r)(1 - \cos\alpha)}}.
\end{aligned}$$

Thus

$$\cos\beta = \frac{1 - \frac{R}{r}(1 - \cos\alpha)}{\sqrt{1 + 2R(\frac{R}{r^2} - \frac{1}{r})(1 - \cos\alpha)}}.$$

By (67) we have $1 - \cos\alpha \le \frac{\Delta_r(s)}{r}$ and therefore

$$\cos\beta \geq \frac{1 - \frac{R\Delta_r(s)}{r^2}}{\sqrt{1 + 2R(\frac{R}{r^2} - \frac{1}{r})\frac{\Delta_r(s)}{r}}} \geq \frac{1 - \frac{R\Delta_r(s)}{r^2}}{1 + R(\frac{R}{r^2} - \frac{1}{r})\frac{\Delta_r(s)}{r}}$$

$$= 1 - \frac{\frac{R^2}{r^3}\Delta_r(s)}{1 + R(\frac{R}{r^2} - \frac{1}{r})\frac{\Delta_r(s)}{r}} \geq 1 - \frac{R^2}{r^3}\Delta_r(s).$$

Thus we have proved (69). From (69) it follows now easily that for all s with $0 < s < s_0$ and all $y \in \partial K \cap H^-(x_s, N_{\partial K_s}(x_s))$

$$1 - \Delta_r(s)\frac{R^2}{r^3} \leq \left\langle N_{\partial K_s}(x_s), \frac{y - (x_0 - rN_{\partial K}(x_0))}{\|y - (x_0 - rN_{\partial K}(x_0))\|} \right\rangle. \tag{70}$$

This follows because the cap $K \cap H^-(x_s, N_{\partial K_s}(x_s))$ is contained in the cap $B_2^n(x_0 - RN_{\partial K}(x_0), R) \cap H^-(x_s, N_{\partial K_s}(x_s))$. Using now (66) ·

$$1 - \frac{R^2}{r^4}\left(\frac{s}{\min_{x \in \partial K} f(x)\mathrm{vol}_{n-1}(B_2^{n-1})}\right)^{\frac{2}{n-1}} \tag{71}$$
$$\leq \left\langle N_{\partial K_s}(x_s), \frac{y - (x_0 - rN_{\partial K}(x_0))}{\|y - (x_0 - rN_{\partial K}(x_0))\|} \right\rangle.$$

For all s with $0 < s < s_0$ and all $y \in \partial K \cap H^-(x_s, N_{\partial K_s}(x_s))$ the angle between $y - (x_0 - rN_{\partial K}(x_0))$ and $N_{\partial K}(y)$ cannot be greater than the angle between $y - (x_0 - rN_{\partial K}(x_0))$ and $N_{\partial K}(x_0)$. This follows from Figure 4.10.7.

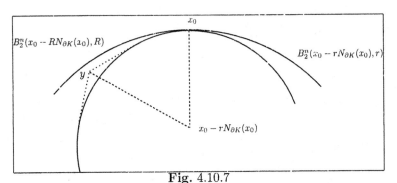

Fig. 4.10.7

A supporting hyperplane of K through y cannot intersect $B_2^n(x_0 - rN_{\partial K}(x_0), r)$. Therefore the angle between $y - (x_0 - rN_{\partial K}(x_0))$ and $N_{\partial K}(y)$ is smaller than the angle between $y - (x_0 - rN_{\partial K}(x_0))$ and the normal of a supporting hyperplane of $B_2^n(x_0 - rN_{\partial K}(x_0), r)$ that contains y.

Let α_1 denote the angle between $N_{\partial K}(x_0)$ and $N_{\partial K_s}(x_s)$, α_2 the angle between $N_{\partial K_s}(x_s)$ and $y - (x_0 - rN_{\partial K}(x_0))$, and α_3 the angle between $N_{\partial K}(x_0)$ and $y - (x_0 - rN_{\partial K}(x_0))$. Then by (65) and (71) we have

$$\alpha_3 \leq \alpha_1 + \alpha_2 \leq \frac{\pi}{2}\sin\alpha_1 + \frac{\pi}{2}\sin\alpha_2$$

$$\leq \frac{\pi}{2}\sqrt{\frac{8R}{3r^3}}\left(s\frac{\max_{x\in\partial K} f(x)}{(\min_{x\in\partial K} f(x))^2 \operatorname{vol}_{n-1}(B_2^{n-1})}\right)^{\frac{1}{n-1}}$$

$$+\frac{\pi}{\sqrt{2}}\frac{R}{r^2}\left(\frac{s}{\min_{x\in\partial K} f(x)\operatorname{vol}_{n-1}(B_2^{n-1})}\right)^{\frac{1}{n-1}}$$

$$\leq 10\frac{R}{r^2}\left(s\frac{\max_{x\in\partial K} f(x)}{(\min_{x\in\partial K} f(x))^2 \operatorname{vol}_{n-1}(B_2^{n-1})}\right)^{\frac{1}{n-1}}.$$

Let α_4 be the angle between $N_{\partial K}(y)$ and $y - (x_0 - rN_{\partial K}(x_0))$. By the above consideration $\alpha_4 \leq \alpha_3$. Thus

$$\alpha_4 \leq 10\frac{R}{r^2}\left(s\frac{\max_{x\in\partial K} f(x)}{(\min_{x\in\partial K} f(x))^2 \operatorname{vol}_{n-1}(B_2^{n-1})}\right)^{\frac{1}{n-1}}.$$

Let α_5 be the angle between $N_{\partial K_s}(x_s)$ and $N_{\partial K}(y)$. Then

$$\sin\alpha_5 \leq \alpha_5 \leq \alpha_2 + \alpha_4$$

$$\leq 10\frac{R}{r^2}\left(s\frac{\max_{x\in\partial K} f(x)}{(\min_{x\in\partial K} f(x))^2 \operatorname{vol}_{n-1}(B_2^{n-1})}\right)^{\frac{1}{n-1}}$$

$$+\frac{\pi}{\sqrt{2}}\frac{R}{r^2}\left(\frac{s}{\min_{x\in\partial K} f(x)\operatorname{vol}_{n-1}(B_2^{n-1})}\right)^{\frac{1}{n-1}}$$

$$\leq 30\frac{R}{r^2}\left(s\frac{\max_{x\in\partial K} f(x)}{(\min_{x\in\partial K} f(x))^2 \operatorname{vol}_{n-1}(B_2^{n-1})}\right)^{\frac{1}{n-1}}.$$

$\square$

Lemma 4.11. *Let K be a convex body in $\mathbb{R}^n$ and $x_0 \in \partial K$. Let $f : \partial K \to \mathbb{R}$ be a strictly positive, continuous function with $\int_{\partial K} f d\mu = 1$. Assume that for all t with $0 < t \leq T$ we have $K_t \subseteq \overset{\circ}{K}$. Let $x_s \in \partial K_s$ be given by the equation $\{x_s\} = [x_0, x_T] \cap \partial K_s$. Suppose that there are r, R with $0 < r, R < \infty$ and*

$$B_2^n(x_0 - rN_{\partial K}(x_0), r) \subseteq K \subseteq B_2^n(x_0 - RN_{\partial K}(x_0), R).$$

Let the normals $N_{\partial K_s}(x_s)$ be such that

$$s = \mathbb{P}_f(\partial K \cap H^-(x_s, N_{\partial K_s}(x_s))).$$

Let s_0 be as in Lemma 4.10. Then we have for all s with $0 < s < s_0$

$$\int_{\partial K \cap H_s} \frac{1}{\sqrt{1- <N_{\partial K_s}(x_s). N_{\partial K}(y)>^2}} d\mu_{\partial K \cap H_s}(y)$$

$$\geq c^n \frac{r^n \left(\min_{x \in \partial K} f(x)\right)^{\frac{2}{n-1}} \left(\text{vol}_{n-1}(B_2^{n-1})\right)^{\frac{2}{n-1}}}{R^{n-1} \max_{x \in \partial K} f(x)} s^{\frac{n-3}{n-1}}$$

where c is an absolute constant and $H_s = H(x_s, N_{\partial K_s}(x_s))$.

Proof. By Lemma 4.10 we have

$$\int_{\partial K \cap H_s} \frac{1}{\sqrt{1- <N_{\partial K_s}(x_s), N_{\partial K}(y)>^2}} d\mu_{\partial K \cap H_s}(y)$$

$$\geq \frac{r^2}{30R} \left(\frac{(\min_{x \in \partial K} f(x))^2 \text{vol}_{n-1}(B_2^{n-1})}{s \max_{x \in \partial K} f(x)}\right)^{\frac{1}{n-1}} \text{vol}_{n-2}(\partial K \cap H_s)$$

$$\geq \frac{r^2}{30R} \left(\frac{(\min_{x \in \partial K} f(x))^2 \text{vol}_{n-1}(B_2^{n-1})}{s \max_{x \in \partial K} f(x)}\right)^{\frac{1}{n-1}}$$

$$\times \text{vol}_{n-2}(\partial B_2^n(x_0 - rN_{\partial K}(x_0), r) \cap H_s). \tag{72}$$

Now we estimate the radius of the $n-1$-dimensional Euclidean ball $B_2^n(x_0 - rN_{\partial K}(x_0), r) \cap H_s$ from below. As in Lemma 4.10 $\Delta_r(s)$ is the height of the cap

$$B_2^n(x_0 - rN_{\partial K}(x_0), r) \cap H^-(x_s, N_{\partial K_s}(x_s))$$

and $\Delta_R(s)$ the one of

$$B_2^n(x_0 - RN_{\partial K}(x_0), R) \cap H^-(x_s, N_{\partial K_s}(x_s)).$$

By (68) we have $\Delta_R(s) \leq \frac{R}{r}\Delta_r(s)$. Moreover,

$$s = \mathbb{P}_f(\partial K \cap H^-(x_s, N_{\partial K_s}(x_s)))$$

$$= \int_{\partial K \cap H_s^-} f(x) d\mu_{\partial K}(x) \leq \max_{x \in \partial K} f(x) \text{vol}_{n-1}(\partial K \cap H_s^-). \tag{73}$$

Since $K \cap H_s^- \subseteq B_2^n(x_0 - RN_{\partial K}(x_0), R) \cap H_s^-$ we have

$$\text{vol}_{n-1}(\partial K \cap H_s^-) \leq \text{vol}_{n-1}(\partial(K \cap H_s^-))$$

$$\leq \text{vol}_{n-1}(\partial(B_2^n(x_0 - RN_{\partial K}(x_0), R) \cap H_s^-))$$

$$\leq 2\text{vol}_{n-1}(\partial B_2^n(x_0 - RN_{\partial K}(x_0), R) \cap H_s^-).$$

By Lemma 1.3 we get

$$\text{vol}_{n-1}(\partial K \cap H_s^-) \leq 2\sqrt{1+ \frac{2\Delta_R(s)R}{(R - \Delta_R(s))^2}} \text{vol}_{n-1}(B_2^{n-1})(2R\Delta_R(s))^{\frac{n-1}{2}}.$$

As we have seen in the proof of Lemma 4.10 we have $\Delta_r(s) \leq \frac{1}{2}r$. Together with $\Delta_R(s) \leq \frac{R}{r}\Delta_r(s)$ we get $\Delta_R(s) \leq \frac{1}{2}R$. This gives us

$$\mathrm{vol}_{n-1}(\partial K \cap H_s^-) \leq 2\sqrt{5}\mathrm{vol}_{n-1}(B_2^{n-1})(2R\Delta_R(s))^{\frac{n-1}{2}}$$

and

$$\frac{R}{r}\Delta_r(s) \geq \Delta_R(s) \geq \frac{1}{2R}\left(\frac{\mathrm{vol}_{n-1}(\partial K \cap H_s^-)}{2\sqrt{5}\mathrm{vol}_{n-1}(B_2^{n-1})}\right)^{\frac{2}{n-1}}$$

$$\geq \frac{1}{2R}\left(\frac{s}{2\sqrt{5}\mathrm{vol}_{n-1}(B_2^{n-1})\max_{x\in\partial K}f(x)}\right)^{\frac{2}{n-1}}.$$

By this and by $\Delta_r(s) \leq \frac{1}{2}r$ the radius of $B_2^n(x_0 - rN_{\partial K}(x_0), r) \cap H_s$ is greater than

$$\sqrt{2r\Delta_r(s) - \Delta_r(s)^2} \geq \sqrt{r\Delta_r(s)}$$

$$\geq \frac{r}{\sqrt{2R}}\left(\frac{s}{2\sqrt{5}\mathrm{vol}_{n-1}(B_2^{n-1})\max_{x\in\partial K}f(x)}\right)^{\frac{1}{n-1}}.$$

Therefore, by (72)

$$\int_{\partial K \cap H_s}\frac{1}{\sqrt{1 - <N_{\partial K_s}(x_s), N_{\partial K}(y)>^2}}\mathrm{d}\mu_{\partial K \cap H_s}(y)$$

$$\geq \frac{r^2}{30R}\left(\frac{(\min_{x\in\partial K}f(x))^2\,\mathrm{vol}_{n-1}(B_2^{n-1})}{s\,\max_{x\in\partial K}f(x)}\right)^{\frac{1}{n-1}}\mathrm{vol}_{n-2}(\partial B_2^{n-1})$$

$$\left(\frac{r}{\sqrt{2R}}\right)^{n-2}\left(\frac{s}{2\sqrt{5}\mathrm{vol}_{n-1}(B_2^{n-1})\max_{x\in\partial K}f(x)}\right)^{\frac{n-2}{n-1}}.$$

By (73) the latter expression is greater than or equal to

$$c^n\frac{r^n\,(\min_{x\in\partial K}f(x))^{\frac{2}{n-1}}\,(\mathrm{vol}_{n-1}(B_2^{n-1}))^{\frac{2}{n-1}}}{R^{n-1}\max_{x\in\partial K}f(x)}\,s^{\frac{n-3}{n-1}}$$

where c is an absolute constant. □

Lemma 4.12. *Let K be a convex body in $\mathbb{R}^n$ and $x_0 \in \partial K$. Let $f : \partial K \to \mathbb{R}$ be a strictly positive, continuous function with $\int_{\partial K} f\mathrm{d}\mu = 1$. Assume that for all t with $0 < t \leq T$ we have $K_t \subseteq \overset{\circ}{K}$. Let $x_s \in \partial K_s$ be given by the equation $\{x_s\} = [x_0, x_T] \cap \partial K_s$. Suppose that there are r, R with $0 < r, R < \infty$ and*

$$B_2^n(x_0 - rN_{\partial K}(x_0), r) \subseteq K \subseteq B_2^n(x_0 - RN_{\partial K}(x_0), R).$$

Let the normals $N_{\partial K_s}(x_s)$ be such that

$$s = \mathbb{P}_f(\partial K \cap H^-(x_s, N_{\partial K_s}(x_s))).$$

Let s_0 be as in Lemma 4.10. Let β be such that $B_2^n(x_T, \beta) \subseteq K_{s_0} \subseteq K \subseteq B_2^n(x_T, \frac{1}{\beta})$ and let $H_s = H(x_s, N_{\partial K_s}(x_s))$. Then there are constants a and b with $0 \le a, b < 1$ that depend only on r, R, and f such that we have for all N

$$N^{\frac{2}{n-1}} \int_0^{s_0} \frac{\mathbb{P}_f^N\{(x_1, \ldots, x_N)| \ x_s \notin [x_1, \ldots, x_N]\}}{\int_{\partial(K \cap H_s)} \frac{f(y) \mathrm{d}\mu_{\partial(K \cap H_s)}(y)}{(1 - <N_{\partial K_s}(x_s), N_{\partial K}(y)>^2)^{\frac{1}{2}}}}$$

$$\times \left(\frac{\|x_s - x_T\|}{\|x_0 - x_T\|} \right)^n \frac{< x_0 - x_T, N_{\partial K}(x_0) >}{< x_s - x_T, N_{\partial K_s}(x_s) >} \mathrm{d}s$$

$$\le c_n \frac{R^{n-1} \max_{x \in \partial K} f(x) \left[(1-a)^{-\frac{2}{n-1}} + (1-b)^{-\frac{2}{n-1}} \right]}{\beta^2 r^n \left(\min_{x \in \partial K} f(x) \right)^{\frac{n+1}{n-1}}}$$

where c_n is a constant that depends only on the dimension n. The constants a and b are the same as in Lemma 4.8. They depend only on n, r, R and f.

Lemma 4.12 provides an uniform estimate. The constants do not depend on the boundary point x_0.

Proof. As in Lemma 4.10 Θ denotes the angle between the vectors $N_{\partial K}(x_0)$ and $x_0 - x_T$. Θ_s is the angle between the vectors $N_{\partial K_s}(x_s)$ and $x_s - x_T$ which is the same as the angle between $N_{\partial K_s}(x_s)$ and $x_0 - x_T$. Thus $< \frac{x_0 - x_T}{\|x_0 - x_T\|}, N_{\partial K}(x_0) > = \cos \Theta$ and $< \frac{x_s - x_T}{\|x_s - x_T\|}, N_{\partial K_s}(x_s) > = \cos \Theta_s$. By Lemma 2.3.(ii) K_s has volume strictly greater than 0 if we choose s small enough. Since $K_t \subseteq \overset{\circ}{K}$ the point x_T is an interior point of K. For small enough s_0 the set K_{s_0} has nonempty interior and therefore there is a $\beta > 0$ such that

$$B_2^n(x_T, \beta) \subseteq K_{s_0} \subseteq K \subseteq B_2^n(x_T, \tfrac{1}{\beta}).$$

Then for all s with $0 < s \le s_0$

$$\beta^2 \le \left\langle \frac{x_0 - x_T}{\|x_0 - x_T\|}, N_{\partial K}(x_0) \right\rangle \le 1 \quad \text{and} \quad \beta^2 \le \left\langle \frac{x_s - x_T}{\|x_s - x_T\|}, N_{\partial K_s}(x_s) \right\rangle \le 1.$$

Thus

$$\frac{\|x_s - x_T\| < x_0 - x_T, N_{\partial K}(x_0) >}{\|x_0 - x_T\| < x_s - x_T, N_{\partial K_s}(x_s) >} \le \frac{1}{\beta^2}.$$

As $\frac{\|x_s - x_T\|}{\|x_0 - x_T\|} \le 1$,

$$N^{\frac{2}{n-1}} \int_0^{s_0} \frac{\mathbb{P}_f^N\{(x_1,\ldots,x_N)|\ x_s \notin [x_1,\ldots,x_N]\}}{\int_{\partial(K\cap H_s)} \frac{f(y)\mathrm{d}\mu_{\partial(K\cap H_s)}(y)}{(1-<N_{\partial K_s}(x_s),N_{\partial K}(y)>^2)^{\frac{1}{2}}}}$$

$$\times \left(\frac{\|x_s - x_T\|}{\|x_0 - x_T\|}\right)^n \frac{<x_0 - x_T, N_{\partial K}(x_0)>}{<x_s - x_T, N_{\partial K_s}(x_s)>}\,\mathrm{d}s$$

$$\leq N^{\frac{2}{n-1}} \frac{1}{\beta^2} \int_0^{s_0} \frac{\mathbb{P}_f^N\{(x_1,\ldots,x_N)|\ x_s \notin [x_1,\ldots,x_N]\}}{\int_{\partial(K\cap H_s)} \frac{f(y)\mathrm{d}\mu_{\partial(K\cap H_s)}(y)}{(1-<N_{\partial K_s}(x_s),N_{\partial K}(y)>^2)^{\frac{1}{2}}}}\,\mathrm{d}s.$$

By Lemma 4.8 and Lemma 4.11 the last expression is less than

$$N^{\frac{2}{n-1}} \frac{R^{n-1}\max_{x\in\partial K} f(x)}{\beta^2 c^n r^n \left(\min_{x\in\partial K} f(x)\right)^{\frac{2}{n-1}} (\mathrm{vol}_{n-1}(B_2^{n-1}))^{\frac{2}{n-1}}} \tag{74}$$

$$\times \int_0^{s_0} \left[2^n\,(a-as+s)^N + 2^n(1-s+bs)^N\right]\,s^{-\frac{n-3}{n-1}}\mathrm{d}s.$$

We estimate now the integral

$$\int_0^{s_0} \left[2^n\,(a-as+s)^N + 2^n(1-s+bs)^N\right]\,s^{-\frac{n-3}{n-1}}\mathrm{d}s$$

$$= 2^n \int_0^{s_0} [1-(1-a)(1-s)]^N s^{-\frac{n-3}{n-1}} + [1-(1-b)s]^N s^{-\frac{n-3}{n-1}}\mathrm{d}s.$$

For $s_0 \leq \frac{1}{2}$ (we may assume this) we have $1-(1-a)(1-s) \leq 1-(1-a)s$. Therefore the above expression is smaller than

$$2^n \int_0^{s_0} [1-(1-a)s]^N s^{-\frac{n-3}{n-1}} + [1-(1-b)s]^N s^{-\frac{n-3}{n-1}}\mathrm{d}s$$

$$= 2^n(1-a)^{-\frac{2}{n-1}} \int_0^{(1-a)s_0} [1-s]^N s^{-\frac{n-3}{n-1}}\mathrm{d}s$$

$$+ 2^n(1-b)^{-\frac{2}{n-1}} \int_0^{(1-b)s_0} [1-s]^N s^{-\frac{n-3}{n-1}}\mathrm{d}s.$$

Since $s_0 \leq \frac{1}{2}$ and $0 < a, b < 1$ the last expression is smaller than

$$2^n \left[(1-a)^{-\frac{2}{n-1}} + (1-b)^{-\frac{2}{n-1}}\right] B\left(N+1, \tfrac{2}{n-1}\right)$$

where B denotes the Beta function. We have

$$\lim_{x\to\infty} \frac{\Gamma(x+\alpha)}{\Gamma(x)} x^{-\alpha} = 1.$$

Thus

$$\lim_{N \to \infty} B(N+1, \tfrac{2}{n-1})(N+1)^{\frac{2}{n-1}}$$

$$= \lim_{N \to \infty} \frac{\Gamma(N+1)\Gamma(\tfrac{2}{n-1})}{\Gamma(N+1+\tfrac{2}{n-1})}(N+1)^{\frac{2}{n-1}} = \Gamma(\tfrac{2}{n-1})$$

and

$$B(N+1, \tfrac{2}{n-1}) \leq 2^{2+\frac{2}{n-1}}\frac{\Gamma(\tfrac{2}{n-1})}{N^{\frac{2}{n-1}}}.$$

We get

$$\int_0^{s_0} \left[2^n (a - as + s)^N + 2^n (1 - s + bs)^N \right] s^{-\frac{n-3}{n-1}}ds$$

$$\leq 2^n \left[(1-a)^{-\frac{2}{n-1}} + (1-b)^{-\frac{2}{n-1}} \right] 2^{2+\frac{2}{n-1}}\frac{\Gamma(\tfrac{2}{n-1})}{N^{\frac{2}{n-1}}}.$$

Therefore, by (74)

$$N^{\frac{2}{n-1}} \int_0^{s_0} \frac{\mathbb{P}_f^N\{(x_1,\ldots,x_N)| \; x_s \notin [x_1,\ldots,x_N]\}}{\int_{\partial(K \cap H_s)} \frac{f(y)d\mu_{\partial(K \cap H_s)}(y)}{(1-(<N(x_s),N(y)>)^2)^{\frac{1}{2}}}}$$

$$\times \left(\frac{\|x_s - x_T\|}{\|x_0 - x_T\|} \right)^n \frac{< x_0 - x_T, N_{\partial K}(x_0) >}{< x_s - x_T, N_{\partial K_s}(x_s) >}ds$$

$$\leq N^{\frac{2}{n-1}} \frac{R^{n-1}\max_{x \in \partial K} f(x)}{\beta^2 c^n r^n (\min_{x \in \partial K} f(x))^{\frac{2}{n-1}} (\mathrm{vol}_{n-1}(B_2^{n-1}))^{\frac{2}{n-1}}}$$

$$2^n \left[(1-a)^{-\frac{2}{n-1}} + (1-b)^{-\frac{2}{n-1}} \right] 2^{2+\frac{2}{n-1}}\frac{\Gamma(\tfrac{2}{n-1})}{N^{\frac{2}{n-1}}}.$$

With a new constant c_n that depends only on the dimension n the last expression is less than

$$c_n \frac{R^{n-1}\max_{x \in \partial K} f(x) \left[(1-a)^{-\frac{2}{n-1}} + (1-b)^{-\frac{2}{n-1}} \right]}{\beta^2 r^n (\min_{x \in \partial K} f(x))^{\frac{2}{n-1}}}.$$

$\square$

Lemma 4.13. Let K be a convex body in $\mathbb{R}^n$ and $x_0 \in \partial K$. Let $f : \partial K \to \mathbb{R}$ be a strictly positive, continuous function with $\int_{\partial K} f d\mu = 1$. Assume that for all t with $0 < t \leq T$ we have $K_t \subseteq \overset{\circ}{K}$. Let $x_s \in \partial K_s$ be given by the equation $\{x_s\} = [x_0, x_T] \cap \partial K_s$. Suppose that there are r, R with $0 < r, R < \infty$ and

$$B_2^n(x_0 - rN_{\partial K}(x_0), r) \subseteq K \subseteq B_2^n(x_0 - RN_{\partial K}(x_0), R).$$

Let the normals $N_{\partial K_s}(x_s)$ be such that

$$s = \mathbb{P}_f(\partial K \cap H^-(x_s, N_{\partial K_s}(x_s))).$$

Let s_0 be as in Lemma 4.10. Then there are $c_1, c_2, c_3 > 0$, N_0, and u_0 such that we have for all $u > u_0$ and $N > N_0$

$$N^{\frac{2}{n-1}} \int_{\frac{u}{N}}^{T} \frac{\mathbb{P}_f^N\{(x_1,\ldots,x_N)\mid x_s \notin [x_1,\ldots,x_N]\}}{\int_{\partial K \cap H_s} \frac{f(y)}{\sqrt{1 - <N_{\partial K_s}(x_s), N_{\partial K}(y)>^2}} d\mu_{\partial K \cap H_s}(y)} ds \leq c_1 e^{-u} + c_2 e^{-c_3 N}$$

where $H_s = H(x_s, N_{\partial K_s}(x_s))$. The constants u_0, N_0, c_1, c_2 and c_3 depend only on n, r, R and f.

Proof. First we estimate the integral from s_0 to $\frac{u}{N}$. As in the proof of Lemma 4.12 we show

$$N^{\frac{2}{n-1}} \int_{\frac{u}{N}}^{s_0} \frac{\mathbb{P}_f^N\{(x_1,\ldots,x_N)\mid x_s \notin [x_1,\ldots,x_N]\}}{\int_{\partial K \cap H_s} \frac{f(y)}{\sqrt{1 - <N_{\partial K_s}(x_s), N_{\partial K}(y)>^2}} d\mu_{\partial K \cap H_s}(y)} ds$$

$$\leq N^{\frac{2}{n-1}} \frac{R^{n-1} \max_{x \in \partial K} f(x)}{\beta^2 c^n r^n \left(\min_{x \in \partial K} f(x)\right)^{\frac{2}{n-1}} \left(\mathrm{vol}_{n-1}(B_2^{n-1})\right)^{\frac{2}{n-1}}}$$
$$2^n \left[(1-a)^{-\frac{2}{n-1}} + (1-b)^{-\frac{2}{n-1}}\right] \int_{\frac{u}{N}}^{s_0} [1-s]^N s^{-\frac{n-3}{n-1}} ds.$$

We estimate the integral

$$\int_{\frac{u}{N}}^{s_0} [1-s]^N s^{-\frac{n-3}{n-1}} ds \leq \int_{\frac{u}{N}}^{s_0} e^{-sN} s^{-\frac{n-3}{n-1}} ds = N^{-\frac{2}{n-1}} \int_{u}^{s_0 N} e^{-s} s^{-\frac{n-3}{n-1}} ds.$$

If we require that $u_0 \geq 1$ then the last expression is not greater than

$$N^{-\frac{2}{n-1}} \int_{u}^{s_0 N} e^{-s} ds \leq N^{-\frac{2}{n-1}} \int_{u}^{\infty} e^{-s} ds = N^{-\frac{2}{n-1}} e^{-u}.$$

Thus

$$N^{\frac{2}{n-1}} \int_{\frac{u}{N}}^{s_0} \frac{\mathbb{P}_f^N\{(x_1,\ldots,x_N)\mid x_s \notin [x_1,\ldots,x_N]\}}{\int_{\partial K \cap H_s} \frac{f(y)}{\sqrt{1 - <N_{\partial K_s}(x_s), N_{\partial K}(y)>^2}} d\mu_{\partial K \cap H_s}(y)} ds$$

$$\leq \frac{R^{n-1} \max_{x \in \partial K} f(x)}{\beta^2 c^n r^n \left(\min_{x \in \partial K} f(x)\right)^{\frac{2}{n-1}} \left(\mathrm{vol}_{n-1}(B_2^{n-1})\right)^{\frac{2}{n-1}}}$$
$$\times 2^n \left[(1-a)^{-\frac{2}{n-1}} + (1-b)^{-\frac{2}{n-1}}\right] e^{-u}.$$

Now we estimate the integral from s_0 to T

$$N^{\frac{2}{n-1}} \int_{s_0}^{T} \frac{\mathbb{P}_f^N\{(x_1,\ldots,x_N)\mid x_s \notin [x_1,\ldots,x_N]\}}{\int_{\partial K \cap H_s} \frac{f(y)}{\sqrt{1 - <N_{\partial K_s}(x_s), N_{\partial K}(y)>^2}} d\mu_{\partial K \cap H_s}(y)} ds.$$

The same arguments that we have used in the proof of Lemma 4.9 in order
to show formula (62) give that the latter expression is less than

$$\frac{2^n N^{\frac{2}{n-1}} \mathrm{vol}_{n-1}(\partial K)(T - s_0)}{c_2 \min_{x \in \partial K} f(x)} \left(1 - \min_{x \in \partial K} f(x) c_1^{n-1} 2^{-n} \mathrm{vol}_{n-1}(\partial B_2^n))\right)^N$$

where c_1 is the distance between ∂K and ∂K_{s_0}. Choosing now new constants
c_1 and c_2 finishes the proof. $\square$

Lemma 4.14. *Let H be a hyperplane in $\mathbb{R}^n$ that contains 0. Then in both
halfspaces there is a 2^n-tant i.e. there is a sequence of signs θ such that*

$$\{x | \forall i, 1 \le i \le n : sgn(x_i) = \theta_i\}.$$

Moreover, if H^+ is the halfspace that contains the above set then

$$H^+ \subset \bigcup_{i=1}^n \{x | sgn(x_i) = \theta_i\}.$$

The following lemma is an extension of a localization principle introduced
by Bárány [Ba1] for random polytopes whose vertices are chosen from the inside of the convex body. The measure in that case is the normalized Lebesgue
measure on the convex body.

For large numbers N of chosen points the probability that a point is
an element of a random polytope is almost 1 provided that this point is
not too close to the boundary. So it leaves us to compute the probability
for those points that are in the vicinity of the boundary. The localization
principle now says that in order to compute the probability that a point
close to the boundary is contained in a random polytope it is enough to
consider only those points that are in a small neighborhood of the point
under consideration. As a neighborhood we choose a cap of the convex body.
The arguments are similar to the ones used in [Sch1].

Lemma 4.15. *Let K be a convex body in $\mathbb{R}^n$ and $x_0 \in \partial K$. Suppose that the
indicatrix of Dupin exists at x_0 and is an ellipsoid (and not a cylinder with
a base that is an ellipsoid). Let $f : \partial K \to \mathbb{R}$ be a continuous, strictly positive
function with $\int_{\partial K} f d\mu_{\partial K} = 1$. Assume that for all t with $0 < t \le T$ we have
$K_t \subseteq \overset{\circ}{K}$. We define the point x_s by $\{x_s\} = [x_T, x_0] \cap \partial K_s$ and*

$$\Delta(s) = < N_{\partial K}(x_0), x_0 - x_s >$$

*is the distance between the planes $H(x_0, N_{\partial K}(x_0))$ and $H(x_s, N_{\partial K}(x_0))$. Sup-
pose that there are r, R with $0 < r, R < \infty$ and*

$$B_2^n(x_0 - rN_{\partial K}(x_0), r) \subseteq K \subseteq B_2^n(x_0 - RN_{\partial K}(x_0), R).$$

Then, there is c_0 such that for all c with $c \geq c_0$ and b with $b > 2$ there is $s_{c,b} > 0$ such that we have for all s with $0 < s \leq s_{c,b}$ and for all $N \in \mathbb{N}$ with

$$N \geq \tfrac{1}{bs}\mathrm{vol}_{n-1}(\partial K)$$

that

$$\left| \mathbb{P}_f^N\{(x_1,\ldots,x_N) \mid x_s \notin [x_1,\ldots,x_N]\} - \right.$$
$$\mathbb{P}_f^N\{(x_1,\ldots,x_N) \mid x_s \notin [\{x_1,\ldots,x_N\} \cap H^-]\}\Big|$$
$$\leq 2^{n-1}\exp(-\tfrac{c_1}{b}\sqrt{c})$$

where $H = H(x_0 - c\Delta(s)N_{\partial K}(x_0), N_{\partial K}(x_0))$ and $c_1 = c_1(n)$ is a constant that only depends on the dimension n.

In particular, for all $\epsilon > 0$ and all $k \in \mathbb{N}$ there is $N_0 \in \mathbb{N}$ such that we have for all $N \geq N_0$ and all $x_s \in [x_0, x_T]$

$$\left| \mathbb{P}_f^N\{(x_1,\ldots,x_N) \mid x_s \notin [x_1,\ldots,x_N]\} - \right.$$
$$\mathbb{P}_f^{N+k}\{(x_1,\ldots,x_{N+k}) \mid x_s \notin [x_1,\ldots,x_{N+k}]\}\Big| \leq \epsilon.$$

The numbers $s_{c,b}$ may depend on the boundary points x_0.

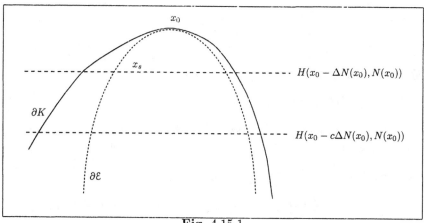

Fig. 4.15.1

Subsequently we apply Lemma 4.15 to a situation where b is already given and we choose c sufficiently big so that

$$2^{n-1}\exp(-\tfrac{c_1}{b}\sqrt{c})$$

is as small as we desire.

Proof. Let c and b be given. Since f is continuous for any given $\epsilon > 0$ we can choose $s_{c,b}$ so small that we have for all s with $0 < s \le s_{c,b}$ and all $x \in \partial K \cap H^-(x_0 - c\Delta(s)N_{\partial K}(x_0), N_{\partial K}(x_0))$

$$|f(x) - f(x_0)| < \epsilon.$$

We may assume that $x_0 = 0$, $N_{\partial K}(x_0) = -e_n$. Let

$$\mathcal{E} = \left\{ x \in \mathbb{R}^n \left| \sum_{i=1}^{n-1} \left| \frac{x_i}{a_i} \right|^2 + \left| \frac{x_n}{a_n} - 1 \right|^2 \le 1 \right. \right\}$$

be the standard approximating ellipsoid at x_0 (see Lemma 1.2). Thus the principal axes are multiples of e_i, $i = 1, \ldots, n$.

We define the operator $T_\eta : \mathbb{R}^n \to \mathbb{R}^n$

$$T_\eta(x_1, \ldots, x_n) = (\eta x_1, \ldots, \eta x_{n-1}, x_n).$$

By Lemma 1.2 for any $\epsilon > 0$ we may choose $s_{c,b}$ so small that we have

$$\begin{aligned}
T_{1-\epsilon}(\mathcal{E} \cap H^-(x_0 - c\Delta(s_{c,b})N_{\partial K}(x_0), N_{\partial K}(x_0))) & \\
\subseteq K \cap H^-(x_0 - c\Delta(s_{c,b})N_{\partial K}(x_0), N_{\partial K}(x_0)) & \quad (75) \\
\subseteq T_{1+\epsilon}(\mathcal{E} \cap H^-(x_0 - c\Delta(s_{c,b})N_{\partial K}(x_0), N_{\partial K}(x_0))). &
\end{aligned}$$

For s with $0 < s \le s_{c,b}$ we denote the lengths of the principal axes of the $n-1$-dimensional ellipsoid

$$T_{1+\epsilon}(\mathcal{E}) \cap H(x_0 - c\Delta(s)N_{\partial K}(x_0), N_{\partial K}(x_0)))$$

by λ_i, $i = 1, \ldots, n-1$, so that the principal axes are $\lambda_i e_i$, $i = 1, \ldots, n-1$. We may assume (for technical reasons) that for all s with $0 < s \le s_{c,b}$

$$x_0 - c\Delta(s)N_{\partial K}(x_0) \pm \lambda_i e_i \notin K \qquad i = 1, \ldots, n-1. \qquad (76)$$

This is done by choosing (if necessary) a slightly bigger ϵ.

For any sequence $\Theta = (\Theta_i)_{i=1}^n$ of signs $\Theta_i = \pm 1$ we put

$$\mathrm{corn}_K(\Theta) = \partial K \cap H^+(x_s, N_{\partial K}(x_0)) \qquad (77)$$

$$\cap \left\{ \bigcap_{i=1}^{n-1} H^-(x_s, (\Theta_i < x_s, e_i > -\lambda_i)e_n + \Theta_i(c-1)\Delta(s)e_i) \right\}.$$

We have

$$\mathrm{corn}_K(\Theta) \subseteq H^-(x_0 - c\Delta(s)N_{\partial K}(x_0), N_{\partial K}(x_0)). \qquad (78)$$

We refer to these sets as corner sets (see Figure 4.15.2). The hyperplanes

$$H(x_s, (\Theta_i < x_s, e_i > -\lambda_i)e_n + \Theta_i(c-1)\Delta(s)e_i)$$

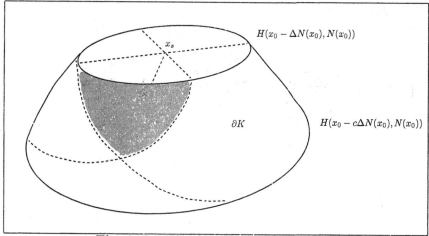

Fig. 4.15.2: The shaded area is $\mathrm{corn}_K(\Theta)$.

$\Theta_i = \pm 1$ and $i = 1, \ldots, n-1$ are chosen in such a way that x_s and

$$x_0 + \Theta_i\lambda_i e_i + c\Delta(s)e_n = \Theta_i\lambda_i e_i + c\Delta(s)e_n$$

($x_0 = 0$) are elements of the hyperplanes. We check this. By definition x_s is an element of this hyperplane. We have

$$< x_s, (\Theta_i < x_s, e_i > -\lambda_i)e_n + \Theta_i(c-1)\Delta(s)e_i >$$
$$= (\Theta_i < x_s, e_i > -\lambda_i) < x_s, e_n > +\Theta_i(c-1)\Delta(s) < x_s, e_i > .$$

Since $N_{\partial K}(x_0) = -e_n$ we have $\Delta(s) = < x_s, e_n >$ and

$$< x_s, (\Theta_i < x_s, e_i > -\lambda_i)e_n + \Theta_i(c-1)\Delta(s)e_i >$$
$$= (\Theta_i < x_s, e_i > -\lambda_i)\Delta(s) + \Theta_i(c-1)\Delta(s) < x_s, e_i >$$
$$= \Delta(s)\{(\Theta_i < x_s, e_i > -\lambda_i) + \Theta_i(c-1) < x_s, e_i >\}$$
$$= \Delta(s)\{-\lambda_i + \Theta_i c < x_s, e_i >\}$$

and

$$< \Theta_i\lambda_i e_i + c\Delta(s)e_n, (\Theta_i < x_s, e_i > -\lambda_i)e_n + \Theta_i(c-1)\Delta(s)e_i >$$
$$= \lambda_i(c-1)\Delta(s) + c\Delta(s)(\Theta_i < x_s, e_i > -\lambda_i)$$
$$= -\lambda_i\Delta(s) + \Theta_i c\Delta(s) < x_s, e_i > .$$

These two equalities show that for all i with $i = 1, \ldots, n-1$

$$\Theta_i\lambda_i e_i + c\Delta(s)e_n \in H(x_s, (\Theta_i < x_s, e_i > -\lambda_i)e_n + \Theta_i(c-1)\Delta(s)e_i).$$

We conclude that for all i with $i = 1, \ldots, n-1$ and all s, $0 < s \leq s_{c,b}$,

$$K \cap H^+(x_0 - c\Delta(s)N_{\partial K}(x_0), N_{\partial K}(x_0)) \tag{79}$$
$$\cap H^-(x_s, (\Theta_i < x_s, e_i > -\lambda_i)e_n + \Theta_i(c-1)\Delta(s)e_i) = \emptyset.$$

We verify this. Since

$$x_0 + \Theta_i\lambda_i e_i + c\Delta(s)e_n \in H(x_s, (\Theta_i < x_s, e_i > -\lambda_i)e_n + \Theta_i(c-1)\Delta(s)e_i)$$

we have

$$H(x_0 - c\Delta(s)N_{\partial K}(x_0), N_{\partial K}(x_0))$$
$$\cap H(x_s, (\Theta_i < x_s, e_i > -\lambda_i)e_n + \Theta_i(c-1)\Delta(s)e_i)$$
$$= \left\{ x_0 + \Theta_i\lambda_i e_i + c\Delta(s)e_n + \sum_{j \neq i,n} a_j e_j \,\middle|\, a_j \in \mathbb{R} \right\}.$$

On the other hand, by (75)

$$K \cap H^-(x_0 - c\Delta(s_{c,b})N_{\partial K}(x_0), N_{\partial K}(x_0))$$
$$\subseteq T_{1+\epsilon}(\mathcal{E} \cap H^-(x_0 - c\Delta(s_{c,b})N_{\partial K}(x_0), N_{\partial K}(x_0)))$$

and by (76)

$$x_0 - c\Delta(s)N_{\partial K}(x_0) + \lambda_i e_i \notin K \qquad i = 1, \ldots, n-1.$$

From this we conclude that

$$H(x_0 - c\Delta(s)N_{\partial K}(x_0))$$
$$\cap H(x_s, (\Theta_i < x_s, e_i > -\lambda_i)e_n + \Theta_i(c-1)\Delta(s)e_i) \cap K = \emptyset.$$

Using this fact and the convexity of K we deduce (78).

We want to show now that we have for all s with $0 < s \leq s_{c,b}$ and $H = H(x_0 - c\Delta(s)N_{\partial K}(x_0), N_{\partial K}(x_0))$

$$\{(x_1, \ldots, x_N)| \; x_s \notin [\{x_1, \ldots, x_N\} \cap H^-]\} \tag{80}$$
$$\setminus \{(x_1, \ldots, x_N)| \; x_s \notin [x_1, \ldots, x_N]\}$$
$$= \{(x_1, \ldots, x_N)| \; x_s \notin [\{x_1, \ldots, x_N\} \cap H^-] \text{ and } x_s \in [x_1, \ldots, x_N]\}$$
$$\subseteq \bigcup_{\Theta} \{(x_1, \ldots, x_N)| \; x_1, \ldots, x_N \in \partial K \setminus \mathrm{corn}_K(\Theta)\}.$$

In order to do this we show first that for $H = H(x_0 - c\Delta(s)N_{\partial K}(x_0), N_{\partial K}(x_0))$ we have

$$\{(x_1, \ldots, x_N)| \; x_s \notin [\{x_1, \ldots, x_N\} \cap H^-] \text{ and } x_s \in [x_1, \ldots, x_N]\} \tag{81}$$
$$\subseteq \{(x_1, \ldots, x_N)|\exists H_{x_s}, \text{hyperplane} : x_s \in H_{x_s}, H_{x_s}^- \cap K \cap H^+ \neq \emptyset$$
$$\text{and } \{x_1, \ldots, x_N\} \cap H^- \subseteq \overset{\circ}{H^+}_{x_s}\}.$$

We show this now. We have $x_s \notin [\{x_1, \ldots, x_N\} \cap H^-]$ and $x_s \in [x_1, \ldots, x_N]$. We observe that there is $z \in K \cap \overset{\circ}{H^+} (x_0 - c\Delta(s)N_{\partial K}(x_0), N_{\partial K}(x_0))$ such that

$$[z, x_s] \cap [\{x_1, \ldots, x_N\} \cap H^-(x_0 - c\Delta(s)N_{\partial K}(x_0), N_{\partial K}(x_0))] = \emptyset. \quad (82)$$

We verify this. Assume that $x_1 \ldots, x_k \in H^-(x_0 - c\Delta(s)N_{\partial K}(x_0), N_{\partial K}(x_0))$ and $x_{k+1} \ldots, x_N \in \overset{\circ}{H^+} (x_0 - c\Delta(s)N_{\partial K}(x_0), N_{\partial K}(x_0))$. Since $x_s \in [x_1, \ldots, x_N]$ there are nonnegative numbers a_i, $i = 1, \ldots, N$, with $\sum_{i=1}^{N} a_i = 1$ and

$$x_s = \sum_{i=1}^{N} a_i x_i.$$

Since $x_s \notin [\{x_1, \ldots, x_N\} \cap H^-]$ we have $\sum_{i=k+1}^{N} a_i > 0$ and since $x_s \in H^-(x_0 - \Delta(s)N_{\partial K}(x_0), N_{\partial K}(x_0))$ we have $\sum_{i=1}^{k} a_i > 0$. Now we choose

$$y = \frac{\sum_{i=1}^{k} a_i x_i}{\sum_{i=1}^{k} a_i} \quad \text{and} \quad z = \frac{\sum_{i=k+1}^{N} a_i x_i}{\sum_{i=k+1}^{N} a_i}.$$

Thus we have $y \in [x_1, \ldots, x_k]$, $z \in [x_{k+1}, \ldots, x_N]$, and

$$x_s = \alpha y + (1 - \alpha)z$$

where $\alpha = \sum_{i=1}^{k} a_i$.

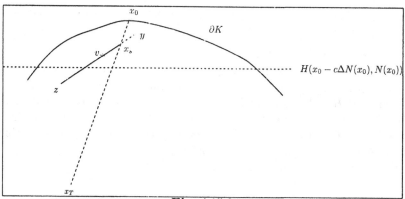

Fig. 4.15.3

We claim that $[z, x_s] \cap [x_1, \ldots, x_k] = \emptyset$. Suppose this is not the case. Then there is $v \in [z, x_s]$ with $v \in [x_1, \ldots, x_k]$. We have $v \neq z$ and $v \neq x_s$. Thus there is β with $0 < \beta < 1$ and $v = \beta z + (1 - \beta)x_s$. Therefore we get

$$v = \beta z + (1 - \beta)x_s = \beta(\tfrac{1}{1-\alpha}x_s - \tfrac{\alpha}{1-\alpha}y) + (1 - \beta)x_s = \tfrac{1-\alpha+\alpha\beta}{1-\alpha}x_s - \tfrac{\alpha\beta}{1-\alpha}y$$

and thus

$$x_s = \frac{1-\alpha}{1-\alpha+\alpha\beta}v + \frac{\alpha\beta}{1-\alpha+\alpha\beta}y.$$

Thus x_s is a convex combination of y and v. Since $v \in [x_1, \ldots, x_k]$ and $y \in [x_1, \ldots, x_k]$ we conclude that $x_s \in [x_1, \ldots, x_k]$ which is not true. Therefore we have reached a contradiction and

$$[z, x_s] \cap [x_1, \ldots, x_k] = \emptyset.$$

We have verified (82).

Now we conclude that

$$\{x_s + t(z - x_s) \mid t \geq 0\} \cap [x_1, \ldots, x_k] = \emptyset.$$

We have

$$\{x_s + t(z - x_s) \mid t \geq 0\} = [z, x_s] \cup \{x_s + t(z - x_s) \mid t > 1\}.$$

We know already that $[z, x_s]$ and $[x_1, \ldots, x_k]$ are disjoint. On the other hand we have

$$\{x_s + t(z - x_s) \mid t > 1\} \subseteq \overset{\circ}{H}{}^+ (x_0 - c\Delta(s)N_{\partial K}(x_0), N_{\partial K}(x_0)).$$

This is true since $x_s \in \overset{\circ}{H}{}^- (x_0 - c\Delta(s)N_{\partial K}(x_0), N_{\partial K}(x_0))$ and

$$z \in \overset{\circ}{H}{}^+ (x_0 - c\Delta(s)N_{\partial K}(x_0), N_{\partial K}(x_0)). \tag{83}$$

Since $\{x_1, \ldots, x_k\} \subseteq H^-(x_0 - c\Delta(s)N_{\partial K}(x_0), N_{\partial K}(x_0))$ we conclude that the sets

$$\{x_s + t(z - x_s) \mid t > 1\} \qquad \text{and} \qquad [x_1, \ldots, x_k]$$

are disjoint. Now we apply the theorem of Hahn-Banach to the convex, closed set $\{x_s + t(z - x_s) \mid t \geq 0\}$ and the compact, convex set $[x_1, \ldots, x_k]$. There is a hyperplane H_{x_s} that separates these two sets strictly. We pass to a parallel hyperplane that separates these two sets and is a support hyperplane of $\{x_s + t(z - x_s) \mid t \geq 0\}$. Let us call this new hyperplane now H_{x_s}. We conclude that $x_s \in H_{x_s}$. We claim that H_{x_s} satisfies (81).

We denote the halfspace that contains z by $H^-_{x_s}$. Then

$$[x_1, \ldots, x_k] \subseteq \overset{\circ}{H}{}^+_{x_s} .$$

Thus we have $x_s \in H_{x_s}$, $H^-_{x_s} \cap K \cap H^+(x_0 - c\Delta(s)N_{\partial K}(x_0), N_{\partial K}(x_0)) \supset \{z\} \neq \emptyset$, and

$$[x_1, \ldots, x_k] \subseteq \overset{\circ}{H}{}^+_{x_s} .$$

Therefore we have shown (81)

$$\{(x_1,\ldots,x_N)| \ x_s \notin [\{x_1,\ldots,x_N\} \cap H^-] \text{ and } x_s \in [x_1,\ldots,x_N]\}$$
$$\subseteq \{(x_1,\ldots,x_N)|\exists H_{x_s} : x_s \in H_{x_s}, H_{x_s}^- \cap K \cap H^+ \neq \emptyset$$
$$\text{and } \{x_1,\ldots,x_N\} \cap H^- \subseteq \overset{\circ}{H_{x_s}^+}\}$$

where $H = H(x_0 - c\Delta(s)N_{\partial K}(x_0), N_{\partial K}(x_0))$. Now we show that

$$\{(x_1,\ldots,x_N)|\exists H_{x_s} : x_s \in H_{x_s}, H_{x_s}^- \cap K \cap H^+ \neq \emptyset \tag{84}$$
$$\text{and } \{x_1,\ldots,x_N\} \cap H^- \subseteq \overset{\circ}{H_{x_s}^+}\}$$
$$\subseteq \bigcup_\Theta \{(x_1,\ldots,x_N)|x_1,\ldots,x_N \in \partial K \setminus \text{corn}_K(\Theta)\}$$

which together with (81) gives us (80).

We show that for every H_{x_s} with $x_s \in H_{x_s}$ and $H_{x_s}^- \cap K \cap H^+ \neq \emptyset$ there is a sequence of signs Θ so that we have

$$\text{corn}_K(\Theta) \subseteq H_{x_s}^- \quad \text{and} \quad \text{corn}_K(-\Theta) \subseteq H_{x_s}^+. \tag{85}$$

This implies that for all sequences $(x_1,\ldots,x_N)$ that are elements of the left hand side set of (4.15.5) there is a Θ such that for all $k = 1,\ldots,N$

$$x_k \notin \text{corn}_K(\Theta).$$

Indeed,

$$\{x_1,\ldots,x_N\} \cap H^-(x_0 - c\Delta(s)N_{\partial K}(x_0), N_{\partial K}(x_0)) \subseteq \overset{\circ}{H_{x_s}^+}$$
$$\text{corn}_K(\Theta) \cap H^+(x_0 - c\Delta(s)N_{\partial K}(x_0), N_{\partial K}(x_0)) = \emptyset.$$

This proves (84). We choose Θ so that (85) is fulfilled. We have for all $i = 1,\ldots,n-1$

$$H(x_s, N_{\partial K}(x_0)) \cap H^-(x_s, (\Theta_i < x_s, e_i > -\lambda_i)e_n + \Theta_i(c-1)\Delta(s)e_i)$$
$$= \{x \in \mathbb{R}^n| < x, e_n >=< x_s, e_n > \text{ and } < x - x_s, \Theta_i e_i >\geq 0\}.$$

Indeed, $N_{\partial K}(x_0) = -e_n$ and

$$H(x_s, N_{\partial K}(x_0)) = \{x \in \mathbb{R}^n| < x, e_n >=< x_s, e_n >\}$$

and

$$H^-(x_s, (\Theta_i < x_s, e_i > -\lambda_i)e_n + \Theta_i(c-1)\Delta(s)e_i)$$
$$= \{x \in \mathbb{R}^n| < x - x_s, (\Theta_i < x_s, e_i > -\lambda_i)e_n + \Theta_i(c-1)\Delta(s)e_i >\geq 0\}.$$

On the intersection of the two sets we have $< x - x_s, e_n >= 0$ and thus

$$0 \leq < x - x_s, (\Theta_i < x_s, e_i > -\lambda_i)e_n + \Theta_i(c-1)\Delta(s)e_i >$$
$$=< x - x_s, \Theta_i(c-1)\Delta(s)e_i >.$$

Since $c - 1$ and $\Delta(s)$ are positive we can divide and get

$$0 \leq \; < x - x_s, \Theta_i e_i > .$$

Therefore, the hyperplanes

$$H(x_s, (\Theta_i < x_s, e_i > -\lambda_i)e_n + \Theta_i(c - 1)\Delta(s)e_i) \qquad i = 1, \ldots, n - 1$$

divide the hyperplane $H(x_s, N_{\partial K}(x_0))$ into 2^{n-1}-tants, i.e. 2^{n-1} sets of equal signs. x_s is considered as the origin in the hyperplane $H(x_s, N_{\partial K}(x_0))$. By Lemma 4.14 there is Θ such that

$$\begin{aligned} &H(x_s, N_{\partial K}(x_0)) \cap H_{x_s}^+ \\ &\supseteq H(x_s, N_{\partial K}(x_0)) \\ &\quad \cap \left\{ \bigcap_{i=1}^{n-1} H^-(x_s, (\Theta_i < x_s, e_i > -\lambda_i)e_n + \Theta_i(c - 1)\Delta(s)e_i) \right\} \end{aligned}$$

and

$$\begin{aligned} &H(x_s, N(x_0)) \cap H_{x_s}^- \\ &\supseteq H(x_s, N(x_0)) \\ &\quad \cap \left\{ \bigcap_{i=1}^{n-1} H^-(x_s, (-\Theta_i < x_s, e_i > -\lambda_i)e_n - \Theta_i(c - 1)\Delta(s)e_i) \right\}. \end{aligned}$$

For a given H_{x_s} we choose this Θ and claim that

$$\mathrm{corn}_K(\Theta) \subseteq H_{x_s}^-. \tag{86}$$

Suppose this is not the case. We consider the hyperplane $\tilde{H}_{x_s}$ with

$$H_{x_s} \cap H(x_s, N_{\partial K}(x_0)) = \tilde{H}_{x_s} \cap H(x_s, N_{\partial K}(x_0))$$

and

$$\bigcap_{i=1}^{n-1} H(x_s, (\Theta_i < x_s, e_i > -\lambda_i)e_n + \Theta_i(c - 1)\Delta(s)e_i) \subseteq \tilde{H}_{x_s}.$$

The set on the left hand side is a 1-dimensional affine space. We obtain $\tilde{H}_{x_s}$ from H_{x_s} by rotating H_{x_s} around the "axis" $H_{x_s} \cap H(x_s, N_{\partial K}(x_0))$. Then we have

$$H^+(x_s, N_{\partial K}(x_0)) \cap H_{x_s}^- \subseteq H^+(x_s, N_{\partial K}(x_0)) \cap \tilde{H}_{x_s}^-.$$

Indeed, from the procedure by which we obtain $\tilde{H}_{x_s}$ from H_{x_s} it follows that one set has to contain the other. Moreover, since $\mathrm{corn}_K(\Theta) \subseteq \tilde{H}_{x_s}^-$, but $\mathrm{corn}_K(\Theta) \not\subseteq H_{x_s}^-$ we verify the above inclusion. On the other hand, by our choice of Θ and by Lemma 4.14

$$\tilde{H}^-_{x_s} \subseteq \bigcup_{i=1}^{n-1} H^-(x_s, (\Theta_i < x_s, e_i > -\lambda_i)e_n + \Theta_i(c-1)\Delta(s)e_i).$$

By (76) none of the halfspaces

$$H^+(x_s, (\Theta_i < x_s, e_i > -\lambda_i)e_n + \Theta_i(c-1)\Delta(s)e_i) \qquad i = 1, \ldots, n-1$$

contains an element of

$$K \cap H^+(x_0 - c\Delta(s)N_{\partial K}(x_0), N_{\partial K}(x_0))$$

and therefore $H^-_{x_s}$ also does not contain such an element. But we know that H_{x_s} contains such an element by (83) giving a contradiction. Altogether we have shown (80) with $H = H(x_0 - c\Delta(s)N_{\partial K}(x_0), N_{\partial K}(x_0))$

$$\{(x_1, \ldots, x_N)| \; x_s \notin [\{x_1, \ldots, x_N\} \cap H^-] \text{ and } x_s \in [x_1, \ldots, x_N]\}$$
$$\subseteq \bigcup_{\Theta} \{(x_1, \ldots, x_N)| \; x_1, \ldots, x_N \in \partial K \setminus \mathrm{corn}_K(\Theta)\}.$$

This gives us

$$\mathbb{P}^N_f\{(x_1, \ldots, x_N)| \; x_s \notin [\{x_1, \ldots, x_N\} \cap H^-] \text{ and } x_s \in [x_1, \ldots, x_N]\}$$
$$\leq \sum_{\Theta} \mathbb{P}^N_f\{(x_1, \ldots, x_N)| \; x_1, \ldots, x_N \in \partial K \setminus \mathrm{corn}_K(\Theta)\}$$
$$= \sum_{\Theta} \left(1 - \int_{\mathrm{corn}_K(\Theta)} f(x)d\mu(x)\right)^N$$
$$\leq \sum_{\Theta} (1 - (f(x_0) - \epsilon)\mathrm{vol}_{n-1}(\mathrm{corn}_K(\Theta)))^N. \qquad (87)$$

Now we establish an estimate for $\mathrm{vol}_{n-1}(\mathrm{corn}_K(\Theta))$. Let p be the orthogonal projection onto the hyperplane $H(x_0, N_{\partial K}(x_0)) = H(0, -e_n)$. By the definition (77) of the set $\mathrm{corn}_K(\Theta)$

$$p\Big(K \cap H^+(x_s, N_{\partial K}(x_0)))$$
$$\cap \Big\{\bigcap_{i=1}^{n-1} H^-(x_s, (\Theta_i < x_s, e_i > -\lambda_i)e_n + \Theta_i(c-1)\Delta(s)e_i)\Big\}\Big)$$
$$= p\Big(\partial\Big(K \cap H^+(x_s, N_{\partial K}(x_0)))$$
$$\cap \Big\{\bigcap_{i=1}^{n-1} H^-(x_s, (\Theta_i < x_s, e_i > -\lambda_i)e_n + \Theta_i(c-1)\Delta(s)e_i)\Big\}\Big)\Big)$$
$$\subseteq p(\mathrm{corn}_K(\Theta)) \cup p(K \cap H(x_s, N_{\partial K}(x_0))). \qquad (88)$$

This holds since $u \in H(x_0, N_{\partial K}(x_0))$ can only be the image of a point

$$w \in \partial\bigg(K \cap H^+(x_s, N_{\partial K}(x_0)))$$
$$\cap\bigg\{\bigcap_{i=1}^{n-1} H^-(x_s, (\Theta_i < x_s, e_i > -\lambda_i)e_n + \Theta_i(c-1)\Delta(s)e_i)\bigg\}\bigg)$$

if $< N(w), N_{\partial K}(x_0) > = < N(w), -e_n > \geq 0$. This holds only for $w \in \mathrm{corn}_K(\Theta)$ or $w \in H(x_s, N_{\partial K}(x_0)) \cap K$. Indeed, the other normals are

$$-(\Theta_i < x_s, e_i > -\lambda_i)e_n - \Theta_i(c-1)\Delta(s)e_i \qquad i = 1, \ldots, n-1$$

and for $i = 1, \ldots, n-1$

$$< -(\Theta_i < x_s, e_i > -\lambda_i)e_n - \Theta_i(c-1)\Delta(s)e_i, -e_n > = \Theta_i < x_s, e_i > -\lambda_i.$$

By (76) we have for all $i = 1, \ldots, n-1$ that $| < x_s, e_i > | < \lambda_i$. This implies that $\Theta_i < x_s, e_i > -\lambda_i < 0$.

Since

$$\mathrm{vol}_{n-1}(p(\mathrm{corn}_K(\Theta))) \leq \mathrm{vol}_{n-1}(\mathrm{corn}_K(\Theta))$$

and

$$\mathrm{vol}_{n-1}(p(K \cap H(x_s, N_{\partial K}(x_0)))) = \mathrm{vol}_{n-1}(K \cap H(x_s, N_{\partial K}(x_0)))$$

we get from (88)

$$\mathrm{vol}_{n-1}(\mathrm{corn}_K(\Theta))) \qquad (89)$$
$$\geq \mathrm{vol}_{n-1}\bigg(p\bigg(K \cap H^+(x_s, N_{\partial K}(x_0)))$$
$$\cap\bigg\{\bigcap_{i=1}^{n-1} H^-(x_s, (\Theta_i < x_s, e_i > -\lambda_i)e_n + \Theta_i(c-1)\Delta(s)e_i)\bigg\}\bigg)\bigg)$$
$$-\mathrm{vol}_{n-1}(K \cap H(x_s, N_{\partial K}(x_0))).$$

Now we use that the indicatrix of Dupin at x_0 exists. Let $\mathcal{E}$ be the standard approximating ellipsoid (Lemma 1.2) whose principal axes have lengths a_i, $i = 1, \ldots, n$. By Lemma 1.2 and Lemma 1.3 for all $\epsilon > 0$ there is s_0 such that for all s with $0 < s \leq s_0$ the set

$$K \cap H(x_s, N_{\partial K}(x_0))$$

is contained in an $n-1$-dimensional ellipsoid whose principal axes have lengths less than

$$(1+\epsilon)a_i\sqrt{\frac{2\Delta(s)}{a_n}} \qquad i = 1, \ldots, n-1.$$

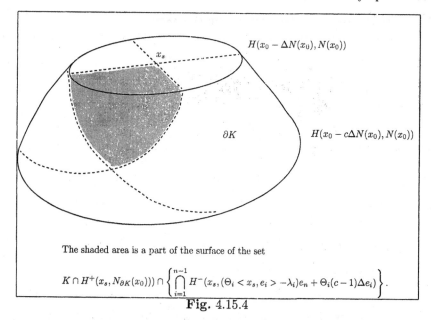

$$H(x_0 - \Delta N(x_0), N(x_0))$$

$$x_s$$

$$\partial K$$

$$H(x_0 - c\Delta N(x_0), N(x_0))$$

The shaded area is a part of the surface of the set

$$K \cap H^+(x_s, N_{\partial K}(x_0))) \cap \left\{ \bigcap_{i=1}^{n-1} H^-(x_s, (\Theta_i < x_s, e_i > -\lambda_i)e_n + \Theta_i(c-1)\Delta e_i) \right\}.$$

Fig. 4.15.4

We choose $s_{c,b}$ to be smaller than this s_0. Therefore for all s with $0 < s \le s_{c,b}$

$$\mathrm{vol}_{n-1}(K \cap H(x_s, N_{\partial K}(x_0)))$$

$$\le (1 + \epsilon)^{n-1} \left(\frac{2\Delta(s)}{a_n} \right)^{\frac{n-1}{2}} \left(\prod_{i=1}^{n-1} a_i \right) \mathrm{vol}_{n-1}(B_2^{n-1}).$$

Thus we deduce from (89)

$$\mathrm{vol}_{n-1}(\mathrm{corn}_K(\Theta))) \qquad\qquad (90)$$

$$\ge \mathrm{vol}_{n-1}\left(p\bigg(K \cap H^+(x_s, N_{\partial K}(x_0)) \right)$$

$$\cap \left\{ \bigcap_{i=1}^{n-1} H^-(x_s, (\Theta_i < x_s, e_i > -\lambda_i)e_n + \Theta_i(c-1)\Delta(s)e_i) \right\} \bigg) \bigg)$$

$$-(1 + \epsilon)^{n-1} \left(\frac{2\Delta(s)}{a_n} \right)^{\frac{n-1}{2}} \left(\prod_{i=1}^{n-1} a_i \right) \mathrm{vol}_{n-1}(B_2^{n-1}).$$

Now we get an estimate for the first summand of the right hand side. Since $\mathcal{E}$ is an approximating ellipsoid we have by Lemma 1.2 that for all $\epsilon > 0$ there is s_0 such that we have for all s with $0 < s \le s_0$

$$x_0 - \Delta(s)N_{\partial K}(x_0) + (1 - \epsilon)\Theta_i a_i \sqrt{\frac{2\Delta(s)}{a_n}} e_i \in K \qquad i = 1, \dots, n-1.$$

Again, we choose $s_{c,b}$ to be smaller than this s_0.

Let θ be the angle between $N_{\partial K}(x_0) = -e_n$ and $x_0 - x_T = -x_T$. Then

$$\|x_s\| = \Delta(s)(\cos\theta)^{-1}. \tag{91}$$

Consequently,

$$\|(x_0 - \Delta(s)N_{\partial K}(x_0)) - x_s\| = \Delta(s)\tan\theta.$$

Therefore, for all $\epsilon > 0$ there is s_0 such that we have for all s with $0 < s \le s_0$

$$x_s + (1-\epsilon)\Theta_i a_i \sqrt{\tfrac{2\Delta(s)}{a_n}} e_i \in K \qquad i = 1,\ldots,n-1.$$

Moreover, for $i = 1,\ldots,n-1$

$$x_s + (1-\epsilon)\Theta_i a_i \sqrt{\tfrac{2\Delta(s)}{a_n}} e_i \in K \cap H^+(x_s, N_{\partial K}(x_0)) \tag{92}$$

$$\cap \left\{ \bigcap_{i=1}^{n-1} H^-(x_s, (\Theta_i < x_s, e_i > -\lambda_i)e_n + \Theta_i(c-1)\Delta(s)e_i) \right\}.$$

Indeed, by the above these points are elements of K. Since $N_{\partial K}(x_0) = -e_n$

$$x_s + (1-\epsilon)\Theta_i a_i \sqrt{\tfrac{2\Delta(s)}{a_n}} e_i \in K \cap H(x_s, N_{\partial K}(x_0)).$$

For $i \ne j$

$$\left\langle x_s + (1-\epsilon)\Theta_j a_j \sqrt{\tfrac{2\Delta(s)}{a_n}} e_j, (\Theta_i < x_s, e_i > -\lambda_i)e_n + \Theta_i(c-1)\Delta(s)e_i \right\rangle$$
$$= \langle x_s, (\Theta_i < x_s, e_i > -\lambda_i)e_n + \Theta_i(c-1)\Delta(s)e_i \rangle$$

and for $i = j$

$$\left\langle x_s + (1-\epsilon)\Theta_i a_i \sqrt{\tfrac{2\Delta(s)}{a_n}} e_i, (\Theta_i < x_s, e_i > -\lambda_i)e_n + \Theta_i(c-1)\Delta(s)e_i \right\rangle$$
$$= \langle x_s, (\Theta_i < x_s, e_i > -\lambda_i)e_n + \Theta_i(c-1)\Delta(s)e_i \rangle$$
$$+ (1-\epsilon)(c-1)a_i \sqrt{\tfrac{2\Delta(s)}{a_n}} \Delta(s).$$

Since the second summand is nonnegative we get for all j with $j = 1,\ldots,n-1$

$$x_s + (1-\epsilon)\Theta_j a_j \sqrt{\frac{2\Delta(s)}{a_n}} e_j \in$$

$$\bigcap_{i=1}^{n-1} H^-(x_s, (\Theta_i < x_s, e_i > -\lambda_i)e_n + \Theta_i(c-1)\Delta(s)e_i).$$

There is a unique point z in $H^+(x_s, N_{\partial K}(x_0))$ with

$$\{z\} = \partial K \cap \left\{ \bigcap_{i=1}^{n-1} H(x_s, (\Theta_i < x_s, e_i > -\lambda_i)e_n + \Theta_i(c-1)\Delta(s)e_i) \right\}. \quad (93)$$

This holds since the intersection of the hyperplanes is 1-dimensional. We have that

$$\mathrm{vol}_{n-1}\left(\left[p(z), p(x_s) + \left((1-\epsilon)\Theta_1 a_1 \sqrt{\tfrac{2\Delta(s)}{a_n}} e_1\right), \dots,\right.\right.$$

$$\left.\left. p(x_s) + \left((1-\epsilon)\Theta_{n-1} a_{n-1} \sqrt{\tfrac{2\Delta(s)}{a_n}} e_{n-1}\right)\right]\right)$$

$$= \mathrm{vol}_{n-1}\left(p\left[z, x_s + \left((1-\epsilon)\Theta_1 a_1 \sqrt{\tfrac{2\Delta(s)}{a_n}} e_1\right), \dots,\right.\right.$$

$$\left.\left. x_s + \left((1-\epsilon)\Theta_{n-1} a_{n-1} \sqrt{\tfrac{2\Delta(s)}{a_n}} e_{n-1}\right)\right]\right)$$

$$\leq \mathrm{vol}_{n-1}\left(p\left(K \cap H^+(x_s, N_{\partial K}(x_0))\right.\right.$$

$$\left.\left. \cap \left\{ \bigcap_{i=1}^{n-1} H^-(x_s, (\Theta_i < x_s, e_i > -\lambda_i)e_n + \Theta_i(c-1)\Delta(s)e_i) \right\}\right)\right).$$

$$(94)$$

The $(n-1)$-dimensional volume of the simplex

$$\left[p(z), p(x_s) + \left((1-\epsilon)\Theta_1 a_1 \sqrt{\tfrac{2\Delta(s)}{a_n}} e_1\right), \dots,\right.$$

$$\left. p(x_s) + \left((1-\epsilon)\Theta_{n-1} a_{n-1} \sqrt{\tfrac{2\Delta(s)}{a_n}} e_{n-1}\right)\right]$$

equals

$$\frac{d}{n-1} \mathrm{vol}_{n-2}\left(\left[(1-\epsilon)a_1 \sqrt{\tfrac{2\Delta(s)}{a_n}} e_1, \dots, (1-\epsilon)a_{n-1} \sqrt{\tfrac{2\Delta(s)}{a_n}} e_{n-1}\right]\right)$$

where d is the distance of $p(z)$ from the plane spanned by

$$p(x_s) + (1-\epsilon)\Theta_i a_i \sqrt{\frac{2\Delta(s)}{a_n}} e_i \qquad i = 1, \dots, n-1$$

in the space $\mathbb{R}^{n-1}$. We have

$$\mathrm{vol}_{n-2}\left(\left[(1-\epsilon)a_1 \sqrt{\tfrac{2\Delta(s)}{a_n}} e_1, \dots, (1-\epsilon)a_{n-1} \sqrt{\tfrac{2\Delta(s)}{a_n}} e_{n-1}\right]\right)$$

$$= (1 - \epsilon)^{n-2} \left(\frac{2\Delta(s)}{a_n} \right)^{\frac{n-2}{2}} \mathrm{vol}_{n-2} \left([a_1 e_1, \ldots, a_{n-1} e_{n-1}] \right)$$

$$= \frac{1}{(n-2)!} (1 - \epsilon)^{n-2} \left(\frac{2\Delta(s)}{a_n} \right)^{\frac{n-2}{2}} \prod_{i=1}^{n-1} a_i \left(\sum_{i=1}^{n-1} |a_i|^{-2} \right)^{\frac{1}{2}}.$$

From this and (94)

$$\frac{d}{(n-1)!} (1 - \epsilon)^{n-2} \left(\frac{2\Delta(s)}{a_n} \right)^{\frac{n-2}{2}} \prod_{i=1}^{n-1} a_i \left(\sum_{i=1}^{n-1} |a_i|^{-2} \right)^{\frac{1}{2}}$$

$$\leq \mathrm{vol}_{n-1} \left(p \left(K \cap H^+(x_s, N_{\partial K}(x_0)) \right. \right.$$

$$\left. \left. \cap \left\{ \bigcap_{i=1}^{n-1} H^-(x_s, (\Theta_i < x_s, e_i > -\lambda_i) e_n + \Theta_i (c-1)\Delta(s) e_i) \right\} \right) \right).$$

From this inequality and (90)

$$\mathrm{vol}_{n-1}(\mathrm{corn}_K(\Theta)) \tag{95}$$

$$\geq \frac{d}{(n-1)!} (1 - \epsilon)^{n-2} \left(\frac{2\Delta(s)}{a_n} \right)^{\frac{n-2}{2}} \prod_{i=1}^{n-1} a_i \left(\sum_{i=1}^{n-1} |a_i|^{-2} \right)^{\frac{1}{2}}$$

$$- (1 + \epsilon)^{n-1} \left(\frac{2\Delta(s)}{a_n} \right)^{\frac{n-1}{2}} \prod_{i=1}^{n-1} a_i \, \mathrm{vol}_{n-1}(B_2^{n-1}).$$

We claim that there is a constant c_2 that depends only on K (and not on s and c) such that we have for all c and s with $0 < s \leq s_{c,b}$

$$d \geq c_2 \sqrt{c\Delta(s)}. \tag{96}$$

d equals the distance of $p(z)$ from the hyperplane that passes through 0 and that is parallel to the one spanned by

$$p(x_s) + (1 - \epsilon)\Theta_i a_i \sqrt{\frac{2\Delta(s)}{a_n}} e_i \qquad\qquad i = 1, \ldots, n-1$$

in $\mathbb{R}^{n-1}$ minus the distance of 0 to the hyperplane spanned by

$$p(x_s) + (1 - \epsilon)\Theta_i a_i \sqrt{\frac{2\Delta(s)}{a_n}} e_i \qquad\qquad i = 1, \ldots, n-1.$$

Clearly, the last quantity is smaller than

$$\|p(x_s)\| + \sqrt{\frac{2\Delta(s)}{a_n}} \max_{1 \leq i \leq n-1} a_i$$

which can be estimated by (91)

$$\|p(x_s)\| + \sqrt{\tfrac{2\Delta(s)}{a_n}} \max_{1\le i\le n-1} a_i \;\le\; \|x_s\| + \sqrt{\tfrac{2\Delta(s)}{a_n}} \max_{1\le i\le n-1} a_i$$
$$= \Delta(s)(\cos\theta)^{-1} + \sqrt{\tfrac{2\Delta(s)}{a_n}} \max_{1\le i\le n-1} a_i.$$

It is left to show that the distance of $p(z)$ to the hyperplane that passes through 0 and that is parallel to the one spanned by

$$p(x_s) + (1-\epsilon)\Theta_i a_i \sqrt{\tfrac{2\Delta(s)}{a_n}} e_i \qquad\qquad i = 1,\ldots,n-1$$

is greater than a constant times $\sqrt{c\Delta(s)}$. Indeed, there is c_0 such that for all c with $c > c_0$ the distance d is of the order $\sqrt{c\Delta(s)}$.

Since z is an element of all hyperplanes

$$H(x_s,(\Theta_i x_s(i) - \lambda_i)e_n + \Theta_i(c-1)\Delta(s)e_i) \qquad i = 1,\ldots,n-1$$

we have for all $i = 1,\ldots,n-1$

$$< z - x_s, (\Theta_i x_s(i) - \lambda_i)e_n + \Theta_i(c-1)\Delta(s)e_i >\; = 0$$

which implies that we have for all $i = 1,\ldots,n-1$

$$z(i) - x_s(i) = (z(n) - x_s(n))\frac{\lambda_i - \Theta_i x_s(i)}{\Theta_i(c-1)\Delta(s)}. \qquad (97)$$

Instead of z we consider $\tilde{z}$ given by

$$\{\tilde{z}\} = \partial T_{1-\epsilon}(\mathcal{E}) \cap \left\{ \bigcap_{i=1}^{n-1} H(x_s,(\Theta_i x_s(i) - \lambda_i)e_n + \Theta_i(c-1)\Delta(s)e_i) \right\}. \qquad (98)$$

We also have

$$\tilde{z}(i) - x_s(i) = (\tilde{z}(n) - x_s(n))\frac{\lambda_i - \Theta_i x_s(i)}{\Theta_i(c-1)\Delta(s)}. \qquad (99)$$

By (75)

$$T_{1-\epsilon}(\mathcal{E} \cap H^-(x_0 - c\Delta(s)N_{\partial K}(x_0), N_{\partial K}(x_0)))$$
$$\subseteq K \cap H^-(x_0 - c\Delta(s)N_{\partial K}(x_0), N_{\partial K}(x_0)).$$

Therefore we have for all $i = 1,\ldots,n$ that $|\tilde{z}(i)| \le |z(i)|$. We will show that we have for all $i = 1,\ldots,n-1$ that $c_3\sqrt{c\Delta(s)} \le |\tilde{z}(i)|$. (We need this estimate for one coordinate only, but get it for all $i = 1,\ldots,n-1$. $\tilde{z}(n)$ is of the order $\Delta(s)$.)

We have

$$1 = \sum_{i=1}^{n-1} \left| \frac{\tilde{z}(i)}{a_i(1-\epsilon)} \right|^2 + \left| \frac{\tilde{z}(n)}{a_n} - 1 \right|^2$$

and equivalently

$$2\frac{\tilde{z}(n)}{a_n} = \sum_{i=1}^{n-1} \left| \frac{\tilde{z}(i)}{a_i(1-\epsilon)} \right|^2 + \left| \frac{\tilde{z}(n)}{a_n} \right|^2$$

$$= \sum_{i=1}^{n-1} \left| \frac{\tilde{z}(i) - x_c(i) + x_s(i)}{a_i(1-\epsilon)} \right|^2 + \left| \frac{\tilde{z}(n) - x_s(n) + x_s(n)}{a_n} \right|^2.$$

By triangle-inequality

$$\sqrt{2\frac{\tilde{z}(n)}{a_n}} - \sqrt{\sum_{i=1}^{n-1} \left| \frac{x_s(i)}{a_i(1-\epsilon)} \right|^2 + \left| \frac{x_s(n)}{a_n} \right|^2}$$

$$\leq \sqrt{\sum_{i=1}^{n-1} \left| \frac{\tilde{z}(i) - x_s(i)}{a_i(1-\epsilon)} \right|^2 + \left| \frac{\tilde{z}(n) - x_s(n)}{a_n} \right|^2}.$$

By (99)

$$\sqrt{2\frac{\tilde{z}(n)}{a_n}} - \sqrt{\sum_{i=1}^{n-1} \left| \frac{x_s(i)}{a_i(1-\epsilon)} \right|^2 + \left| \frac{x_s(n)}{a_n} \right|^2}$$

$$\leq |\tilde{z}(n) - x_s(n)| \sqrt{\sum_{i=1}^{n-1} \left| \frac{\lambda_i - \Theta_i x_s(i)}{(c-1)\Delta(s)a_i(1-\epsilon)} \right|^2 + \left| \frac{1}{a_n} \right|^2}.$$

Since $\tilde{z} \in H^+(x_s, N_{\partial K}(x_0))$ we have $\tilde{z}(n) \geq \Delta(s)$. By (91) we have for all $i = 1, \ldots, n$ that $|x_s(i)| \leq \|x_s\| \leq \Delta(s)(\cos\theta)^{-1}$. Therefore, for small enough s

$$\sqrt{\frac{\tilde{z}(n)}{a_n}} \leq |\tilde{z}(n) - x_s(n)| \sqrt{\sum_{i=1}^{n-1} \left| \frac{\lambda_i - \Theta_i x_s(i)}{(c-1)\Delta(s)a_i(1-\epsilon)} \right|^2 + \left| \frac{1}{a_n} \right|^2}.$$

Since $\tilde{z}(n) \geq x_s(n) \geq 0$

$$\frac{1}{a_n} \leq |\tilde{z}(n) - x_s(n)| \left(\sum_{i=1}^{n-1} \left| \frac{\Theta_i < x_s, e_i > -\lambda_i}{(c-1)\Delta(s)a_i(1-\epsilon)} \right|^2 + \frac{1}{a_n} \right).$$

For sufficiently small s we have $|\tilde{z}(n) - x_s(n)| \leq \frac{1}{2}$ and therefore

$$\frac{1}{2a_n} \leq |\tilde{z}(n) - x_s(n)| \sum_{i=1}^{n-1} \left| \frac{\Theta_i < x_s, e_i > -\lambda_i}{(c-1)\Delta(s)a_i(1-\epsilon)} \right|^2$$

and

$$\sqrt{\frac{1}{2a_n}} \le \sqrt{\tilde{z}(n) - x_s(n)} \left(\sum_{i=1}^{n-1} \left| \frac{\Theta_i < x_s, e_i > -\lambda_i}{(c-1)\Delta(s)a_i(1-\epsilon)} \right|^2 \right)^{\frac{1}{2}}$$

$$\le \sqrt{\tilde{z}(n) - x_s(n)} \left\{ \left(\sum_{i=1}^{n-1} \left| \frac{\Theta_i < x_s, e_i >}{(c-1)\Delta(s)a_i(1-\epsilon)} \right|^2 \right)^{\frac{1}{2}} + \left(\sum_{i=1}^{n-1} \left| \frac{\lambda_i}{(c-1)\Delta(s)a_i(1-\epsilon)} \right|^2 \right)^{\frac{1}{2}} \right\}.$$

Therefore

$$\sqrt{\frac{1}{2a_n}} \le \frac{\sqrt{|\tilde{z}(n) - x_s(n)|}}{(c-1)(1-\epsilon)\Delta(s)} \left\{ \left(\sum_{i=1}^{n-1} \left| \frac{x_s(i)}{a_i} \right|^2 \right)^{\frac{1}{2}} + \left(\sum_{i=1}^{n-1} \left| \frac{\lambda_i}{a_i} \right|^2 \right)^{\frac{1}{2}} \right\}$$

$$\le \frac{\sqrt{|\tilde{z}(n) - x_s(n)|}}{(c-1)(1-\epsilon)\Delta(s)} \left\{ \frac{\left(\sum_{i=1}^{n-1} |x_s(i)|^2 \right)^{\frac{1}{2}}}{\min_{1 \le i \le n-1} a_i} + \left(\sum_{i=1}^{n-1} \left| \frac{\lambda_i}{a_i} \right|^2 \right)^{\frac{1}{2}} \right\}.$$

By (91) we have $\|x_s\| = \Delta(s)(\cos\theta)^{-1}$. From the definition of λ_i, $i = 1, \ldots, n-1$, (following formula (75)) and Lemma 1.3 we get $\lambda_i \le (1 + \epsilon)a_i\sqrt{\frac{c\Delta(s)}{a_n}}$. Therefore we get

$$\sqrt{\frac{1}{2a_n}} \le \frac{\sqrt{|\tilde{z}(n) - x_s(n)|}}{(c-1)(1-\epsilon)\Delta(s)} \left\{ \frac{\Delta(s)(\cos\theta)^{-1}}{\min_{1 \le i \le n-1} a_i} + (1+\epsilon)\sqrt{(n-1)\frac{c\Delta(s)}{a_n}} \right\}.$$

Thus there is a constant c_3 such that for all c with $c \ge 2$ and s with $0 < s \le s_{c,b}$

$$\frac{1}{a_n} \le \frac{c_3}{c\Delta(s)} |\tilde{z}(n) - x_s(n)|.$$

By this inequality and (99)

$$|\tilde{z}(i) - x_s(i)| = |\tilde{z}(n) - x_s(n)| \frac{|\Theta_i < x_s, e_i > -\lambda_i|}{(c-1)\Delta(s)} \ge c_4 |\Theta_i < x_s, e_i > -\lambda_i|.$$

By (91) we have $\|x_s\| = \Delta(s)(\cos\theta)^{-1}$ and from the definition of λ_i, $i = 1, \ldots, n-1$, we get $\lambda_i \ge (1-\epsilon)a_i\sqrt{\frac{c\Delta(s)}{a_n}}$. Therefore $\tilde{z}(i)$ is of the order of λ_i which is in turn of the order of $\sqrt{c\Delta(s)}$.

The orthogonal projection p maps $(z_1, \ldots, z_n)$ onto $(z_1, \ldots, z_{n-1}, 0)$. The distance d of $p(z)$ to the $n-2$-dimensional hyperplane that passes through 0 and that is parallel to the one spanned by

$$p(x_s) + (1 - \epsilon)a_i\sqrt{\tfrac{2\Delta(s)}{a_n}}e_i \qquad\qquad i = 1, \ldots, n-1$$

equals $|<p(z),\xi>|$ where ξ is the normal to this plane. We have

$$\xi = \left(\frac{\frac{1}{a_i}}{\left(\sum_{i=1}^{n-1}a_i^{-2}\right)^{\frac{1}{2}}}\right)_{i=1}^{n-1}$$

and get $|<p(z),\xi>| \geq c_4\sqrt{c\Delta(s)}$. Thus we have proved (96). By (95) and (96) there is a constant c_0 such that for all c with $c \geq c_0$

$$\mathrm{vol}_{n-1}(\mathrm{corn}_K(\Theta))$$

$$\geq \frac{c_4\sqrt{c\Delta(s)}}{(n-1)!}(1-\epsilon)^{n-2}\left(\frac{2\Delta(s)}{a_n}\right)^{\frac{n-2}{2}}\prod_{i=1}^{n-1}a_i\left(\sum_{i=1}^{n-1}|a_i|^{-2}\right)^{\frac{1}{2}}$$

$$-(1+\epsilon)^{n-1}\mathrm{vol}_{n-1}(B_2^{n-1})\left(\frac{2\Delta(s)}{a_n}\right)^{\frac{n-1}{2}}\prod_{i=1}^{n-1}a_i$$

$$\geq c_5\sqrt{c}\Delta(s)^{\frac{n-1}{2}}$$

where c_5 depends only on K. Finally, by the latter inequality and by (87)

$$\mathbb{P}_f^N\{(x_1,\ldots,x_N)|\ x_s \notin [\{x_1,\ldots,x_N\}\cap H^-]\ \text{and}\ x_s \in [x_1,\ldots,x_N]\}$$

$$\leq \sum_{\Theta}\left(1 - (f(x_0) - \epsilon)\mathrm{vol}_{n-1}(\mathrm{corn}_K(\Theta))\right)^N$$

$$\leq 2^{n-1}\left(1 - (f(x_0) - \epsilon)c_5\sqrt{c}\Delta(s)^{\frac{n-1}{2}}\right)^N$$

$$\leq 2^{n-1}\exp\left(-N(f(x_0) - \epsilon)c_5\sqrt{c}\Delta(s)^{\frac{n-1}{2}}\right).$$

By hypothesis we have $\frac{1}{bN}\mathrm{vol}_{n-1}(\partial K) \leq s$. We have

$$s \leq \mathbb{P}_f(\partial K \cap H^-(x_s, N_{\partial K}(x_0)))$$
$$\leq (f(x_0) + \epsilon)\mathrm{vol}_{n-1}(\partial K \cap H^-(x_s, N_{\partial K}(x_0))).$$

By Lemma 1.3 we get

$$s \leq c_6 f(x_0)\Delta(s)^{\frac{n-1}{2}}$$

and therefore

$$\frac{N}{\mathrm{vol}_{n-1}(\partial K)} \geq \frac{1}{bs} \geq \frac{1}{c_6 b f(x_0)\Delta(s)^{\frac{n-1}{2}}}.$$

Therefore

$$\mathbb{P}_f^N\{(x_1,\ldots,x_N)|\ x_s \notin [\{x_1,\ldots,x_N\}\cap H^-]\text{ and }x_s \in [x_1,\ldots,x_N]\}$$
$$\leq 2^{n-1}\exp\left(-c_7\frac{\sqrt{c}}{b}\right).$$

Now we derive

$$\Big|\mathbb{P}_f^N\{(x_1,\ldots,x_N)|\ x_s \notin [x_1,\ldots,x_N]\} -$$
$$\mathbb{P}_f^{N+k}\{(x_1,\ldots,x_{N+k})|\ x_s \notin [x_1,\ldots,x_{N+k}]\}\Big| \leq \epsilon.$$

It is enough to show

$$\Big|\mathbb{P}_f^N\{(x_1,\ldots,x_N)|\ x_s \in [\{x_1,\ldots,x_N\}\cap H^-]\} -$$
$$\mathbb{P}_f^{N+k}\{(x_1,\ldots,x_{N+k})|\ x_s \in [\{x_1,\ldots,x_{N+k}\}\cap H^-]\}\Big| \leq \epsilon.$$

We have

$$\{(x_1,\ldots,x_{N+k})|\ x_s \in [\{x_1,\ldots,x_{N+k}\}\cap H^-]\}$$
$$= \{(x_1,\ldots,x_{N+k})|\ x_s \in [\{x_1,\ldots,x_N\}\cap H^-]\}$$
$$\cup\{(x_1,\ldots,x_{N+k})|\ x_s \notin [\{x_1,\ldots,x_N\}\cap H^-]\text{ and }$$
$$x_s \in [\{x_1,\ldots,x_{N+k}\}\cap H^-]\}.$$

Clearly, the above set is contained in

$$\{(x_1,\ldots,x_{N+k})|\ x_s \in [\{x_1,\ldots,x_N\}\cap H^-]\}$$
$$\cup\{(x_1,\ldots,x_{N+k})|\exists i, 1\leq i\leq k: x_{N+i} \in H^-\cap\partial K\}.$$

Therefore we have

$$\mathbb{P}_f^{N+k}\{(x_1,\ldots,x_{N+k})|\ x_s \in [\{x_1,\ldots,x_{N+k}\}\cap H^-]\}$$
$$\leq \mathbb{P}_f^N\{(x_1,\ldots,x_N)|\ x_s \in [\{x_1,\ldots,x_N\}\cap H^-]\}$$
$$+\mathbb{P}_f^k\{(x_{N+1},\ldots,x_{N+k})|\exists i, 1\leq i\leq k: x_{N+i} \in H^-\cap\partial K\}$$
$$= \mathbb{P}_f^N\{(x_1,\ldots,x_N)|\ x_s \in [\{x_1,\ldots,x_N\}\cap H^-]\}$$
$$+k\int_{\partial K\cap H^-} f(x)\mathrm{d}\mu.$$

We choose H so that $k\int_{\partial K\cap H^-}f(x)\mathrm{d}\mu$ is sufficiently small. $\square$

Lemma 4.16. *Let K be a convex body in $\mathbb{R}^n$ and $x_0 \in \partial K$. Let $\mathcal{E}$ be the standard approximating ellipsoid at x_0. Let $f : \partial K \to \mathbb{R}$ be a continuous, strictly positive function with $\int_{\partial K} f\mathrm{d}\mu = 1$ and K_s be the surface body with respect to the measure $f\mathrm{d}\mu_{\partial K}$ and $\mathcal{E}_s$ the surface body with respect to the measure with the constant density $(\mathrm{vol}_{n-1}(\partial\mathcal{E}))^{-1}$ on $\partial\mathcal{E}$. Suppose that the* *indicatrix of Dupin at x_0 exists and is an ellipsoid (and not a cylinder with an ellipsoid as base). We define x_s, y_s and z_s by*

$$\{x_s\} = [x_0, x_T] \cap \partial K_s \qquad \{z_s\} = [x_0, z_T] \cap \partial \mathcal{E}_s$$

$$\{y_s\} = [x_0, x_T] \cap H(z_s, N_{\partial K}(x_0)).$$

(i) For every $\epsilon > 0$ and all $\ell \in \mathbb{N}$ there are $c_0 > 1$ and $s_0 > 0$ so that we have for all $k \in \mathbb{N}$ with $1 \leq k \leq \ell$, all s and all c with $0 < cs < s_0$ and $c_0 \leq c$, and all hyperplanes H that are orthogonal to $N_{\partial K}(x_0)$ and that satisfy $\mathrm{vol}_{n-1}(\partial K \cap H^-) = cs$

$$\left| \mathbb{P}^k_{f, \partial K \cap H^-} \{(x_1, \ldots, x_k) | \quad x_s \in [x_1, \ldots, x_k]\} - \right.$$
$$\left. \mathbb{P}^k_{\partial K \cap H^-} \{(x_1, \ldots, x_k) | \ x_s \in [x_1, \ldots, x_k]\} \right| < \epsilon$$

where $\mathbb{P}_{f, \partial K \cap H^-}$ is the normalized restriction of the measure $\mathbb{P}_f$ to the set $\partial K \cap H^-$.

(ii) For every $\epsilon > 0$ and all $\ell \in \mathbb{N}$ there are $c_0 > 1$ and $s_0 > 0$ so that we have for all $k \in \mathbb{N}$ with $1 \leq k \leq \ell$, all s and all c with $0 < cs < s_0$ and $c_0 \leq c$, and all hyperplanes H that are orthogonal to $N_{\partial K}(x_0)$ and that satisfy $\mathrm{vol}_{n-1}(\partial K \cap H^-) = cs$

$$\left| \mathbb{P}^k_{\partial K \cap H^-} \{(x_1, \ldots, x_k) | \quad x_s \in [x_1, \ldots, x_k]\} - \right.$$
$$\left. \mathbb{P}^k_{\partial \mathcal{E} \cap H^-} \{(z_1, \ldots, z_k) | \ x_s \in [z_1, \ldots, z_k]\} \right| < \epsilon.$$

(iii) For every $\epsilon > 0$ and all $\ell \in \mathbb{N}$ there are $c_0 > 1$ and $s_0 > 0$ so that we have for all $k \in \mathbb{N}$ with $1 \leq k \leq \ell$, all s and all c with $0 < cs < s_0$ and $c_0 \leq c$, and all hyperplanes H that are orthogonal to $N_{\partial K}(x_0)$ and that satisfy $\mathrm{vol}_{n-1}(\partial K \cap H^-) = cs$

$$\left| \mathbb{P}^k_{\partial \mathcal{E} \cap H^-} \{(z_1, \ldots, z_k) | \quad z_s \in [z_1, \ldots, z_k]\} - \right.$$
$$\left. \mathbb{P}^k_{\partial \mathcal{E} \cap H^-} \{(z_1, \ldots, z_k) | \ y_s \in [z_1, \ldots, z_k]\} \right| < \epsilon.$$

(iv) For every $\epsilon > 0$ and all $\ell \in \mathbb{N}$ there are $c_0 > 1$, $s_0 > 0$, and $\delta > 0$ so that we have for all $k \in \mathbb{N}$ with $1 \leq k \leq \ell$, all s, s' and all c with $0 < cs, cs' < s_0$, $(1-\delta)s \leq s' \leq (1+\delta)s$, and $c_0 \leq c$, and all hyperplanes H_s that are orthogonal to $N_{\partial \mathcal{E}}(x_0)$ and that satisfy $\mathrm{vol}_{n-1}(\partial \mathcal{E} \cap H_s^-) = cs$

$$\left| \mathbb{P}^k_{\partial \mathcal{E} \cap H_s^-} \{(z_1, \ldots, z_k) | \quad z_s \in [z_1, \ldots, z_k]\} - \right.$$
$$\left. \mathbb{P}^k_{\partial \mathcal{E} \cap H_{s'}^-} \{(z_1, \ldots, z_k) | \ z_{s'} \in [z_1, \ldots, z_k]\} \right| < \epsilon.$$

(v) For every $\epsilon > 0$ and all $\ell \in \mathbb{N}$ there are $c_0 > 1$ and $\Delta_0 > 0$ so that we have for all $k \in \mathbb{N}$ with $1 \leq k \leq \ell$, all Δ, all $\gamma \geq 1$ and all c with $0 < c\gamma\Delta < \Delta_0$ and $c_0 \leq c$, and

$$\left| \mathbb{P}^k_{\partial \mathcal{E} \cap H^-_{c\Delta}} \{(x_1, \ldots, x_k)| \ x_0 - \Delta N_{\partial K}(x_0) \in [x_1, \ldots, x_k]\} - \right.$$

$$\left. \mathbb{P}^k_{\partial \mathcal{E} \cap H^-_{c\gamma \Delta}} \{(x_1, \ldots, x_k)| \ x_0 - \gamma \Delta N_{\partial K}(x_0) \in [x_1, \ldots, x_k]\} \right| < \epsilon$$

where $H_{c\Delta} = H_{c\Delta}(x_0 - c\Delta N_{\partial K}(x_0), N_{\partial K}(x_0))$.

(vi) For every $\epsilon > 0$ and all $\ell \in \mathbb{N}$ there are $c_0 > 1$ and $s_0 > 0$ so that we have for all $k \in \mathbb{N}$ with $1 \leq k \leq \ell$, all s with $0 < cs < s_0$, all c with $c_0 \leq c$, and all hyperplanes H and $\tilde{H}$ that are orthogonal to $N_{\partial K}(x_0)$ and that satisfy

$$\mathbb{P}_f(\partial K \cap H^-) = cs \qquad \frac{\mathrm{vol}_{n-1}(\partial \mathcal{E} \cap \tilde{H}^-)}{\mathrm{vol}_{n-1}(\partial \mathcal{E})} = cs$$

that

$$\left| \mathbb{P}^k_{f, \partial K \cap H^-} \{(x_1, \ldots, x_k)| \ x_s \in [x_1, \ldots, x_k]\} - \right.$$

$$\left. \mathbb{P}^k_{\partial \mathcal{E} \cap \tilde{H}^-} \{(z_1, \ldots, z_k)| \ z_s \in [z_1, \ldots, z_k]\} \right| < \epsilon.$$

(The hyperplanes H and $\tilde{H}$ may not be very close, depending on the value $f(x_0)$.)

Proof. (i) This is much simpler than the other cases. We define $\Phi_{x_s} : \partial K \times \cdots \times \partial K \to \mathbb{R}$ by

$$\Phi_{x_s}(x_1, \ldots, x_k) = \begin{cases} 0 & x_s \notin [x_1, \ldots, x_k] \\ 1 & x_s \in [x_1, \ldots, x_k]. \end{cases}$$

Then we have

$$\mathbb{P}^k_{f, \partial K \cap H^-} \{(x_1, \ldots, x_k)| \ x_s \in [x_1, \ldots, x_k]\}$$
$$= (\mathbb{P}_f(\partial K \cap H^-))^{-k} \times$$

$$\int_{\partial K \cap H^-} \cdots \int_{\partial K \cap H^-} \Phi_{x_s}(x_1, \ldots, x_k) \prod_{i=1}^k f(x_i) \mathrm{d}\mu_{\partial K}(x_1) \cdots \mathrm{d}\mu_{\partial K}(x_k).$$

By continuity of f for every $\delta > 0$ we find s_0 so small that we have for all s with $0 < s \leq s_0$ and all $x \in \partial K \cap H^-(x_s, N_{\partial K}(x_0))$

$$|f(x_0) - f(x)| < \delta.$$

(ii) We may suppose that $x_0 = 0$ and that e_n is orthogonal to K at x_0. Let $T_s : \mathbb{R}^n \to \mathbb{R}^n$ be given by

$$T_s(x(1), \ldots, x(n)) = (sx(1), \ldots, sx(n-1), x(n)). \tag{100}$$

Then, by Lemma 1.2, for every $\delta > 0$ there is a hyperplane H orthogonal to e_n such that for

$$\mathcal{E}_1 = T_{\frac{1}{1+\delta}}(\mathcal{E}) \qquad \mathcal{E}_2 = T_{1+\delta}(\mathcal{E})$$

we have

$$\mathcal{E}_1 \cap H^- = T_{\frac{1}{1+\delta}}(\mathcal{E}) \cap H^- \subseteq K \cap H^- \subseteq T_{1+\delta}(\mathcal{E}) \cap H^- = \mathcal{E}_2 \cap H^-.$$

Since the indicatrix of Dupin at x_0 is an ellipsoid and not a cylinder and since f is continuous with $f(x_0) > 0$ we conclude that there is s_0 such that

$$T_{\frac{1}{1+\delta}}(\mathcal{E}) \cap H^-(x_{s_0}, N_{\partial K}(x_0)) \subseteq K \cap H^- \subseteq T_{1+\delta}(\mathcal{E}) \cap H^-(x_{s_0}, N_{\partial K}(x_0)). \tag{101}$$

We have that

$$\mathbb{P}^k_{\partial K \cap H^-}\{(x_1, \ldots, x_k)| \ x_s \in [x_1, \ldots, x_k]\}$$
$$= \mathbb{P}^k_{\partial K \cap H^-}\{(x_1, \ldots, x_k)| \ x_s \in [x_1, \ldots, x_k]^\circ\}.$$

This follows from Lemma 4.2. Therefore it is enough to verify the claim for this set. The set

$$\{(x_1, \ldots, x_k)| \ x_s \in [x_1, \ldots, x_k]^\circ, x_1, \ldots, x_k \in \partial K \cap H^-\}$$

is an open subset of the k-fold product $(\partial K \cap H^-) \times \cdots \times (\partial K \cap H^-)$. Indeed, since x_s is in the interior of the polytope $[x_1, \ldots, x_k]$ we may move the vertices slightly and x_s is still in the interior of the polytope.

Therefore this set is an intersection of $(\partial K \cap H^-) \times \cdots \times (\partial K \cap H^-)$ with an open subset $\mathcal{O}$ of $\mathbb{R}^{kn}$. Such a set $\mathcal{O}$ can be written as the countable union of cubes whose pairwise intersections have measure 0. Cubes are sets $B^n_\infty(x_0, r) = \{x| \max_i |x(i) - x_0(i)| \le r\}$. Thus there are cubes $B^n_\infty(x_i^j, r_i^j)$, $1 \le i \le k$, $j \in \mathbb{N}$, in $\mathbb{R}^n$ such that

$$\mathcal{O} = \bigcup_{j=1}^{\infty} \prod_{i=1}^{k} B^n_\infty(x_i^j, r_i^j) \tag{102}$$

and for $j \neq m$

$$\text{vol}_{kn}\left(\prod_{i=1}^{k} B^n_\infty(x_i^j, r_i^j) \cap \prod_{i=1}^{k} B^n_\infty(x_i^m, r_i^m)\right)$$
$$= \prod_{i=1}^{k} \text{vol}_n(B^n_\infty(x_i^j, r_i^j) \cap B^n_\infty(x_i^m, r_i^m)) = 0.$$

Therefore, for every pair j, m with $j \neq m$ there is i, $1 \le i \le k$, such that

$$B^n_\infty(x_i^j, r_i^j) \cap B^n_\infty(x_i^m, r_i^m) \tag{103}$$

is contained in a hyperplane that is orthogonal to one of the vectors $e_1, \ldots, e_n$. We put

$$W_j = \prod_{i=1}^{k} \left(B_\infty^n(x_i^j, r_i^j) \cap \partial K \cap H^- \right) \qquad j \in \mathbb{N} \qquad (104)$$

and get

$$\{(x_1, \ldots, x_k) | \ x_s \in [x_1, \ldots, x_k]^\circ, x_1, \ldots, x_k \in \partial K \cap H^- \} = \bigcup_{j=1}^{\infty} W_j. \quad (105)$$

Then we have for $j \neq m$ that

$$\mathrm{vol}_{k(n-1)}(W_j \cap W_m) = 0.$$

Indeed,

$$W_j \cap W_m = \left\{ (x_1, \ldots, x_k) | \forall i : x_i \in \partial K \cap B_\infty^n(x_i^j, r_i^j) \cap B_\infty^n(x_i^m, r_i^m) \cap H^- \right\}.$$

There is at least one i_0 such that

$$B_\infty^n(x_{i_0}^j, r_{i_0}^j) \cap B_\infty^n(x_{i_0}^m, r_{i_0}^m)$$

is contained in a hyperplane L that is orthogonal to one of the vectors $e_1, \ldots, e_n$. Therefore

$$\mathrm{vol}_{n-1}(\partial K \cap B_\infty^n(x_{i_0}^j, r_{i_0}^j) \cap B_\infty^n(x_{i_0}^m, r_{i_0}^m)) \leq \mathrm{vol}_{n-1}(\partial K \cap L).$$

The last expression is 0 if the hyperplane is chosen sufficiently close to x_0. Indeed, $\partial K \cap L$ is either a face of K or $\partial K \cap L = \partial(K \cap L)$. In the latter case $\mathrm{vol}_{n-1}(\partial K \cap L) = \mathrm{vol}_{n-1}(\partial(K \cap H)) = 0$. If H is sufficiently close to x_0, then L does not contain a $n-1$-dimensional face of K. This follows from the fact that the indicatrix exists and is an ellipsoid and consequently all normals are close to $N_{\partial K}(x_0) = e_n$ but not equal.

Let $rp : \partial K \to \partial \mathcal{E}$ where $rp(x)$ is the unique point with

$$\{rp(x)\} = \{x_s + t(x - x_s) | t \geq 0\} \cap \partial \mathcal{E}. \qquad (106)$$

For s_0 small enough we have for all s with $0 < s \leq s_0$ that $x_s \in \mathcal{E}$. In this case rp is well defined. $Rp : \partial K \times \cdots \times \partial K \to \partial \mathcal{E} \times \cdots \times \partial \mathcal{E}$ is defined by

$$Rp(x_1, \ldots, x_k) = (rp(x_1), \ldots, rp(x_k)). \qquad (107)$$

There is a map $\alpha : \partial K \to (-\infty, 1)$ such that

$$rp(x) = x - \alpha(x)(x - x_s). \qquad (108)$$

Since x_s is an interior point of K the map α does not attain the value 1. For every $\epsilon > 0$ there is s_0 such that we have for all s and c with $0 < cs \leq s_0$ and $c \geq c_0$ and for all hyperplanes H that are orthogonal to $N_{\partial K}(x_0) = e_n$ and that satisfy $\mathrm{vol}_{n-1}(\partial K \cap H^-) = cs$ and all cubes $B_\infty^n(x_i^j, r_i^j)$ that satisfy (104) and (105)

$$\mathrm{vol}_{n-1}(\partial K \cap B_\infty^n(x_i^j, r_i^j)) \leq (1+\epsilon)\mathrm{vol}_{n-1}(rp(\partial K \cap B_\infty^n(x_i^j, r_i^j))). \qquad (109)$$

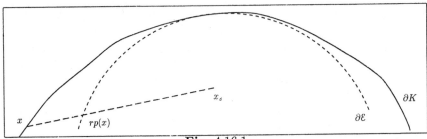

Fig. 4.16.1

To show this we have to establish that there is s_0 such that for all $x \in \partial K \cap H^-(x_{s_0}, N_{\partial K}(x_0))$ and all s with $0 < s \leq s_0$

$$\|x - rp(x)\| \leq \epsilon \|x_s - rp(x)\| \qquad (110)$$

$$(1-\epsilon) \leq \frac{\left\langle N_{\partial K}(x), \frac{x-x_s}{\|x-x_s\|} \right\rangle}{\left\langle N_{\partial \mathcal{E}}(rp(x)), \frac{x-x_s}{\|x-x_s\|} \right\rangle} \leq (1+\epsilon). \qquad (111)$$

Indeed, the volume of a surface element changes under the map rp by the factor

$$\left(\frac{\|rp(x) - x_s\|}{\|x - x_s\|} \right)^{n-1} \frac{\left\langle N_{\partial K}(x), \frac{x-x_s}{\|x-x_s\|} \right\rangle}{\left\langle N_{\partial \mathcal{E}}(rp(x)), \frac{x-x_s}{\|x-x_s\|} \right\rangle}.$$

We establish (111). We have

$$\frac{\langle N_{\partial K}(x), x - x_s \rangle}{\langle N_{\partial \mathcal{E}}(rp(x)), x - x_s \rangle} = 1 + \frac{\langle N_{\partial K}(x) - N_{\partial \mathcal{E}}(rp(x)), x - x_s \rangle}{\langle N_{\partial \mathcal{E}}(rp(x)), x - x_s \rangle}$$

$$\leq 1 + \frac{\|N_{\partial K}(x) - N_{\partial \mathcal{E}}(rp(x))\| \, \|x - x_s\|}{\langle N_{\partial \mathcal{E}}(rp(x)), x - x_s \rangle}.$$

We have

$$\|N_{\partial K}(x) - N_{\partial \mathcal{E}}(rp(x))\| \leq \epsilon \|x - x_0\|.$$

This can be shown in the same way as (33) (Consider the plane $H(x, N_{\partial K}(x_0))$. The distance of this plane to x_0 is of the order $\|x - x_0\|^2$.) Thus we have

$$\frac{\langle N_{\partial K}(x), x - x_s \rangle}{\langle N_{\partial \mathcal{E}}(rp(x)), x - x_s \rangle} \leq 1 + \frac{\epsilon \|x - x_0\| \|x - x_s\|}{\langle N_{\partial \mathcal{E}}(rp(x)), x - x_s \rangle} \leq 1 + \frac{\epsilon c_0 \|x - x_s\|^2}{\langle N_{\partial \mathcal{E}}(rp(x)), x - x_s \rangle}.$$

It is left to show

$$| < N_{\partial \mathcal{E}}(rp(x)), x - x_s > | \geq c_0 \|x - x_s\|^2.$$

If x is close to x_0 then this estimate reduces to $\|x - x_s\| \geq \|x - x_s\|^2$ which is obvious. If x is not close to x_0 then $\|x - x_s\|^2$ is of the order of the height of the cap $\partial \mathcal{E} \cap H^-(rp(x), N_{\partial K}(x_0))$. Therefore, it is enough to show

$$| < N_{\partial \mathcal{E}}(rp(x)), x - x_s > | \geq c_0 | < N_{\partial K}(x_0), rp(x) - x_0 > |.$$

We consider the map $T : \mathbb{R}^n \to \mathbb{R}^n$ that transforms the standard approximating ellipsoid into a Euclidean ball (5)

$$T(x) = \left(\frac{x_1}{a_1} \left(\prod_{i=1}^{n-1} b_i \right)^{\frac{2}{n-1}}, \ldots, \frac{x_{n-1}}{a_{n-1}} \left(\prod_{i=1}^{n-1} b_i \right)^{\frac{2}{n-1}}, x_n \right).$$

Thus it is enough to show

$$| < T^{-1t} N_{\partial \mathcal{E}}(rp(x)), Tx - Tx_s > | \geq c_0 | < N_{\partial K}(x_0), rp(x) - x_0 > |.$$

Since $Tx_0 = x_0 = 0$ and $T^{-1t}(N_{\partial K}(x_0)) = N_{\partial K}(x_0) = e_n$ the above inequality is equivalent to

$$| < T^{-1t} N_{\partial \mathcal{E}}(rp(x)), Tx - Tx_s > | \geq c_0 | < N_{\partial K}(x_0), T(rp(x)) - x_0 > |.$$

Allowing another constant c_0, the following is equivalent to the above

$$\left| \left\langle \frac{T^{-1t} N_{\partial \mathcal{E}}(rp(x))}{\|T^{-1t} N_{\partial \mathcal{E}}(rp(x))\|}, Tx - Tx_s \right\rangle \right| \geq c_0 | < N_{\partial K}(x_0), T(rp(x)) - x_0 > |.$$

Thus we have reduced the estimate to the case of a Euclidean ball.

The hyperplane $H(T(rp(x)), N_{\partial K}(x_0))$ intersects the line

$$\{x_0 + t N_{\partial K}(x_0) | t \in \mathbb{R}\}$$

at the point z with $\|x_0 - z\| = | < N_{\partial K}(x_0), T(rp(x)) - x_0 > |$. Let the radius of $T(\mathcal{E})$ be r. See Figure 4.16.2. We may assume that $< T^{-1t} N_{\partial \mathcal{E}}(rp(x)), N_{\partial K}(x_0) >> \frac{1}{2}$. Therefore we have by Figure 4.16.2 ($h = \|x_0 - z\|$)

$$\left| \left\langle \frac{T^{-1t} N_{\partial \mathcal{E}}(rp(x))}{\|T^{-1t} N_{\partial \mathcal{E}}(rp(x))\|}, T(rp(x)) - x_0 \right\rangle \right|$$

$$= \left(\|x_0 - z\| + \frac{\|x_0 - z\|^2}{r - \|x_0 - z\|} \right) \left\langle \frac{T^{-1t} N_{\partial \mathcal{E}}(rp(x))}{\|T^{-1t} N_{\partial \mathcal{E}}(rp(x))\|}, N_{\partial K}(x_0) \right\rangle$$

$$\geq \|x_0 - z\| \left\langle \frac{T^{-1t} N_{\partial \mathcal{E}}(rp(x))}{\|T^{-1t} N_{\partial \mathcal{E}}(rp(x))\|}, N_{\partial K}(x_0) \right\rangle$$

$$\geq \tfrac{1}{2} | < N_{\partial K}(x_0), T(rp(x)) - x_0 > |$$

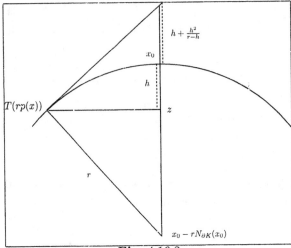

Fig. 4.16.2

where r is the radius of $T(\mathcal{E})$. Since there is a constant c_0 such that

$$\left|\left\langle \frac{T^{-1t}N_{\partial\varepsilon}(rp(x))}{\|T^{-1t}N_{\partial\varepsilon}(rp(x))\|}, T(x) - T(x_s) \right\rangle\right|$$
$$\geq c_0 \left|\left\langle \frac{T^{-1t}N_{\partial\varepsilon}(rp(x))}{\|T^{-1t}N_{\partial\varepsilon}(rp(x))\|}, T(rp(x)) - x_0 \right\rangle\right|$$

we get

$$\left|\left\langle \frac{T^{-1t}N_{\partial\varepsilon}(rp(x))}{\|T^{-1t}N_{\partial\varepsilon}(rp(x))\|}, T(x) - T(x_s) \right\rangle\right| \geq \tfrac{1}{2}c_0| < N_{\partial K}(x_0), T(rp(x)) - x_0 > |.$$

The left hand inequality of (111) is shown in the same way.

Now we verify (110).

Again we apply the affine transform T to K that transforms the indicatrix of Dupin at x_0 into a Euclidean sphere (5). T leaves x_0 and $N_{\partial K}(x_0)$ invariant.

An affine transform maps a line onto a line and the factor by which a segment of a line is stretched is constant. We have

$$\frac{\|x - rp(x)\|}{\|x_s - rp(x)\|} = \frac{\|T(x) - T(rp(x))\|}{\|T(x_s) - T(rp(x))\|}.$$

Thus we have

$$B_2^n(x_0 - rN_{\partial K}(x_0), r) \cap H^-(T(x_{s_0}), N_{\partial K}(x_0))$$
$$\subseteq T(K) \cap H^-(T(x_{s_0}), N_{\partial K}(x_0))$$
$$\subseteq B_2^n(x_0 - (1 + \epsilon)rN_{\partial K}(x_0), (1 + \epsilon)r) \cap H^-(T(x_{s_0}), N_{\partial K}(x_0)).$$

The center of the $n - 1$-dimensional sphere

$$B_2^n(x_0 - rN_{\partial K}(x_0), r) \cap H(T(rp(x)), N_{\partial K}(x_0))$$

is

$$x_0 - \ <x_0 - T(rp(x)), N_{\partial K}(x_0)>\ N_{\partial K}(x_0)$$

and the height of the cap

$$B_2^n(x_0 - rN_{\partial K}(x_0), r) \cap H^-(T(rp(x)), N_{\partial K}(x_0))$$

is

$$|<x_0 - T(rp(x)), N_{\partial K}(x_0)>|.$$

Therefore, for sufficiently small s_0 and all s with $0 < s \leq s_0$ we get that the radius of the cap $\|T(rp(x)) - (x_0 - <x_0 - T(rp(x)), N_{\partial K}(x_0)> N_{\partial K}(x_0))\|$ satisfies

$$\sqrt{r|<x_0 - T(rp(x)), N_{\partial K}(x_0)>|} \tag{112}$$
$$\leq \|T(rp(x)) - (x_0 - <x_0 - T(rp(x)), N_{\partial K}(x_0)> N_{\partial K}(x_0))\|.$$

We show that there is a constant $c_0 > 0$ so that we have for all s with $0 < s \leq s_0$ and all $x \in \partial K \cap H^-(x_{s_0}, N_{\partial K}(x_0))$

$$\|T(rp(x)) - T(x_s)\| \geq c_0 \sqrt{r|<x_0 - T(rp(x)), N_{\partial K}(x_0)>|}. \tag{113}$$

Let α be the angle between $N_{\partial K}(x_0)$ and $x_0 - T(x_T)$. We first consider the case

$$\|T(rp(x)) - (x_0 - <x_0 - T(rp(x)), N_{\partial K}(x_0)> N_{\partial K}(x_0))\|$$
$$\geq 2(1 + (\cos\alpha)^{-1})|<x_0 - T(x_s), N_{\partial K}(x_0)>|. \tag{114}$$

(This case means: x_0 is not too close to $T(rp(x))$.) Then we have

$$\|T(rp(x)) - T(x_s)\|$$
$$\geq \|T(rp(x)) - (x_0 - <x_0 - T(rp(x)), N_{\partial K}(x_0)> N_{\partial K}(x_0))\|$$
$$\quad - \|x_0 - T(x_s)\| - |<x_0 - T(rp(x)), N_{\partial K}(x_0)>|$$
$$= \|T(rp(x)) - (x_0 - <x_0 - T(rp(x)), N_{\partial K}(x_0)> N_{\partial K}(x_0))\|$$
$$\quad - (\cos\alpha)^{-1}|<x_0 - T(x_s), N_{\partial K}(x_0)>|$$
$$\quad - |<x_0 - T(rp(x)), N_{\partial K}(x_0)>|.$$

By the assumption (114)

$$\|T(rp(x)) - T(x_s)\|$$
$$\geq \tfrac{1}{2}\|T(rp(x)) - (x_0 - <x_0 - T(rp(x)), N_{\partial K}(x_0)> N_{\partial K}(x_0))\|$$
$$\quad + (1 + (\cos\alpha)^{-1})|<x_0 - T(x_s), N_{\partial K}(x_0)>|$$
$$\quad - (\cos\alpha)^{-1}|<x_0 - T(x_s), N_{\partial K}(x_0)>|$$
$$\quad - |<x_0 - T(rp(x)), N_{\partial K}(x_0)>|$$
$$= \tfrac{1}{2}\|T(rp(x)) - (x_0 - <x_0 - T(rp(x)), N_{\partial K}(x_0)> N_{\partial K}(x_0))\|$$
$$\quad + |<x_0 - T(x_s), N_{\partial K}(x_0)>| - |<x_0 - T(rp(x)), N_{\partial K}(x_0)>|.$$

By (112)

$$\|T(rp(x)) - T(x_s)\|$$
$$\geq \tfrac{1}{2}\sqrt{r|<x_0 - T(rp(x)), N_{\partial K}(x_0)>|}$$
$$+|<x_0 - T(x_s), N_{\partial K}(x_0)>| - |<x_0 - T(rp(x)), N_{\partial K}(x_0)>|$$
$$\geq \tfrac{1}{2}\sqrt{r|<x_0 - T(rp(x)), N_{\partial K}(x_0)>|}$$
$$-|<x_0 - T(rp(x)), N_{\partial K}(x_0)>|.$$

We get for sufficiently small s_0 that for all s with $0 < s \leq s_0$

$$\|T(rp(x)) - T(x_s)\| \geq \tfrac{1}{4}\sqrt{r|<x_0 - T(rp(x)), N_{\partial K}(x_0)>|}.$$

The second case is

$$\|T(rp(x)) - (x_0 - <x_0 - T(rp(x)), N_{\partial K}(x_0)> N_{\partial K}(x_0))\| \qquad (115)$$
$$< 2(1 + (\cos\alpha)^{-1})|<x_0 - T(x_s), N_{\partial K}(x_0)>|.$$

(In this case, x_0 is close to $T(rp(x))$.) $\|T(rp(x)) - T(x_s)\|$ can be estimated from below by the least distance of $T(x_s)$ to the boundary of $B_2^n(x_0 - rN_{\partial K}(x_0), r)$. This, in turn, can be estimated from below by

$$c'|<x_0 - T(x_s), N_{\partial K}(x_0)>|.$$

Thus we have

$$\|T(rp(x)) - T(x_s)\| \geq c'|<x_0 - T(x_s), N_{\partial K}(x_0)>|.$$

On the other hand, by our assumption (115)

$$\|T(rp(x)) - T(x_s)\|$$
$$\geq \frac{c'}{2(1 + (\cos\alpha)^{-1})} \times$$
$$\|T(rp(x)) - (x_0 - <x_0 - T(rp(x)), N_{\partial K}(x_0)> N_{\partial K}(x_0))\|.$$

By (112)

$$\|T(rp(x)) - T(x_s)\| \geq \frac{c'}{2(1 + (\cos\alpha)^{-1})}\sqrt{r|<x_0 - T(rp(x)), N_{\partial K}(x_0)>|}.$$

This establishes (113).

Now we show that for s_0 sufficiently small we have for all s with $0 < s \leq s_0$ and all x

$$\|T(x) - T(rp(x))\| \leq 2\sqrt{2\epsilon(1 + \epsilon)r|<x_0 - T(rp(x)), N_{\partial K}(x_0)>|}. \qquad (116)$$

Instead of $T(x)$ we consider the points z and z' with

$\{z\} = B_2^n(x_0 - (1+\epsilon)rN_{\partial K}(x_0), (1+\epsilon)r) \cap \{T(x_s) + t(T'(x) - T(x_s)) | t \geq 0\}$

$\{z'\} = B_2^n(x_0 - (1-\epsilon)rN_{\partial K}(x_0), (1-\epsilon)r) \cap \{T(x_s) + t(T(x) - T(x_s)) | t \geq 0\}.$

We have

$$\|T(x) - T(rp(x))\| \leq \max\{\|z - T(rp(x))\|, \|z' - T(rp(x))\|\}.$$

We may assume that $\|x - x_s\| \geq \|rp(x) - x_s\|$. This implies $\|T(x) - T(rp(x))\| \leq \|z - T(rp(x))\|$. $\|z - T(rp(x))\|$ is smaller than the diameter of the cap

$$B_2^n(x_0 - (1+\epsilon)rN_{\partial K}(x_0), (1+\epsilon)r)$$
$$\cap H^-(T(rp(x)), N_{\partial B_2^n(x_0 - rN_{\partial K}(x_0), r)}(T(rp(x))))$$

because z and $T(rp(x))$ are elements of this cap. See Figure 4.16.3. We compute the radius of this cap. The two triangles in Figure 4.16.3 are homothetic with respect to the point x_0. The factor of homothety is $1 + \epsilon$. The distance between the two tangents to $B_2^n(x_0 - (1+\epsilon)rN_{\partial K}(x_0), (1+\epsilon)r)$ and $B_2^n(x_0 - rN_{\partial K}(x_0), r)$ is

$$\epsilon| < x_0 - T(rp(x)), N_{\partial K}(x_0) > |.$$

Consequently the radius is less than

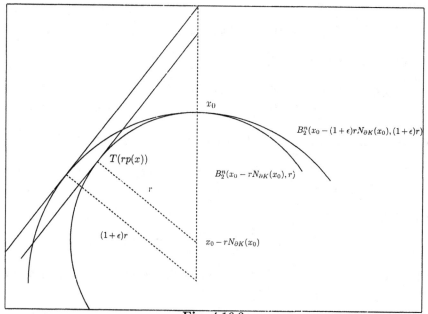

Fig. 4.16.3

$$\sqrt{2\epsilon(1+\epsilon)}r| < \overline{x_0 - T(rp(x)), N_{\partial K}(x_0) > |}.$$

Thus we have established (116). The inequalities (113) and (116) give (110).

From the inequalities (110) and (111) we get for $x \in \partial K \cap B_\infty^n(x_i^j, r_i^j)$ and r_i^j sufficiently small

$$\text{vol}_{n-1}(\partial K \cap B_\infty^n(x_i^j, r_i^j))$$

$$\leq \frac{\left|\frac{\|x_s-x\|}{\|x_s-rp(x)\|}\right|^{n-1}}{\langle N_{\partial K}(x), N_{\partial \mathcal{E}}(rp(x))\rangle}\text{vol}_{n-1}(rp(\partial K \cap B_\infty^n(x_i^j, r_i^j)))$$

$$\leq (1+\epsilon)\frac{(1+\epsilon)^{n-1}}{1-\epsilon}\text{vol}_{n-1}(rp(\partial K \cap B_\infty^n(x_i^j, r_i^j))).$$

It follows that for a new s_0

$$\text{vol}_{k(n-1)}(W_j) = \prod_{i=1}^{k}\text{vol}_{n-1}(\partial K \cap B_\infty^n(x_i^j, r_i^j))$$

$$\leq (1+\epsilon)^k \prod_{i=1}^{k}\text{vol}_{n-1}(rp(\partial K \cap B_\infty^n(x_i^j, r_i^j)))$$

$$= (1+\epsilon)^k\text{vol}_{k(n-1)}(Rp(W_j)).$$

And again with a new s_0

$$\text{vol}_{k(n-1)}(W_j) \leq (1+\epsilon)\text{vol}_{k(n-1)}(Rp(W_j)). \tag{117}$$

We also have for all $x_i \in \partial K$, $i = 1, \ldots, k$

$$Rp(\{(x_1, \ldots, x_k)| \quad x_s \in [x_1, \ldots, x_k]^\circ \text{ and } x_i \in \partial K\}) \tag{118}$$

$$\subseteq \{(z_1, \ldots, z_k)| \quad x_s \in [z_1, \ldots, z_k] \text{ and } z_i \in \partial\mathcal{E}\}.$$

We verify this. Let a_i, $i = 1, \ldots, k$, be nonnegative numbers with $\sum_{i=1}^{k} a_i = 1$ and

$$x_s = \sum_{i=1}^{k} a_i x_i.$$

We choose

$$b_i = \frac{a_i}{(1-\alpha(x_i))(1+\sum_{j=1}^{k}\frac{\alpha(x_j)a_j}{1-\alpha(x_j)})}$$

where $\alpha(x_i)$, $i = 1, \ldots, k$, are defined by (108). We claim that $\sum_{i=1}^{k} b_i = 1$ and

$$x_s = \sum_{i=1}^{k} b_i rp(x_i).$$

We have

$$\sum_{i=1}^{k} b_i = \sum_{i=1}^{k} \frac{a_i}{(1-\alpha(x_i))(1+\sum_{j=1}^{k}\frac{\alpha(x_j)a_j}{1-\alpha(x_j)})}$$

$$= \sum_{i=1}^{k} \frac{a_i(1+\frac{\alpha(x_i)}{1-\alpha(x_i)})}{1+\sum_{j=1}^{k}\frac{\alpha(x_j)a_j}{1-\alpha(x_j)}} = 1.$$

Moreover, by (108) we have $rp(x_i) = x_i - \alpha(x_i)(x_i - x_s)$

$$\sum_{i=1}^{k} b_i rp(x_i) = \sum_{i=1}^{k} b_i(x_i - \alpha(x_i)(x_i - x_s))$$

$$= \sum_{i=1}^{k} \frac{a_i(x_i - \alpha(x_i)(x_i - x_s))}{(1-\alpha(x_i))(1+\sum_{j=1}^{k}\frac{\alpha(x_j)a_j}{1-\alpha(x_j)})}$$

$$= \sum_{i=1}^{k} \frac{a_i x_i}{1+\sum_{j=1}^{k}\frac{\alpha(x_j)a_j}{1-\alpha(x_j)}} + \sum_{i=1}^{k} \frac{a_i\alpha(x_i)x_s}{(1-\alpha(x_i))(1+\sum_{j=1}^{k}\frac{\alpha(x_j)a_j}{1-\alpha(x_j)})}$$

$$= \frac{x_s}{1+\sum_{j=1}^{k}\frac{\alpha(x_j)a_j}{1-\alpha(x_j)}} + \frac{\sum_{i=1}^{k}\frac{a_i\alpha(x_i)}{1-\alpha(x_i)}x_s}{1+\sum_{j=1}^{k}\frac{\alpha(x_j)a_j}{1-\alpha(x_j)}} = x_s.$$

Thus we have established (118)

$$Rp(\{(x_1,\ldots,x_k)|\ x_s \in [x_1,\ldots,x_k]^\circ \text{ and } x_i \in \partial K\})$$
$$\subseteq \{(z_1,\ldots,z_k)|\ x_s \in [z_1,\ldots,z_k] \text{ and } z_i \in \partial\mathcal{E}\}.$$

Next we verify that there is a hyperplane $\tilde{H}$ that is parallel to H and such that

$$\text{vol}_{n-1}(\partial K \cap \tilde{H}^-) \le (1+\epsilon)\text{vol}_{n-1}(\partial K \cap H^-) \tag{119}$$

and

$$Rp(\{(x_1,\ldots,x_k)|\ x_s \in [x_1,\ldots,x_k]^\circ, x_i \in \partial K \cap H^-\}) \tag{120}$$
$$\subseteq \{(z_1,\ldots,z_k)|\ x_s \in [z_1,\ldots,z_k] \text{ and } z_i \in \partial\mathcal{E} \cap \tilde{H}^-\}.$$

This is done by arguments similar to the ones above. Thus we get with a new s_0

$$\mathbb{P}^k_{\partial K \cap H^-}\{(x_1,\ldots,x_k)|\ x_s \in [x_1,\ldots,x_k]\} = \frac{\text{vol}_{k(n-1)}\left(\bigcup_{j=1}^{\infty} W_j\right)}{(\text{vol}_{n-1}(\partial K \cap H^-))^k}$$

$$\le (1+\epsilon)\frac{\text{vol}_{k(n-1)}\left(\bigcup_{j=1}^{\infty} Rp(W_j)\right)}{(\text{vol}_{n-1}(\partial K \cap H^-))^k}$$

$$\le (1+\epsilon)\frac{\text{vol}_{k(n-1)}\{(z_1,\ldots,z_k)|\ x_s \in [z_1,\ldots,z_k] \text{ and } z_i \in \partial\mathcal{E} \cap \tilde{H}^-\}}{(\text{vol}_{n-1}(\partial K \cap H^-))^k}$$

$$\le (1+\epsilon)\frac{\text{vol}_{k(n-1)}\{(z_1,\ldots,z_k)|\ x_s \in [z_1,\ldots,z_k] \text{ and } z_i \in \partial\mathcal{E} \cap H^-\}}{(\text{vol}_{n-1}(\partial K \cap H^-))^k} + k\epsilon.$$

$\text{vol}_{n-1}(\partial K \cap H^-)$ and $\text{vol}_{n-1}(\partial \mathcal{E} \cap H^-)$ differ only by a factor between $1 - \epsilon$ and $1 + \epsilon$ if we choose s_0 small enough. Therefore, for sufficiently small s_0 we have for all s with $0 < s \le s_0$

$$\mathbb{P}^k_{\partial K \cap H^-}\{(x_1, \ldots, x_k)| \ x_s \in [x_1, \ldots, x_k]\}$$
$$\le (1 + \epsilon)\mathbb{P}^k_{\partial \mathcal{E} \cap H^-}\{(z_1, \ldots, z_k)| \ x_s \in [z_1, \ldots, z_k]\} + \epsilon.$$

(iii) We show now that for sufficiently small s_0 we have

$$|\mathbb{P}^k_{\partial \mathcal{E} \cap H^-}\{(z_1, \ldots, z_k)| \ y_s \in [z_1, \ldots, z_k]\}$$
$$-\mathbb{P}^k_{\partial \mathcal{E} \cap H^-}\{(z_1, \ldots, z_k)| \ z_s \in [z_1, \ldots, z_k]\}| < \epsilon.$$

The arguments are very similar to those for the first inequality. We consider the standard approximating ellipsoid $\mathcal{E}$ and the map $tp : \partial \mathcal{E} \to \partial \mathcal{E}$ mapping $x \in \partial \mathcal{E}$ onto the unique point $tp(x)$ with

$$\{tp(x)\} = \partial \mathcal{E} \cap \{y_s + t(x - z_s)| \ t \ge 0\}.$$

See Figure 4.16.4.

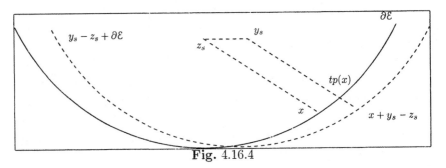

Fig. 4.16.4

We define $Tp : \partial \mathcal{E} \times \cdots \partial \mathcal{E} \to \partial \mathcal{E} \times \cdots \times \partial \mathcal{E}$ by $Tp(z_1, \ldots, z_k) = (tp(z_1), \ldots, tp(z_k))$. Then we have

$$Tp(\{(z_1, \ldots, z_k)| \ z_s \in [z_1, \ldots, z_k] \text{ and } z_i \in \partial \mathcal{E}\})$$
$$\subseteq \{(y_1, \ldots, y_k)| \ y_s \in [y_1, \ldots, y_k] \text{ and } y_i \in \partial \mathcal{E}\}.$$

The calculation is the same as for the inequality (ii). The map tp changes the volume of a surface-element at the point x by the factor

$$\left(\frac{\|y_s - tp(x)\|}{\|y_s - (x + y_s - z_s)\|}\right)^{n-1} \frac{\left\langle \frac{y_s - tp(x)}{\|y_s - tp(x)\|}, N_{\partial \mathcal{E}}(x) \right\rangle}{\left\langle \frac{y_s - tp(x)}{\|y_s - tp(x)\|}, N_{\partial \mathcal{E}}(tp(x)) \right\rangle} \qquad (121)$$

$$= \left(\frac{\|y_s - tp(x)\|}{\|x - z_s\|}\right)^{n-1} \frac{\left\langle \frac{y_s - tp(x)}{\|y_s - tp(x)\|}, N_{\partial \mathcal{E}}(x) \right\rangle}{\left\langle \frac{y_s - tp(x)}{\|y_s - tp(x)\|}, N_{\partial \mathcal{E}}(tp(x)) \right\rangle}.$$

We have to show that this expression is arbitrarily close to 1 provided that s is sufficiently small. Since we consider an ellipsoid

$$\frac{\left\langle \frac{y_s - tp(x)}{\|y_s - tp(x)\|}, N_{\partial \mathcal{E}}(x) \right\rangle}{\left\langle \frac{y_s - tp(x)}{\|y_s - tp(x)\|}, N_{\partial \mathcal{E}}(tp(x)) \right\rangle} \tag{122}$$

is sufficiently close to 1 provided that s is sufficiently small. We check this. We have

$$\frac{\left\langle \frac{y_s - tp(x)}{\|y_s - tp(x)\|}, N_{\partial \mathcal{E}}(x) \right\rangle}{\left\langle \frac{y_s - tp(x)}{\|y_s - tp(x)\|}, N_{\partial \mathcal{E}}(tp(x)) \right\rangle} = 1 + \frac{\left\langle \frac{y_s - tp(x)}{\|y_s - tp(x)\|}, N_{\partial \mathcal{E}}(tp(x)) - N_{\partial \mathcal{E}}(x) \right\rangle}{\left\langle \frac{y_s - tp(x)}{\|y_s - tp(x)\|}, N_{\partial \mathcal{E}}(tp(x)) \right\rangle}.$$

We show that (122) is close to 1 first for the case that $\mathcal{E}$ is a Euclidean ball. We have $\|N_{\partial \mathcal{E}}(x) - N_{\partial \mathcal{E}}(tp(x))\| \leq c_0 \|x_0 - z_s\|$ for some constant c_0 because $\|N_{\partial \mathcal{E}}(x) - N_{\partial \mathcal{E}}(tp(x))\| \leq \|y_s - z_s\|$ and $\|y_s - z_s\| \leq c_0 \|x_0 - z_s\|$. The inequality $\|y_s - z_s\| \leq c_0 \|x_0 - z_s\|$ holds because $\{z_s\} = [x_0, z_T] \cap \partial \mathcal{E}_s$ and $\{y_s\} = [x_0, x_T] \cap H(z_s, N_{\partial K}(x_0))$.

On the other hand, there is a constant c_0 such that for all s

$$\left\langle \frac{y_s - tp(x)}{\|y_s - tp(x)\|}, N_{\partial \mathcal{E}}(tp(x)) \right\rangle \geq c_0 \sqrt{\|x_0 - z_s\|}.$$

These two inequalities give that (122) is close to 1 in the case that $\mathcal{E}$ is a Euclidean ball. In order to obtain these inequalities for the case of an ellipsoid we apply the diagonal map A that transforms the Euclidean ball into the ellipsoid. A leaves e_n invariant. Lemma 2.6 gives the first inequality and the second inequality gives

$$\left\langle A\left(\frac{y_s - tp(x)}{\|y_s - tp(x)\|} \right), A^{-1t}(N_{\partial \mathcal{E}}(tp(x))) \right\rangle \geq c_0 \sqrt{\|x_0 - z_s\|}.$$

This gives that (122) is close to 1 for ellipsoids. Therefore, in order to show that the expression (121) converges to 1 for s to 0 it is enough to show that for all x

$$\left(\frac{\|y_s - tp(x)\|}{\|x - z_s\|} \right)^{n-1} \tag{123}$$

is arbitrarily close to 1 provided that s is small. In order to prove this we show for all x

$$1 - c_1 \|z_s - x_0\|^{\frac{1}{6}} \leq \frac{\|y_s - tp(x)\|}{\|y_s - (x + y_s - z_s)\|} \leq 1 + c_2 \|z_s - x_0\|^{\frac{1}{6}} \tag{124}$$

or, equivalently, that there is a constant c_3 such that

$$\frac{\|tp(x) - (x + y_s - z_s)\|}{\|y_s - tp(x)\|} \leq c_3\|z_s - x_0\|^{\frac{1}{6}}. \tag{125}$$

We verify the equivalence. By triangle inequality

$$1 + c_3\|z_s - x_0\|^{\frac{1}{6}} \geq \frac{\|y_s - tp(x)\| + \|tp(x) - (x + y_s - z_s)\|}{\|y_s - tp(x)\|}$$

$$\geq \frac{\|y_s - (x + y_s - z_s)\|}{\|y_s - tp(x)\|}$$

which gives the left hand inequality of (124). Again, by triangle inequality

$$1 - c_3\|z_s - x_0\|^{\frac{1}{6}} \leq \frac{\|y_s - tp(x)\| - \|tp(x) - (x + y_s - z_s)\|}{\|y_s - tp(x)\|}$$

$$\leq \frac{\|y_s - (x + y_s - z_s)\|}{\|y_s - tp(x)\|}$$

which gives the right hand inequality of (124).

We show (125). We begin by showing that

$$\frac{\|tp(x_0) - (x_0 + y_s - z_s)\|}{\|y_s - tp(x_0)\|} \leq c_3\|z_s - x_0\|^{\frac{1}{2}}. \tag{126}$$

See Figure 4.16.5.

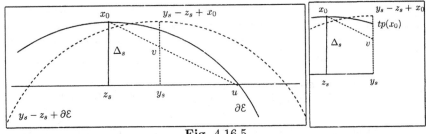

Fig. 4.16.5

Clearly, by Figure 4.16.5

$$\|tp(x_0) - (x_0 + y_s - z_s)\| \leq \|v - (x_0 + y_s - z_s)\|.$$

There is ρ such that for all s with $0 < s \leq s_0$ we have $\|z_s - u\| \geq \rho\sqrt{\|x_0 - z_s\|}$. Let θ be the angle between $x_0 - x_T$ and $N_{\partial K}(x_0)$. By this and $\|z_s - y_s\| = (\tan\theta)\|x_0 - z_s\|$

$$\|tp(x_0) - (x_0 + y_s - z_s)\| \leq \|v - (x_0 + y_s - z_s)\|$$

$$= \|z_s - y_s\|\frac{\|x_0 - z_s\|}{\|z_s - u\|} \leq \frac{\tan\theta}{\rho}\|x_0 - z_s\|^{\frac{3}{2}}.$$

It follows

$$\|y_s - tp(x_0)\| = \|z_s - x_0\| - \|tp(x_0) - (x_0 + y_s - z_s)\| \geq \|z_s - x_0\| - \frac{\tan\theta}{\rho}\|x_0 - z_s\|^{\frac{3}{2}}.$$

This proves (126) which is the special case $x = x_0$ for (125).

Now we treat the general case of (125). We consider three cases: One case being $x \in H^-(z_s, N_{\partial K}(x_0))$ and $\|y_s - w_1\| \leq \|x_0 - z_s\|^{\frac{2}{3}}$, another $x \in H^-(z_s, N_{\partial K}(x_0))$ and $\|y_s - w_1\| \geq \|x_0 - z_s\|^{\frac{2}{3}}$ and the last $x \in H^+(z_s, N_{\partial K}(x_0))$. First we consider the case that $x \in H^-(z_s, N_{\partial K}(x_0))$ and

$$\|y_s - w_1\| \leq \|x_0 - z_s\|^{\frac{2}{3}}.$$

We observe that (see Figure 4.16.5 and 4.16.7)

$$\|y_s - tp(x)\| \geq \|y_s - w_2\|$$
$$\|tp(x) - (x + y_s - z_s)\| \leq \|w_2 - (x + y_s - z_s)\| \|w_2 - w_5\|.$$

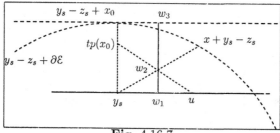

Fig. 4.16.6

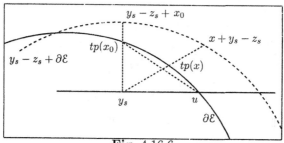

Fig. 4.16.7

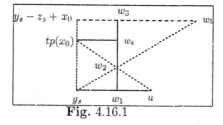

Fig. 4.16.1

Thus we get

$$\frac{\|tp(x) - (x + y_s - z_s)\|}{\|y_s - tp(x)\|} \leq \frac{\|w_2 - w_5\|}{\|y_s - w_2\|} = \frac{\|w_3 - w_2\|}{\|w_1 - w_2\|}. \tag{127}$$

Comparing the triangles $(tp(x_0), w_4, w_2)$ and $(tp(x_0), u, y_s)$ we get

$$\frac{\|w_2 - w_4\|}{\|tp(x_0) - w_4\|} = \frac{\|tp(x_0) - y_s\|}{\|y_s - u\|}.$$

Since $\|tp(x_0) - w_4\| = \|y_s - w_1\|$

$$\|w_2 - w_4\| = \|y_s - w_1\| \frac{\|tp(x_0) - y_s\|}{\|y_s - u\|}.$$

By the assumption $\|y_s - w_1\| \leq \|x_0 - z_s\|^{\frac{2}{3}}$, by $\|tp(x_0) - y_s\| \leq \|x_0 - z_s\|$ and by $\|y_s - u\| \geq c_0 \sqrt{\|x_0 - z_s\|}$ we get with a new constant c_0

$$\|w_2 - w_4\| \leq c_0 \|x_0 - z_s\|^{\frac{7}{6}}$$

and with a new c_0

$$\begin{aligned}
\|w_2 - w_3\| &= \|w_2 - w_4\| + \|w_3 - w_4\| \\
&= \|w_2 - w_4\| + \|tp(x_0) - (y_s - z_s + x_0)\| \\
&\leq c_0(\|x_0 - z_s\|^{\frac{7}{6}} + \|z_s - x_0\|^{\frac{3}{2}}).
\end{aligned}$$

From this and $\|w_1 - w_3\| = \|z_s - x_0\|$ we conclude

$$\|w_1 - w_2\| \geq \|z_s - x_0\| - c_0 \|x_0 - z_s\|^{\frac{7}{6}}.$$

The inequality (127) gives now

$$\frac{\|tp(x) - (x + y_s - z_s)\|}{\|y_s - tp(x)\|} \leq \frac{\|w_3 - w_2\|}{\|w_1 - w_2\|} \leq \frac{c\|x_0 - z_s\|^{\frac{7}{6}}}{\|z_s - x_0\| - c\|x_0 - z_s\|^{\frac{7}{6}}}.$$

The second case is that $tp(x) \in H^-(z_s, N_{\partial K}(x_0))$ and

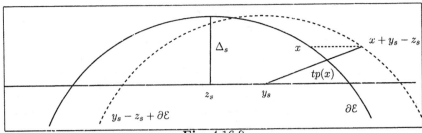

Fig. 4.16.9

$$\|y_s - w_1\| \geq \|x_0 - z_s\|^{\frac{2}{3}}.$$

Compare Figure 4.16.9. Since $\|y_s - w_1\| \geq \|x_0 - z_s\|^{\frac{2}{3}}$ we get

$$\|y_s - tp(x)\| \geq \|x_0 - z_s\|^{\frac{2}{3}}.$$

We have $\|tp(x) - (x + y_s - z_s)\| \leq \|y_s - z_s\|$ because $x \in H^-(z_s, N_{\partial K}(x_0))$ (see Figure 4.16.9). Since $\|z_s - y_s\| \leq c_0\|x_0 - z_s\|$ we deduce $\|tp(x) - (x + y_s - z_s)\| \leq c_0\|x_0 - z_s\|$. Thus we get

$$\frac{\|tp(x) - (x + y_s - z_s)\|}{\|y_s - tp(x)\|} \leq \frac{c_0\|x_0 - z_s\|}{\|x_0 - z_s\|^{\frac{2}{3}}} = c_0\|x_0 - z_s\|^{\frac{1}{3}}.$$

The last case is $tp(x) \in H^+(z_s, N_{\partial K}(x_0))$ (See Figure 4.16.10). We have

$$\|y_s - tp(x)\| \geq \|y_s - u\| \geq \|z_s - u\| - \|y_s - z_s\|.$$

There are constants c_0 and ρ such that

$$\|y_s - tp(x)\| \geq \rho\sqrt{\|x_0 - z_s\|} - c_0\|x_0 - z_s\|$$
$$\|tp(x) - (x + y_s - z_s)\| \leq c_0\|x_0 - z_s\|. \tag{128}$$

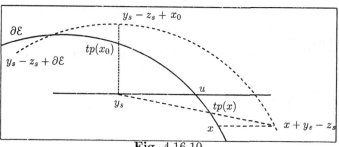

Fig. 4.16.10

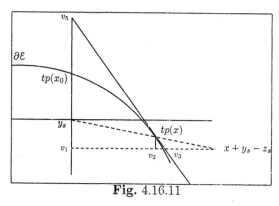

Fig. 4.16.11   **Fig.** 4.16.12

The first inequality is apparent, the second is not. We show the second inequality. We know that the distance between v_3 and $x + y_s - z_s$ is less than $\|y_s - z_s\|$ which is less than $c_0\|x_0 - z_s\|$ (See Figure 4.16.11). The angles α and β are given in Figure 4.16.12. We show that there is a constant c_0 such that $\beta \geq c_0\alpha$. We have

$$\tan\alpha = \frac{\|y_s - v_4\|}{\|v_1 - v_2\|} \qquad \tan(\alpha + \beta) = \frac{\|v_4 - v_5\|}{\|v_1 - v_2\|}.$$

We have

$$\frac{1}{1 - \tan\alpha\tan\beta} + \frac{\frac{\tan\beta}{\tan\alpha}}{1 - \tan\alpha\tan\beta} = \frac{\tan(\alpha+\beta)}{\tan\alpha} = \frac{\|v_4 - v_5\|}{\|y_s - v_4\|} = 1 + \frac{\|y_s - v_5\|}{\|y_s - v_4\|}$$

which gives

$$\frac{\tan\beta}{\tan\alpha} = -\tan\alpha\tan\beta + (1 - \tan\alpha\tan\beta)\frac{\|y_s - v_5\|}{\|y_s - v_4\|}.$$

It is not difficult to show that there is a constant c such that for all s with $0 < s \leq s_0$

$$\|y_s - v_5\| \geq c\|y_s - v_4\|.$$

This gives

$$\frac{\tan\beta}{\tan\alpha} \geq -\tan\alpha\tan\beta + c(1 - \tan\alpha\tan\beta).$$

For s_0 sufficiently small α and β will be as small as we require. Therefore, the right hand side is positive. Since the angles are small we have $\tan\alpha \sim \alpha$ and $\tan\beta \sim \beta$. From $\beta \geq c_0\alpha$ we deduce now that

$$\|tp(x) - (x + y_s - z_s)\| \leq c_0\|v_3 - (x + y_s - z_s)\| \leq c\|y_s - z_s\|.$$

We obtain by (128)

$$\frac{\|tp(x) - (x + y_s - z_s)\|}{\|y_s - tp(x)\|} \leq \frac{c\|y_s - z_s\|}{\rho\sqrt{\|x_0 - z_s\|} - c_0\|x_0 - z_s\|}.$$

There is a constant c such that $\|y_s - z_s\| \leq c_0\|x_0 - z_s\|$.

(iv) First we show

$$\left| \mathbb{P}^k_{\partial\mathcal{E}\cap H_s^-}\{(z_1,\ldots,z_k)|\ z_s \in [z_1,\ldots,z_k]\} - \right.$$

$$\left. \mathbb{P}^k_{\partial\mathcal{E}\cap H_s^-}\{(z_1,\ldots,z_k)|\ z_{s'} \in [z_1,\ldots,z_k]\}\right| < \epsilon.$$

Here the role of the maps rp and tp used in (ii) and (iii) is played by the map that maps $x \in \partial\mathcal{E}$ onto the element $[z_s, x + z_s - z_{s'}] \cap \partial\mathcal{E}$. See Figure 4.16.13.

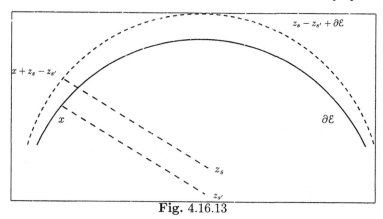

Fig. 4.16.13

Then we show

$$\left| \mathbb{P}^k_{\partial\mathcal{E}\cap H_s^-}\{(z_1,\ldots,z_k)| \ z_{s'} \in [z_1,\ldots,z_k]\} - \right.$$

$$\left. \mathbb{P}^k_{\partial\mathcal{E}\cap H_{s'}^-}\{(z_1,\ldots,z_k)| \ z_{s'} \in [z_1,\ldots,z_k]\} \right| < \epsilon.$$

This is easy to do. It is enough to choose δ small enough so that the probability that a random point z_i is chosen from $\partial\mathcal{E} \cap H_s^- \cap H_{s'}^+$ is very small, e.g. $\delta = \ell^{-2}$ suffices.

(v) We assume that $x_0 = 0$, $N_{\partial K}(x_0) = e_n$, and $\gamma \geq 1$. We consider the transform dil $: \partial\mathcal{E} \rightarrow \partial(\frac{1}{\gamma}\mathcal{E})$ defined by $dil(x) = \frac{1}{\gamma}x$. Then

$$\mathrm{dil}(\partial\mathcal{E}\cap H_{c\gamma\Delta}^-) = \partial(\tfrac{1}{\gamma}\mathcal{E})\cap H_{c\Delta}^- \qquad \mathrm{dil}(x_0 - \gamma\Delta N_{\partial K}(x_0)) = x_0 - \Delta N_{\partial K}(x_0)$$

where $H_\Delta = H(x_0 - \Delta N_{\partial K}(x_0), N_{\partial K}(x_0))$. A surface element on $\partial\mathcal{E}$ is mapped onto one of $\partial(\frac{1}{\gamma}\mathcal{E})$ whose volume is smaller by the factor γ^{-n+1}. Therefore we get

$$\left| \mathbb{P}^k_{\partial(\frac{1}{\gamma}\mathcal{E})\cap H_{c\Delta}^-}\{(x_1,\ldots,x_k)| \ x_0 - \Delta N_{\partial K}(x_0) \in [x_1,\ldots,x_k]\} - \right. \tag{129}$$

$$\left. \mathbb{P}^k_{\partial\mathcal{E}\cap H_{c\gamma\Delta}^-}\{(x_1,\ldots,x_k)| \ x_0 - \gamma\Delta N_{\partial K}(x_0) \in [x_1,\ldots,x_k]\} \right| < \epsilon.$$

Now we apply the map $pd : \mathbb{R}^n \rightarrow \mathbb{R}^n$ with

$$pd(x) = (tx(1),\ldots,tx(n-1),x(n)).$$

We choose t such that the lengths of the principal radii of curvature of $pd(\partial(\frac{1}{\gamma}\mathcal{E}))$ at x_0 coincide with those of $\partial\mathcal{E}$ at x_0. Thus $pd(\partial(\frac{1}{\gamma}\mathcal{E}))$ approximates $\partial\mathcal{E}$ well at x_0 and we can apply Lemma 1.2. See Figure 4.16.14. The relation

$$x_0 - \Delta N_{\partial K}(x_0) \in [x_1,\ldots,x_k]$$

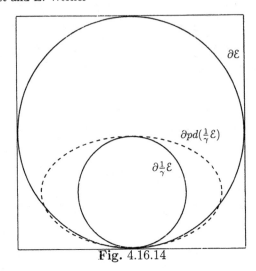

$\partial\mathcal{E}$

$\partial pd(\frac{1}{\gamma}\mathcal{E})$

$\partial\frac{1}{\gamma}\mathcal{E}$

Fig. 4.16.14

holds if and only if

$$x_0 - \Delta N_{\partial K}(x_0) \in [pd(x_1), \dots, pd(x_k)].$$

Indeed, this follows from

$$x_0 - \Delta N_{\partial K}(x_0) = pd(x_0 - \Delta N_{\partial K}(x_0))$$

and

$$pd([x_1, \dots, x_k]) = [pd(x_1), \dots, pd(x_k)].$$

Let $x \in \partial(\frac{1}{\gamma}\mathcal{E})$ and let $N_{\partial(\frac{1}{\gamma}\mathcal{E})\cap H}(x)$ with $H = H(x, N_{\partial K}(x_0)) = H(x_0 - \Delta N_{\partial K}(x_0), N_{\partial K}(x_0))$ be the normal in H to $\partial(\frac{1}{\gamma}\mathcal{E}) \cap H$. Let α be the angle between $N_{\partial(\frac{1}{\gamma}\mathcal{E})}(x)$ and $N_{\partial(\frac{1}{\gamma}\mathcal{E})\cap H}(x)$.

Then a $n - 2$-dimensional surface element in $\partial(\frac{1}{\gamma}\mathcal{E}) \cap H$ at x is mapped onto one in $\partial pd(\frac{1}{\gamma}\mathcal{E}) \cap H$ and the volume changes by a factor t^{n-2}. A $n-1$-dimensional surface element of $\partial(\frac{1}{\gamma}\mathcal{E})$ at x has the volume of a surface element of $\partial(\frac{1}{\gamma}\mathcal{E}) \cap H$ times $(\cos\alpha)^{-1}d\Delta$. When applying the map pd the tangent $\tan\alpha$ changes by the factor t (see Figure 4.16.15). Thus a $n - 1$-dimensional surface element of $\partial(\frac{1}{\gamma}\mathcal{E})$ at x is mapped by pd onto one in $\partial pd(\frac{1}{\gamma}\mathcal{E})$ and its $n - 1$-dimensional volume changes by the factor

$$t^{n-2}\cos\alpha\sqrt{1 + t^2\tan^2\alpha} = t^{n-2}\sqrt{\cos^2\alpha + t^2\sin^2\alpha}.$$

See Figure 4.16.15. If we choose Δ_0 sufficiently small then for all Δ with $0 < \Delta \leq \Delta_0$ the angle α will be very close to $\frac{\pi}{2}$. Thus, for every δ there is Δ_0 such that for all $x \in \partial(\frac{1}{\gamma}\mathcal{E}) \cap H^-(x_0 - \Delta_0 N_{\partial K}(x_0), N_{\partial K}(x_0))$

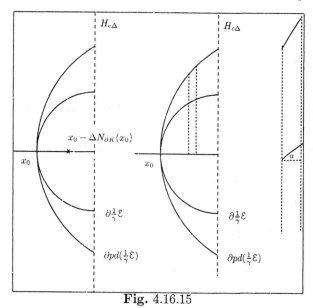

Fig. 4.16.15

$$(1 - \delta)t^{n-1} \le t^{n-2}\sqrt{\cos^2\alpha + t^2\sin^2\alpha} \le (1+\delta)t^{n-1}.$$

Therefore, the image measure of the surface measure on $\partial(\frac{1}{\gamma}\mathcal{E})$ under the map pd has a density that deviates only by a small number from a constant function. More precisely, for every δ there is Δ_0 so that the density function differs only by δ from a constant function. By (i) of this lemma

$$\left| \mathbb{P}^k_{\partial(\frac{1}{\gamma}\mathcal{E})\cap H^-_{c\Delta}} \{(x_1,\ldots,x_k)|\ x_0 - \Delta N_{\partial K}(x_0) \in [x_1,\ldots,x_k]\} - \right.$$

$$\left. \mathbb{P}^k_{\partial pd(\frac{1}{\gamma}\mathcal{E})\cap H^-_{c\Delta}} \{(x_1,\ldots,x_k)|\ x_0 - \Delta N_{\partial K}(x_0) \in [x_1,\ldots,x_k]\} \right| < \epsilon.$$

(In fact, we need only the continuity of this density function at x_0.) $\partial pd(\frac{1}{\gamma}\mathcal{E})$ and $\partial\mathcal{E}$ have the same principal curvature radii at x_0. Therefore, we can apply (ii) of this lemma and get

$$\left| \mathbb{P}^k_{\partial\mathcal{E}\cap H^-_{c\Delta}} \{(x_1,\ldots,x_k)|\ x_0 - \Delta N_{\partial K}(x_0) \in [x_1,\ldots,x_k]\} - \right.$$

$$\left. \mathbb{P}^k_{\partial(\frac{1}{\gamma}\mathcal{E})\cap H^-_{c\Delta}} \{(x_1,\ldots,x_k)|\ x_0 - \Delta N_{\partial K}(x_0) \in [x_1,\ldots,x_k]\} \right| < \epsilon.$$

By (129)

$$\left| \mathbb{P}^k_{\partial\mathcal{E}\cap H^-_{c\Delta}} \{(x_1,\ldots,x_k)|\ x_0 - \Delta N_{\partial K}(x_0) \in [x_1,\ldots,x_k]\} - \right.$$

$$\left. \mathbb{P}^k_{\partial\mathcal{E}\cap H^-_{c\gamma\Delta}} \{(x_1,\ldots,x_k)|\ x_0 - \gamma\Delta N_{\partial K}(x_0) \in [x_1,\ldots,x_k]\} \right| < \epsilon.$$

(vi) By (i) and (ii) of this lemma

$$\left| \mathbb{P}^k_{f,\partial K \cap H^-}\{(x_1,\ldots,x_k)| \quad x_s \in [x_1,\ldots,x_k]\} - \right.$$
$$\left. \mathbb{P}^k_{\partial \mathcal{E} \cap H^-}\{(z_1,\ldots,z_k)| \ x_s \in [z_1,\ldots,z_k]\}\right| < \epsilon$$

where H satisfies $\mathrm{vol}_{n-1}(\partial K \cap H^-) = cs$ and H is orthogonal to $N_{\partial K}(x_0)$. We choose $\tilde{s}$ so that

$$\{z_{\tilde{s}}\} = [x_0, z_T] \cap H(x_s, N_{\partial K}(x_0)) \qquad \{z_{\tilde{s}}\} = [x_0, z_T] \cap \partial \mathcal{E}_{\tilde{s}}.$$

We have $(1-\epsilon)\tilde{s} \leq \dfrac{s}{f(x_0)\mathrm{vol}_{n-1}(\partial \mathcal{E})} \leq (1+\epsilon)\tilde{s}$. We verify this. For sufficiently small s_0 we have for all s with $0 < s \leq s_0$ and $H_s = H(x_s, N_{\partial K}(x_0))$

$$(1-\epsilon)s \leq \int_{\partial K \cap H_s} f(x)\mathrm{d}\mu_{\partial K} \leq (1+\epsilon)s.$$

(H and H_s are generally different.) By the continuity of f at x_0 we get for a new s_0 and all s with $0 < s \leq s_0$

$$(1-\epsilon)s \leq f(x_0)\mathrm{vol}_{n-1}(\partial \mathcal{E} \cap H_s^-) \leq (1+\epsilon)s.$$

Since

$$\tilde{s} = \frac{\mathrm{vol}_{n-1}(\partial \mathcal{E} \cap H_s^-)}{\mathrm{vol}_{n-1}(\partial \mathcal{E})}$$

we get the estimates on $\tilde{s}$.

By (iii) of this lemma

$$\left| \mathbb{P}^k_{f,\partial K \cap H^-}\{(x_1,\ldots,x_k)| \quad x_s \in [x_1,\ldots,x_k]\} - \right.$$
$$\left. \mathbb{P}^k_{\partial \mathcal{E} \cap H^-}\{(z_1,\ldots,z_k)| \ z_{\tilde{s}} \in [z_1,\ldots,z_k]\}\right| < \epsilon.$$

A perturbation argument allows us to assume that $\tilde{s} = \dfrac{s}{f(x_0)\mathrm{vol}_{n-1}(\partial \mathcal{E})}$. By (iv) we get for H with $\mathrm{vol}_{n-1}(\partial K \cap H^-) = cs$

$$\left| \mathbb{P}^k_{f,\partial K \cap H^-}\{(x_1,\ldots,x_k)| \quad x_s \in [x_1,\ldots,x_k]\} - \right.$$
$$\left. \mathbb{P}^k_{\partial \mathcal{E} \cap H^-}\{(z_1,\ldots,z_k)| \ z_{\tilde{s}} \in [z_1,\ldots,z_k]\}\right| < \epsilon.$$

Let L and $\tilde{L}$ be hyperplanes orthogonal to $N_{\partial K}(x_0)$ with $\mathrm{vol}_{n-1}(\partial \mathcal{E} \cap L^-) = cs$ and $\mathrm{vol}_{n-1}(\partial \mathcal{E} \cap \tilde{L}^-) = cs f(x_0)\mathrm{vol}_{n-1}(\partial \mathcal{E})$. By (v) of this lemma

$$\left| \mathbb{P}^k_{\partial \mathcal{E} \cap \tilde{L}^-}\{(z_1,\ldots,z_k)| \ z_s \in [z_1,\ldots,z_k]\} \right.$$
$$\left. -\mathbb{P}^k_{\partial \mathcal{E} \cap L^-}\{(z_1,\ldots,z_k)| \ z_{\tilde{s}} \in [z_1,\ldots,z_k]\}\right| < \epsilon.$$

In order to verify this it is enough to check that the quotient of the height of the cap $\partial \mathcal{E} \cap L^-$ and the distance of $z_{\tilde{s}}$ to x_0 equals up to a small error

$(cf(x_0)\mathrm{vol}_{n-1}(\partial\mathcal{E}))^{\frac{2}{n-1}}$. Indeed, by Lemma 1.3 the height of the cap $\partial\mathcal{E}\cap\tilde{L}^-$ resp. the distance of z_s to x_0 equal up to a small error

$$\frac{1}{2}\left(\frac{csf(x_0)\mathrm{vol}_{n-1}(\partial\mathcal{E})\sqrt{\kappa}}{\mathrm{vol}_{n-1}(B_2^{n-1})}\right)^{\frac{2}{n-1}} \qquad \text{resp.} \qquad \frac{1}{2}\left(\frac{s\sqrt{\kappa}}{\mathrm{vol}_{n-1}(B_2^{n-1})}\right)^{\frac{2}{n-1}}.$$

For the height of the cap $\partial\mathcal{E}\cap L^-$ and the distance of $z_{\tilde{s}}$ to x_0

$$\frac{1}{2}\left(\frac{cs\sqrt{\kappa}}{\mathrm{vol}_{n-1}(B_2^{n-1})}\right)^{\frac{2}{n-1}} \qquad \text{resp.} \qquad \frac{1}{2}\left(\frac{s\sqrt{\kappa}}{f(x_0)\mathrm{vol}_{n-1}(\partial\mathcal{E})\mathrm{vol}_{n-1}(B_2^{n-1})}\right)^{\frac{2}{n-1}}.$$

Therefore the quotients are the same.

Since $\mathrm{vol}_{n-1}(\partial K\cap H^-) = cs$ and $\mathrm{vol}_{n-1}(\partial\mathcal{E}\cap L^-) = cs$ and $\mathcal{E}$ is the standard approximating ellipsoid of K at x_0 we have

$$(1-\epsilon)cs \le \mathrm{vol}_{n-1}(\partial\mathcal{E}\cap H^-) \le (1+\epsilon)cs$$

and

$$\left|\mathbb{P}^k_{\partial\mathcal{E}\cap H^-}\{(z_1,\ldots,z_k)|\ z_{\tilde{s}}\in[z_1,\ldots,z_k]\}\right.$$
$$\left.-\mathbb{P}^k_{\partial\mathcal{E}\cap L^-}\{(z_1,\ldots,z_k)|\ z_{\tilde{s}}\in[z_1,\ldots,z_k]\}\right| < \epsilon.$$

Therefore

$$\left|\mathbb{P}^k_{f,\partial K\cap H^-}\{(x_1,\ldots,x_k)|\ x_s\in[x_1,\ldots,x_k]\} - \right.$$
$$\left.\mathbb{P}^k_{\partial\mathcal{E}\cap\tilde{L}^-}\{(z_1,\ldots,z_k)|\ z_s\in[z_1,\ldots,z_k]\}\right| < \epsilon$$

with $\mathrm{vol}_{n-1}(\partial K\cap H^-) = cs$ and $\mathrm{vol}_{n-1}(\partial\mathcal{E}\cap\tilde{L}^-) = csf(x_0)\mathrm{vol}_{n-1}(\partial\mathcal{E})$. Introducing the constant $c' = cf(x_0)$

$$\mathrm{vol}_{n-1}(\partial K\cap H^-) = \frac{c's}{f(x_0)} \qquad\qquad \frac{\mathrm{vol}_{n-1}(\partial\mathcal{E}\cap\tilde{L}^-)}{\mathrm{vol}_{n-1}(\partial\mathcal{E})} = c's.$$

Since

$$(1-\epsilon)\mathbb{P}_f(\partial K\cap H^-) \le f(x_0)\mathrm{vol}_{n-1}(\partial K\cap H^-) \le (1+\epsilon)\mathbb{P}_f(\partial K\cap H^-)$$

we get the result. $\square$

Lemma 4.17. *Let K be a convex body in $\mathbb{R}^n$ and $x_0\in\partial K$. Suppose that the indicatrix of Dupin exists at x_0 and is an ellipsoid (and not a cylinder with a base that is an ellipsoid). Let $\mathcal{E}$ be the standard approximating ellipsoid at x_0. Let $f:\partial K\to\mathbb{R}$ be a continuous, positive function with $\int_{\partial K}f d\mu = 1$. Let K_s be the surface body with respect to the density f and $\mathcal{E}_s$ the surface body* *with respect to the measure with the constant density* $(\mathrm{vol}_{n-1}(\partial\mathcal{E}))^{-1}$ *on $\partial\mathcal{E}$. Let x_s and z_s be defined by*

$$\{x_s\} = [x_T, x_0] \cap \partial K_s \qquad and \qquad \{z_s\} = [z_T, z_0] \cap \partial\mathcal{E}_s.$$

Then for all $\epsilon > 0$ there is s_ϵ such for all $s \in [0, s_\epsilon]$ and for all $N \in \mathbb{N}$

$$\left| \mathbb{P}_f^N \{(x_1, \ldots, x_N) \mid x_s \notin [x_1, \ldots, x_N]\} - \right.$$
$$\left. \mathbb{P}_{\partial\mathcal{E}}^N \{(z_1, \ldots, z_N) \mid z_s \notin [z_1, \ldots, z_N]\} \right| < \epsilon.$$

Moreover, for all $\epsilon > 0$ there is a $\delta > 0$ such that we have for all s and s' with $0 < s, s' \le s_\epsilon$ and $(1 - \delta)s \le s' \le (1 + \delta)s$

$$\left| \mathbb{P}_f^N \{(x_1, \ldots, x_N) \mid x_s \notin [x_1, \ldots, x_N]\} - \right.$$
$$\left. \mathbb{P}_{\partial\mathcal{E}}^N \{(z_1, \ldots, z_N) \mid z_{s'} \notin [z_1, \ldots, z_N]\} \right| < \epsilon.$$

Proof. For all $\alpha \ge 1$, for all s with $0 < s \le T$ and all $N \in \mathbb{N}$ with

$$N \le \frac{1}{\alpha s}$$

we have

$$1 \ge \mathbb{P}_f^N \{(x_1, \ldots, x_N) \mid x_s \notin [x_1, \ldots, x_N]\}$$
$$\ge \mathbb{P}_f^N \{(x_1, \ldots, x_N) \mid x_1, \ldots, x_N \in (H^-(x_s, N_{\partial K_s}(x_s)) \cap \partial K)^\circ\}$$
$$\ge (1 - s)^N \ge \left(1 - \frac{1}{\alpha N}\right)^N \ge 1 - \frac{1}{\alpha}$$

and

$$1 \ge \mathbb{P}_{\partial\mathcal{E}}^N \{(z_1, \ldots, z_N) \mid z_s \notin [z_1, \ldots, z_N]\}$$
$$\ge \mathbb{P}_{\partial\mathcal{E}}^N \{(z_1, \ldots, z_N) \mid z_1, \ldots, z_N \in (H^-(z_s, N_{\partial\mathcal{E}_s}(z_s)) \cap \partial\mathcal{E})^\circ\}$$
$$\ge (1 - s)^N \ge 1 - sN \ge 1 - \frac{1}{\alpha}.$$

Therefore, if we choose $\alpha \ge \frac{1}{\epsilon}$ we get for all N with $N < \frac{1}{\alpha s}$

$$\left| \mathbb{P}_f^N \{(x_1, \ldots, x_N) \mid x_s \notin [x_1, \ldots, x_N]\} \right.$$
$$\left. - \mathbb{P}_{\partial\mathcal{E}}^N \{(z_1, \ldots, z_N) \mid z_s \notin [z_1, \ldots, z_N]\} \right| \le \epsilon.$$

By Lemma 4.8 for a given x_0 there are constants a, b with $0 \le a, b < 1$, and s_ϵ such that we have for all s with $0 < s \le s_\epsilon$

$$\mathbb{P}_f^N \{(x_1, \ldots, x_N) \mid x_s \notin [x_1, \ldots, x_N]\}$$
$$\le 2^n (a - as + s)^N + 2^n (1 - s + bs)^N$$
$$\le 2^n \exp(N(\ln a + s(\tfrac{1}{a} - 1))) + 2^n \exp(-Ns(1 - b)).$$

We choose s_ϵ so small that $|\ln a| \geq 2s_\epsilon(\frac{1}{2} - 1)$. Thus

$$\mathbb{P}_f^N\{(x_1, \ldots, x_N)|\ x_s \notin [x_1, \ldots, x_N]\}$$
$$\leq 2^n \exp(-\tfrac{1}{2}sN|\ln a|) + 2^n \exp(-Ns(1 - b)).$$

Now we choose β so big that

$$2^n e^{-\beta(1-b)} < \tfrac{1}{2}\epsilon \qquad \text{and} \qquad 2^n e^{-\frac{1}{2}\beta|\ln a|} < \tfrac{1}{2}\epsilon.$$

Thus, for sufficiently small s_ϵ and all N with $N \geq \frac{\beta}{s}$ we get

$$\mathbb{P}_f^N\{(x_1, \ldots, x_N)|\ x_s \notin [x_1, \ldots, x_N]\} \leq \epsilon$$

and

$$\mathbb{P}_{\partial\mathcal{E}}^N\{(z_1, \ldots, z_N)|\ z_s \notin [z_1, \ldots, z_N]\} \leq \epsilon.$$

Please note that β depends only on a, b, n and ϵ. This leaves us with the case $\frac{1}{\alpha s} \leq N \leq \frac{\beta}{s}$.

We put $\gamma = \alpha \operatorname{vol}_{n-1}(\partial K)$. By Lemma 4.15 for all c with $c \geq c_0$ and γ there is $s_{c,\gamma}$ such that for all s with $0 < s \leq s_{c,\gamma}$ and for all $N \in \mathbb{N}$ with

$$N \geq \frac{1}{\gamma s}\operatorname{vol}_{n-1}(\partial K) = \frac{1}{\alpha s}$$

that

$$\left|\mathbb{P}_f^N\{(x_1, \ldots, x_N)|\ x_s \notin [x_1, \ldots, x_N]\} - \right.$$
$$\left. \mathbb{P}_f^N\{(x_1, \ldots, x_N)|\ x_s \notin [\{x_1, \ldots, x_N\} \cap H^-]\}\right|$$
$$\leq 2^{n-1} \exp(-\tfrac{c_1}{\gamma}\sqrt{c}) = 2^{n-1} \exp\left(-\frac{c_1\sqrt{c}}{\alpha\operatorname{vol}_{n-1}(\partial K)}\right)$$

where $H = H(x_0 - c\Delta N_{\partial K}(x_0), N_{\partial K}(x_0))$ and $\Delta = \Delta(s)$ as in Lemma 4.15. We choose c so big that

$$2^{n-1} \exp(-\tfrac{c_1}{\gamma}\sqrt{c}) < \epsilon.$$

Thus for all ϵ there are c and s_ϵ such that for all s with $0 < s \leq s_\epsilon$

$$\left|\mathbb{P}_f^N\{(x_1, \ldots, x_N)|\quad x_s \notin [x_1, \ldots, x_N]\} - \right.$$
$$\left. \mathbb{P}_f^N\{(x_1, \ldots, x_N)|\ x_s \notin [\{x_1, \ldots, x_N\} \cap H^-]\}\right| \leq \epsilon$$

and in the same way that

$$\left|\mathbb{P}_{\partial\mathcal{E}}^N\{(x_1, \ldots, x_N)|\quad z_s \notin [x_1, \ldots, x_N]\} - \right.$$
$$\left. \mathbb{P}_{\partial\mathcal{E}}^N\{(x_1, \ldots, x_N)|\ z_s \notin [\{x_1, \ldots, x_N\} \cap H^-]\}\right| \leq \epsilon.$$

By Lemma 1.3 there are constants c_1 and c_2 such that

$$c_1 \Delta^{\frac{n-1}{2}} \leq \mathrm{vol}_{n-1}(H^-(x_0 - c\Delta N_{\partial K}(x_0), N_{\partial K}(x_0)) \cap \partial\mathcal{E}) \leq c_2 \Delta^{\frac{n-1}{2}}$$

where Δ is the height of the cap. Now we adjust the cap that will allow us to apply Lemma 4.16. There is $d > 0$ such that for all s with $0 < s \leq s_\epsilon$ there are hyperplanes H_{ds} and $\tilde{H}_{ds}$ that are orthogonal to $N_{\partial K}(x_0)$ and that satisfy

$$\mathbb{P}_f(\partial K \cap H_{ds}^-) = ds \qquad \frac{\mathrm{vol}_{n-1}(\partial\mathcal{E} \cap \tilde{H}_{ds}^-)}{\mathrm{vol}_{n-1}(\partial\mathcal{E})} = ds$$

and

$$\partial K \cap H^-(x_0 - c\Delta N_{\partial K}(x_0), N_{\partial K}(x_0)) \subseteq \partial K \cap H_{ds}^-$$
$$\partial\mathcal{E} \cap H^-(x_0 - c\Delta N_{\partial K}(x_0), N_{\partial K}(x_0)) \subseteq \partial\mathcal{E} \cap \tilde{H}_{ds}^-.$$

Thus we have for all s with $0 < s \leq s_\epsilon$

$$|\mathbb{P}_f^N\{(x_1, \ldots, x_N)| \quad x_s \notin [x_1, \ldots, x_N]\} - \tag{130}$$
$$\mathbb{P}_f^N\{(x_1, \ldots, x_N)| \ x_s \notin [\{x_1, \ldots, x_N\} \cap H_{ds}^-]\}| \leq \epsilon$$

and

$$|\mathbb{P}_{\partial\mathcal{E}}^N\{(x_1, \ldots, x_N)| \quad z_s \notin [x_1, \ldots, x_N]\} - \tag{131}$$
$$\mathbb{P}_{\partial\mathcal{E}}^N\{(x_1, \ldots, x_N)| \ z_s \notin [\{x_1, \ldots, x_N\} \cap \tilde{H}_{ds}^-]\}| \leq \epsilon.$$

We choose ℓ so big that

$$\sum_{k=\ell}^{\infty} \frac{(d\beta)^k}{k!} < \epsilon.$$

By Lemma 4.16.(vi) we can choose s_ϵ so small that we have for all k with $1 \leq k \leq \ell$

$$\left| \mathbb{P}_{f, \partial K \cap H_{ds}^-}^k\{(x_1, \ldots, x_k)| \quad x_s \in [x_1, \ldots, x_k]\} - \right. \tag{132}$$
$$\left. \mathbb{P}_{\partial\mathcal{E} \cap \tilde{H}_{ds}^-}^k\{(z_1, \ldots, z_k)| \ z_s \in [z_1, \ldots, z_k]\} \right| < \epsilon.$$

We have

$$|\mathbb{P}_f^N\{(x_1, \ldots, x_N)| \ x_s \notin [x_1, \ldots, x_N]\} \tag{133}$$
$$-\mathbb{P}_{\partial\mathcal{E}}^N\{(z_1, \ldots, z_N)| \ z_s \notin [z_1, \ldots, z_N]\}|$$
$$\leq |\mathbb{P}_f^N\{(x_1, \ldots, x_N)| \ x_s \notin [x_1, \ldots, x_N]\}$$
$$-\mathbb{P}_f^N\{(x_1, \ldots, x_N)| \ x_s \notin [\{x_1, \ldots, x_N\} \cap H_{ds}^-]\}|$$
$$+|\mathbb{P}_f^N\{(x_1, \ldots, x_N)| \ x_s \notin [\{x_1, \ldots, x_N\} \cap H_{ds}^-]\}$$
$$-\mathbb{P}_{\partial\mathcal{E}}^N\{(z_1, \ldots, z_N)| \ z_s \notin [\{z_1, \ldots, z_N\} \cap \tilde{H}_{ds}^-]\}|$$
$$+|\mathbb{P}_{\partial\mathcal{E}}^N\{(z_1, \ldots, z_N)| \ z_s \notin [z_1, \ldots, z_N]\}$$
$$-\mathbb{P}_{\partial\mathcal{E}}^N\{(z_1, \ldots, z_N)| \ z_s \notin [\{z_1, \ldots, z_N\} \cap \tilde{H}_{ds}^-]\}|.$$

By (130) and (131) the first and third summand are smaller than ϵ. It remains to estimate the second summand. We do this now. We have

$$\mathbb{P}_f^N \{(x_1,\ldots,x_N)|\ x_s \notin [\{x_1,\ldots,x_N\} \cap H_{ds}^-]\}$$

$$= \sum_{k=0}^N \binom{N}{k} \mathbb{P}_f^N \{(x_1,\ldots,x_N)|\ x_s \notin [x_1,\ldots,x_k],\ x_1,\ldots,x_k \in H_{ds}^-,$$

$$x_{k+1},\ldots,x_N \in H_{ds}^+\}$$

$$= \sum_{k=0}^N \binom{N}{k} (1-ds)^{N-k} (ds)^k \mathbb{P}_{f,\partial K \cap H_{ds}^-}^k \{(x_1,\ldots,x_k)|\ x_s \notin [x_1,\ldots,x_k]\}.$$

Moreover, since $N \le \frac{\beta}{s}$ we have

$$\sum_{k=\ell}^N \binom{N}{k} (1-ds)^{N-k} (ds)^k \mathbb{P}_{f,\partial K \cap H_{ds}^-}^k \{(x_1,\ldots,x_k)|\ x_s \notin [x_1,\ldots,x_k]\}$$

$$\le \sum_{k=\ell}^N \binom{N}{k} \left(\frac{d\beta}{N}\right)^k \le \sum_{k=\ell}^N \frac{(d\beta)^k}{k!} < \epsilon.$$

Thus we have

$$\left| \mathbb{P}_f^N \{(x_1,\ldots,x_N)|\ x_s \notin [\{x_1,\ldots,x_N\} \cap H_{ds}^-]\} - \right.$$

$$\left. \sum_{k=0}^{\ell-1} \binom{N}{k} (1-ds)^{N-k} (ds)^k \mathbb{P}_{f,\partial K \cap H_{ds}^-}^k \{(x_1,\ldots,x_k)|\ x_s \notin [x_1,\ldots,x_k]\} \right| < \epsilon.$$

In the same way we get

$$\left| \mathbb{P}_{\partial\mathcal{E}}^N \{(z_1,\ldots,z_N)|\ z_s \notin [\{z_1,\ldots,z_N\} \cap \tilde{H}_{ds}^-]\} \right.$$

$$\left. - \sum_{k=0}^{\ell-1} \binom{N}{k} (1-ds)^{N-k} (ds)^k \mathbb{P}_{\partial\mathcal{E} \cap \tilde{H}_{ds}^-}^k \{(z_1,\ldots,z_k)|\ z_s \notin [z_1,\ldots,z_k]\} \right| < \epsilon.$$

From these two inequalities we get

$$\left| \mathbb{P}_f^N \{(x_1,\ldots,x_N)|\ x_s \notin [\{x_1,\ldots,x_N\} \cap H_{ds}^-]\} \right.$$

$$\left. - \mathbb{P}_{\partial\mathcal{E}}^N \{(z_1,\ldots,z_N)|\ z_s \notin [\{z_1,\ldots,z_N\} \cap \tilde{H}_{ds}^-]\} \right|$$

$$\le 2\epsilon +$$

$$\left| \sum_{k=0}^{\ell-1} \binom{N}{k} (1-ds)^{N-k} (ds)^k \mathbb{P}_{f,\partial K \cap H_{ds}^-}^k \{(x_1,\ldots,x_k)|\ x_s \notin [x_1,\ldots,x_k]\} \right.$$

$$-\sum_{k=0}^{\ell-1}\binom{N}{k}(1-ds)^{N-k}(ds)^k\,\mathbb{P}^k_{\partial\mathcal{E}\cap\tilde{H}_{ds}^-}\{(z_1,\ldots,z_k)|\;z_s\notin[z_1,\ldots,z_k]\}\Bigg|$$

$$=2\epsilon+$$

$$\Bigg|\sum_{k=0}^{\ell-1}\binom{N}{k}(1-ds)^{N-k}(ds)^k\,\big[\mathbb{P}^k_{f,\partial K\cap H_{ds}^-}\{(x_1,\ldots,x_k)|\;x_s\notin[x_1,\ldots,x_k]\}$$

$$-\mathbb{P}^k_{\partial\mathcal{E}\cap\tilde{H}_{ds}^-}\{(z_1,\ldots,z_k)|\;z_s\notin[z_1,\ldots,z_k]\}\big]\Bigg|.$$

By (132) the last expression is less than

$$2\epsilon+\epsilon\sum_{k=0}^{\ell-1}\binom{N}{k}(1-ds)^{N-k}(ds)^k\le 3\epsilon.$$

Together with (133) this gives the first inequality of the lemma.

We show now that for all $\epsilon>0$ there is a $\delta>0$ such that we have for all s and s' with $0<s,s'\le s_\epsilon$ and $(1-\delta)s\le s'\le(1+\delta)s$

$$\Big|\mathbb{P}_f^N\{(x_1,\ldots,x_N)|\;x_s\notin[x_1,\ldots,x_N]\}-$$

$$\mathbb{P}_{\partial\mathcal{E}}^N\{(z_1,\ldots,z_N)|\;z_{s'}\notin[z_1,\ldots,z_N]\}\Big|<\epsilon.$$

Using the first inequality we see that it is enough to show that for all $\epsilon>0$ there is a $\delta>0$ such that we have for all s and s' with $0<s,s'\le s_\epsilon$ and $(1-\delta)s\le s'\le(1+\delta)s$

$$\Big|\mathbb{P}_{\partial\mathcal{E}}^N\{(z_1,\ldots,z_N)|\;z_s\notin[z_1,\ldots,z_N]\}-$$

$$\mathbb{P}_{\partial\mathcal{E}}^N\{(z_1,\ldots,z_N)|\;z_{s'}\notin[z_1,\ldots,z_N]\}\Big|<\epsilon.$$

As in the proof of the first inequality we show that we just have to consider the case $\frac{1}{\alpha\,s}\le N\le\frac{\beta}{s}$. We choose $\delta=\frac{\epsilon}{\ell}$. Thus δ depends on ℓ, but ℓ depends only on β and c. In particular, ℓ does not depend on N. As above, we write

$$\mathbb{P}_{\partial\mathcal{E}}^N\{(z_1,\ldots,z_N)|\;z_s\notin[\{z_1,\ldots,z_N\}\cap\tilde{H}_{ds}^-]\}$$

$$=\sum_{k=0}^N\binom{N}{k}(1-ds)^{N-k}(ds)^k\,\mathbb{P}^k_{\partial\mathcal{E}\cap\tilde{H}_{ds}^-}\{(z_1,\ldots,z_k)|\;z_s\notin[z_1,\ldots,z_k]\}.$$

We get as above

$$\Big|\mathbb{P}_{\partial\mathcal{E}}^N\{(z_1,\ldots,z_N)|\;z_s\notin[\{z_1,\ldots,z_N\}\cap\tilde{H}_{ds}^-]\}$$

$$-\mathbb{P}_{\partial\mathcal{E}}^N\{(z_1,\ldots,z_N)|\;z_{s'}\notin[\{z_1,\ldots,z_N\}\cap\tilde{H}_{ds'}^-]\}\Big|$$

$$\le\Bigg|\sum_{k=0}^{\ell}\binom{N}{k}(1-ds)^{N-k}(ds)^k\,\mathbb{P}^k_{\partial\mathcal{E}\cap\tilde{H}_{ds}^-}\{(z_1,\ldots,z_k)|\;z_s\notin[z_1,\ldots,z_k]\}$$

$$-\sum_{k=0}^{\ell}\binom{N}{k}(1-ds')^{N-k}(ds')^k\,\mathbb{P}^k_{\partial\mathcal{E}\cap\tilde{H}_{ds'}^-}\{(z_1,\ldots,z_k)|\;z_{s'}\notin[z_1,\ldots,z_k]\}\Bigg|.$$

This expression is not greater than

$$\sum_{k=0}^{\ell} \binom{N}{k} \left[(1-ds)^{N-k}(ds)^k - (1-ds')^{N-k}(ds')^k\right]$$

$$\mathbb{P}^k_{\partial\mathcal{E}\cap\tilde{H}_{ds}^-}\{(z_1,\ldots,z_k)|\ z_s \notin [z_1,\ldots,z_k]\}$$

$$+\sum_{k=0}^{\ell} \binom{N}{k}(1-ds')^{N-k}(ds')^k \left|\mathbb{P}^k_{\partial\mathcal{E}\cap\tilde{H}_{ds'}^-}\{(z_1,\ldots,z_k)|\ z_{s'} \notin [z_1,\ldots,z_k]\}\right.$$

$$\left.-\mathbb{P}^k_{\partial\mathcal{E}\cap\tilde{H}_{ds}^-}\{(z_1,\ldots,z_k)|\ z_s \notin [z_1,\ldots,z_k]\}\right|.$$

By Lemma 4.16.(iv) the second summand is smaller than

$$\epsilon\sum_{k=0}^{\ell} \binom{N}{k}(1-ds')^{N-k}(ds')^k \leq \epsilon.$$

The first summand can be estimated by (we may assume that $s > s'$)

$$\sum_{k=0}^{\ell} \binom{N}{k} \left[(1-ds)^{N-k}(ds)^k - (1-ds')^{N-k}(ds')^k\right]$$

$$=\sum_{k=0}^{\ell} \binom{N}{k}(1-ds)^{N-k}(ds)^k \left[1 - \left(\frac{1-ds'}{1-ds}\right)^{N-k}\left(\frac{s'}{s}\right)^k\right].$$

Since $s > s'$ we have $1 - ds' \geq 1 - ds$ and the above expression is smaller than

$$\sum_{k=0}^{\ell} \binom{N}{k}(1-ds)^{N-k}(ds)^k \left[1 - (1-\delta)^k\right]$$

$$\leq \sum_{k=0}^{\ell} \binom{N}{k}(1-ds)^{N-k}(ds)^k k\delta \leq \ell\delta.$$

$\square$

4.2 Probabilistic Estimates for Ellipsoids

Lemma 4.18. *Let $x_0 \in \partial B_2^n$ and let $(B_2^n)_s$ be the surface body with respect to the measure $\mathbb{P}_f$ with constant density $f = (\mathrm{vol}_{n-1}(\partial B_2^n))^{-1}$. We have*

$$\lim_{N\to\infty} N^{\frac{2}{n-1}} \int_0^{\frac{1}{2}} \frac{\mathbb{P}^N_{\partial B_2^n}\{(x_1,\ldots,x_N)|\ x_s \notin [x_1,\ldots,x_N]\}}{\int_{\partial(B_2^n\cap H_s)} \frac{(\mathrm{vol}_{n-1}(\partial B_2^n))^{-1}}{(1-<N_{\partial(B_2^n)_s}(x_s),N_{\partial B_2^n}(y)>^2)^{\frac{1}{2}}} d\mu_{\partial(B_2^n\cap H_s)}(y)} ds$$

$$= (n-1)^{\frac{n+1}{n-1}} \left(\frac{\mathrm{vol}_{n-1}(\partial B_2^n)}{\mathrm{vol}_{n-2}(\partial B_2^{n-1})}\right)^{\frac{2}{n-1}} \frac{\Gamma\left(n+1+\frac{2}{n-1}\right)}{2(n+1)!}$$

where $H_s = H(x_s, N_{\partial(B_2^n)_s}(x_s))$ and $\{x_s\} = [0, x_0] \cap \partial(B_2^n)_s$. (*Let us note that* $N_{\partial(B_2^n)_s}(x_s) = x_0$ *and* $N_{\partial B_2^n}(y) = y$.)

Proof. Clearly, for all s with $0 \leq s < \frac{1}{2}$ the surface body $(B_2^n)_s$ is homothetic to B_2^n. We have

$$\mathrm{vol}_n(B_2^n) - \mathbb{E}(\partial B_2^n, N) = \int_{B_2^n} \mathbb{P}_{\partial B_2^n}^N \{(x_1, \ldots, x_N) | x \notin [x_1, \ldots, x_N]\} \mathrm{d}x.$$

We pass to polar coordinates

$$\mathrm{vol}_n(B_2^n) - \mathbb{E}(\partial B_2^n, N)$$
$$= \int_0^1 \int_{\partial B_2^n} \mathbb{P}_{\partial B_2^n}^N \{(x_1, \ldots, x_N) | r\xi \notin [x_1, \ldots, x_N]\} r^{n-1} \mathrm{d}\xi \mathrm{d}r$$

where $\mathrm{d}\xi$ is the surface measure on ∂B_2^n. Since B_2^n is rotationally invariant

$$\mathbb{P}_{\partial B_2^n}^N \{(x_1, \ldots, x_N) | r\xi \notin [x_1, \ldots, x_N]\}$$

is independent of ξ. We get that the last expression equals

$$\mathrm{vol}_{n-1}(\partial B_2^n) \int_0^1 \mathbb{P}_{\partial B_2^n}^N \{(x_1, \ldots, x_N) | r\xi \notin [x_1, \ldots, x_N]\} r^{n-1} \mathrm{d}r$$

for all $\xi \in \partial B_2^n$. Now we perform a change of variable. We define the function $s : [0, 1] \to [0, \frac{1}{2}]$ by

$$s(r) = \frac{\mathrm{vol}_{n-1}(\partial B_2^n \cap H^-(r\xi, \xi))}{\mathrm{vol}_{n-1}(\partial B_2^n)}.$$

The function is continuous, strictly decreasing, and invertible. We have by Lemma 2.11.(iii)

$$\frac{\mathrm{d}s}{\mathrm{d}r} = -\int_{\partial(B_2^n \cap H_s)} \frac{(\mathrm{vol}_{n-1}(\partial B_2^n))^{-1}}{(1 - \langle N_{\partial(B_2^n)_s}(x_s), N_{\partial B_2^n}(y) \rangle^2)^{\frac{1}{2}}} \mathrm{d}\mu_{\partial(B_2^n \cap H_s)}(y).$$

We have $r(s)\xi = x_s$. Thus we get

$$\frac{\mathrm{vol}_n(B_2^n) - \mathbb{E}(\partial B_2^n, N)}{\mathrm{vol}_{n-1}(\partial B_2^n)}$$
$$= \int_0^{\frac{1}{2}} \frac{\mathbb{P}_{\partial B_2^n}^N \{(x_1, \ldots, x_N) | x_s \notin [x_1, \ldots, x_N]\} (r(s))^{n-1} \mathrm{d}s}{\int_{\partial(B_2^n \cap H_s)} \frac{(\mathrm{vol}_{n-1}(\partial B_2^n))^{-1}}{(1 - \langle N_{\partial(B_2^n)_s}(x_s), N_{\partial B_2^n}(y) \rangle^2)^{\frac{1}{2}}} \mathrm{d}\mu_{\partial(B_2^n \cap H_s)}(y)}.$$

Now we apply Proposition 3.1 and obtain

$$\lim_{N\to\infty} N^{\frac{2}{n-1}} \int_0^{\frac{1}{2}} \frac{\mathbb{P}^N_{\partial B_2^n}\{(x_1,\dots,x_N)|\ x_s \notin [x_1,\dots,x_N]\}(r(s))^{n-1}\mathrm{d}s}{\int_{\partial(B_2^n \cap H_s)} \frac{(\mathrm{vol}_{n-1}(\partial B_2^n))^{-1}}{(1-<N_{\partial(B_2^n)_s}(x_s),N_{\partial B_2^n}(y)>^2)^{\frac{1}{2}}} \mathrm{d}\mu_{\partial(B_2^n \cap H_s)}(y)}$$

$$= (n-1)^{\frac{n+1}{n-1}} \left(\frac{\mathrm{vol}_{n-1}(\partial B_2^n)}{\mathrm{vol}_{n-2}(\partial B_2^{n-1})} \right)^{\frac{2}{n-1}} \frac{\Gamma\left(n+1+\frac{2}{n-1}\right)}{2(n+1)!}.$$

By Lemma 4.13 it follows that we have for all s_0 with $0 < s_0 \leq \frac{1}{2}$

$$\lim_{N\to\infty} N^{\frac{2}{n-1}} \int_0^{s_0} \frac{\mathbb{P}^N_{\partial B_2^n}\{(x_1,\dots,x_N)|\ x_s \notin [x_1,\dots,x_N]\}(r(s))^{n-1}\mathrm{d}s}{\int_{\partial(B_2^n \cap H_s)} \frac{(\mathrm{vol}_{n-1}(\partial B_2^n))^{-1}}{(1-<N_{\partial(B_2^n)_s}(x_s),N_{\partial B_2^n}(y)>^2)^{\frac{1}{2}}} \mathrm{d}\mu_{\partial(B_2^n \cap H_s)}(y)}$$

$$= (n-1)^{\frac{n+1}{n-1}} \left(\frac{\mathrm{vol}_{n-1}(\partial B_2^n)}{\mathrm{vol}_{n-2}(\partial B_2^{n-1})} \right)^{\frac{2}{n-1}} \frac{\Gamma\left(n+1+\frac{2}{n-1}\right)}{2(n+1)!}.$$

By this and since $r(s)$ is a continuous function with $\lim_{s\to 0} r(s) = 1$ we get

$$\lim_{N\to\infty} N^{\frac{2}{n-1}} \int_0^{\frac{1}{2}} \frac{\mathbb{P}^N_{\partial B_2^n}\{(x_1,\dots,x_N)|\ x_s \notin [x_1,\dots,x_N]\}\mathrm{d}s}{\int_{\partial(B_2^n \cap H_s)} \frac{(\mathrm{vol}_{n-1}(\partial B_2^n))^{-1}}{(1-<N_{\partial(B_2^n)_s}(x_s),N_{\partial B_2^n}(y)>^2)^{\frac{1}{2}}} \mathrm{d}\mu_{\partial(B_2^n \cap H_s)}(y)}$$

$$= (n-1)^{\frac{n+1}{n-1}} \left(\frac{\mathrm{vol}_{n-1}(\partial B_2^n)}{\mathrm{vol}_{n-2}(\partial B_2^{n-1})} \right)^{\frac{2}{n-1}} \frac{\Gamma\left(n+1+\frac{2}{n-1}\right)}{2(n+1)!}.$$

$\square$

Lemma 4.19. *Let K be a convex body in $\mathbb{R}^n$ and $x_0 \in \partial K$. Suppose that the indicatrix of Dupin exists at x_0 and is an ellipsoid (and not a cylinder with a base that is an ellipsoid). Let $f, g : \partial K \to \mathbb{R}$ be continuous, strictly positive functions with*

$$\int_{\partial K} f \mathrm{d}\mu = \int_{\partial K} g \mathrm{d}\mu = 1.$$

Let

$$\mathbb{P}_f = f \mathrm{d}\mu_{\partial K} \quad \text{and} \quad \mathbb{P}_g = g \mathrm{d}\mu_{\partial K}.$$

Then for all $\epsilon > 0$ there is s_ϵ such that we have for all $0 < s < s_\epsilon$, all x_s with $\{x_s\} = [0,x_0] \cap \partial K_{f,s}$, all $\{y_s\} = [0,x_0] \cap \partial K_{g,s}$, and all $N \in \mathbb{N}$

$$|\mathbb{P}_f^N\{(x_1,\dots,x_N)|x_s \notin [x_1,\dots,x_N]\} - \mathbb{P}_g^N\{(x_1,\dots,x_N)|y_s \notin [x_1,\dots,x_N]\}| < \epsilon.$$

Proof. By Lemma 4.17

$$|\mathbb{P}_f^N\{(x_1,\dots,x_N)|\ x_s \notin [x_1,\dots,x_N]\} -$$
$$\mathbb{P}_{\partial \mathcal{E}}^N\{(z_1,\dots,z_N)|\ z_s \notin [z_1,\dots,z_N]\}| < \epsilon,$$

and

$$\left| \mathbb{P}_g^N \{(x_1, \ldots, x_N)| \quad y_s \notin [x_1, \ldots, x_N]\} - \right.$$
$$\left. \mathbb{P}_{\partial \mathcal{E}}^N \{(z_1, \ldots, z_N)| \ z_s \notin [z_1, \ldots, z_N]\} \right| < \epsilon.$$

The result follows by triangle-inequality. □

Lemma 4.20. *Let* $a_1, \ldots, a_n > 0$ *and let* $A : \mathbb{R}^n \to \mathbb{R}^n$ *be defined by* $Ax = (a_i x(i))_{i=1}^n$. *Let* $\mathcal{E} = A(B_2^n)$, *i.e.*

$$\mathcal{E} = \left\{ x \left| \sum_{i=1}^n \left| \frac{x(i)}{a_i} \right|^2 \le 1 \right. \right\}.$$

Let $f : \partial \mathcal{E} \to \mathbb{R}$ *be given by*

$$f(x) = \left(\left(\prod_{i=1}^n a_i \right) \sqrt{\sum_{i=1}^n \frac{x(i)^2}{a_i^4}} \operatorname{vol}_{n-1}(\partial B_2^n) \right)^{-1}.$$

Then we have $\int_{\partial \mathcal{E}} f d\mu_{\partial \mathcal{E}} = 1$ *and for all* $x \in B_2^n$

$$\mathbb{P}_{\partial B_2^n}^N \{(x_1, \ldots, x_N)| x \notin [x_1, \ldots, x_N]\} = \mathbb{P}_f^N \{(z_1, \ldots, z_N)| A(x) \notin [z_1, \ldots, z_N]\}.$$

Proof. We have that

$$x \notin [x_1, \ldots, x_N] \qquad \text{if and only if} \qquad Ax \notin [Ax_1, \ldots, Ax_N].$$

For all subsets M of $\partial \mathcal{E}$ such that $A^{-1}(M)$ is measurable we put

$$\nu(M) = \mathbb{P}_{\partial B_2^n}(A^{-1}(M))$$

and get

$$\mathbb{P}_{\partial B_2^n}^N \{(x_1, \ldots, x_N)| x \notin [x_1, \ldots, x_N]\} = \nu^N \{(z_1, \ldots, z_N)| Ax \notin [z_1, \ldots, z_N]\}.$$

We want to apply the Theorem of Radon-Nikodym. ν is absolutely continuous with respect to the surface measure $\mu_{\partial \mathcal{E}}$. We check this.

$$\nu(M) = \mathbb{P}_{\partial B_2^n}(A^{-1}(M)) = \frac{h_{n-1}(A^{-1}(M))}{\operatorname{vol}_{n-1}(\partial B_2^n)}$$

where h_{n-1} is the $n-1$-dimensional Hausdorff-measure. By elementary properties of the Hausdorff-measure ([EvG], p. 75) we get

$$\nu(M) \le (\operatorname{Lip}(A))^{n-1} \frac{h_{n-1}(M)}{\operatorname{vol}_{n-1}(\partial B_2^n)} = (\operatorname{Lip}(A))^{n-1} \frac{1}{\operatorname{vol}_{n-1}(\partial B_2^n)} \mu_{\partial \mathcal{E}}(M)$$

where $\mathrm{Lip}(A)$ is the Lipschitz-constant of A. Thus $\nu(M) = 0$ whenever $\mu_{\partial\mathcal{E}}(M) = 0$.

Therefore, by the Theorem of Radon-Nikodym there is a density f such that $d\nu = f d\mu_{\partial\mathcal{E}}$. The density is given by

$$f(x) = \left(\left(\prod_{i=1}^{n} a_i \right) \sqrt{\sum_{i=1}^{n} \frac{x(i)^2}{a_i^4}} \mathrm{vol}_{n-1}(\partial B_2^n) \right)^{-1}.$$

We show this. We may assume that $x(n) \geq \frac{a_n}{\sqrt{n}}$ (there is at least one coordinate $x(i)$ with $|x(i)| \geq \frac{a_i}{\sqrt{n}}$). Let U be a small neighborhood of x in $\partial\mathcal{E}$. We may assume that for all $y \in U$ we have $y(n) \geq \frac{a_n}{2\sqrt{n}}$. Thus the orthogonal projection p_{e_n} onto the subspace orthogonal to e_n is injective on U. Since $x \in \partial\mathcal{E}$ we have $(\frac{x(i)}{a_i})_{i=1}^{n} \in \partial B_2^n$ and $N_{\partial B_2^n}(A^{-1}(x)) = (\frac{x(i)}{a_i})_{i=1}^{n}$. Then we have up to a small error

$$\nu(U) = \mathbb{P}_{\partial B_2^n}(A^{-1}(U))$$
$$\sim \frac{\mathrm{vol}_{n-1}(p_{e_n}(A^{-1}(U)))}{< e_n, N_{\partial B_2^n}(A^{-1}(x)) > \mathrm{vol}_{n-1}(\partial B_2^n)} = \frac{a_n \mathrm{vol}_{n-1}(p_{e_n}(A^{-1}(U)))}{x(n)\, \mathrm{vol}_{n-1}(\partial B_2^n)}.$$

Moreover, since

$$N_{\partial\mathcal{E}}(x) = \left(\sum_{i=1}^{n} \frac{x(i)^2}{a_i^4} \right)^{-\frac{1}{2}} \left(\frac{x(i)}{a_i^2} \right)_{i=1}^{n}$$

we have

$$\mu_{\partial\mathcal{E}}(U) \sim \frac{\mathrm{vol}_{n-1}(p_{e_n}(U))}{< e_n, N_{\partial\mathcal{E}}(x) >} = a_n^2 \sqrt{\sum_{i=1}^{n} \frac{x(i)^2}{a_i^4}} \left(\frac{\mathrm{vol}_{n-1}(p_{e_n}(U))}{x(n)} \right).$$

We also have that

$$\mathrm{vol}_{n-1}(p_{e_n}(U)) = \left(\prod_{i=1}^{n-1} a_i \right) \mathrm{vol}_{n-1}(p_{e_n}(A^{-1}(U))).$$

Therefore we get

$$\mu_{\partial\mathcal{E}}(U) \sim a_n \left(\prod_{i=1}^{n} a_i \right) \sqrt{\sum_{i=1}^{n} \frac{x(i)^2}{a_i^4}} \left(\frac{\mathrm{vol}_{n-1}(p_{e_n}(A^{-1}(U)))}{x(n)} \right)$$
$$\sim \left(\prod_{i=1}^{n} a_i \right) \sqrt{\sum_{i=1}^{n} \frac{x(i)^2}{a_i^4}} \mathrm{vol}_{n-1}(\partial B_2^n)\nu(U).$$

$\square$

Lemma 4.21. *Let $a_1, \ldots, a_n > 0$ and*

$$\mathcal{E} = \left\{ x \left| \sum_{i=1}^{n} \left| \frac{x(i)}{a_i} \right|^2 \leq 1 \right. \right\}$$

Let $\mathcal{E}_s$, $0 < s \leq \frac{1}{2}$, be the surface body with respect to the measure $\mathbb{P}_g$ with constant density $g = (\mathrm{vol}_{n-1}(\partial\mathcal{E}))^{-1}$. Moreover, let $\lambda_\mathcal{E} : [0, \frac{1}{2}] \to [0, a_n]$ be such that $\lambda_\mathcal{E}(s)e_n \in \partial\mathcal{E}_s$. Then we have for all t with $0 \leq t \leq \frac{1}{2}$

$$\lim_{N \to \infty} N^{\frac{2}{n-1}} \int_0^t \frac{\mathbb{P}_{\partial\mathcal{E}}^N \{(x_1, \ldots, x_N) | \lambda_\mathcal{E}(s)e_n \notin [x_1, \ldots, x_N]\}}{\int_{\partial(\mathcal{E} \cap H_s)} \frac{(\mathrm{vol}_{n-1}(\partial\mathcal{E}))^{-1}}{(1 - <N_{\partial\mathcal{E}_s}(\lambda_\mathcal{E}(s)e_n), N_{\partial\mathcal{E}}(y)>^2)^{\frac{1}{2}}} d\mu_{\partial(\mathcal{E} \cap H_s)}(y)} ds$$

$$= a_n \left(\prod_{i=1}^{n-1} a_i \right)^{-\frac{2}{n-1}} \left(\frac{\mathrm{vol}_{n-1}(\partial\mathcal{E})}{\mathrm{vol}_{n-2}(\partial B_2^{n-1})} \right)^{\frac{2}{n-1}} \frac{\Gamma\left(n + 1 + \frac{2}{n-1}\right)}{2(n+1)!} (n-1)^{\frac{n+1}{n-1}}$$

where $H_s = H(\lambda_\mathcal{E}(s)e_n, N_{\partial\mathcal{E}_s}(\lambda_\mathcal{E}(s)e_n))$. (Please note that $N_{\partial\mathcal{E}_s}(\lambda_\mathcal{E}(s)e_n) = e_n$.)

Proof. $(B_2^n)_t$, $0 < t \leq \frac{1}{2}$, are the surface bodies with respect to the constant density $(\mathrm{vol}_{n-1}(\partial B_2^n))^{-1}$. $\lambda_B : [0, \frac{1}{2}] \to [0, 1]$ is the map defined by $\lambda_B(t)e_n \in \partial(B_2^n)_t$.

By Lemma 4.18

$$\lim_{N \to \infty} N^{\frac{2}{n-1}} \int_0^{\frac{1}{2}} \frac{\mathbb{P}_{\partial B_2^n}^N \{(x_1, \ldots, x_N) | \lambda_B(s)e_n \notin [x_1, \ldots, x_N]\}}{\int_{\partial(B_2^n \cap H_s)} \frac{(\mathrm{vol}_{n-1}(\partial B_2^n))^{-1}}{(1 - <N_{\partial(B_2^n)_s}(\lambda_B(s)e_n), N_{\partial B_2^n}(y)>^2)^{\frac{1}{2}}} d\mu_{\partial(B_2^n \cap H_s)}(y)} ds$$

$$= \left(\frac{\mathrm{vol}_{n-1}(\partial B_2^n)}{\mathrm{vol}_{n-2}(\partial B_2^{n-1})} \right)^{\frac{2}{n-1}} \frac{\Gamma\left(n + 1 + \frac{2}{n-1}\right)}{2(n+1)!} (n-1)^{\frac{n+1}{n-1}}$$

where $\lambda_B(s)e_n \in \partial(B_2^n)_s$ and $H_s = H(\lambda_B(s)e_n, e_n)$. By Lemma 4.13 for c with $c_0 < c$ and N with $N_0 < N$

$$\left| N^{\frac{2}{n-1}} \int_0^{\frac{c}{N}} \frac{\mathbb{P}_{\partial B_2^n}^N \{(x_1, \ldots, x_N) | \lambda_B(s)e_n \notin [x_1, \ldots, x_N]\}}{\int_{\partial B_2^n \cap H_s)} \frac{(\mathrm{vol}_{n-1}(\partial B_2^n))^{-1}}{(1 - <N_{\partial(B_2^n)_s}(\lambda_B(s)e_n), N_{\partial B_2^n}(y)>^2)^{\frac{1}{2}}} d\mu_{\partial(B_2^n \cap H_s)}(y)} ds \right.$$

$$\left. - \left(\frac{\mathrm{vol}_{n-1}(\partial B_2^n)}{\mathrm{vol}_{n-2}(\partial B_2^{n-1})} \right)^{\frac{2}{n-1}} \frac{\Gamma\left(n + 1 + \frac{2}{n-1}\right)}{2(n+1)!} (n-1)^{\frac{n+1}{n-1}} \right| \leq c_1 e^{-c} + c_2 e^{-c_3 N}.$$

Let A be the diagonal operator with $A(x) = (a_i x_i)_{i=1}^n$ such that $A(B_2^n) = \mathcal{E}$. By Lemma 4.20 we have

$$\mathbb{P}_{\partial B_2^n}^N \{(x_1, \ldots, x_N) | \quad A^{-1}(x) \notin [x_1, \ldots, x_N]\}$$

$$= \mathbb{P}_f^N \{(z_1, \ldots, z_N) | \quad x \notin [z_1, \ldots, z_N]\}$$

where $f : \partial \mathcal{E} \to (0, \infty)$

$$f(x) = \left(\left(\prod_{i=1}^{n} a_i \right) \sqrt{\sum_{i=1}^{n} \frac{x(i)^2}{a_i^4}} \mathrm{vol}_{n-1}(\partial B_2^n) \right)^{-1}.$$

For all c with $c_0 < c$ and N with $N_0 < N$

$$\left| N^{\frac{2}{n-1}} \int_0^{\frac{c}{N}} \frac{\mathbb{P}_f^N \{(z_1, \ldots, z_N) | A(\lambda_B(s)e_n) \notin [z_1, \ldots, z_N]\}}{\int_{\partial(B_2^n \cap H_s)} \frac{(\mathrm{vol}_{n-1}(\partial B_2^n))^{-1}}{(1 - <N_{\partial(B_2^n)_s}(\lambda_B(s)e_n), N_{\partial B_2^n}(y)>^2)^{\frac{1}{2}}} \mathrm{d}\mu_{\partial(B_2^n \cap H_s)}(y)} \mathrm{d}s \right.$$

$$\left. - \left(\frac{\mathrm{vol}_{n-1}(\partial B_2^n)}{\mathrm{vol}_{n-2}(\partial B_2^{n-1})} \right)^{\frac{2}{n-1}} \frac{\Gamma\left(n + 1 + \frac{2}{n-1}\right)}{2(n+1)!} (n-1)^{\frac{n+1}{n-1}} \right| \leq c_1 e^{-c} + c_2 e^{-c_3 N}.$$

The functions λ_B and $\lambda_\mathcal{E}$ are strictly decreasing, bijective, continuous functions. Therefore, the function $s : [0, a_n] \to [0, 1]$

$$s(t) = \lambda_B^{-1}\left(\frac{\lambda_\mathcal{E}(t)}{a_n} \right)$$

exists, is continuous and has $t : [0, 1] \to [0, a_n]$

$$t(s) = \lambda_\mathcal{E}^{-1}(a_n \lambda_B(s))$$

as its inverse function. Clearly, $a_n \lambda_B(s(t)) = \lambda_\mathcal{E}(t)$ and $A(\lambda_B(s(t))e_n) = \lambda_\mathcal{E}(t)e_n$. Thus

$$\left| N^{\frac{2}{n-1}} \int_0^{\frac{c}{N}} \frac{\mathbb{P}_f^N \{(z_1, \ldots, z_N) | \lambda_\mathcal{E}(t(s))e_n \notin [z_1, \ldots, z_N]\}}{\int_{\partial(B_2^n \cap H_s)} \frac{(\mathrm{vol}_{n-1}(\partial B_2^n))^{-1}}{(1 - <N_{\partial(B_2^n)_s}(\lambda_B(s)e_n), N_{\partial B_2^n}(y)>^2)^{\frac{1}{2}}} \mathrm{d}\mu_{\partial(B_2^n \cap H_s)}(y)} \mathrm{d}s \right.$$

$$\left. - \left(\frac{\mathrm{vol}_{n-1}(\partial B_2^n)}{\mathrm{vol}_{n-2}(\partial B_2^{n-1})} \right)^{\frac{2}{n-1}} \frac{\Gamma\left(n + 1 + \frac{2}{n-1}\right)}{2(n+1)!} (n-1)^{\frac{n+1}{n-1}} \right| \leq c_1 e^{-c} + c_2 e^{-c_3 N}.$$

Now we perform a change of variable. By Lemma 2.11.(iii) and $a_n \lambda_B(s(t)) = \lambda_\mathcal{E}(t)$

$$\frac{\mathrm{d}s}{\mathrm{d}t} = \frac{1}{a_n} \cdot \frac{\frac{\mathrm{d}\lambda_\mathcal{E}}{\mathrm{d}t}(t)}{\frac{\mathrm{d}\lambda_B}{\mathrm{d}s}(s(t))}$$

$$= \frac{1}{a_n} \frac{\mathrm{vol}_{n-1}(\partial \mathcal{E})}{\mathrm{vol}_{n-1}(\partial B_2^n)} \frac{\int_{\partial B_2^n \cap H(\lambda_B(s(t))e_n, e_n)} \frac{\mathrm{d}\mu_{\partial B_2^n \cap H(\lambda_B(s(t))e_n, e_n)}(y)}{\sqrt{1 - <e_n, N(y)>^2}}}{\int_{\partial \mathcal{E} \cap H(\lambda_\mathcal{E}(t)e_n, e_n)} \frac{\mathrm{d}\mu_{\partial \mathcal{E} \cap H(\lambda_\mathcal{E}(t)e_n, e_n)}(y)}{\sqrt{1 - <e_n, N(y)>^2}}}.$$

Therefore we get for all c with $c_0 < c$ and N with $N_0 < N$

$$\left| N^{\frac{2}{n-1}} \int_0^{t(\frac{c}{N})} \frac{\mathbb{P}_f^N\{(z_1,\ldots,z_N) | \; \lambda_{\mathcal{E}}(t)e_n \notin [z_1,\ldots,z_N]\}}{\int_{\partial(\mathcal{E}\cap H_t)} \frac{(\mathrm{vol}_{n-1}(\partial\mathcal{E}))^{-1}}{(1-<N_{\partial\mathcal{E}_t}(\lambda_{\mathcal{E}}(t)e_n),N_{\partial\mathcal{E}}(y)>^2)^{\frac{1}{2}}}d\mu_{\partial(\mathcal{E}\cap H_t)}(y)} dt \right.$$

$$\left. -a_n \left(\frac{\mathrm{vol}_{n-1}(\partial B_2^n)}{\mathrm{vol}_{n-2}(\partial B_2^{n-1})}\right)^{\frac{2}{n-1}} \frac{\Gamma\left(n+1+\frac{2}{n-1}\right)}{2(n+1)!}(n-1)^{\frac{n+1}{n-1}} \right|$$

$$\leq \frac{1}{a_n}\left[c_1 e^{-c} + c_2 e^{-c_3 N}\right]$$

where H_t now denotes $H(\lambda_{\mathcal{E}}(t)e_n, N(\lambda_{\mathcal{E}}(t)e_n))$. Since $a_n\lambda_B(s(t)) = \lambda_{\mathcal{E}}(t)$ we get that for sufficiently small t the quantities t and s are up to a small error directly proportional. We have

$$t(s) \sim s\frac{c_n a_n^{\frac{n-1}{2}}}{\kappa(a_n e_n)^{\frac{n-1}{4}}}.$$

Therefore, with a constant α and new constants c_1, c_2 we can substitute $t(\frac{c}{N})$ by $\frac{c}{N}$.

$$\left| N^{\frac{2}{n-1}} \int_0^{\frac{c}{N}} \frac{\mathbb{P}_f^N\{(z_1,\ldots,z_N) | \; \lambda_{\mathcal{E}}(t)e_n \notin [z_1,\ldots,z_N]\}}{\int_{\partial(\mathcal{E}\cap H_t)} \frac{(\mathrm{vol}_{n-1}(\partial\mathcal{E}))^{-1}}{(1-<N_{\partial\mathcal{E}_t}(\lambda_{\mathcal{E}}(t)e_n),N_{\partial\mathcal{E}}(y)>^2)^{\frac{1}{2}}}d\mu_{\partial(\mathcal{E}\cap H_t)}(y)} dt \right.$$

$$\left. -a_n \left(\frac{\mathrm{vol}_{n-1}(\partial B_2^n)}{\mathrm{vol}_{n-2}(\partial B_2^{n-1})}\right)^{\frac{2}{n-1}} \frac{\Gamma\left(n+1+\frac{2}{n-1}\right)}{2(n+1)!}(n-1)^{\frac{n+1}{n-1}} \right|$$

$$\leq c_1 e^{-\alpha c} + c_2 e^{-c_3 N}$$

We have $\lambda_{\mathcal{E}}(tf(a_n e_n)\mathrm{vol}_{n-1}(\partial\mathcal{E}))e_n \in \partial\mathcal{E}_{t'}$ with $t' = tf(a_n e_n)\mathrm{vol}_{n-1}(\partial\mathcal{E})$. By Lemma 2.7.(i) for every $\delta > 0$ there is t'' with $\lambda_{\mathcal{E}}(t)e_n \in \partial\mathcal{E}_{f,t''}$ and

$$(1-\delta)tf(a_n e_n)\mathrm{vol}_{n-1}(\partial\mathcal{E}) \leq t'' \leq (1+\delta)tf(a_n e_n)\mathrm{vol}_{n-1}(\partial\mathcal{E})$$

i.e.

$$(1-\delta)t' \leq t'' \leq (1+\delta)t'.$$

Applying Lemma 4.17 gives

$$\left| \mathbb{P}_f^N\{(x_1,\ldots,x_N) | \; \lambda_{\mathcal{E}}(t)e_n \notin [x_1,\ldots,x_N]\} - \right.$$
$$\left. \mathbb{P}_{\partial\mathcal{E}}^N\{(z_1,\ldots,z_N) | \; \lambda_{\mathcal{E}}(tf(a_n e_n)\mathrm{vol}_{n-1}(\partial\mathcal{E}))e_n \notin [z_1,\ldots,z_N]\} \right| < \epsilon.$$

Therefore

$$\left| N^{\frac{2}{n-1}} \int_0^{\frac{c}{N}} \frac{\mathbb{P}_{\partial\mathcal{E}}^N\{(z_1,\ldots,z_N) | \; \lambda_{\mathcal{E}}(tf(a_n e_n)\mathrm{vol}_{n-1}(\partial\mathcal{E}))e_n \notin [z_1,\ldots,z_N]\}}{\int_{\partial(\mathcal{E}\cap H_t)} \frac{(\mathrm{vol}_{n-1}(\partial\mathcal{E}))^{-1}}{(1-<N_{\partial\mathcal{E}_t}(\lambda_{\mathcal{E}}(t)e_n),N_{\partial\mathcal{E}}(y)>^2)^{\frac{1}{2}}}d\mu_{\partial(\mathcal{E}\cap H_t)}(y)} dt \right.$$

$$\left. -a_n \left(\frac{\mathrm{vol}_{n-1}(\partial B_2^n)}{\mathrm{vol}_{n-2}(\partial B_2^{n-1})}\right)^{\frac{2}{n-1}} \frac{\Gamma\left(n+1+\frac{2}{n-1}\right)}{2(n+1)!}(n-1)^{\frac{n+1}{n-1}} \right|$$

$$\leq \left| N^{\frac{2}{n-1}} \int_0^{\frac{c}{N}} \frac{\epsilon}{\int_{\partial(\mathcal{E}\cap H_t)} \frac{(\mathrm{vol}_{n-1}(\partial\mathcal{E}))^{-1}}{(1-<N_{\partial\mathcal{E}_t}(\lambda_\mathcal{E}(t)e_n),N_{\partial\mathcal{E}}(y)>^2)^{\frac{1}{2}}} \, \mathrm{d}\mu_{\partial(\mathcal{E}\cap H_t)}(y)} \, \mathrm{d}t \right|$$
$$+ c_1 \mathrm{e}^{-\alpha c} + c_2 \mathrm{e}^{-c_3 N}.$$

By Lemma 4.11

$$\int_{\partial\mathcal{E}\cap H_t} (1- <N_{\partial\mathcal{E}_t}(\lambda_\mathcal{E}(t)e_n),N_{\partial\mathcal{E}}(y)>^2)^{-\frac{1}{2}} \mathrm{d}\mu_{\partial(\mathcal{E}\cap H_t)}(y) \geq \gamma t^{\frac{n-3}{n-1}}.$$

Therefore we have

$$\int_0^{\frac{c}{N}} \frac{\epsilon}{\int_{\partial\mathcal{E}\cap H_t}(1- <N_{\partial\mathcal{E}_t}(\lambda_\mathcal{E}(t)e_n),N_{\partial\mathcal{E}}(y)>^2)^{-\frac{1}{2}} \mathrm{d}\mu_{\partial(\mathcal{E}\cap H_t)}(y)} \, \mathrm{d}t$$
$$\leq \frac{\epsilon}{\gamma} \int_0^{\frac{c}{N}} t^{-\frac{n-3}{n-1}} \mathrm{d}t = \frac{\epsilon}{\gamma} \frac{n-1}{2} \left(\frac{c}{N}\right)^{\frac{2}{n-1}}.$$

Therefore

$$\left| N^{\frac{2}{n-1}} \int_0^{\frac{c}{N}} \frac{\mathbb{P}_{\partial\mathcal{E}}^N\{(z_1,\dots,z_N)| \ \lambda_\mathcal{E}(tf(a_ne_n)\mathrm{vol}_{n-1}(\partial\mathcal{E}))e_n \notin [z_1,\dots,z_N]\}}{\int_{\partial(\mathcal{E}\cap H_t)} \frac{(\mathrm{vol}_{n-1}(\partial\mathcal{E}))^{-1}}{(1-<N_{\partial\mathcal{E}_t}(\lambda_\mathcal{E}(t)e_n),N_{\partial\mathcal{E}}(y)>^2)^{\frac{1}{2}}} \mathrm{d}\mu_{\partial(\mathcal{E}\cap H_t)}(y)} \mathrm{d}t \right.$$
$$\left. -a_n \left(\frac{\mathrm{vol}_{n-1}(\partial B_2^n)}{\mathrm{vol}_{n-2}(\partial B_2^{n-1})}\right)^{\frac{2}{n-1}} \frac{\Gamma\left(n+1+\frac{2}{n-1}\right)}{2(n+1)!}(n-1)^{\frac{n+1}{n-1}} \right|$$
$$\leq \frac{\epsilon}{\gamma} \frac{n-1}{2} \left(\frac{c}{N}\right)^{\frac{2}{n-1}} + c_1 \mathrm{e}^{-\alpha c} + c_2 \mathrm{e}^{-c_3 N}.$$

We perform another transform, $u = tf(a_ne_n)\mathrm{vol}_{n-1}(\partial\mathcal{E})$. With a new constant α

$$\left| N^{\frac{2}{n-1}} \int_0^{\frac{c}{N}} \frac{\mathbb{P}_{\partial\mathcal{E}}^N\{(z_1,\dots,z_N)| \ \lambda_\mathcal{E}(u)e_n \notin [z_1,\dots,z_N]\}}{\int_{\partial(\mathcal{E}\cap H_{t(u)})} \frac{(\mathrm{vol}_{n-1}(\partial\mathcal{E}))^{-1}}{(1-<N_{\partial\mathcal{E}_{t(u)}}(\lambda_\mathcal{E}(t(u))e_n),N_{\partial\mathcal{E}}(y)>^2)^{\frac{1}{2}}} \mathrm{d}\mu_{\partial(\mathcal{E}\cap H_{t(u)})}(y)} \right.$$
$$\left. \times \frac{\mathrm{d}u}{f(a_ne_n)\mathrm{vol}_{n-1}(\partial\mathcal{E})} - a_n \left(\frac{\mathrm{vol}_{n-1}(\partial B_2^n)}{\mathrm{vol}_{n-2}(\partial B_2^{n-1})}\right)^{\frac{2}{n-1}} \frac{\Gamma\left(n+1+\frac{2}{n-1}\right)}{2(n+1)!(n-1)^{-\frac{n+1}{n-1}}} \right|$$
$$\leq \frac{\epsilon}{\gamma} \frac{n-1}{2} \left(\frac{c}{N}\right)^{\frac{2}{n-1}} + c_1 \mathrm{e}^{-\alpha c} + c_2 \mathrm{e}^{-c_3 N}.$$

By Lemma 2.10.(iii)

$$\int_{\partial\mathcal{E}\cap H_u} \frac{1}{\sqrt{1- <N_{\partial\mathcal{E}_u}(x_u),N_{\partial\mathcal{E}}(y)>^2}} \mathrm{d}\mu_{\partial\mathcal{E}\cap H_u}(y)$$
$$\leq (1+\epsilon)(\tfrac{u}{t})^{\frac{n-3}{n-1}} \int_{\partial\mathcal{E}\cap H_t} \frac{1}{\sqrt{1- <N_{\partial\mathcal{E}_t}(x_t),N_{\partial\mathcal{E}}(y)>^2}} \mathrm{d}\mu_{\partial\mathcal{E}\cap H_t}(y)$$

and the inverse inequality. Thus

$$\left| N^{\frac{2}{n-1}} \int_0^{\frac{c}{N}} \frac{\mathbb{P}_{\partial\mathcal{E}}^N \{(z_1,\ldots,z_N) | \; \lambda_{\mathcal{E}}(u)e_n \notin [z_1,\ldots,z_N]\}}{\int_{\partial(\mathcal{E}\cap H_u)} \frac{(\mathrm{vol}_{n-1}(\partial\mathcal{E}))^{-1}}{(1-<N_{\partial\mathcal{E}_u}(\lambda_{\mathcal{E}}(u)e_n),N_{\partial\mathcal{E}}(y)>^2)^{\frac{1}{2}}} d\mu_{\partial(\mathcal{E}\cap H_u)}(y)} \times \right.$$

$$\left. \frac{du}{(f(a_n e_n)\mathrm{vol}_{n-1}(\partial\mathcal{E}))^{\frac{2}{n-1}}} - a_n \left(\frac{\mathrm{vol}_{n-1}(\partial B_2^n)}{\mathrm{vol}_{n-2}(\partial B_2^{n-1})}\right)^{\frac{2}{n-1}} \frac{\Gamma\left(n+1+\frac{2}{n-1}\right)}{2(n+1)!(n-1)^{-\frac{n+1}{n-1}}} \right|$$

$$\leq \epsilon + \frac{\epsilon}{\gamma}\frac{n-1}{2}\left(\frac{c}{N}\right)^{\frac{2}{n-1}} + c_1 e^{-\alpha c} + c_2 e^{-c_3 N}.$$

Since $f(a_n e_n) = ((\prod_{i=1}^{n-1} a_i)\mathrm{vol}_{n-1}(\partial B_2^n))^{-1}$

$$\left| N^{\frac{2}{n-1}} \int_0^{\frac{c}{N}} \frac{\mathbb{P}_{\partial\mathcal{E}}^N \{(z_1,\ldots,z_N) | \; \lambda_{\mathcal{E}}(u)e_n \notin [z_1,\ldots,z_N]\}}{\int_{\partial(\mathcal{E}\cap H_u)} \frac{(\mathrm{vol}_{n-1}(\partial\mathcal{E}))^{-1}}{(1-<N_{\partial\mathcal{E}_u}(\lambda_{\mathcal{E}}(u)e_n),N_{\partial\mathcal{E}}(y)>^2)^{\frac{1}{2}}} d\mu_{\partial(\mathcal{E}\cap H_u)}(y)} du \right.$$

$$\left. -a_n \left(\prod_{i=1}^{n-1} a_i\right)^{-\frac{2}{n-1}} \left(\frac{\mathrm{vol}_{n-1}(\partial\mathcal{E})}{\mathrm{vol}_{n-2}(\partial B_2^{n-1})}\right)^{\frac{2}{n-1}} \frac{\Gamma\left(n+1+\frac{2}{n-1}\right)}{2(n+1)!}(n-1)^{\frac{n+1}{n-1}} \right|$$

$$\leq \left(\frac{\mathrm{vol}_{n-1}(\partial B_2^n)}{\mathrm{vol}_{n-1}(\partial\mathcal{E})}\prod_{i=1}^{n-1} a_i\right)^{\frac{2}{n-1}} \left(\epsilon + \frac{\epsilon}{\gamma}\frac{n-1}{2}\left(\frac{c}{N}\right)^{\frac{2}{n-1}} + c_1 e^{-\alpha c} + c_2 e^{-c_3 N}\right).$$

By choosing first c sufficiently big and then ϵ sufficiently small we get the above expression as small as possible provided that N is sufficiently large. By this and Lemma 4.13

$$\lim_{N\to\infty} N^{\frac{2}{n-1}} \int_0^{t_0} \frac{\mathbb{P}_{\partial\mathcal{E}}^N \{(z_1,\ldots,z_N) | \; \lambda_{\mathcal{E}}(t)e_n \notin [z_1,\ldots,z_N]\}}{\int_{\partial(\mathcal{E}\cap H_t)} \frac{(\mathrm{vol}_{n-1}(\partial\mathcal{E}))^{-1}}{(1-<N_{\partial\mathcal{E}_t}(\lambda_{\mathcal{E}}(t)e_n),N_{\partial\mathcal{E}}(y)>^2)^{\frac{1}{2}}} d\mu_{\partial(\mathcal{E}\cap H_t)}(y)} dt$$

$$= a_n \left(\prod_{i=1}^{n-1} a_i\right)^{-\frac{2}{n-1}} \left(\frac{\mathrm{vol}_{n-1}(\partial\mathcal{E})}{\mathrm{vol}_{n-2}(\partial B_2^{n-1})}\right)^{\frac{2}{n-1}} \frac{\Gamma\left(n+1+\frac{2}{n-1}\right)}{2(n+1)!}(n-1)^{\frac{n+1}{n-1}}.$$

$\square$

5 Proof of the Theorem

Lemma 5.1. *Let K be a convex body in $\mathbb{R}^n$ such that the generalized Gauß-curvature exists at $x_0 \in \partial K$ and is not 0. Let $f : \partial K \to \mathbb{R}$ be a continuous, strictly positive function with $\int_{\partial K} f d\mu = 1$. Let K_s be the surface body with respect to the measure $f d\mu$. Let $\{x_s\} = [x_T, x_0] \cap K_s$ and $H_s = H(x_s, N_{\partial K_s}(x_s))$. Assume that there are r and R with $0 < r, R < \infty$ and*

$$B_2^n(x_0 - rN_{\partial K}(x_0), r) \subseteq K \subseteq B_2^n(x_0 - RN_{\partial K}(x_0), R).$$

Then for all s_0 with $0 < s_0 \leq T$

$$\lim_{N \to \infty} N^{\frac{2}{n-1}} \int_0^{s_0} \frac{\mathbb{P}_f^N\{(x_1, \ldots, x_N) \mid x_s \notin [x_1, \ldots, x_N]\}}{\int_{\partial(K \cap H_s)} \frac{f(y) d\mu_{\partial(K \cap H_s)}(y)}{(1 - <N_{\partial K_s}(x_s), N_{\partial K}(y)>^2)^{\frac{1}{2}}}} ds = c_n \frac{\kappa(x_0)^{\frac{1}{n-1}}}{f(x_0)^{\frac{2}{n-1}}}$$

where

$$c_n = \frac{(n-1)^{\frac{n+1}{n-1}} \Gamma(n + 1 + \frac{2}{n-1})}{2(n+1)!(\mathrm{vol}_{n-2}(\partial B_2^{n-1}))^{\frac{2}{n-1}}}.$$

We can recover Lemma 4.21 from Lemma 5.1 by choosing $K = \mathcal{E}$ and $f = (\mathrm{vol}_{n-1}(\partial \mathcal{E}))^{-1}$.

Proof. Let $\mathcal{E}$ be the standard approximating ellipsoid at x_0 with principal axes having the lengths a_i, $i = 1, \ldots, n-1$. Then we have (4)

$$\kappa(x_0) = \prod_{i=1}^{n-1} \frac{a_n}{a_i^2}.$$

Therefore, by Lemma 4.21 we get for all s_0 with $0 < s_0 \leq \frac{1}{2}$

$$\lim_{N \to \infty} N^{\frac{2}{n-1}} \int_0^{s_0} \frac{\mathbb{P}_{\partial \mathcal{E}}^N\{(z_1, \ldots, z_N) \mid \lambda_{\mathcal{E}}(s)e_n \notin [z_1, \ldots, z_N]\}}{\int_{\partial(\mathcal{E} \cap H_s)} \frac{(\mathrm{vol}_{n-1}(\partial \mathcal{E}))^{-1}}{(1 - <N_{\partial \mathcal{E}_s}(\lambda_{\mathcal{E}}(s)e_n), N_{\partial \mathcal{E}}(y)>^2)^{\frac{1}{2}}} d\mu_{\partial(\mathcal{E} \cap H_s)}(y)} ds$$

$$= a_n \left(\prod_{i=1}^{n-1} a_i\right)^{-\frac{2}{n-1}} \left(\frac{\mathrm{vol}_{n-1}(\partial \mathcal{E})}{\mathrm{vol}_{n-2}(\partial B_2^{n-1})}\right)^{\frac{2}{n-1}} \frac{\Gamma\left(n + 1 + \frac{2}{n-1}\right)}{2(n+1)!}(n-1)^{\frac{n+1}{n-1}}$$

$$= c_n \kappa^{\frac{1}{n-1}}(x_0)(\mathrm{vol}_{n-1}(\partial \mathcal{E}))^{\frac{2}{n-1}}$$

where

$$c_n = \frac{(n-1)^{\frac{n+1}{n-1}} \Gamma(n + 1 + \frac{2}{n-1})}{2(n+1)!(\mathrm{vol}_{n-2}(\partial B_2^{n-1}))^{\frac{2}{n-1}}}$$

and $H_s = H(\lambda_{\mathcal{E}}(s)e_n, e_n)$. H_s is a tangent hyperplane to the surface body $\mathcal{E}_s$ with respect to the constant density $(\mathrm{vol}_{n-1}(\partial \mathcal{E}))^{-1}$.

By this for all $\epsilon > 0$ and sufficiently big N

$$\left| N^{\frac{2}{n-1}} \int_0^{s_0} \frac{\mathbb{P}_f^N\{(x_1, \ldots, x_N) \mid x_s \notin [x_1, \ldots, x_N]\}}{\int_{\partial(K \cap H(x_s, N(x_s)))} \frac{f(y) d\mu_{\partial(K \cap H(x_s, N(x_s)))}(y)}{(1 - <N_{\partial K_s}(x_s), N_{\partial K}(y)>^2)^{\frac{1}{2}}}} ds - c_n \frac{\kappa(x_0)^{\frac{1}{n-1}}}{f(x_0)^{\frac{2}{n-1}}} \right|$$

$$\leq \epsilon +$$

$$\left| N^{\frac{2}{n-1}} \int_0^{s_0} \frac{\mathbb{P}_f^N\{(x_1, \ldots, x_N) \mid x_s \notin [x_1, \ldots, x_N]\}}{\int_{\partial(K \cap H(x_s, N(x_s)))} \frac{f(y) d\mu_{\partial(K \cap H(x_s, N(x_s)))}(y)}{(1 - <N_{\partial K_s}(x_s), N_{\partial K}(y)>^2)^{\frac{1}{2}}}} ds \right|$$

$$-\left(\frac{N}{f(x_0)\mathrm{vol}_{n-1}(\partial\mathcal{E})}\right)^{\frac{2}{n-1}}\times$$

$$\left.\int_0^{s_0}\frac{\mathbb{P}_{\partial\mathcal{E}}^N\{(z_1,\dots,z_N)|\ \lambda_\mathcal{E}(s)e_n\notin[z_1,\dots,z_N]\}}{\int_{\partial(\mathcal{E}\cap H_s)}\frac{(\mathrm{vol}_{n-1}(\partial\mathcal{E}))^{-1}}{(1-<N_{\partial\mathcal{E}_s}(\lambda_\mathcal{E}(s)e_n),N_{\partial\mathcal{E}}(y)>^2)^{\frac{1}{2}}}d\mu_{\partial(\mathcal{E}\cap H_s)}(y)}ds\right|.$$

By Lemma 4.13 there are constants b_1,b_2,b_3 such that for all sufficiently big c the latter expression is smaller than

$$\epsilon+2(b_1e^{-c}+b_2e^{-b_3N})$$

$$+\left|N^{\frac{2}{n-1}}\int_0^{\frac{c}{N}}\frac{\mathbb{P}_f^N\{(x_1,\dots,x_N)|\ x_s\notin[x_1,\dots,x_N]\}}{\int_{\partial(K\cap H(x_s.N(x_s)))}\frac{f(y)d\mu_{\partial(K\cap H(x_s,N_{\partial K_s}(x_s)))}(y)}{(1-<N_{\partial K_s}(x_s),N_{\partial K}(y)>^2)^{\frac{1}{2}}}}ds\right.$$

$$\left.-N^{\frac{2}{n-1}}\int_0^{\frac{c}{N}}\frac{\mathbb{P}_{\partial\mathcal{E}}^N\{(z_1,\dots,z_N)|\ \lambda_\mathcal{E}(s)e_n\notin[z_1,\dots,z_N]\}}{\int_{\partial(\mathcal{E}\cap H_s)}\frac{f(x_0)^{\frac{2}{n-1}}(\mathrm{vol}_{n-1}(\partial\mathcal{E}))^{-\frac{n-3}{n-1}}}{(1-<N_{\partial\mathcal{E}_s}(\lambda_\mathcal{E}(s)e_n),N_{\partial\mathcal{E}}(y)>^2)^{\frac{1}{2}}}d\mu_{\partial(\mathcal{E}\cap H_s)}(y)}ds\right|.$$

By triangle-inequality this is smaller than

$$\epsilon+2(b_1e^{-c}+b_2e^{-b_3N})$$

$$+\left|N^{\frac{2}{n-1}}\int_0^{\frac{c}{N}}\frac{\mathbb{P}_f^N\{(x_1,\dots,x_N)|\ x_s\notin[x_1,\dots,x_N]\}}{\int_{\partial(K\cap H(x_s,N_{\partial K_s}(x_s)))}\frac{f(y)d\mu_{\partial(K\cap H(x_s,N_{\partial K_s}(x_s)))}(y)}{(1-<N_{\partial K_s}(x_s),N_{\partial K}(y)>^2)^{\frac{1}{2}}}}\right.$$

$$\left.-\frac{\mathbb{P}_f^N\{(x_1,\dots,x_N)|\ x_s\notin[x_1,\dots,x_N]\}}{\int_{\partial(\mathcal{E}\cap H_s)}\frac{f(x_0)^{\frac{2}{n-1}}(\mathrm{vol}_{n-1}(\partial\mathcal{E}))^{-\frac{n-3}{n-1}}}{(1-<N_{\partial\mathcal{E}_s}(\lambda_\mathcal{E}(s)e_n),N_{\partial\mathcal{E}}(y)>^2)^{\frac{1}{2}}}d\mu_{\partial(\mathcal{E}\cap H_s)}(y)}ds\right|$$

$$+\left|N^{\frac{2}{n-1}}\int_0^{\frac{c}{N}}\frac{\mathbb{P}_{\partial\mathcal{E}}^N\{(z_1,\dots,z_N)|\ \lambda_\mathcal{E}(s)e_n\notin[z_1,\dots,z_N]\}}{\int_{\partial(\mathcal{E}\cap H_s)}\frac{f(x_0)^{\frac{2}{n-1}}(\mathrm{vol}_{n-1}(\partial\mathcal{E}))^{-\frac{n-3}{n-1}}}{(1-<N_{\partial\mathcal{E}_s}(\lambda_\mathcal{E}(s)e_n),N_{\partial\mathcal{E}}(y)>^2)^{\frac{1}{2}}}d\mu_{\partial(\mathcal{E}\cap H_s)}(y)}\right.$$

$$\left.-\frac{\mathbb{P}_f^N\{(x_1,\dots,x_N)|\ x_s\notin[x_1,\dots,x_N]\}}{\int_{\partial(\mathcal{E}\cap H_s)}\frac{f(x_0)^{\frac{2}{n-1}}(\mathrm{vol}_{n-1}(\partial\mathcal{E}))^{-\frac{n-3}{n-1}}}{(1-<N_{\partial\mathcal{E}_s}(\lambda_\mathcal{E}(s)e_n),N_{\partial\mathcal{E}}(y)>^2)^{\frac{1}{2}}}d\mu_{\partial(\mathcal{E}\cap H_s)}(y)}ds\right|.$$

By Lemma 4.17

$$\left|\mathbb{P}_f^N\{(x_1,\dots,x_N)|\quad x_s\notin[x_1,\dots,x_N]\}-\right.$$
$$\left.\mathbb{P}_{\partial\mathcal{E}}^N\{(z_1,\dots,z_N)|\ \lambda_\mathcal{E}(s)e_n\notin[z_1,\dots,z_N]\}\right|<\epsilon.$$

Therefore, the above quantity is less than

$$\epsilon+2(b_1e^{-c}+b_2e^{-b_3N})$$

$$+\left|N^{\frac{2}{n-1}}\int_0^{\frac{c}{N}}\frac{\mathbb{P}_f^N\{(x_1,\dots,x_N)|\ x_s\notin[x_1,\dots,x_N]\}}{\int_{\partial(K\cap H(x_s,N_{\partial K_s}(x_s)))}\frac{f(y)d\mu_{\partial(K\cap H(x_s,N_{\partial K_s}(x_s)))}(y)}{(1-<N_{\partial K_s}(x_s),N_{\partial K}(y)>^2)^{\frac{1}{2}}}}\right.$$

$$
- \frac{\mathbb{P}_f^N\{(x_1,\dots,x_N)\mid x_s \notin [x_1,\dots,x_N]\}}{\int_{\partial(\mathcal{E}\cap H_s)} \frac{f(x_0)^{\frac{2}{n-1}}(\mathrm{vol}_{n-1}(\partial\mathcal{E}))^{-\frac{n-3}{n-1}}}{(1-<N_{\partial\mathcal{E}_s}(\lambda_{\mathcal{E}}(s)e_n),N_{\partial\mathcal{E}}(y)>^2)^{\frac{1}{2}}}\, d\mu_{\partial(\mathcal{E}\cap H_s)}(y)}\, ds \Bigg|
$$

$$
+ \Bigg| N^{\frac{2}{n-1}} \int_0^{\frac{c}{N}} \frac{\epsilon}{\int_{\partial(\mathcal{E}\cap H_s)} \frac{f(x_0)^{\frac{2}{n-1}}(\mathrm{vol}_{n-1}(\partial\mathcal{E}))^{-\frac{n-3}{n-1}}}{(1-<N_{\partial\mathcal{E}_s}(\lambda_{\mathcal{E}}(s)e_n),N_{\partial\mathcal{E}}(y)>^2)^{\frac{1}{2}}}\, d\mu_{\partial(\mathcal{E}\cap H_s)}(y)}\, ds \Bigg|.
$$

By Lemma 4.11 we have

$$
\int_0^{\frac{c}{N}} \frac{1}{\int_{\partial(\mathcal{E}\cap H_s)} \frac{d\mu_{\partial(\mathcal{E}\cap H_s)}(y)}{(1-<N_{\partial\mathcal{E}_s}(\lambda_{\mathcal{E}}(s)e_n),N_{\partial\mathcal{E}}(y)>^2)^{\frac{1}{2}}}}\, ds
$$

$$
\leq c_0^n \frac{R^{n-1}}{r^n}(\mathrm{vol}_{n-1}(B_2^{n-1}))^{-\frac{2}{n-1}}(\mathrm{vol}_{n-1}(\partial\mathcal{E}))^{-\frac{n-3}{n-1}}\int_0^{\frac{c}{N}} s^{-\frac{n-3}{n-1}}\, ds
$$

$$
= c_0^n \frac{R^{n-1}}{r^n}(\mathrm{vol}_{n-1}(B_2^{n-1}))^{-\frac{2}{n-1}}(\mathrm{vol}_{n-1}(\partial\mathcal{E}))^{-\frac{n-3}{n-1}}\frac{n-1}{2}\left(\frac{c}{N}\right)^{\frac{2}{n-1}}.
$$

Therefore, the above expression is not greater than

$$
\epsilon + b_1 e^{-c} + b_2 e^{-b_3 N} + b_4 \epsilon\, c^{\frac{2}{n-1}}
$$

$$
+ \Bigg| N^{\frac{2}{n-1}} \int_0^{\frac{c}{N}} \frac{\mathbb{P}_f^N\{(x_1,\dots,x_N)\mid x_s \notin [x_1,\dots,x_N]\}}{\int_{\partial(K\cap H(x_s,N_{\partial K_s}(x_s)))} \frac{f(y)d\mu_{\partial(K\cap H(x_s,N_{\partial K_s}(x_s)))}(y)}{(1-<N_{\partial K_s}(x_s),N_{\partial K}(y)>^2)^{\frac{1}{2}}}}
$$

$$
- \frac{\mathbb{P}_f^N\{(x_1,\dots,x_N)\mid x_s \notin [x_1,\dots,x_N]\}}{\int_{\partial(\mathcal{E}\cap H_s)} \frac{f(x_0)^{\frac{2}{n-1}}(\mathrm{vol}_{n-1}(\partial\mathcal{E}))^{-\frac{n-3}{n-1}}}{(1-<N_{\partial\mathcal{E}_s}(\lambda_{\mathcal{E}}(s)e_n),N_{\partial\mathcal{E}}(y)>^2)^{\frac{1}{2}}}\, d\mu_{\partial(\mathcal{E}\cap H_s)}(y)}\, ds \Bigg|
$$

for some constant b_4. Let z_s be defined by

$$
\{z_s\} = \{x_0 + tN_{\partial K}(x_0)\mid t \in \mathbb{R}\} \cap H(x_s,N_{\partial K}(x_0)).
$$

By Lemma 2.7 there is a sufficiently small s_ϵ such that we have for all s with $0 < s \leq s_\epsilon$

$$
s \leq \mathbb{P}_f(\partial K \cap H^-(z_s,N_{\partial K}(x_0))) \leq (1+\epsilon)s.
$$

Because f is continuous at x_0 and because $\mathcal{E}$ is the standard approximating ellipsoid at x_0 we have for all s with $0 < s \leq s_\epsilon$

$$
(1-\epsilon)s \leq f(x_0)\mathrm{vol}_{n-1}(\partial\mathcal{E}\cap H^-(z_s,N_{\partial K}(x_0))) \leq (1+\epsilon)s.
$$

Since $s = \frac{\mathrm{vol}_{n-1}(\partial\mathcal{E}\cap H_s^-)}{\mathrm{vol}_{n-1}(\partial\mathcal{E})}$ we get by Lemma 2.10.(iii) for a new s_ϵ that for all s with $0 < s \leq s_\epsilon$

$$
\frac{(1-\epsilon)}{(f(x_0)\mathrm{vol}_{n-1}(\partial\mathcal{E}))^{\frac{n-3}{n-1}}}\int_{\partial\mathcal{E}\cap H_s} \frac{d\mu_{\partial\mathcal{E}\cap H_s}(y)}{\sqrt{1-<N_{\partial\mathcal{E}_s}(x_0),N_{\partial\mathcal{E}}(y)>^2}}
$$

$$\leq \int_{\partial \mathcal{E} \cap H^{-}(z_s, N_{\partial K}(x_0))} \frac{\mathrm{d}\mu_{\partial \mathcal{E} \cap H^{-}(z_s, N_{\partial K}(x_0))}(y)}{\sqrt{1 - <N_{\partial \mathcal{E}_t}(x_0), N_{\partial \mathcal{E}}(y)>^2}}$$

$$\leq \frac{(1+\epsilon)}{(f(x_0)\mathrm{vol}_{n-1}(\partial \mathcal{E}))^{\frac{n-3}{n-1}}} \int_{\partial \mathcal{E} \cap H_s} \frac{\mathrm{d}\mu_{\partial \mathcal{E} \cap H_s}(y)}{\sqrt{1 - <N_{\partial \mathcal{E}_s}(x_0), N_{\partial \mathcal{E}}(y)>^2}}$$

where $t \sim s(f(x_0)\mathrm{vol}_{n-1}(\partial \mathcal{E}))^{\frac{n-3}{n-1}}$. Please note that $N_{\partial K}(x_0) = N_{\partial \mathcal{E}_s}(z_s)$. Therefore, if we pass to another s_c the above expression is not greater than

$$\epsilon + b_1 e^{-c} + b_2 e^{-b_3 N} + b_4 \epsilon c^{\frac{2}{n-1}}$$

$$+ \left| N^{\frac{2}{n-1}} \int_0^{\frac{c}{N}} \frac{\mathbb{P}_f^N\{(x_1,\ldots,x_N)| \; x_s \notin [x_1,\ldots,x_N]\}}{\int_{\partial(K \cap H(x_s, N(x_s)))} \frac{f(y)}{(1 - <N(x_s), N(y)>^2)^{\frac{1}{2}}} \mathrm{d}\mu_{\partial(K \cap H(x_s, N(x_s)))}(y)} \right.$$

$$\left. - \frac{\mathbb{P}_f^N\{(x_1,\ldots,x_N)| \; x_s \notin [x_1,\ldots,x_N]\}}{\int_{\partial(\mathcal{E} \cap H(z_s, N_{\partial K}(x_0)))} \frac{f(x_0)}{(1 - <N(\lambda_{\mathcal{E}}(s)e_n), N(y)>^2)^{\frac{1}{2}}} \mathrm{d}\mu_{\partial(\mathcal{E} \cap H(z_s, N_{\partial K}(x_0)))}(y)} \mathrm{d}s \right|.$$

Now we apply Lemma 2.10.(i). Choosing another s_ϵ the above expression is less than $\epsilon + b_1 e^{-c} + b_2 e^{-b_3 N} + b_4 \epsilon c^{\frac{2}{n-1}}$. We choose c and N sufficiently big and ϵ sufficiently small. □

Proof. (Proof of Theorem 1.1) We assume here that $x_T = 0$. For $x_0 \in \partial K$ the point x_s is given by $\{x_s\} = [x_T, x_0] \cap \partial K_s$.

$$\mathrm{vol}_n(K) - \mathbb{E}(f, N) = \int_K \mathbb{P}_f^N\{(x_1,\ldots,x_N)|x \notin [x_1,\ldots,x_N]\}\mathrm{d}x.$$

By Lemma 2.1.(iv) we have that $K_0 = K$ and by Lemma 2.4.(iii) that K_T consists of one point only. Since $\mathbb{P}_f^N\{(x_1,\ldots,x_N)|x_s \notin [x_1,\ldots,x_N]\}$ is a continuous functions of the variable x_s we get by Lemma 2.12

$$\mathrm{vol}_n(K) - \mathbb{E}(f, N)$$
$$= \int_0^T \int_{\partial K_s} \frac{\mathbb{P}_f^N\{(x_1,\ldots,x_N)|x_s \notin [x_1,\ldots,x_N]\}}{\int_{\partial(K \cap H_s)} \frac{f(y)\mathrm{d}\mu_{\partial(K \cap H_s)}(y)}{(1 - <N_{\partial K_s}(x_s), N_{\partial K}(y)>^2)^{\frac{1}{2}}}} \mathrm{d}\mu_{\partial K_s}(x_s)\mathrm{d}s$$

where $H_s = H(x_s, N_{\partial K_s}(x_s))$. By Lemma 4.9 for all s_0 with $0 < s_0 \leq T$

$$\lim_{N \to \infty} N^{\frac{2}{n-1}} \int_{s_0}^T \int_{\partial K_s} \frac{\mathbb{P}_f^N\{(x_1,\ldots,x_N)| \; x_s \notin [x_1,\ldots,x_N]\}\mathrm{d}\mu_{\partial K_s}(x_s)\mathrm{d}s}{\int_{\partial K \cap H_s} \frac{f(y)}{\sqrt{1 - <N_{\partial K_s}(x_s), N_{\partial K}(y)>^2}}\mathrm{d}\mu_{\partial K \cap H_s}(y)} = 0.$$

We get for all s_0 with $0 < s_0 \leq T$

$$\lim_{N \to \infty} \frac{\mathrm{vol}_n(K) - \mathbb{E}(f, N)}{N^{-\frac{2}{n-1}}} =$$

$$\lim_{N \to \infty} N^{\frac{2}{n-1}} \int_0^{s_0} \int_{\partial K_s} \frac{\mathbb{P}_f^N\{(x_1,\ldots,x_N)| \; x_s \notin [x_1,\ldots,x_N]\}\mathrm{d}\mu_{\partial K_s}(x_s)\mathrm{d}s}{\int_{\partial K \cap H_s} \frac{f(y)}{\sqrt{1 - <N_{\partial K_s}(x_s), N_{\partial K}(y)>^2}}\mathrm{d}\mu_{\partial K \cap H_s}(y)}.$$

We apply now the bijection between ∂K and ∂K_s mapping an element $x \in \partial K$ to x_s given by $\{x_s\} = [x_T, x_0] \cap \partial K_s$. The ratio of the volumes of a surface element in ∂K and its image in ∂K_s is

$$\frac{\|x_s\|^n <x_0, N_{\partial K}(x_0)>}{\|x_0\|^n <x_s, N_{\partial K_s}(x_s)>}.$$

Thus we get

$$\int_{\partial K_s} \frac{\mathbb{P}_f^N\{(x_1,\ldots,x_N)|\ x_s \notin [x_1,\ldots,x_N]\}}{\int_{\partial K \cap H_s} \frac{f(y)}{\sqrt{1-<N_{\partial K_s}(x_s),N_{\partial K}(y)>^2}} d\mu_{\partial K \cap H_s}(y)} d\mu_{\partial K_s}(x_s)$$

$$= \int_{\partial K} \frac{\mathbb{P}_f^N\{(x_1,\ldots,x_N)|\ x_s \notin [x_1,\ldots,x_N]\}}{\int_{\partial K \cap H_s} \frac{f(y)}{\sqrt{1-<N_{\partial K_s}(x_s),N_{\partial K}(y)>^2}} d\mu_{\partial K \cap H_s}(y)} \times$$

$$\frac{\|x_s\|^n <x, N_{\partial K}(x)>}{\|x\|^n <x_s, N_{\partial K_s}(x_s)>} d\mu_{\partial K}(x).$$

We get for all s_0 with $0 < s_0 \le T$

$$\lim_{N\to\infty} \frac{\mathrm{vol}_n(K) - \mathbb{E}(f,N)}{N^{-\frac{2}{n-1}}} =$$

$$\lim_{N\to\infty} N^{\frac{2}{n-1}} \int_0^{s_0} \int_{\partial K} \frac{\mathbb{P}_f^N\{(x_1,\ldots,x_N)|\ x_s \notin [x_1,\ldots,x_N]\}}{\int_{\partial K \cap H_s} \frac{f(y)}{\sqrt{1-<N_{\partial K_s}(x_s),N_{\partial K}(y)>^2}} d\mu_{\partial K \cap H_s}(y)} \times$$

$$\frac{\|x_s\|^n <x, N_{\partial K}(x)>}{\|x\|^n <x_s, N_{\partial K_s}(x_s)>} d\mu_{\partial K}(x) ds.$$

By the theorem of Tonelli

$$\lim_{N\to\infty} \frac{\mathrm{vol}_n(K) - \mathbb{E}(f,N)}{N^{-\frac{2}{n-1}}} =$$

$$\lim_{N\to\infty} N^{\frac{2}{n-1}} \int_{\partial K} \int_0^{s_0} \frac{\mathbb{P}_f^N\{(x_1,\ldots,x_N)|\ x_s \notin [x_1,\ldots,x_N]\}}{\int_{\partial K \cap H_s} \frac{f(y)}{\sqrt{1-<N_{\partial K_s}(x_s),N_{\partial K}(y)>^2}} d\mu_{\partial K \cap H_s}(y)} \times$$

$$\frac{\|x_s\|^n <x_0, N_{\partial K}(x_0)>}{\|x_0\|^n <x_s, N_{\partial K_s}(x_s)>} ds d\mu_{\partial K}(x).$$

Now we want to apply the dominated convergence theorem in order to change the limit and the integral over ∂K. By Lemma 5.1 for all s_0 with $0 < s_0 \le T$

$$\lim_{N\to\infty} N^{\frac{2}{n-1}} \int_0^{s_0} \frac{\mathbb{P}_f^N\{(x_1,\ldots,x_N)|\ x_s \notin [x_1,\ldots,x_N]\}}{\int_{\partial(K\cap H_s)} \frac{f(y)d\mu_{\partial(K\cap H_s)}(y)}{(1-<N_{\partial K_s}(x_s),N_{\partial K}(y)>^2)^{\frac{1}{2}}}} ds = c_n \frac{\kappa(x_0)^{\frac{1}{n-1}}}{f(x_0)^{\frac{2}{n-1}}}.$$

Clearly, we have $\lim_{s\to 0} \|x_s\| = \|x\|$ and by Lemma 2.5

$$\lim_{s \to 0} < x_s, N_{\partial K_s}(x_s) > = < x, N_{\partial K}(x) > .$$

By this and since the above formula holds for all s_0 with $0 < s_0 \leq T$

$$\lim_{N \to \infty} N^{\frac{2}{n-1}} \int_0^{s_0} \frac{\mathbb{P}_f^N\{(x_1, \ldots, x_N) | \ x_s \notin [x_1, \ldots, x_N]\}}{\int_{\partial(K \cap H_s)} \frac{f(y) d\mu_{\partial(K \cap H_s)}(y)}{(1 - < N_{\partial K_s}(x_s), N_{\partial K}(y) >^2)^{\frac{1}{2}}}} \frac{\|x_s\|^n < x, N(x) >}{\|x\|^n < x_s, N(x_s) >} ds$$

$$= c_n \frac{\kappa(x_0)^{\frac{1}{n-1}}}{f(x_0)^{\frac{2}{n-1}}}.$$

By Lemma 4.12 the functions with variable $x_0 \in \partial K$

$$N^{\frac{2}{n-1}} \int_0^{s_0} \frac{\mathbb{P}_f^N\{(x_1, \ldots, x_N) | \ x_s \notin [x_1, \ldots, x_N]\}}{\int_{\partial(K \cap H_s)} \frac{f(y) d\mu_{\partial(K \cap H_s)}(y)}{(1 - < N_{\partial K_s}(x_s), N_{\partial K}(y) >^2)^{\frac{1}{2}}}} \frac{\|x_s\|^n < x_0, N(x_0) >}{\|x_0\|^n < x_s, N(x_s) >} ds$$

are uniformly bounded. Thus we can apply the dominated convergence theorem.

$$\lim_{N \to \infty} \frac{\text{vol}_n(K) - \mathbb{E}(f, N)}{N^{-\frac{2}{n-1}}} = c_n \int_{\partial K} \frac{\kappa(x)^{\frac{1}{n-1}}}{f(x)^{\frac{2}{n-1}}} d\mu_{\partial K}(x)$$

□

References

[Al]   Aleksandrov A.D. (1939): Almost everywhere existence of the second differential of a convex function and some properties of convex surfaces connected with it. Uchenye Zapiski Leningrad Gos. Univ., Math. Ser., **6**, 3–35

[Ban]   Bangert V. (1979): Analytische Eigenschaften konvexer Funktionen auf Riemannschen Mannigfaltigkeiten. Journal für die Reine und Angewandte Mathematik, **307**, 309–324

[Ba1]   Bárány I. (1992): Random polytopes in smooth convex bodies. Mathematika, **39**, 81–92

[Ba2]   Bárány I. (1997): Approximation by random polytopes is almost optimal. II International Conference in "Stochastic Geometry, Convex Bodies and Empirical Measures"(Agrigento 1996). Rend. Circ. Mat. Palermo (2) Suppl. No., **50**, 43–50

[BCP]   Bianchi G., Colesanti A., Pucci C. (1996): On the second differentiability of convex surfaces. Geometriae Dedicata, **60**, 39–48

[Bil]   Billingsley P. (1968): Convergence of Probability Measures. John Wiley& Sons

[Bla1] Blaschke W. (1935): Integralgeometrie 2: Zu Ergebnissen von M.W. Crofton. Bull. Math. Soc. Roumaine Sci., **37**, 3–11

[Bla2] Blaschke W. (1956): Kreis und Kugel. Walter de Gruyter, Berlin

[BrI]   Bronshteyn E.M., Ivanov L.D. (1975): The approximation of convex sets by polyhedra. Siberian Math. J., **16**, 852–853

[BuR]   Buchta C., Reitzner M. (2001): The convex hull of random points in a tetra-
hedron: Solution of Blaschke's problem and more general results. Journal
für die Reine und Angewandte Mathematik, **536**, 1–29

[E]   Edelsbrunner H. (1993): Geometric algorithms. In: P.M. Gruber, J.M.
Wills (eds) Handbook of Convex Geometry. North-Holland, 699–735

[EvG]   Evans L.C., Gariepy R.F. (1992): Measure Theory and Fine Properties of
Functions. CRC Press

[Fra]   Fradelizi M. (1999): Hyperplane sections of convex bodies in isotropic position. Beiträge zur Algebra und Geometrie/Contributions to Algebra and
Geometry **40**, 163–183

[Ga]   Gardner R.J. (1995): Tomography. Encyclopedia of Mathematics and its
Applications. Cambridge University Press, Cambridge

[GRS1]   Gordon Y., Reisner S., Schütt C. (1997): Umbrellas and polytopal approximation of the Euclidean ball, Journal of Approximation Theory **90**, 9–22

[GRS2]   Gordon Y., Reisner S., Schütt C. (1998): Erratum. Journal of Approximation Theory **95**, 331

[Gr1]   Gruber P.M. (1983): Approximation of convex bodies. In: P.M. Gruber,
J.M. Wills (eds.) Convexity and its Applications. Birkhäuser, 131–162

[Gr2]   Gruber P.M. (1993): Asymptotic estimates for best and stepwise approximation of convex bodies II. Forum Mathematicum **5**, 521–538

[Gr3]   Gruber P.M. (1993): Aspects of approximation of convex bodies. In: P.M.
Gruber, J.M. Wills (eds.) Handbook of Convex Geometry, North-Holland,
319–345

[Gr4]   Gruber P.M. (1997): Comparisons of best and random approximation of
convex bodies by polytopes. Rend. Circ. Mat. Palermo, II. Ser., Suppl. **50**,
189–216

[Hu]   Hug D. (1996): Contributions to Affine Surface Area. Manuscripta Mathematica **91**, 283–301

[J]   John F. (1948): Extremum problems with inequalities as subsidiary conditions. R. Courant Anniversary Volume. Interscience New York, 187–204

[Ki]   Kingman J.F.C. (1969): Random secants of a convex body. Journal of
Applied Probability **6**, 660–672

[Lei]   Leichtweiss K. (1993): Convexity and differential geometry. In: P.M. Gruber and J.M. Wills (eds.) Handbook of Convex Geometry, 1045–1080

[Lu]   Lutwak E. (1996): The Brunn-Minkowski-Fiery Theory II: Affine and Ge-
ominimal Surface Areas. Advances in Mathematics **118**, 194–244

[Ma]   Macbeath A. M. (1951): An extremal property of the hypersphere. Proc.
Cambridge Philos. Soc., **47**, 245–247

[MaS1]   Mankiewicz P., Schütt C. (2000): A simple proof of an estimate for the
approximation of the Euclidean ball and the Delone triangulation numbers.
Journal of Approximation Theory, **107**, 268–280

[MaS2]   Mankiewicz P., Schütt C. (2001): On the Delone triangulations numbers.
Journal of Approximation Theory, **111**, 139–142

[McV]   McClure D.E., Vitale R. (1975): Polygonal approximation of plane convex
bodies. J. Math. Anal. Appl., **51**, 326–358

[MW1]   Meyer M., Werner E. (1998): The Santaló-regions of a convex body. Trans-
actions of the AMS, **350**, 4569–4591

[MW2]   Meyer M., Werner E. (2000): On the p-affine surface area. Advances in
Mathematics, **152**, 288–313

422   C. Schütt and E. Werner

[Mil]   Miles R.E. (1971): Isotropic random simplices. Advances in Appl. Probability, **3**, 353–382.

[Mü]   Müller J.S. (1990): Approximation of the ball by random polytopes. Journal of Approximation Theory, **63**, 198–209

[Pe]   Petkantschin B. (1936): Zusammenhänge zwischen den Dichten der linearen Unterräume im n-dimensionalen Raume. Abhandlungen des Mathematischen Seminars der Universität Hamburg, **11**, 249–310

[ReS1] Rényi A., Sulanke R. (1963): Über die konvexe Hülle von n zufällig gewählten Punkten, Zeitschrift für Wahrscheinlichkeitstheorie und verwandte Gebiete, **2**, 75–84

[ReS2] Rényi A., Sulanke R. (1964): Über die konvexe Hülle von n zufällig gewählten Punkten II. Zeitschrift für Wahrscheinlichkeitstheorie und verwandte Gebiete, **3**, 138–147

[Ro]   Rockafellar R.T. (1970): Convex Analysis. Princeton University Press

[San]   Santaló L.A. (1976): Integral Geometry and Geometric Probability. Encyclopedia of Mathematics and its Applications, Addison-Wesley

[SaT1] Sapiro G., Tannenbaum A. (1994): On affine plane curve evolution. Journal of Functional Analysis, **119**, 79–120

[SaT2] Sapiro G., Tannenbaum A. (1994): On invariant curve evolution and image analysis. Indiana University Journal of Mathematics, **42**, 985–1009

[Sch1] Schütt C. (1994): Random polytopes and affine surface area. Mathematische Nachrichten, **170**, 227–249

[Sch2] Schütt C. (1999): Floating body, illumination body, and polytopal approximation. In: K.M. Ball and V.D. Milman (eds.) Convex Geometric Analysis, Mathematical Science Research Institute Publications **34**, 203–230

[SW1]   Schütt C., Werner E. (1990): The convex floating body. Mathematica Scandinavica **66**, 275–290

[SW2]   Schütt C., Werner E.: Surface bodies and p-affine surface area, preprint

[Ta]   Tannenbaum A. (1996): Three snippets of curve evolution theory in computer vision. Mathematical and Computer Modelling Journal **24**, 103–119

[Wie]   Wieacker J.A. (1978): Einige Probleme der polyedrischen Approximation. Diplomarbeit, Freiburg im Breisgau

Israel Seminar on Geometric Aspects of Functional Analysis (GAFA) 2000-2001

Friday, November 3, 2000

1. *Mikhail Sodin* (Tel Aviv): Dimension free estimates of polynomials and analytic functions (joint with F. Nazarov and A. Volberg)
2. *Boris Tsirelson* (Tel Aviv): Logarithm of a Hilbert space

Friday, November 10, 2000

1. *Michael Krivelevich* (Tel Aviv): On the concentration of eigenvalues of random symmetric matrices
2. *Gideon Schechtman* (Rehovot): MAX CUT and an isoperimetric inequality on the sphere

Friday, November 24, 2000

1. *Michael Entov* (Rehovot): A symplectic proof of Schaffer's conjecture in convex geometry (after J.C. Alvarez Paiva)
2. *Jean-Michel Bismut* (Orsay): Secondary invariants in real and complex geometry

Friday, December 15, 2000

1. *William B. Johnson* (College Station): Non linear quotients vs. non linear quotients (joint work with J. Lindenstrauss, D. Preiss and G. Schechtman)
2. *Marianna Csornyei* (London): The visibility of invisible sets

Friday, December 29, 2000

1. *David Preiss* (London): Deformation with finitely many gradients
2. *Peter Sarnak* (Princeton): L-functions, arithmetic, and semiclassics: L^p norms of eigenfunctions on surfaces

Friday, January 26, 2001

1. *Pavel Shvartsman* (Haifa): Extension of Lipschitz mappings and the K-divisibility theorem (joint work with Yu. Brudnyi)
2. *Roman Vershynin* (Rehovot): Coordinate restrictions of operators

Friday, March 9, 2001

1. *Elisabeth Werner* (Cleveland): An analysis of completely-positive trace-preserving maps on M_2 (joint work with M.B. Ruskai and S.J. Szarek)
2. *Carsten Schutt* (Kiel): Orlicz norms of sequences of random variables (joint work with Y. Gordon, A. Litvak and E. Werner)

Friday, March 30, 2001

1. *Noga Alon* (Tel Aviv): Testing subgraphs in large graphs
2. *Béla Bollobás* (Memphis and Cambridge): Large subgraphs of random graphs

Friday, May 11, 2001

1. *Nicole Tomczak-Jaegermann* (Edmonton): Families of random projections of symmetric convex bodies (joint work with P. Mankiewicz)
2. *Vitali Milman* (Tel Aviv): Some old problems in a new presentation

Friday, November 16, 2001

1. *Marcel Berger* (Bures-sur-Yvette): Metric geometry from Blumenthal to Gromov
2. *Boaz Klartag* (Tel Aviv): $5n$ Minkowski symmetrizations enough to approximate a Euclidean ball starting any convex body

Friday, December 7, 2001

1. *Anatolij Plichko* (Lviv): Superstrictly singular operators
2. *Assaf Naor* (Jerusalem): Girth and Euclidean distortion

Israel Mathematical Union – Functional Analysis Meeting

(Organized by J. Lindenstrauss and G. Schechtman)

Friday, June 8, 2001

1. *Assaf Naor* (Hebrew University): Hyperplane projections of the unit ball in ℓ_p^n

2. *Joram Lindenstrauss* (Hebrew University): On the work of Yaki Sternfeld in functional analysis

3. *Mark Rudelson* (University of Missouri): Embeddings of Levy families in Banach spaces

4. *David Shoikhet* (Technion and Ort Braude Karmiel): A non-linear analogue of the Lumer–Phillips theorem for holomorphic maps and applications to the geometry of domains in Banach spaces

5. *Gideon Schechtman* (Weizmann Institute): Block bases of the Haar system as complemented subspaces of L_p

6. *Vladimir Fonf* (Ben Gurion University): The stochastic approximation property

7. *Shiri Artstein* (Tel Aviv University): Asymptotic behaviors of neighborhoods of sections of S^n with applications to local theory

8. *Boris Rubin* (Hebrew University): Radon transforms and fractional integrals on hyperbolic spaces

Workshop on Convex Geometric Analysis
Anogia Academic Village, Crete (August 2001)

(Organized by A. Giannopoulos, V. Milman, R. Schneider and S. Szarek)

Sunday, August 19

1. *Keith Ball*: The complex plank problem
2. *Imre Bárány*: 0-1 polytopes with many facets
3. *Daniel Hug*: Almost transversal intersections of convex surfaces and translative integral formulae
4. *Markus Kiderlen*: Determination of a convex body from Crofton-type averages of surface area measures
5. *Ulrich Brehm*: Moment inequalities and central limit properties of isotropic convex bodies
6. *Yehoram Gordon*: Local theory of convex bodies between zonotopes and polytopes
7. *Boaz Klartag*: Minkowski symmetrizations of convex bodies
8. *Piotr Mankiewicz*: Average diameters of projections of symmetric convex bodies
9. *Assaf Naor*: The cone measure on the sphere of ℓ_p^n

Monday, August 20

1. *Peter Gruber*: Recent results on asymptotic best approximation of convex bodies
2. *Gideon Schechtman*: A non-standard isoperimetric inequality with applications to the complexity of approximating MAX CUT
3. *Shiri Artstein*: Proportional concentration phenomena on the sphere
4. *Roman Vershynin*: Restricted invertibility of linear operators and applications
5. *Monika Ludwig*: L_p floating bodies
6. *Matthias Reitzner*: Stochastical approximation of smooth convex bodies
7. *Karoly Böröczky*: Polytopal approximation if the number of edges is restricted
8. *Shlomo Reisner*: Linear time approximation of three dimensional polytopes
9. *Szilard Revesz*: A generalized Minkowski distance function and applications in approximation theory

Tuesday, August 21

1. *Alexander Koldobsky*: Applications of the Fourier transform to sections of convex bodies
2. *Hermann König*: Sharp constants for Khintchine type inequalities
3. *Alex Iosevich*: The notion of a dimension of a convex planar set and applications to lattice points and irregularity of distribution
4. *Yossi Lonke*: Curvature via the q-cosine transform
5. *Mihail Kolountzakis*: Orthogonal bases of exponentials for convex bodies

Wednesday, August 22

1. *Stanislaw Szarek*: Duality of metric entropy
2. *Richard Vitale*: Convex bodies in Hilbert space: some metric issues, open problems
3. *Marianna Csornyei*: Absolutely continuous functions of several variables
4. *Yves Martinez-Maure*: Examples of analytical problems related to hedgehogs (differences of convex bodies)
5. *Matthieu Fradelizi*: Some inequalities about mixed volumes
6. *Antonis Tsolomitis*: Volume radius of a random polytope in a convex body
7. *Miguel Romance*: Extremal positions for dual mixed volumes
8. *Aljoša Volčič*: Determination of convex bodies and reconstruction of polytopes by certain section functions
9. *Apostolos Giannopoulos*: On the volume ratio of two convex bodies

Thursday, August 23

1. *Rolf Schneider*: On the mixed convex bodies of Goodey and Weil
2. *Olivier Guedon*: Supremum of a process in terms of the geometry of the set
3. *Krzysztof Oleszkiewicz*: On ℓ_p^n-ball slicing and pseudo-p-stable random variables
4. *Boris Kashin*: N-term approximation
5. *Vitali Milman*: Random cotype-2 of normed spaces

Conference on Geometric and Topological Aspects of Functional Analysis

Haifa, Israel (May, 2002)

(In memory of Yaki Sternfeld. Organized by J. Arazy, Y. Benyamini, Y. Gordon, V. Harnik, S. Reich and S. Reisner)

Sunday, May 19

1. *Vitali Milman* (Tel Aviv, Israel): Can we recognize in a reasonable time that a convex body in a high-dimensional space is very far from an ellipsoid?

2. *Marianna Csornyei* (London, UK): Some periodic and non-periodic recursions

3. *Michael Levin* (Beer-Sheva, Israel): The Chogoshvili-Pontrjagin conjecture

4. *Joram Lindenstrauss* (Jerusalem, Israel): The work of Yaki Sternfeld in Functional Analysis

5. *Jim Hagler* (Denver, Colorado): The structure of hereditarily indecomposable continua

6. *Arkady Leiderman* (Beer-Sheva, Israel): Basic families of functions and embeddings of free locally convex spaces

Monday, May 20

1. *Nicole Tomczak-Jaegermann* (Edmonton, Alberta): Families of random sections of convex bodies

2. *Wieslaw Kubis* (Beer-Sheva, Israel): Hyperspaces of separable Banach spaces with the Wijsman topology

3. *Henryk Torunczyk* (Warsaw, Poland): Equilibria in a class of games: geometric and topological aspects

4. *Paolo Terenzi* (Milan, Italy): The solution of the basis problem

5. *Boaz Klartag* (Tel Aviv, Israel): Isomorphic Steiner symmetrization

6. *Assaf Naor* (Jerusalem, Israel): Entropy production and the Brunn-Minkowski inequality

7. *Olga Maleva* (Rehovot, Israel): On ball non-collapsing mappings of the plane

8. *Michael Megrelishvili* (Ramat Gan, Israel): Reflexively and unitarily representable topological groups

Tuesday, May 21

1. *Edward W. Odell* (Austin, Texas): Asymptotic structures in Banach spaces
2. *Mark Rudelson* (Columbia, Missouri): Phase transitions for sections of convex bodies
3. *Haskell Rosenthal* (Austin, Texas): Invariant subspaces for certain algebras of operators
4. *Alexander Litvak* (Edmonton, Alberta): Projections of quasi-convex bodies
5. *David Preiss* (London, UK): Measure and category do not mix or do they?

Wednesday, May 22

1. *William B. Johnson* (College Station, Texas): Lipschitz quotients
2. *Eva Matouskova* (Prague, Czech Republic): Bilipschitz mappings of nets
3. *Tadeusz Dobrowolski* (Pittsburg, Kansas): The simplicial approximation and fixed-point properties
4. *Vladimir Fonf* (Beer-Sheva, Israel): On the set of functionals that do not attain their norms
5. *Gideon Schechtman* (Rehovot, Israel): ℓ_p^n, $1 < p < 2$, well embed in ℓ_1^{an} for any $a > 1$
6. *Aleksander Pelczynski* (Warsaw, Poland): Elliptic sections of convex bodies

Recent Reprints and New Editions